With hope
that Jason reach
his best potential

[signature]
Sep/26/00

Quantitative Genetics in Maize Breeding

Quantitative Genetics in Maize Breeding

SECOND EDITION

ARNEL R. HALLAUER

J. B. MIRANDA, FO.

Iowa State University Press / Ames

Arnel R. Hallauer is research geneticist, Agricultural Research, Science and Education Administration, U.S. Department of Agriculture, and professor of plant breeding, Iowa State University.

J. B. Miranda, Fo., is professor of genetics, E.S.A. "Luiz de Queiroz," University of São Paulo, Brazil.

Frontispiece: Courtesy Funk Seeds International, Bloomington, Ill.

First edition, 1981
Second printing, 1982
Third printing, 1985
Fourth printing, 1986
Second edition, 1988

Library of Congress Cataloging-in-Publication Data

Hallauer, Arnel R.,   1932–
    Quantitative genetics in maize breeding / Arnel R. Hallauer, J. B. Miranda, Fo. – 2nd ed.
        p.   cm.
    Includes bibliographies and index.
    ISBN 0–8138–1522–3
    1. Corn – Breeding. 2. Corn – Genetics. 3. Quantitative genetics.
    I. Miranda, J. B., Filho, 1945-   . II. Title
    SB191.M2H29 1988   88–13452
    633.1′52 – dc19   CIP

To Jan, Beth, Paul, and the memory of my mother

—ARNEL R. HALLAUER

CONTENTS

Contents

PREFACE

Maize breeding has been effective in developing improved varieties and hybrids to meet the rapidly changing cultural conditions of the past 100 years. Maize is an economically important crop in the world economy and is an ingredient in manufactured items that affect a large proportion of the world's population. To meet the increasing and divergent uses of maize, breeding methods have evolved to increase the effectiveness and efficiency of selection for many traits. Modern maize breeding methods are primarily a twentieth century phenomenon. Successful development of the inbred line-hybrid concept of maize to a useful form is still considered one of plant breeding's greatest achievements. Development of the commercial seed industry is testimony of breeding methods that have evolved for the economical production of high-quality hybrid seed that is accepted and demanded by the modern farmer.

Although the basic breeding methods for development of maize inbred lines and hybrids were described by 1910, significant contributions have been visualized and tested for modernizing the basic breeding methods. Our main concern in this volume is to describe how the principles of quantitative genetics and cyclical selection schemes have been used in maize breeding research. In addition to the principles discussed, data are summarized from reported studies. It is not intended that the breeding methods discussed in this volume will, or should, replace those currently used; they are intended to supplement those currently used. The integration of the breeding methods discussed herein with those currently used should enhance future breeding efforts to maintain the level of genetic gain of the past 50 years in future years.

This volume was not written with any particular audience in mind. It is hoped that all or portions of the material included will be of interest to graduate students in plant breeding, to maize breeders conducting basic research on breeding and selection methods, and to maize breeders whose main concerns are developing improved inbred lines and hybrids. Some of the elementary principles of quantitative genetics and selection are given to provide guidelines in interpreting the summaries of empirical data. For students, the chapters describing the principles of quantitative genetics, estimation of genetic variability, and selection may be of greater interest than the empirical data; for maize breeders, the summaries of the empirical data may be of greater interest to assist in planning their breeding programs. Because of rapid development of the inbred line-hybrid concept in maize, some chapters pertain more to maize than to other crop species.

Maize breeders have been very active in developing and testing breeding methods, conducting quantitative genetic studies, and conducting cyclical selection studies for the improvement of populations. Consequently, a large volume of literature has accumulated for these three aspects of maize breed-

ing. No attempt has been made to include a complete listing of all the available references. Only those that are considered key references or illustrate
specific principles are included; obviously, others would have a different
listing. Many of the references are not limited to a specific topic and are
repeated for the different chapters. There also is some repetition among
chapters, which in most instances is intentional. It is desirable to have
each chapter as complete as possible and, because of the relation of certain
topics to others, some repetition is inevitable.

The authors wish to acknowledge the contributions others have provided
in the preparation of this volume. The following have read either portions
or all of the manuscript used in the first four printings: C. W. Crum,
T. C. Hoegemeyer, G. R. Johnson, E. Paterniani, W. A. Russell, O. S. Smith,
and R. Vencovsky. A special thanks is given to Wyman E. Nyquist: his
patience and attention to details in reviewing the volume have contributed
immeasurably to the revisions included in this printing. The content is
essentially the same as the previous printings, but corrections of errors
detected by the authors and others have been made, some changes were made
in notation for consistency, some rewriting was done for clarity, and some
recent references on the subject matter were added. The authors
appreciated receiving from others the notes and comments that were used in
the revision. Finally, the authors gratefully acknowledge the time,
patience, and stellar abilities of Mary A. Lents in the typing of the
original manuscript and preparing the revisions.

Arnel R. Hallauer
J. B. Miranda, Fo.

Quantitative Genetics in Maize Breeding

1

Introduction

Maize (*Zea mays* L.) breeding and improvement started when people realized the potential of the species for food, feed, and fuel. Although the ancestral pedigree of maize has not been resolved, the early maize breeders certainly played an important role in developing the species as we know it today. Maize, however, is known to be indigenous to the Western Hemisphere.

The ancestral form of cultivated maize would be weedlike for survival, and the transition from a weed species to a cultivated crop species that depends on care for survival required years and patience of the early maize breeders. Depending on the preferences and environments of the different Indian settlements, different selection pressures would have been applied. Growers could effectively select and fix traits such as kernel color and texture for a range of environments. Because all harvesting was on an individual plant basis, variation among plants and ears for easily recognizable traits would be noted. Nature, on the other hand, would play an important role in the development of maize strains that were resistant to pests; day-length sensitivity; and drought, heat, and cold tolerance. The traits that people could select and fix were transferable from one local area to another; people and nature would cooperatively develop maize populations adapted for wide differences in environments. Local settlements probably used certain traits as genetic markers for their particular populations. Accidental mixtures and cross-pollination would contribute to the wide variation of maize varieties before the landing of Columbus in the Western Hemisphere.

Selection procedures used by early maize breeders would seem primitive compared with present-day breeding methods, but they recognized the traits needed to sustain their civilizations. The effectiveness of selection by early maize breeders is evident from the 250 New World races and thousands of varieties that have been collected and described. There is overlap in the genes among the races and varieties, but maize can be grown throughout almost the complete range of altitudes and latitudes around the world. Certainly the greatest maize breeding advance was made by converting a wild weed species of relatively low yield to a cultivated crop species that can yield over 100 q/ha (quintals per hectare) under modern cultural practices.

Maize breeding also has evolved through several stages during the past 100 years. Because maize was indigenous to the Western Hemisphere, European settlers soon were introduced to and depended on it for survival. As the European migration expanded along the coastlines and into the interiors of the Western Hemisphere, it would have come into contact with a great diversity of maize germplasm. Movement of the people and exchange of maize germplasm permitted the introgression of divergent varieties and complexes of the Western Hemisphere. In the Northern Hemisphere, expansion of the cultivation of the

3

Northern Flint complex and the Southern Dent complex led to the development of
the highly productive Corn Belt Dent varieties. The Northern Flints and
Southern Dents were distinct complexes, but the expansion in maize cultivation
southward (for the Northern Flints) and northward (for the Southern Dents) resulted in a reciprocal introgression of the two complexes. Seeds of the two
distinct landrace complexes were carried by people, but the crossing between
the two complexes probably occurred more by contamination than by planned
crosses. Crossing of the two distinct complexes created a vast reservoir of
genetic variability for plant and ear traits. Simple mass selection, therefore, was an effective breeding method for developing varieties that possessed
traits appealing to the fancies of the growers and early colonial maize breeders. The hybrid swarms arising from crossing the Northern Flint and Southern
Dent complexes would not have been as extensive as in pre-Columbian times, but
the range in variability was great enough for the development of varieties
having distinctive plant and ear traits.

Development of the Corn Belt Dent varieties was not a planned breeding
program. As the interior of the United States was developed, seeds of settlers' varieties were brought westward from the Atlantic seacoast. Each individual seed lot would have been subjected to different selection pressures,
depending on the individuals selecting seed to propagate the crop the following season. In some instances, careful selection was given to ear traits (row
number, flint vs. dent, color, prolificacy, etc.), whereas others selected for
early maturity, shorter plants, freedom from tillers, and plant type. Selection developed varieties with distinctive traits and adapted to specific environments. As cultivation of maize became more extensive, introgression among
the selected strains occurred because isolation was reduced. The amount of
introgression among selected strains depended on the topography of the areas
and the amount of interchange of seed among pioneer settlers. Genetic divergence among selected strains was sufficiently great that crosses among them
suggested that variety crosses were superior to growing the selected varieties. Beal (1880) suggested the potential of variety crosses from seed produced from controlled crossing of two varieties; although variety crosses were
superior to the parental varieties, they apparently were never used to any
great extent. A brief summary of the early history of maize breeding was giv-
en by Goodman (1976).

Two developments in the early part of the twentieth century had a pro-
found effect on modern maize breeding: (1) development of the maize show card
for exhibiting maize ears had an effect on selection procedures of farmers,
producers, and breeders for selecting ear types that conformed to show card
standards; and (2) research reported by Shull (1908, 1909) outlined the basis
of the breeding methods for developing and producing modern maize hybrids.
The development of the maize show card had only a temporary effect on maize
breeding, whereas Shull's research formed the foundation of modern maize
breeding.

Development of the maize show card created further interest in selection
within maize varieties because intense selection pressure was given to ears
that conformed to the standards of the show cards (see example in Table 1.1).
Individual ear and small-ear sample winners of shows often commanded high pre-
miums when sold as a seed source. Although fields of maize were open pollinated, bottlenecks for genetic variability often would have resulted from the
small number of ears representing the winning samples of maize shows. Usually
the winning samples resulted from the time and patience the exhibitor spent
inspecting fields for ears that were uniform in appearance and conformed to
show card standards. Time and patience in selecting ear types resulted in a
number of named varieties that had distinctive traits. Often several sub-

Table 1.1. Score card adopted by the Iowa Corn Growers' Association in 1908 (from
Corn by Bowman and Crossley 1908).

Traits	Points	
I. General appearance	25	
1. Size and shape of ear		10
2. Filling of butts and tips		5
3. Straightness of rows		5
4. Uniformity of kernels		5
II. Productiveness	60	
1. Maturity		25
2. Vitality		25
3. Shelling percentage		10
III. Breed type	15	
1. Size and shape of ear		5
2. Size, shape, and dent of kernel		5
3. Color of grain		2
4. Color of cob		2
5. Arrangement of rows		1

strains of a limited sample that had won top honors at a prestigious maize
show were developed. Although maize shows were very popular in the early part
of the twentieth century, critical comparisons of prize-winning samples that
met score card standards with samples that deviated from them showed that es-
thetic ear traits were not an indicator of yield performance. Results of comparative yield tests quickly led to the demise of the maize shows when it was
demonstrated that a Krug variety that was not selected to meet score card
standards was equal to a Reid variety that was selected to conform with the
score card.

The research of East (1908), Shull (1908, 1909), and Jones (1918), contrary to the development of show score cards as selection criteria, provided
the framework for maize breeding that is still used today. Although the essential features of modern hybrid maize breeding were outlined before 1920,
several distinct phases have been recorded during the past 50 years (Fig.
1.1). Shull (1908) and East showed that an open-pollinated maize variety con-
sisted of a range of heterogeneous genotypes from which homozygous and homogeneous pure lines (inbred lines) could be developed by five to seven generations of selfing; i.e., controlled pollinations of placing the pollen (male
gametes) on the silks (female gametes) of the same plant. Although inbred
lines that were uniform and repeatable for phenotype were obtained by self-
pollination, the inbred lines often were weak and difficult to propagate.
However, when certain of the weak inbred lines were crossed, vigor was re-
stored and the yield of the inbred line crosses (or single crosses) usually
exceeded the original open-pollinated variety from which the lines were developed. The single crosses were uniform and seemed to be potentially useful.
These observations of inbreeding and crossing led Shull (1909) to describe a
method for producing maize hybrids. His remarkable conclusions gave a lucid
outline of the subsequent course of maize breeding--conclusions that were

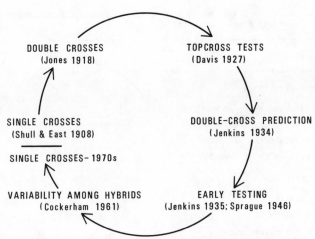

Fig. 1.1.   Sequence of breeding methods developed for production of hybrids in
            the United States.

based on the limited data and experience available at that time.

East was conducting research similar to that of Shull (1908). Because of
the poor vigor and seed production of the inbred lines, East was somewhat pessimistic that the procedure would have practical usefulness. It was not until
Jones (1918) suggested that two single crosses be used as parents in the production of hybrids that it seemed hybrid maize would become a reality; i.e.,
use the two single crosses that were highly productive to produce double-cross
hybrid seed, which in turn could be distributed and grown by farmers. Removal
of the concern for producing hybrid seed on a practical basis stimulated research to develop inbred lines for use in the production of double-cross hybrids during the 1920s. No good estimate is available of the number of inbred
lines developed during the 1920s, but it soon became evident that it was simpler to develop lines than it was to determine their worth in hybrids.

Not all lines in hybrid combinations were superior to the original open-
pollinated varieties. The variation among double-cross hybrids showed that
only certain combinations of inbred lines gave a satisfactory double-cross hybrid. Preliminary screening techniques by Davis (1927) and Jenkins and Brunson (1932) showed that crossing lines with a common tester (topcross) was effective for discarding lines that did not have satisfactory performance in hybrids. Many lines could be discarded on the preliminary topcross performance
information. Often a sizable number of lines that had potential for use in
hybrids still were available. Even if only 20 lines were available for testing in hybrids, 190 different single crosses and 14,535 different double
crosses were possible. Jenkins (1934) developed and tested four methods for
predicting double-cross hybrids from use of single-cross hybrid data. Experimental data showed the reliability of method B (average performance of four
nonparental single crosses) for reducing the amount of testing needed to determine better performing double-cross hybrids. Topcross tests for preliminary screening of lines and prediction methods for reducing the number of double-cross hybrids for testing were widely used and contributed greatly to the
efficiency of maize breeding.

Several generations (five to seven) of selfing are required to obtain
lines that are nearly homozygous. Topcross tests and prediction methods aided
testing of lines in hybrids after they were inbred to near homozygosity, i.e.,
isolation of pure genotypes. The next logical step was to establish whether

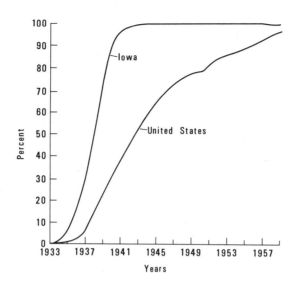

Fig. 1.2. Rate of acceptance of double-cross hybrids by farmers in the United
          States and Iowa.

yielding ability of hybrid lines could be determined before they approached
homozygosity. Selection during inbreeding can be effective for some traits
(e.g., plant characteristics, ability to shed pollen, resistance to attack by
pests, kernel color, etc.), but yield in hybrids was usually determined by
producing crosses and testing when lines were nearly homozygous. If yield potential of lines in crosses could be determined before homozygosity, many
lines could be discarded to reduce the number continued by selfing. Jenkins
(1935) and Sprague (1946) presented data showing that yield potential of lines
in crosses could be determined in early generations of inbreeding. Consequently, lines that lacked yield contributions to crosses could be discarded
in early generations of inbreeding; this permitted better sampling of populations for superior lines, and the number of lines continued by inbreeding
could be substantially reduced before they were homozygous.

     Because of the development of maize breeding methodology, formal programs
expanded rapidly in scope and number during the 1930s and 1940s. Hundreds of
inbred lines were developed and tested. Cooperation and interchange of inbred
lines among maize breeders identified elite inbred lines that were used extensively in the production and growing of double-cross hybrids. Acceptance of
double-cross hybrids by farmers depended on the availability of seed and their
knowledge of new seeds. Figure 1.2 shows that nearly 100% of the maize acre-
age was planted to hybrids in Iowa by 1943 and in the United States by 1960.

     Although double-cross hybrids had essentially replaced open-pollinated
varieties by the 1950s, there was some concern that a yield plateau had been
attained. Double-cross hybrids were used because it seemed the only feasible
method of producing maize hybrids from the experiences of East (1908), Shull
(1908, 1909), and Jones (1918). During the late 1950s and early 1960s a few
single-cross hybrids were produced and grown. Improvement in management (use
of herbicides, pesticides, and fertilizers) and inbred lines (improved by selection, in many instances, from earlier developed lines) permitted the production of single-cross hybrids that were economically feasible for producers
and farmers. Double-cross hybrids require four parental inbred lines, whereas
single-cross hybrids require two. Selection pressure, therefore, would be 100%

greater among inbred lines for use in single-cross hybrids. Although elite
inbred lines that were extensively tested were used for producing hybrids,
trade-offs usually were necessary in the selection of lines to compensate for
their known weaknesses; this would be easier for two than for four elite inbred lines. Cockerham (1961) also showed theoretically that expected genetic
variability and expected genetic gain among single-cross hybrids within a population would be at least double that among double-cross hybrids. From field
experience and theoretical considerations, the change from production and
growing of double-cross hybrids to single-cross hybrids has increased in the
United States from about zero in 1960 to about 85% at the present time. Because of the development of field husbandry and maize breeding methodology,
the circuit was completed from Shull's first suggestion of single-cross hybrids in 1908 to the present (Fig. 1.1).

One other development that stimulated basic research to formalize maize
breeding procedures was the introduction of quantitative genetic theory to
breeders. Researchers at North Carolina State, Raleigh, were responsible primarily for applying quantitative genetic theory to maize breeding during the
latter part of the 1940s. Quantitative genetics had been used extensively by
animal breeders, but plant breeders had used only limited quantitative genetic
theory for guidance in the development of breeding plans. Perhaps because
large numbers of genotypes with a relatively short generation interval could
be evaluated and the emphasis given to inbred line and hybrid development,
plant breeders generally ignored the potential of quantitative genetics.
Since 1950 quantitative genetics has strongly influenced the thinking in maize
breeding. Population structure and improvement, selection theory, types of
gene action, predicted gain, and mating designs were concepts that became
known to the maize breeder and served as guides in developing breeding programs. In addition to the principles and terminology of quantitative genetics, an important aspect (previously generally neglected) was the importance
of improving the basic breeding populations, i.e., population improvement.
The first population improvement programs usually were initiated to compare
empirical results with theoretical predictions. Because of the seeming yield

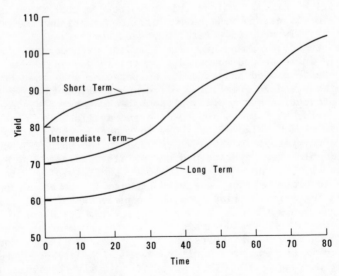

Fig. 1.3. Objectives of breeding programs to meet present and anticipated
           goals.

plateau in the 1950s, population improvement programs seemed logical for upgrading the breeding populations used as source materials for extraction of inbred lines. Estimates of genetic components of variance showed that a preponderance of total genetic variance was additive. Hence, cyclical population improvement (generally called recurrent selection) programs seemed to be an obvious method for increasing the frequency of superior lines.

General acceptance of the principles of quantitative genetics in maize breeding was relatively slow by applied maize breeders, whose major responsibilities were for development of elite inbred lines for use in hybrids. Most applied breeders, however, are cognizant of quantitative genetics research and usually appreciate its importance for quantifying the genetic parameters they frequently have observed empirically. Recent reviews by Dudley and Moll (1969) and Moll and Stuber (1974) have summarized and interpreted data obtained from quantitative genetic studies and how they are related to plant breeding. More extensive quantitative genetic studies have been conducted with maize than with most other crop species. In maize one has greater flexibility of mating, and seed supplies usually are not limiting, which makes maize amenable for development of progenies for testing and recombination.

The methods of maize breeding depend on the objectives of the breeder, demands of the seed industry, and the level of agricultural development of the areas in which the final product is to be used. The objectives of the breeder may be long-, intermediate-, and short-term, in which case the objectives may vary at different stages of the breeding program (Fig. 1.3). Breeding programs need flexibility to meet future contingencies that are either predictable or nonpredictable, e.g., the southern leaf blight (*Helminthosporium maydis*) epidemic in the United States in 1970. In developing countries, an inbred line and single-cross hybrid breeding program may have low priority because of the problems associated with costs for the production and distribution of seed. Breeding programs in developing countries may have objectives of developing, evaluating, and improving populations and testing population crosses; these objectives may have been initiated in more advanced programs several years ago and may have little interest for breeders emphasizing inbred line and hybrid development. All programs, however, should have long-, intermediate-, and short-term objectives to have current relevance and long-term viability. The breeder should have germplasm identified for present objectives and be working on germplasm for intermediate and future objectives.

Several distinct methods of maize breeding techniques have evolved with the transition from the growing of open-pollinated varieties to hybrids (Fig. 1.1). The chronological phases are not distinct because of considerable overlapping, personal preferences, data and information available as maize breeding developed, and objectives of the breeding programs. Any listing of the methods of maize breeding becomes arbitrary because they may or may not be applicable for short- vs. long-term objectives, intra- vs. inter-population selection procedures, and population improvement vs. inbred line and hybrid development. A brief discussion of the different methods, however, will be given to provide a general framework of the methods suggested for use in maize breeding. The methods are listed under two broad categories: (1) inbred line development for eventual use in hybrids, and (2) improvement of populations for use either as breeding populations or by farmers (Table 1.2). The stage of development of maize breeding and husbandry will determine which method(s) in either category is used. Obviously, some of the methods in each category and more than one method in one category are used in specific breeding programs. Programs that have short- and long-term objectives would include methods from both.

Table 1.2. Methods of maize selection procedures that have been suggested and used for inbred
            line development and population improvement.

Population improvement		Inbred line development
Intra-	Inter-	
1. Mass	1. Half-sib - Reciprocal	1. Pedigree
2. Half-sib	recurrent (HS-RSS)	2. Backcross
a. Ear-to-row	2. Full-sib - Reciprocal	3. Single-seed descent
b. Testcross	full-sib (FS-RRS)	4. Gamete selection
3. Full-sib	3. Testcross	5. Monoploids
4. Selfed progeny		a. Homozygous diploids - maternal
		b. Androgenesis - paternal
		c. Gametophyte factor
		d. Pollen culture

1.1 INBRED LINE DEVELOPMENT

Because of the acceptance of the tenets of Shull (1908, 1909) for use of
inbred lines in production of hybrids, breeding methods that emphasize line
development are used extensively. Other breeding methods often were neglected
and emphasis was given to isolating inbred lines and testing in hybrids. All
methods have been developed during the interval shown in Fig. 1.1.

1.1.1 Pedigree Selection

Pedigree selection was and is the most commonly used selection method of
line development. Pedigree selection became common after Shull (1908, 1909)
outlined the principles of inbreeding and hybridization. Progenies are developed from source populations by some form of inbreeding, with selfing the most
commonly used. Although pedigree selection often implies selection in an F_2
population produced from a planned cross of two genotypes (often, in maize
breeding between two or more elite inbred lines), pedigree selection could be
practiced with progenies developed in open-pollinated, composite, and synthetic varieties; backcross populations; and mixtures of germplasm as well as F_2
populations. In each instance, progenies are established that have an assigned designation (pedigree) to identify the origin, numerical sequence, and
generation of development. Detailed record keeping is required to properly
identify the progenies selected in each generation of inbreeding. The selection unit usually is a combination of progeny row and individual plants within
a progeny row. Seed of selected plants within selected progeny rows is planted ear-to-row the following season. Progeny rows usually are nonreplicated,
but because of controlled pollinations, selection is effective for traits of
relatively high heritability. For traits such as yield of progenies themselves and in crosses, replicated yield tests become necessary at some stage
of the pedigree selection; in some instances testing is done in the early generations of inbreeding and in others as the selected progenies approach homozygosity. Pedigree selection is extensively used in recycling of lines that
have known strengths and weaknesses for specific traits. Pedigrees developed
by recycling of lines can become quite complex (Fig. 1.4), but parentage con-

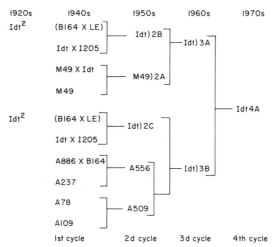

Fig. 1.4. Example of four cycles of pedigree selection for the development of
           inbred lines. Idt and I205 are from Reid's Iodent (sister lines);
           B164 is from Indiana Reid variety; LE is from the open-pollinated
           variety, Long Ear; and M49 is the Minnesota culture of 1949.
           (Courtesy Raymond Baker, Pioneer Hi-Bred International, Inc.,
           Johnston, Iowa)

trol permits development of lines to meet specific requirements. Techniques
of pedigree selection are described in greater detail by Harrington (1952) and
Allard (1960) and used extensively in maize breeding programs (Bauman 1981).

1.1.2 Backcross Selection

The backcross method is a special case of pedigree selection. Planned
crosses are made, as with pedigree selection, but usually it is desired to incorporate a specific gene(s) into an otherwise desirable genotype. In maize
breeding the objective of backcross selection often is to transfer a specific
trait (e.g., Ht, o_2, fl_2, lg_2, Lg_3, etc.) from a donor parent (nonrecurrent)
to an elite inbred line (recurrent parent); the breeder wants to recover the
genotype of the elite inbred line in successive backcrosses and incorporate
the desired trait from the donor parent. As for pedigree selection, detailed
records are needed to identify recurrent parent, donor parent, generation of
backcrossing, and specific trait under selection. Because the trait transferred by backcrossing usually can be classified in discrete classes, the selection unit is a combination of individual plants within nonreplicated progeny rows. To transfer a one-gene trait from one parent to another, backcross
selection has an advantage over pedigree selection because 50% of the plants
for backcross selection would be homozygous for the trait compared to only 25%
of the plants for pedigree selection in the first generation. Backcrossing
also may be used for traits that are not controlled by one or two genes, e.g.,
incorporation of exotic germplasm into adapted populations. In these instances larger plant numbers are needed to sample the array of genotypes in
the recurrent and nonrecurrent parental populations. Next to pedigree selection, backcross selection is probably the most commonly used method in maize
breeding for line development. Backcrossing is a convenient method for either
the transfer or the improvement of a specific trait of an otherwise elite inbred line. Richey (1927) developed the concept of convergent improvement,
which involves the reciprocal addition to each of two inbred lines the domi-

nant favorable genes lacking in one parent but present in the other. Convergent improvement, as for backcrossing, has been successful for introducing simply inherited traits into established lines; but for complex traits such as yield the method is less efficient. The techniques and principles of backcrossing are discussed in greater detail by Hayes et al. (1955) and Allard (1960), and backcrossing is an important component in maize breeding (Bauman 1981).

1.1.3 Single-Seed Descent

The single-seed descent method is not used extensively in maize breeding, but Jones and Singleton (1934) suggested a modification called the single-hill method, which so far has not been used to any great extent. The single-seed descent method is not as commonly used in maize breeding as in self-pollinated species. This method is used when a population of genotypes (say 500) are sampled and at least one seed is saved from each genotype in successive generations of inbreeding until approximate homozygosity is reached. Except for natural selection a minimum of selection pressure is applied. At some level of homozygosity, seed supplies of the progenies obtained by single-seed descent are increased for evaluation in replicated tests. For obligate self-pollinated species, self-pollinations are made naturally.

The pedigree and backcross methods, however, use the principles of single-seed descent because selected plants in selected progenies are propagated to the next generation. The distinction between pedigree and backcross selection and single-seed descent is that artificial selection is maximized in the pedigree and backcross methods and minimized in single-seed descent. Natural selection, of course, is operative in all methods. The single-seed descent method has been used in maize for quantitative genetic studies estimating parameters in reference to a specific population. Because controlled pollinations are required to effect inbreeding, the single-seed descent method is not as amenable for maize as for self-pollinated species.

1.1.4 Gamete Selection

Gamete selection was suggested by Stadler (1944) and has had limited use compared with the pedigree and backcross methods in maize breeding. Stadler theorized that if the frequency of superior zygotes was p^2, the frequency of superior gametes was p; i.e., if p^2 is 0.25, p is 0.5. Because the theoretical frequencies of superior gametes were greater than for superior zygotes, gametic sampling should be more efficient than zygotic sampling. Briefly, gamete selection includes: crossing a bulk sample of pollen from a source population with an elite inbred line in season 1, self-fertilizing the F_1 plants (each F_1 plant differing only by the gametic complement contributed by the source population) and crossing the F_1 plants and an elite inbred line to a common tester in season 2, and growing the testcrosses in replicated yield trials in season 3. Therefore, any testcross that exceeded the elite line × tester combination presumably received a superior gamete from the source population. Although the testcrosses identify the superior gametes, the major disadvantage is that the superior gamete cannot be fixed in homozygous inbred lines. Because individual plants by elite line F_1 plants were self-fertilized, only 25% are homozygous for individual loci; this would increase to 37.5% over all S_1 progenies, but not all progenies would be retained from the testcross data. Selection units are individual plants from the source population but would include progeny row and individual plants within the selfed progenies of the F_1 plants. Some positive data have been reported from use of gamete selection, but the method is not used generally. The method seems to have merit and with some modifications could be useful for identifying superior genotypes in a source population.

Zygote selection was suggested (Hallauer 1970) as one modification of gamete selection. Plants in a source population are selfed and crossed to an elite line used as a tester. The gametic array of the plant used in testcrosses would be similar to that of the S_1 progenies. Although recombination would prevent the recovery of identical superior gametes expressed in the testcrosses, the opportunities would be greater than by gamete selection. In all instances the primary objectives of these methods are either to improve an elite inbred line or to select a companion line for use in hybrids.

1.1.5 Monoploids

Several interesting methods for rapid development of inbred lines have been suggested. In most instances none of the methods has been adequately evaluated to determine its relative merits. Development of inbred lines, however, generally is not a serious problem in maize breeding. Determining the relative merit of the lines in hybrids is the more difficult problem. Also, tremendous selection pressure can be applied during inbreeding for the phenotypic expression of plant and ear traits, disease and insect resistance, and yield in crosses by use of early testing. Techniques for rapid development of inbred lines yield a group of lines that are a random sample for most traits; after the inbred lines are available, they will have to be grown and evaluated for the traits necessary for use as parental seed stocks in hybrids.

Homozygous diploids. Production of homozygous diploids of maize from monoploids was outlined by Chase (1952a, 1952b) as a new technique for inbred line development. Use of the method depends on the occurrence of monoploids, and it was found in several populations that maternal monoploids occurred spontaneously at a high enough rate (1/1000) to yield sufficient numbers of monoploid plants. Successful production of homozygous diploids from monoploids depends on (1) production and recognition of monoploids, and (2) deriving homozygous, diploid progeny from the isolated monoploids. Spontaneously occurring monoploids were identified by Chase in the seedling stage by use of a genetic marker system. Many of the monoploids survived to maturity to produce normal gametophytes. Successful self-pollination of the mature monoploids yields homozygous, diploid lines at a frequency of one homozygous diploid line for every ten isolated monoploids. At this time, opportunities for selection have occurred only for the self-pollinated monoploid plants.

Relative merits of the homozygous diploid method for practical breeding were tested by Thompson (1954) by comparing the combining ability of yield of selected and unselected lines developed by the usual inbreeding system of self-pollination with lines developed from monoploids; all lines originated from the same source population. Generally, Thompson found that the lines derived from monoploids were a random sample with respect to yield in topcrosses because the differences between means were nonsignificant and the frequency distributions were similar. So far the method has not been used extensively by maize breeders. Theoretically, the homozygous diploid method has the same advantage as Stadler's (1944) gamete selection method--the frequency of superior gametes is greater than the frequency of superior zygotes. Contrary to the gamete selection method, however, the superior gametes of monoploids can be fixed in the homozygous condition in one generation of self-pollination. The difficulty is in identifying the superior gametes, which can be done only in observations repeated over environments.

Androgenesis. Goodsell (1961) described a technique whereby paternal monoploids could be identified and selfed in one generation to produce homozygous, diploid lines. The breeding procedure is similar to the one used by Chase

(1952a, 1952b) for maternal monoploids and requires a genetic marker system to identify the monoploids (Goodsell used a genetic marker that caused purple coloration in the seedling root tips, which also was similar to Chase's system). Nonpurple, fertile males were crossed to purple, male-sterile females. The F_1 seedlings of the crosses were examined for seedlings failing to exhibit the purple coloration of the root tips. From 400,000 progenies of four crosses Goodsell found 4 seedlings having white roots, which is lower than the frequency (1/1000) reported for maternal monoploids. Pollinations of the monoploids resulted in plants similar to the male pollinator and male sterile; i.e., the male gamete failed to unite with the female gamete but acquired the cytoplasm of the female nucleus. This method has the same advantages and disadvantages of maternal monoploids and so far has not been used to any extent. The method also seems promising for quick conversion of lines to male-sterile cytoplasms, provided one can increase the frequency of occurrence of paternal monoploids.

Indeterminate gametophyte. Kermicle (1969) reported on a spontaneous mutation, designated indeterminate gametophyte (ig), that produced unusual effects in the embryo sacs of the plants carrying the mutant gene. Because of the influence of the ig gene on female gametophyte development, some of the effects associated with ig are: homozygous mutant plants (igig) are generally male sterile; about 50% of the kernels produced on igig plants and 25% of those on Igig plants either abort or are defective; about 6% of the seeds with normal endosperm that received ig from either Igig or igig females are polyembryonic; and when igig plants are crossed as females about 3% of the progeny are monoploids, which occurred as maternal and paternal monoploids in a ratio of about 1:2. Hence, the action of the ig gene reflects the loss of normal functions in the female gametophyte development.

It seems that production of monoploids by the ig gene may have useful advantages in maize breeding programs for (1) rapid isolation of new inbred lines by doubling of the paternal monoploids and (2) rapid conversion of existing elite lines with normal cytoplasms to versions with male-sterile cytoplasms. These two advantages are the same as those given by Goodsell (1961) except the occurrence of monoploids is at a higher frequency, which was emphasized by Goodsell if the method of androgenesis was to be useful in breeding programs. Kermicle (1969) found that the anthocyanin pigmentation characteristics of the R-Navajo (R^{nj}) gene and its acyanic allele r^g were useful genetic markers for the detection of monoploids. Reciprocal crosses of $r^g r^g$ and $R^{nj}R^{nj}$ stocks were used to detect monoploids of maternal and paternal origin, respectively. So far as known, the indeterminate gametophyte is not used extensively in maize breeding, but the higher frequency of monoploids would make it more attractive than the other previously described systems.

Pollen culture. Recent developments of tissue and cell culture techniques offer tremendous opportunities for the visionary plant breeder. If one assumes a maize plant produces 1 to 10 million pollen grains, large numbers of haploid plants could be developed from cultured microspores. Presently, however, techniques are not available to culture microspores of maize to differentiated haploid plants that are doubled to homozygosity and fertile. Such techniques, if available, would provide for thousands of monoploid plants that could be doubled to homozygous, diploid inbreds. To be useful, some monitoring tests could perhaps be developed that would determine the relative merit of the doubled monoploids for disease and insect resistance, grain quality, and yield potential as homozygous inbred lines and in hybrids; development of these tests remains for the distant future. Although culture techniques are of in-

terest to maize breeders, they are not available at the present. The possibilities of pollen culture are intriguing and could add to breeders' choices
if practical methods for pollen culture are developed. Pollen culture would
solve the limitation of frequency of occurrence of monoploids, which is the
limitation of the other three methods.

1.2 POPULATION IMPROVEMENT

Population improvement provides for the cyclical upgrading of open-pollinated varieties, synthetic varieties, and composites formed from a mixture of
races, varieties, and inbred lines. Improvement of populations may be either
for use as source populations for new inbred lines or for use by the farmer.
The former is for use in more advanced programs, whereas the latter is for use
in developing areas or countries that do not have the means for production,
distribution, and growing of hybrids. Hence, population improvement has always been an important objective from the most primitive to the most advanced
maize breeding programs. Intrapopulation improvement is as old as maize
breeding, whereas interpopulation improvement was stimulated by the expression
of heterosis.

Methods of selection for population improvement have evolved from the
simplest type of mass selection for intrapopulation improvement to the complex
procedures of reciprocal recurrent selection for interpopulation improvement.
Obviously, all methods have been successful for some traits at some stage of
maize breeding development. Some evidence has been presented for the relative
efficiency of the different methods of selection, but additional data are
still needed to determine their long-term effects. Some of the methods are
applicable only to intrapopulation improvement (e.g., mass selection), whereas
others are useful for either intra- or interpopulation improvement (e.g.,
half-sib). The choice of any one method of selection depends on the breeder,
stage of the breeding program, stage of germplasm development, stage of knowledge of the populations, and objectives of the breeding program. Usually each
breeding program will include more than one method of selection either for one
population or for more than one.

1.2.1 Mass Selection

Mass selection is undoubtedly the oldest method used in maize breeding.
The selection unit comprises the nonreplicated phenotypes of individual
plants. Ears of the selected plants are harvested and shelled and equal quantities of the selected seed are composited for planting the following season.
Mass selection was effective in fixing some traits, but it generally was not
effective for yield improvement (Sprague 1955)--not because of the method but
because of the techniques frequently used. Factors that contributed to the
seeming ineffectiveness of mass selection were poor isolation, no environmental control, genotype-environment interaction, no parental control, and poor
plot technique.

The distinctive differences among the many different populations of
maize, however, show that mass selection was effective for fixing traits that
were not affected by the environment to the same extent as, say, yield. Mass
selection, therefore, was effective for traits of high heritability. Some of
the disadvantages of mass selection as an effective breeding method were cor-
rected by the suggestions of Gardner (1961). Although mass selection is the
oldest method of maize breeding, it is still used for the improvement of
breeding populations; its effectiveness is dependent on the plant trait selected, adequate isolation, and the precision of the experimental techniques
used by the breeder.

1.2.2 Half-sib Selection

Half-sib selection is effected in all instances by use of a common tester parent. The tester parent could be either the population that is under selection (intra) or a population unrelated to the population under selection (inter); one that is either a narrow genetic base (inbred line) or a broad genetic base (open-pollinated, synthetic, and composite varieties); or one that either has poor or has good combining ability. Because of possible choice of testers, half-sib selection probably has been used more frequently than any other method of selection of population improvement.

Choice of a tester parent for use in half-sib selection depends on the objective of the breeding program and knowledge of the types of gene action operative in yield heterosis. In all instances individual plants of the population under selection are crossed to a common tester; this may be done either by hand pollination or in an isolation plot where the tester is the male parent. If the tester is not an inbred line, adequate sampling of the gametic array of the tester also is necessary. The half-sib progenies (all progenies that have a common parent) are evaluated in replicated yield trials to determine the superior performing genotypes. Either remnant seed of the half-sib progenies (when the parental population is used as the tester) or selfed (S_1) seed of plants used in making the crosses are recombined to form the next cycle population. Half-sib selection can be used for the improvement of either one population or two, from the suggestion of Comstock et al. (1949) of reciprocal recurrent selection.

Ear-to-row. The ear-to-row selection method was devised and introduced by Hopkins (1899) at Illinois for improvement of the chemical composition of maize. Ears from a population are planted ear-to-row and evaluated in a progeny-row test. In contrast to mass selection, the selection unit of ear-to-row is the progeny (a half-sib family of an ear) rather than individual plants. (It is conjectural whether mass selection also may have been effective for improvement of the chemical composition because environmental effects probably were not as great as for some traits being improved by mass selection.) Ear-to-row selection also requires adequate isolation to prevent contamination from other varieties. Ear-to-row selection was used to some extent by maize breeders until about 1925; it was effective in some instances and not in others depending to some extent on the trait selected and how it was affected by the environment. Two modifications of ear-to-row selection were made by Lonnquist (1964) and Compton and Comstock (1976) to make it useful. The modified ear-to-row selection method has progeny-row tests replicated in different environments, one of which is isolated from other maize fields for recombination; these two suggestions increased the precision of progeny-row tests and prevented contamination from other maize populations of selected progenies. Consequently, the ear-to-row selection method is receiving renewed interest and, with adequate isolation, is a form of half-sib selection.

Testcross. Selection based on testcrosses is commonly used in maize improvement, but it obviously is a method based on half-sib family selection. Because evaluation of testcrosses is commonly used, the terminology may be more obvious than that of half-sib family selection. Expected genetic advance is the same for testcross and half-sib selections when the tester is the parental population (Sprague and Eberhart 1977). A common tester parent is used to produce the testcross (or half-sib) progenies for evaluation and selection. Remnant seed of selected progenies is used for recombination if the tester parent is used as the male. If individual plants in the population under selection are used as males, they can be selfed and S_1 seed used for recombination to form the succeeding cycle population.

Testcross selection usually was used for intrapopulation improvement, but Russell and Eberhart (1975) proposed use of inbred testers as a modified form of reciprocal recurrent selection. In place of using the opposing populations as testers, they suggested use of elite inbred lines, one from each of the two populations under selection, as testers. The suggestion of currently usable elite inbred lines as testers was predicated on the evidence that the greatest proportion of the total genetic variation arises from additive genetic effects. Because additive gene action seems to be of major importance, substitution of inbred lines as testers should not hinder future progress from selection. Also, such use as testers to produce the testcrosses (or half-sibs) for two opposing populations would be simpler than the half-sib selection of reciprocal recurrent selection. (Testcrosses using inbred lines as testers are here referred to as half-sibs; they can, however, be referred to as full-sibs if one considers that the tester population is a constant genotype.) Testcross selection has been used more extensively for intrapopulation improvement but seems to have promise for interpopulation improvement as well.

1.2.3 Full-sib Selection

Full-sib selection is effected by crosses in which the progenies have the same two parents, whereas for half-sib selection the progenies had only one parent in common. Full-sib progenies usually are developed by crossing two individuals that are either in the same population (intra) or in two separate populations (inter). Full-sib progenies are evaluated in replicated yield tests and remnant seed is used for recombination to form the next cycle population. Full-sib selection, however, is not used by maize breeders to the same extent as half-sib selection. Hallauer and Eberhart (1970) suggested use of prolific plants for producing full-sib progenies; full-sib progenies are produced reciprocally on one ear of each plant, and the other ear on each plant is selfed and used for recombination to form the next cycle population.

1.2.4 Selfed Progeny Selection

The selfed progeny method of selection has not been used and tested as extensively as mass, half-sib, and full-sib selection for population improvement. Selfed progeny selection includes systems that evaluate selfed progenies themselves, uses the information obtained to determine the superior progenies, and recombines remnant seed of the selfed progenies to form the next cycle population. The generation of inbreeding of the progenies evaluated usually is in S_1 or S_2, although any level of inbreeding could be used. The advantages of selfed progeny selection are increased variability among the progenies evaluated and an exposing of the deleterious recessive genes that can be eliminated. Disadvantages may be longer cycle intervals (which can be minimized by off-season nurseries), greater experimental errors in progeny evaluation (but this often is not the case), and possible linkage and inbreeding effects if advanced generation lines are recombined. Selfed progeny selection seems to have a place in maize selection programs and will probably receive more attention in the future.

All the methods discussed briefly and listed in Table 1.2 have a place in maize breeding programs. Some such as pedigree selection and backcrossing have been used extensively in the past, are being used extensively at the present, and will continue to be used extensively in the future. On the other hand, other methods, such as mass selection and ear-to-row, will only be used in more specific instances. The choice of breeding methods used in a specific program will be dictated by the circumstances at a given site. For instance, in the Corn Belt of the United States, conditions dictate that hybrids will be used by the farmer; consequently, breeding methods that enhance the development of superior inbred lines for use in the production of hybrids are paramount.

We will concern ourselves primarily with the breeding methods classified under population improvement, which, regardless of the circumstances and methods selected for use, is an important facet of any breeding program. Funds, facilities, and objectives will determine the proportion of the effort a given maize breeding program will allot for population improvement. Historically, this has always been a part of maize breeding regardless of the training and knowledge of the breeder. However, techniques have been developed to assist the maize breeder in estimating and quantifying empirical data for comparison with predicted response. This type of information provides guidelines for the maize breeder to evaluate selection programs for relative gain and efficiency of response to selection. Hence changes in experimental technique and allocation of resources may be necessary based on estimated parameters. It is hoped the information presented will be of value to assist maize breeders in the estimation of genetic parameters and the selection of population improvement procedures to meet their short-, intermediate-, and long-term objectives.

REFERENCES

Allard, R. W. 1960. *Principles of Plant Breeding*. Wiley, New York.

Bauman, L. F. 1981. Review of methods used by breeders to develop superior corn inbreds. *Proc. Annu. Corn Sorghum Res. Conf.* 36:199-208.

Beal, W. J. 1880. Indian Corn Rep., Mich. Board Agric. 19:279-89.

Bowman, M. L., and B. W. Crossley. 1908. *Corn*. Kenyon, Des Moines, Iowa.

Chase, S. S. 1952a. Production of homozygous diploids of maize from monoploids. *Agron. J.* 44:263-67.

————. 1952b. Monoploids in maize. In *Heterosis,* J. W. Gowen, ed., pp. 389-99. Iowa State Univ. Press, Ames.

Cockerham, C. C. 1961. Implications of genetic variances in a hybrid breeding program. *Crop Sci.* 1:47-52.

Compton, W. A., and R. E. Comstock. 1976. More on modified ear-to-row selection. *Crop Sci.* 16:122.

Comstock, R. E.; H. F. Robinson; and P. H. Harvey. 1949. A breeding procedure designed to make maximum use of both general and specific combining ability. *Agron. J.* 41:360-67.

Davis, R. L. 1927. Report of the plant breeder. Rep. Puerto Rico Agric. Exp. Stn. Pp. 14-15.

Dudley, J. W., and R. H. Moll. 1969. Interpretation and use of estimates of heritability and genetic variances to plant breeding. *Crop Sci.* 9:257-62.

East, E. M. 1908. Inbreeding in corn. Connecticut Agric. Exp. Stn. Rep. 1907. Pp. 419-28.

Gardner, C. O. 1961. An evaluation of effects of mass selection and seed irradiation with thermal neutrons on yield of corn. *Crop Sci.* 1:241-45.

Goodman, M. M. 1976. Maize. In *Evolution of Crop Plants,* N. W. Simmonds, ed., pp. 128-36. Longman Group, London.

Goodsell, S. 1961. Male sterility in corn by androgenesis. *Crop Sci.* 1:227-28.

Hallauer, A. R. 1970. Zygote selection for the development of single-cross hybrids in maize. *Adv. Front. Plant Sci.* 25:75-81.

Hallauer, A. R., and S. A. Eberhart. 1970. Reciprocal full-sib selection. *Crop Sci.* 10:315-16.

Harrington, J. B. 1952. Cereal breeding procedures. FAO Pap. 28, Rome.

Hayes, H. K.; F. R. Immer; and D. C. Smith. 1955. *Methods of Plant Breeding.*
    McGraw-Hill, New York.
Hopkins, C. G. 1899. Improvement in the chemical composition of the corn
    kernel. Illinois Agric. Exp. Stn. Bull. 55:205-40.
Jenkins, M. T. 1934. Methods of estimating the performance of double crosses
    in corn. *J. Am. Soc. Agron.* 26:199-204.
————. 1935. The effect of inbreeding and of selection within inbred lines
    of maize upon the hybrids made after successive generations of selfing.
    Iowa State J. Sci. 9:429-50.
Jenkins, M. T., and A. M. Brunson. 1932. Methods of testing inbred lines of
    maize in crossbred combinations. *J. Am. Soc. Agron.* 24:523-30.
Jones, D. F. 1918. The effects of inbreeding and crossbreeding upon develop-
    ment. Connecticut Agric. Exp. Stn. Bull. 207:5-100.
Jones, D. F., and W. R. Singleton. 1934. Crossed sweet corn. Connecticut
    Agric. Exp. Stn. Bull. 361:489-536.
Kermicle, J. L. 1969. Androgenesis conditioned by a mutation in maize. *Sci-
    ence* 166:1422-24.
Lonnquist, J. H. 1964. Modification of the ear-to-row procedures for the im-
    provement of maize populations. *Crop Sci.* 4:227-28.
Moll, R. H., and C. W. Stuber. 1974. Quantitative genetics: Empirical re-
    sults relevant to plant breeding. *Adv. Agron.* 26:277-313.
Richey, F. D. 1927. The convergent improvement of selfed lines of corn. *Am.
    Nat.* 61:430-49.
Russell, W. A., and S. A. Eberhart. 1975. Hybrid performance of selected
    maize lines from reciprocal recurrent and testcross selection programs.
    Crop Sci. 15:1-4.
Shull, G. H. 1908. The composition of a field of maize. Am. Breeders'
    Assoc. Rep. 4:296-301.
————. 1909. A pure-line method of corn breeding. Am. Breeders' Assoc.
    Rep. 5:51-59.
Simmonds, N. W. 1973. Plant breeding. *Philos. Trans. R. Soc. London* B267:
    145-56.
Sprague, G. F. 1946. Early testing of inbred lines of corn. *J. Am. Soc.
    Agron.* 38:108-17.
————. 1955. Corn breeding. In *Corn and Corn Improvement,* G. F. Sprague,
    ed., pp. 221-92. Academic Press, New York.
Sprague, G. F., and S. A. Eberhart. 1977. Corn breeding. In *Corn and Corn
    Improvement,* G. F. Sprague, ed., pp. 305-62. Am. Soc. Agron., Madison,
    Wis.
Stadler, L. J. 1944. Gamete selection in corn breeding. *J. Am. Soc. Agron.*
    36:988-89.
Thompson, D. L. 1954. Combining ability of homozygous diploids of corn rela-
    tive to lines derived by inbreeding. *Agron. J.* 46:133-36.

2

Means and Variances

The concepts of population means and variances in current quantitative genetics theory are based on gene frequencies and gene effects or, in other words, on the genetic structure of the population under study. The population structure, however, depends on several other factors such as ploidy level, linkage, mating system, and a number of environmental and genetic factors. Therefore, either some of these factors must be known or restrictions must be imposed about their effects to be able to establish a theoretical model for study. As pointed out by Cockerham (1963), estimated parameters refer to a specific population from which the experimental material is a sample for a specific set of environmental conditions. Thus one must specify the reference population for both genotypes and environments because inferences cannot generally be translated from one population to another. Our first task, therefore, is to characterize a population of genotypes as a basis for introducing some of the current theories. More detailed descriptions of the population means and variances were given by Kempthorne (1957) and Falconer (1960).

A population of maize can be characterized by the following properties: diploid (2n = 20), panmictic (random mating with more than 95% of cross-pollination), monoecious (both sexes in the same individual but in different inflorescences), and a tendency for protandry. Additional assumptions necessary for a complete characterization of a maize population are no maternal effects, linkage equilibrium, normal fertilization (noncompeting gametes), normal meiosis, and normal segregation. The reference population of genotypes may result from a cross between two homozygous inbred lines, crosses among a set of homozygous inbred lines (synthetic variety), an open-pollinated variety, or a mixture of varieties and races (composites). General theories, however, make no distinction about the origin of the population unless it does not fill some of the basic requirements.

Finally, for the first of our theoretical presentations we will assume the reference population is in Hardy-Weinberg equilibrium.

2.1 HARDY-WEINBERG EQUILIBRIUM

In 1908 Hardy and Weinberg independently demonstrated that in a large random mating population both gene frequencies and genotypic frequencies remain constant from generation to generation in the absence of mutation, migration, and selection. Such a population is said to be in Hardy-Weinberg equilibrium and remains so unless any disturbing force changes its gene or genotypic frequency. It also can be demonstrated that for a single locus any population will attain its equilibrium after one generation of random mating.

21

The Hardy-Weinberg law can be demonstrated by taking one locus with two alleles (A_1 and A_2) in a diploid organism such as maize. Let us consider a population whose genotypic frequencies are as follows:

Genotypes	A_1A_1	A_1A_2	A_2A_2	
Number of individuals	n_1	n_2	n_3	$n_1 + n_2 + n_3 = N$
Frequency	$P = n_1/N$	$Q = n_2/N$	$R = n_3/N$	$P + Q + R = 1$

The total number of genes relative to locus A in this population is 2N, i.e., two genes in each diploid individual. Thus the numbers of A_1 and A_2 alleles are $2n_1 + n_2$ and $2n_3 + n_2$, respectively, and their frequencies are

$$p(A_1) = \frac{2n_1 + n_2}{2N} = \frac{n_1 + (1/2)n_2}{N} = P + \frac{1}{2}Q$$

$$q(A_2) = \frac{2n_3 + n_2}{2N} = \frac{n_3 + (1/2)n_2}{N} = R + \frac{1}{2}Q$$

Because gametes unite at random in a population under random mating, the genotypic array and its frequency in the next generation will be:

Genotypes	Male gametes		Frequencies	Male gametes	
	A_1	A_2		p	q
Female gametes A_1	A_1A_1	A_1A_2	Female gametes p	p^2	pq
A_2	A_1A_2	A_2A_2	q	pq	q^2

So the genotypic frequencies are $p^2(A_1A_1):2pq(A_1A_2):q^2(A_2A_2)$, and this population is said to be in Hardy-Weinberg equilibrium since genotypic frequencies are expected to be unchanged in the next generation.

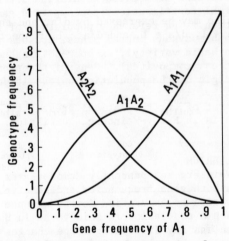

Fig. 2.1. Distributions of genotypic frequencies for gene frequencies ranging from 0 to 1.0 for one locus with two alleles in a population in Hardy-Weinberg equilibrium.

The Hardy-Weinberg law can also be extended to multiple alleles. In general, if p_i is the frequency of the ith allele at a given locus, the genotypic frequency array is given by

$$\sum_i p_i^2 \quad \text{for homozygotes } (A_iA_i)$$

$$2 \sum_{i<i'} p_ip_{i'} \quad \text{for heterozygotes } (A_iA_{i'})$$

With two alleles per locus the gene frequency that gives the maximum frequency of heterozygotes ($Q = 2pq$) is found by taking its first derivative equal to zero; i.e., $dQ/dp = 2 - 4p = 0$, and then $p = 0.5$ for Q maximum. Figure 2.1 shows the variation of genotypic frequencies for gene frequencies in the range from zero to one.

2.2 MEANS OF NONINBRED POPULATIONS AND DERIVED FAMILIES

Considering one locus with two alleles, A_1 and A_2, it is assumed that each locus has a particular effect on the total individual phenotype. Arbitrarily assuming A_1 to be the allele that increases the value, we can denote by $+a$, $-a$, and d the effects of genotypes A_1A_1, A_2A_2, and A_1A_2, respectively. Such effects are taken as deviations from the mean u of the two homozygotes, as shown on the following linear scale:

```
             A₂A₂                    A₁A₂      A₁A₁
  ┴─┴┴┴──────────────────────┴───────────────┴──
  0        -a          u            d        +a
```

where d is related to the level of dominance for

```
d = 0:   no dominance
0 < d < a:   partial dominance
d = a:   complete dominance
d > a:   overdominance
```

The level of dominance also may be expressed by d/a, and the above relations will be $d/a = 0$, $0 < d/a < 1$, $d/a = 1$, and $d/a > 1$, respectively.

The population mean u_0 is thus calculated considering both the genotypic frequencies and genotypic effects, as shown in Table 2.1. Hence, u_0 is determined as

Table 2.1. Genotypic values and frequencies in a population in Hardy-Weinberg equilibrium for one locus with two alleles.

Genotype	Frequency f_i	Coded genotypic value Y_i	Number of A_1 alleles in individuals X_i
A_1A_1	p^2	$+a$	2
A_1A_2	$2pq$	d	1
A_2A_2	q^2	$-a$	0

$$u_0 = p^2(a) + 2pq(d) + q^2(-a) = (p - q)a + 2pqd \qquad (2.1)$$

which expresses the average deviation from the mean of two homozygotes. The population mean can be expressed as

$$u' = u + u_0 = u + (p - q)a + 2pqd \quad \text{for one locus}$$

$$u' = \sum_i [u_i + (p_i - q_i)a_i + 2p_iq_id_i] \quad \text{for all loci}$$

For simplicity we can delete the subscript i. It is sufficient to know that summation $\sum_i$ is over all loci, and then we use p, q, a, and d instead of p_i, q_i, a_i, and d_i.

Since environmental effects are taken as deviations from the general mean over the whole population, they sum to zero, and then u' also expresses the mean phenotypic value.

2.2.1 Half-sib Family Means

A half-sib family is obtained from seeds produced by one plant (female parent) that was pollinated by a random sample of pollen from the population, as shown in Table 2.2.

The mean of the population of half-sib families is

$$u_{HS} = p^2(pa + qd) + 2pq(1/2)[(p - q)a + (1/2)d] + q^2(pd - qa)$$

$$= (p - q)a + 2pqd$$

and the total mean is

$$u'_{HS} = u + (p - q)a + 2pqd \quad \text{for one locus}$$

$$u'_{HS} = \sum_i [u + (p - q)a + 2pqd] \quad \text{for all loci}$$

which is equal to the original population mean.

Table 2.2. Genotypic values and frequencies of half-sib families from a population in Hardy-Weinberg equilibrium for one locus with two alleles.

Female parent	Frequency f_i	Family genotypes[†] A_1A_1	A_1A_2	A_2A_2	Coded half-sib family values, Y_i
A_1A_1	p^2	p	q	---	pa + qd
A_1A_2	2pq	(1/2)p	1/2	(1/2)q	(1/2)[(p-q)a + d]
A_2A_2	q^2	---	p	q	pd - qa

[†]Produced after pollination by $p(A_1)$ and $q(A_2)$ male gametes.

2.2.2 Full-sib Families

A full-sib family is obtained by crossing a random pair of plants from the population. The probability of each cross is obtained by the product of genotypic frequencies, as shown in Table 2.3.

Table 2.3. Genotypic values and frequencies of full-sib families from
a population in Hardy-Weinberg equilibrium for one locus
with two alleles.

Female parent	Male parent	Probability of cross, f_i	Family genotypes			Coded full-sib family values, Y_i
			A_1A_1	A_1A_2	A_2A_2	
A_1A_1	A_1A_1	p^4	1	---	---	a
	A_1A_2	$2p^3q$	1/2	1/2	---	$(1/2)(a+d)$
	A_2A_2	p^2q^2	---	1	---	d
A_1A_2	A_1A_1	$2p^3q$	1/2	1/2	---	$(1/2)(a+d)$
	A_1A_2	$4p^2q^2$	1/4	1/2	1/4	$(1/2)d$
	A_2A_2	$2pq^3$	---	1/2	1/2	$(1/2)(d-a)$
A_2A_2	A_1A_1	p^2q^2	---	1	---	d
	A_1A_2	$2pq^3$	---	1/2	1/2	$(1/2)(d-a)$
	A_2A_2	q^4	---	---	1	-a

The mean of the population for full-sib families is

$$u_{FS} = p^4(a) + 2p^3q(1/2)(a + d) + \ldots + q^4(-a) = (p - q)a + 2pqd$$

and the total mean is

$$u'_{FS} = u + (p - q)a + 2pdq \quad \text{for one locus}$$

$$u'_{FS} = \sum_i [u + (p - q)a + 2pqd] \quad \text{for all loci}$$

Results so far obtained show that the expected value of half-sib families
as well as of full-sib families equals the mean of the reference population.

2.2.3 Inbred Families

Selfing is the most common system of inbreeding used in practical maize
breeding. Considering a noninbred parent population in Hardy-Weinberg equilibrium from which selfed lines will be drawn, we have the family structure as
shown in Table 2.4 for S_1 families, i.e., families developed by one generation
of selfing.

The mean of the population for inbred families is

$$u_{S_1} = p^2a + 2pq(1/2)d + q^2(-a) = (p - q)a + pqd$$

which equals the reference population mean when $d = 0$, i.e., when there are no
dominance effects. If dominance effects are present, the mean is reduced. If
the gene frequencies are the same for the reference and S_1 populations, the
mean of the S_1 population will be halfway between the mean of the S_0 and S_∞
generations.

Under a regular system of selfing the general mean decreases in each generation due to decreases in the frequency of heterozygotes. The general for-

Table 2.4. Genotypic values and frequencies of inbred (S_1) families
from a noninbred population in Hardy-Weinberg equilibrium
for one locus with two alleles.

| Parent genotypes | Frequency, f_i | Family genotypes[†] | | | Coded S_1 family values, Y_i |
		A_1A_1	A_1A_2	A_2A_2	
A_1A_1	p^2	1	---	---	a
A_1A_2	$2pq$	1/4	1/2	1/4	$(1/2)d$
A_2A_2	q^2	---	---	1	$-a$

[†]After one generation of selfing.

mula for the nth generation of inbreeding is

$$u_{S_n} = (p - q)a + (1/2)^{n-1}pqd \quad \text{for one locus}$$

$$u'_{S_n} = \sum_i [u + (p - q)a + (1/2)^{n-1}pqd] \quad \text{for all loci}$$

which equals the noninbred population mean for $n = 0$.

The above formula also may be expressed as a function of F_n, the coefficient of inbreeding, of progenies in the nth generation of selfing:

$$u_{S_n} = \sum_i [u + (p - q)a + 2(1 - F_n)pqd]$$

which equals the noninbred population mean when $F = 0$.

2.3 MEANS OF INBRED POPULATIONS AND DERIVED FAMILIES

The main difference between inbred and noninbred populations is in genotypic frequencies. Gene frequencies remain constant, but genotypic frequencies change under inbreeding because inbreeding decreases the frequency of heterozygotes and consequently increases the frequency of homozygous genotypes. Using Wright's coefficient F as a measure of inbreeding, genotypic frequencies are distributed according to the pattern shown in Table 2.5.

Hence the mean of an inbred population is $u_S = (p - q)a + 2pq(1 - F)d$ or $u'_S = [u + (p - q)a + 2pq(1 - F)d]$. When $F = 0$, $u'_S = u'_0$, the mean of a noninbred population. When $F = 1$ (completely homozygous population), then $u_S =$

Table 2.5. Genotypic values and frequencies in inbred populations
(inbreeding measured by F) for one locus with two alleles.

Genotypes	Frequencies	Coded genotypic values
A_1A_1	$p^2 + Fpq$	a
A_1A_2	$2pq(1-F)$	d
A_2A_2	$q^2 + Fpq$	$-a$

(p − q)a because there will be no dominance effects expressed. If we take F = 1/2, then $u_S = (p − q)a + pqd$, which equals the S_1 family mean, as previously shown in Table 2.4.

The general equation for the mean of an inbred population is

$$u_S' = \sum_i [u + (p − q)a + 2pq(1 − F)d]$$

which equals the mean of families in the same generation of inbreeding, as previously demonstrated.

However, half-sib and full-sib families drawn from an inbred population result in noninbred progenies, and their mean equals that of a noninbred population; i.e., $u_{HS} = u_{FS} = (p − q)a + 2pqd$ because one generation of random mating is involved.

Kempthorne (1956) gives a general formulation for the changes in population mean under inbreeding, including epistatic effects. The general formula is

$$u_S' = u_0 + FD_1 + F^2 D_2 + F^3 D_3 + \ldots$$

where u_S' is the mean of the population under a known degree of inbreeding and u_0 is the mean of the original random mating population. The D_j are defined in terms of dominance deviations in the original random mating population. Note that if there is no dominance ($D_1 = 0$) and dominance types of epistasis ($D_2 = D_3 = \ldots = 0$), the population mean does not change with inbreeding. If there are no dominance types of epistasis, the mean of the inbred population is linearly related to F even in the presence of additive types of epistasis.

2.4 MEAN OF A CROSS BETWEEN TWO POPULATIONS

Let P_1 and P_2 be two populations in Hardy-Weinberg equilibrium. Denoting by p and q the frequencies of both alleles, A_1 and A_2, in population P_1 and by r and s the frequencies of the same alleles in population P_2, we have the following structure in the crossed population (Table 2.6). The population cross mean is $u_{12} = (pr − qs)a + (ps + qr)d$ for one locus.

The cross between two populations also may be obtained according to a family structure. If half-sib families are drawn with, for example, P_1 as female parents, we have the family structure shown in Table 2.7. The mean of half-sib families is then

$$u_{HS_{12}} = p^2(ra + sd) + 2pq[(1/2)(r − s)a + (1/2)d] + q^2(rd − sa)$$

$$= (pr − qs)a + (ps + qr)d$$

which equals the randomly crossed population mean. Note that the mean will be the same whatever population is used as the female parent.

If the crossed population is structured as full-sib families, we have the genotypes and frequencies shown in Table 2.8. The mean becomes

$$u_{FS_{12}} = p^2 r^2 a + 2p^2 rs(1/2)(a + d) + \ldots + q^2 s^2(−a)$$

$$= (pr − qs)a + (ps + qr)d$$

which equals the randomly crossed population mean and has the same value whatever parent is used as female.

Table 2.6. Genotypic values and frequencies in a cross between two
            populations in Hardy–Weinberg equilibrium for one locus
            with two alleles.

Genotypes	Frequency	Coded genotypic values
A_1A_1	pr	a
A_1A_2	ps + qr	d
A_2A_2	qs	-a

Table 2.7. Genotypic values and frequencies in a cross between two
            populations structured as half-sib families for one locus
            with two alleles.

Female parent, P_1	Frequencies	Family genotypes[+] A_1A_1	A_1A_2	A_2A_2	Coded half-sib family values
A_1A_1	p^2	r	s	---	ra + sd
A_1A_2	2pq	(1/2)r	(1/2)(r+s)	(1/2)s	(1/2)(r-s)a + (1/2)d
A_2A_2	q^2	---	r	s	rd - sa

[+]After pollination by $r(A_1)$ and $s(A_2)$ male gametes (from P_2).

Table 2.8. Genotypic values and frequencies in a cross between two
            populations structured as full-sib families for one locus
            with two alleles.

Female parent, P_1	Male parent, P_2	Frequency of crosses	Family genotypes A_1A_1	A_1A_2	A_2A_2	Coded full-sib family values
A_1A_1	A_1A_1	p^2r^2	1	0	0	a
	A_1A_2	$2p^2rs$	1/2	1/2	0	(1/2)(a+d)
	A_2A_2	p^2s^2	0	1	0	d
A_1A_2	A_1A_1	$2pqr^2$	1/2	1/2	0	(1/2)(a+d)
	A_1A_2	4pqrs	1/4	1/2	1/4	(1/2)d
	A_2A_2	$2pqs^2$	0	1/2	1/2	(1/2)(d-a)
A_2A_2	A_1A_1	q^2r^2	0	1	0	d
	A_1A_2	$2q^2rs$	0	1/2	1/2	(1/2)(d-a)
	A_2A_2	q^2s^2	0	0	1	-a

2.5 AVERAGE EFFECT AND BREEDING VALUE

When panmictic populations are under consideration, one must consider that the genotypes of any offspring are not identical to their parents. The relationship between any individual in the offspring and one of its parents is established by the gamete received from that parent. It is known that gametes are haploid entities and carry genes and not genotypes. So for the understanding of the inheritance of a quantitative trait in a panmictic population it is valuable to have an individual measure associated with its genes and not its genotype. Such a value is designated by Falconer (1960) as "breeding value," which is "the value of an individual, judged by the mean value of its progeny."

The concept of "average effect" of a gene is basic to the understanding of breeding value. The average effect of a gene, say A_1, is defined as the mean deviation from the population mean of a group of individuals that received the gene from the same parent, the other gene of such individuals being randomly sampled from the whole population as shown in Table 2.9.

A similar concept is the "average effect of a gene substitution," which is the average deviation due to the substitution of one gene by its allele in each genotype. Consider the A_2 gene being substituted by its gene A_1 at random in the population. Since the A_1A_1, A_1A_2, and A_2A_2 genotypes have frequencies p^2, $2pq$, and q^2, respectively, then genes that will be substituted are found in genotypes A_1A_2 and A_2A_2 with frequency $pq + q^2 = q$. Proportionally, we have $pq/q = p$ A_1A_2 genotypes:$q^2/q = q$ A_2A_2 genotypes; i.e., A_2 genes will be substituted in A_1A_2 and A_2A_2 genotypes at frequencies p and q, respectively. When the substitution is in A_1A_2 genotypes, the change in genotypic value will be from d to a; and when substitution takes place in A_2A_2, the change is from $-a$ to d. The change in the population is $\alpha = p(a - d) + q(d + a) = a + (q - p)d$, which is by definition the average effect of a gene substitution. It can be seen that the average effect of a gene substitution α is the difference between the average effects of genes involved in the substitution; i.e., $\alpha = \alpha_1 - \alpha_2$, as shown in Table 2.9. Both the average effect of a gene and the average effect of a gene substitution depend on gene effects and gene frequency; therefore, both are a property of the population as well as of the gene.

Extending the concept of average effect of genes to the individual genotype gives the concept of breeding value of the individual. At the gene level the breeding value is the sum of average effects of genes, summation being over all alleles and over all loci. Similar to the average effect of a gene, breeding value is a property of the individual as well as of the population; but the breeding value can be measured experimentally. The breeding value is therefore a measurable quantity and is of much relevance in animal breeding

Table 2.9. Genotypes and average effects of progenies having a common parental gamete for one locus (after Falconer 1960).

Gamete	Genotypes in progenies[†] A_1A_1	A_1A_2	A_2A_2	Progeny effects X	Population mean $\bar{Y}$	Average effect of a gene $X-\bar{Y}$
A_1	p	q	--	pa + qd	(p-q)a+2pqd	$\alpha_1 = q[a+(q-p)d]$
A_2	--	p	q	pd - qa	(p-q)a+2pqd	$\alpha_2 = -p[a+(q-p)d]$

[†]After pollination with a random sample of gametes: $p(A_1)$ and $q(A_2)$.

where the individual value is an important criterion. On the other hand, individual values are less important in crop species like maize, since the whole population is concerned. In this case individuals are looked upon as ephemeral representatives of the whole population and its gene pool. The average effect of a gene and individual breeding value concepts, however, are closely related to genotype evaluation procedures like topcross tests in maize.

2.6 VARIANCES: POPULATIONS AND FAMILIES

Fisher (1918) first demonstrated that the hereditary variance in a random mating population can be partitioned into three parts: (1) an additive portion associated with average effects of genes, (2) a dominance portion due to allelic interactions, and (3) a portion due to nonallelic interactions or epistatic effects. A general theory for the partition of hereditary variance was further developed by Cockerham (1954) and Kempthorne (1954). Thus, in general, the total genetic variance σ_G^2 can be partitioned into the following components:

σ_A^2 = additive variance due to the average effects of alleles at the same locus (additive effects)

σ_D^2 = dominance variance due to interaction of average effects of alleles at the same locus (dominance effects)

σ_{AA}^2, σ_{AAA}^2, ... = epistatic variances due to interaction of additive effects of two or more loci

σ_{DD}^2, σ_{DDD}^2, ... = epistatic variances due to interaction of dominance effects of two or more loci

σ_{AD}^2, σ_{AAD}^2, σ_{ADD}^2, ... = epistatic variances due to interaction of additive and dominance effects involving two or more loci

Collecting all components together, the total genetic variance is $\sigma_G^2 = \sigma_A^2 + \sigma_D^2 + \sigma_{AA}^2 + \sigma_{DD}^2 + \sigma_{AD}^2 + \sigma_{AAA}^2 + \sigma_{AAD}^2 + \ldots$.

Details about complex models including epistasis are beyond the scope of this book. Nevertheless, epistasis will be included in the general results. A simple model will be assumed to illustrate some theory commonly used for the study of quantitative traits.

2.6.1 Total Genetic Variance

Total genetic variance of a population in Hardy-Weinberg equilibrium is obtained from Table 2.1 as follows:

$$\sigma_G^2 = \sum_i f_i Y_i^2 - u_0^2 = 2pq[a + (q - p)d]^2 + 4p^2q^2d^2$$

Additive genetic variance, defined as the variance due to linear regression of Y (genotypic values) on X (gene frequencies in individual genotypes), is $\sigma_A^2 = [\text{Cov }(Y, X)]^2/\sigma_X^2$. From Table 2.1 we get

$$\text{Cov }(Y, X) = \sum_i f_i X_i Y_i - u_X u_0 = 2pq[a + (q - p)d]$$

Table 2.10. Gene frequency (p) at maximum values for σ_G^2, σ_A^2, and σ_D^2 for no dominance and complete dominance for one locus with two alleles.

Gene action	σ_G^2	σ_A^2	σ_D^2
No dominance	1/2	1/2	---
Complete dominance	$1 - \sqrt{1/2}$	1/4	1/2

Then $\sigma_X^2 = \Sigma_i f_i X_i^2 - \mu_X^2 = 2pq$ so that $\sigma_A^2 = 2pq[a + (q - p)d]^2 = 2pq\alpha^2$ for one locus, and $\sigma_A^2 = 2\Sigma_i pq[a + (q - p)d]^2 = 2\Sigma_i pq\alpha^2$ for all loci.

Dominance variance is the remainder from the total variance and is calculated by subtraction as $\sigma_D^2 = \sigma_G^2 - \sigma_A^2 = 4 \Sigma_i p^2 q^2 d^2$. Hence the dominance variance is the deviation from the regression of genotypic values on the gene content of the genotypes.

Gene frequencies, which give the maximum values for σ_G^2, σ_A^2, and σ_D^2, are found by taking the first derivative of their respective expressions equal to zero. The simplest cases are those for no dominance and complete dominance, as shown in Table 2.10.

2.6.2 Variance among Noninbred Families

Genetic variance among half-sib families from a population in Hardy-Weinberg equilibrium is obtained from Table 2.2 as

$$\sigma_{HS}^2 = \sum_i f_i Y_i^2 - u_{HS}^2 = (1/2)pq[a + (1 - 2p)d]^2 \quad \text{for one locus}$$

Because $\sigma_A^2 = 2pq[a + (q - p)d]^2$, it is seen that $\sigma_{HS}^2 = (1/4)\sigma_A^2$, where σ_A^2 is the additive genetic variance of the reference population. Theoretically, the total genetic variance (among and within half-sib families) equals the total genetic variance in the reference population; i.e., $\sigma_G^2 = \sigma_A^2 + \sigma_D^2$. From this total variance $(1/4)\sigma_A^2$ is expressed among half-sib families, and so the remainder, $(3/4)\sigma_A^2 + \sigma_D^2$, is expected to be present within half-sib families over the entire population of families.

Genetic variance among full-sib families is obtained from Table 2.3 as

$$\sigma_{FS}^2 = \sum_i f_i Y_i^2 - u_{FS}^2 = pq[a + (q - p)d]^2 + p^2 q^2 d^2 \quad \text{for one locus}$$

Relative to the total genetic variance of a population it is seen that $\sigma_{FS}^2 = (1/2)\sigma_A^2 + (1/4)\sigma_D^2$, which is the portion of the total genetic variance expressed among families. Thus the remainder of the total genetic variance, $(1/2)\sigma_A^2 + (3/4)\sigma_D^2$, is expected to be present within families over the whole population.

If epistasis is considered, the approximation to the real genetic variance is $\sigma_G^2 = \sigma_A^2 + \sigma_D^2 + \sigma_{AA}^2 + \sigma_{AD}^2 + \sigma_{DD}^2 + \sigma_{AAA}^2 + \ldots$, where all components were defined previously.

In the same way, $\sigma_{HS}^2 = (1/4)\sigma_A^2 + (1/16)\sigma_{AA}^2 + (1/64)\sigma_{AAA}^2 + \ldots$, or simply $\sigma_{HS}^2 = (1/4)\sigma_A^2 + \ldots$, indicating a bias due to epistatic components of variance. The variance among full-sib families is $\sigma_{FS}^2 = (1/2)\sigma_A^2 + (1/4)\sigma_D^2 + (1/4)\sigma_{AA}^2 + (1/16)\sigma_{DD}^2 + \ldots$, or simply $\sigma_{FS}^2 = (1/2)\sigma_A^2 + (1/4)\sigma_D^2 + \ldots$, indicating a bias due to epistatic components. The relation between the variance among families and the covariance between relatives within families is presented in Chap. 3.

2.6.3 Variance among Inbred Families

Variance among S_1 families from a noninbred reference population is obtained from Table 2.4 as

$$\sigma_{S_1}^2 = \sum_i f_i Y_i^2 - u_{S_1}^2 = 2pq[a + (1/2)(q - p)d]^2 + p^2q^2d^2 \quad \text{for one locus}$$

A problem that arises with inbreeding is that genetic variance among inbred families is not linearly related to genetic components of variance of the

Table 2.11. Distribution of variances among and within inbred lines under continuous selfing, assuming $p = q = 0.5$.

Generation	F[†]	Among lines σ_A^2	Among lines σ_D^2	Within lines σ_A^2	Within lines σ_D^2	Total σ_A^2[‡]	Total σ_D^2[§]
S_1	1/2	1	1/4	1/2	1/2	3/2	3/4
S_2	3/4	3/2	3/16	1/4	1/4	7/4	7/16
S_3	7/8	7/4	7/64	1/8	1/8	15/8	15/64
S_4	15/16	15/8	15/256	1/16	1/16	31/16	31/256
S_5	31/32	31/16	31/1024	1/32	1/32	63/32	63/1024
S_6	63/64	63/32	63/4096	1/64	1/64	127/64	127/4096
$\vdots$	$\vdots$	$\vdots$	$\vdots$	$\vdots$	$\vdots$	$\vdots$	$\vdots$
S_n	$1 - \dfrac{1}{2^n}$	$\dfrac{2^n-1}{2^{n-1}} = 2F$	$\dfrac{2^n-1}{4^n} = \dfrac{F}{2^n}$	$\dfrac{1}{2^n} = 1 - F$	$\dfrac{1}{2^n} = 1 - F$	$\dfrac{2^{n+1}-1}{2^n} = 1 + F$	$\dfrac{2^{n+1}-1}{4^n} = \dfrac{F(1-2^n)}{2^n} + 1$
$\vdots$							
S_{oo}	1	2	0	0	0	2	0

[†]F is the coefficient of inbreeding.

[‡]Values under σ_A^2 are coefficients of σ_A^2, or

$$\frac{2^n-1}{2^{n-1}} + \frac{1}{2^n} = \frac{2(2^n-1)+1}{2^n} = \frac{2^{n+1}-2+1}{2^n} = \frac{2^{n+1}-1}{2^n}$$

[§]Values under σ_D^2 are coefficients of σ_D^2, or

$$\frac{2^n-1}{4^n} + \frac{1}{2^n} = \frac{2^n-1+2^n}{4^n} = \frac{2^{n+1}-1}{4^n}$$

reference population. From the above expression it can be seen that $\sigma^2_{S_1}$ can be translated into σ^2_A and/or σ^2_D only in the following situations:

No dominance (d_i = 0 for all loci): $\sigma^2_{S_1} = \sigma^2_A$

Gene frequency 1/2 for all loci: $\sigma^2_{S_1} = \sigma^2_A + (1/4)\sigma^2_D$

Otherwise, variance among S_1 families can be expressed by $\sigma^2_{A'} + (1/4)\sigma^2_D$, where $\sigma^2_{A'} = \sigma^2_A + \beta'$, where β' is a deviation from σ^2_A due to dominance effects. For one locus $\beta' = 2pq[p-(1/2)]d[2a + (3/2)(q-p)d]$ and $\sigma^2_{A'} = 2pq[a + (1/2)(q-p)d]^2$. which equals zero for either d = 0 or p = 1/2. If the restriction of no dominance is imposed, it also can be demonstrated that the genetic variance among S_2 families (after two generations of selfing) is $\sigma^2_{S_2} = (3/2)\sigma^2_A$. In the same way, if gene frequencies are assumed to be 1/2, then $\sigma^2_{S_2} = (3/2)\sigma^2_A + (3/16)\sigma^2_D$. A summary of the distribution of σ^2_A and σ^2_D among and within lines for successive generations of inbreeding is given in Table 2.11. Two important features are that (1) the genetic variance among inbred families increases and within inbred families decreases with increased inbreeding; and (2) the total genetic variance doubles from F = 0 to F = 1, and all the genetic variance at F = 1 is additive. General limitations when inbreeding is involved are discussed in Chap. 3.

2.6.4 Variance in Inbred Populations
The effect of inbreeding is to increase total genetic variance in the population, and such an increase depends on the level of inbreeding. Total genetic variance is calculated from Table 2.5 as follows:

$$\sigma^2_{G_S} = (p^2 + Fpq)a^2 + 2pq(1 - F)d^2 + (q^2 + Fpq)a^2 - u^2_S$$

$$= 2pq(1 + F)[a + \frac{1 - F}{1 + F}(q - p)d]^2 + 4pq\frac{1 - F}{1 + F}(p + Fq)(q + Fp)d^2$$

which equals the genetic variance of a noninbred population when F = 0. The additive genetic variance, defined as the variance due to linear regression of Y (genotypic values) on X (gene content in individual genotypes), is

$$\sigma^2_{A_S} = 2pq(1 + F)[a + \frac{1 - F}{1 + F}(q - p)d]^2$$

and the dominance variance is calculated as $\sigma^2_{G_S} - \sigma^2_{A_S}$:

$$\sigma^2_{D_S} = 4pq\frac{1 - F}{1 + F}(p + Fq)(q + Fp)d^2$$

When F = 0, $\sigma^2_{A_S}$ and $\sigma^2_{D_S} = \sigma^2_A$ and σ^2_D, respectively. For F > 0, $\sigma^2_{A_S}$ and $\sigma^2_{D_S}$ cannot be expressed by σ^2_A and σ^2_D nor be translated from one generation of of inbreeding to another. If gene frequency is assumed to be 1/2, then $\sigma^2_{A_S}$ = (1 + F)σ^2_A and $\sigma^2_{D_S}$ = (1 - F^2)σ^2_D. The first expression also is valid when only additive effects are considered and $\sigma_{G_S}^2 = (1 + F)\sigma_A^2$.

When half-sib and full-sib families are drawn from an inbred population, the families are themselves noninbreds and the variance among families is expressed by

$$\sigma_F^2 = \theta_1\sigma_A^2 + \theta_2\sigma_D^2 + \theta_1^2\sigma_{AA}^2 + \theta_2^2\sigma_{DD}^2 + \theta_1\theta_2\sigma_{AD}^2 + \theta_1^3\sigma_{AAA}^2 + \ldots$$

where $\theta_1 = (1 + F)/4$ and $\theta_2 = 0$ for half-sib families and $\theta_1 = (1 + F)/2$ and $\theta_2 = (1 + F)^2/4$ for full-sib families, according to adaptation from Cockerham (1963). In Chap. 3 these covariances are expressed in terms of covariance between relatives.

Under continuous selfing, assuming only additive effects, the total genetic variance is partitioned into among lines and within lines as follows:

Among lines	$2F\sigma_G^2$
Within lines	$(1 - F)\sigma_G^2$
Total	$(1 + F)\sigma_G^2$

where σ_G^2 is the total genetic variance in a random mating noninbred population. Thus when inbreeding is complete (F = 1), the genetic variance among lines is twice the genetic variance of the reference population (noninbred) and no genetic variation is expected within lines. At complete inbreeding (F = 1), the additive model is completely valid (since there is no dominance) and is a good approximation for F slightly less than 1 (highly inbred lines).

2.6.5 Variance in a Cross between Two Populations

The first cross between two distinct populations is not in equilibrium, and its total genetic variance is not linearly related to any of the parent populations. However, total genetic variance can be partitioned into additive and dominance components as follows:

$$\sigma_{A_{(12)}}^2 = pq[a + (s - r)d]^2 + rs[a + (q - p)d]^2 = 1/2(\sigma_{A_{12}}^2 + \sigma_{A_{21}}^2)$$

$$\sigma_{D_{12}}^2 = 4p(1 - p)r(1 - r)d^2 = 4\ pqrsd^2$$

according to notation used in Table 2.6. (See Compton et al. 1965.)

The two components may be called *homologues* of additive and dominance variances as defined for one population, i.e., σ_A^2 and σ_D^2. In fact, they equal σ_A^2 and σ_D^2 when p = r; i.e., both populations have exactly the same gene frequency. In the above notation we used $\sigma_{A_{(12)}}^2$ to denote the total additive genetic variance and either $\sigma_{A_{12}}^2$ or $\sigma_{A_{21}}^2$ to denote the subcomponents when either P_1 or P_2, respectively, is used as the female parent.

When the crossed population is in a half-sib family structure, the variance among families is obtained from Table 2.7 as follows:

$$\sigma_{HS_{12}}^2 = p^2(ra + sd)^2 + \ldots + q^2(rd - sa)^2 - u_{HS_{12}}^2 = (1/2)pq[a + (s - r)d]^2 = (1/4)\sigma_{A_{12}}^2$$

If population P_2 is used as the female parent,

$$\sigma^2_{HS_{21}} = (1/2)rs[a + (q - p)d]^2 = (1/4)\sigma^2_{A_{21}}$$

If both types of families are drawn, their mean variance is $(1/4)\sigma^2_{A_{12}} + (1/4) \cdot \sigma^2_{A_{21}}$; and for $p = r$ this equals $(1/2)[(1/4)\sigma^2_A + (1/4)\sigma^2_A] = (1/4)\sigma^2_A$, which is the variance among half-sib families for one population.

In the same way, genetic variance among full-sib families is obtained from Table 2.8 as follows:

$$\sigma^2_{FS_{(12)}} = p^2r^2a^2 + \ldots + q^2s^2a^2 - u^2_{FS_{(12)}}$$

$$= (1/2)pq[a + (1 - 2r)d]^2 + (1/2)rs[a + (1 - 2p)d]^2 + pqrsd^2$$

$$= (1/2)\sigma^2_{A_{12}} + (1/4)\sigma^2_{D_{12}}$$

Note that this result will be the same whatever parental population is used as female parent. When $p = r$, then $\sigma^2_{FS} = (1/2)\sigma^2_A + (1/4)\sigma^2_D$, the variance among full-sib families as previously demonstrated for one population.

REFERENCES

Cockerham, C. C. 1954. An extension of the concept of partitioning hereditary variance for analysis of covariance among relatives when epistasis
is present. *Genetics* 39:859–82.

————. 1963. Estimation of genetic variances. In *Statistical Genetics and
Plant Breeding*, W. D. Hanson and H. F. Robinson, eds., pp. 53–94. NAS-
NRC Publ. 982.

Compton, W. A.; C. O. Gardner; and J. H. Lonnquist. 1965. Genetic variability in two open-pollinated varieties of corn (*Zea mays* L.) and their F_1
progenies. *Crop Sci.* 5:505–8.

Falconer, D. S. 1960. *Introduction to Quantitative Genetics*. Ronald Press,
New York.

Fisher, R. A. 1918. The correlation between relatives on the supposition of
Mendelian inheritance. *Trans. R. Soc. Edinburgh* 52:399–433.

Kempthorne, O. 1954. The correlations between relatives in a random mating
population. *Proc. R. Soc. London* B143:103–13.

————. 1957. *An Introduction to Genetic Statistics*. Wiley, New York.

Lush, J. L. 1945. *Animal Breeding Plans*. Iowa State Univ. Press, Ames.

3

Covariances between Relatives

Covariance and correlation between relatives are of great importance in plant and animal breeding because nearly all breeding methods deal to some extent with resemblance between relatives. Progress from selection (recurrent or not) is directly proportional to the degree of resemblance between progenies and the selected parent. Besides parent and offspring, other types of relationships are useful in several aspects of quantitative genetics and breeding procedures. Therefore covariance between relatives is important in modern plant breeding for at least two reasons: (1) In most instances covariances between relatives can be expressed in terms of components of genetic variance of the reference population. On the other hand, the variance among families can in some instances be expressed as linear functions of covariance between relatives, thus allowing the estimation of components of genetic variance by using appropriate experimental and breeding designs. (2) The expected progress from selection depends basically on the degree of relationship (i.e., covariance) between the unit of selection (individuals or families) and the individuals descendant from the selected parents.

The first reports of covariation and correlation between relatives were given by Fisher (1918) and Wright (1921), although previous studies were reported in the early part of this century (Pearson 1904; Yule 1906; Weinberg 1908, 1910) as mentioned by Kempthorne (1954). A general theory for the genetic interpretation of covariance between relatives was given by Cockerham (1954) and Kempthorne (1954, 1955) that included epistasis in addition to additive and dominance effects. Both authors used a factorial approach to the partition of genetic variability, assuming no linkage, and Kempthorne's (1954) model is not restricted to the number of alleles per locus. Linkage effects were further presented by Cockerham (1956) and Schnell (1963).

Our objective is to summarize results reported by several authors without showing the exact procedures of derivation. The subject was treated in detail by Cockerham (1954, 1963), Kempthorne (1954, 1957), and Falconer (1960). For a complete review the papers of Fisher (1918) and Wright (1921) are suggested.

3.1 THEORETICAL BASIS OF COVARIANCE

Basically, the covariance between relatives depends on the genetic resemblance between them. Environmental effects are usually expected to be uncorrelated between relatives so that covariance deals with only genotypic values. General theories are limited to some reference population, which is assumed to be in Hardy-Weinberg equilibrium, and the results are extended to any number of loci and any number of alleles per locus. Effects of inbreeding may be included for either the relatives or the reference population or both. As a simple example consider the covariance between parent and offspring.

Parents	Offspring
X_1	Y_1
X_2	Y_2
.	.
X_i	Y_i
.	.
X_n	X_n

First, let us find the causation of resemblance between X_i and Y_i. The X_i genotype may be expressed by $G_x = A_x + D_x$, where G_x is measured as the deviation from the population mean, A_x is the breeding value or additive genotypic value of X_i, and D_x is the dominance deviation from the population mean. Since A and D are taken as deviations from the population mean, specification of a reference population is an obvious requirement.

Consider an offspring Y_i, which represents the expected value or average value of the progeny. It has been seen that the average value of a progeny is one-half the breeding value of the parent; i.e., $G_y = (1/2)A_x$. A dominance effect is not assigned in Y genotypic value because it has no relation with dominance effect in X. In other words, dominance effects are not transferable through gametes (haploid entities) but are recreated at random in the offspring. Hence the parent-offspring covariance is that of individual genotypes with one-half their respective breeding values; i.e., $\text{Cov}(X, Y) = E[(A_x + D_x) \cdot (1/2)A_x] = (1/2)E(A_x^2) = (1/2)\sigma_A^2$ because the expected value E of the squared breeding value is the additive genetic variance. Because A and D are uncorrelated, the expected value of the term AD = 0.

The above calculations of covariance between relatives are known as the *direct method* (Kempthorne 1957). It can be shown that the resemblance between relatives arises from the probability that two relatives have identical alleles (one or two) at a given locus. By *identical* allele we mean *identical by descent;* i.e., an allele present in both relatives is an exact copy of an allele present in a common ancestor. In the case of the parent-offspring relationship, the probability is that an allele in the offspring is identical by descent to the same allele in the parent. In this case the probability that any individual in the offspring has, at a given locus, both alleles identical to those in the parent is zero, since one of these alleles comes from another parent taken at random from the population. This is the reason why a dominance effect in the offspring cannot be identical to that in the parent.

It also can be seen that the probability of both relatives having the same alleles (identical by descent) at a given locus is different from zero only when they have two or more common ancestors. For example, the covariance between full-sibs is $(1/2)\sigma_A^2 + (1/4)\sigma_D^2$ for a noninbred reference population.

The dominance component arises because two full-sibs taken at random from a progeny may have both alleles common at a given locus with a probability of 1/4. However, the possible common dominance effects in both full-sibs is not related to the dominance effects in the parent as far as identity by descent is concerned. For example, consider two parents with genotypes A_iA_j and A_kA_l. The possible genotypes in the full-sib family are:

Genotypes	Probability
A_iA_k	1/4
A_iA_l	1/4
A_jA_k	1/4
A_jA_l	1/4

If a pair of individuals is taken at random, the probability that both have the same alleles at locus A is $P(A_iA_k, A_iA_k) + P(A_iA_l, A_iA_l) + P(A_jA_k, A_jA_k) + P(A_jA_l, A_jA_l) = 1/4$.

When the full-sibs are identical twins, the probability that they have the same alleles at a locus is 1 and their covariance is $\sigma_A^2 + \sigma_D^2$.

There is another case where the dominance effects in the offspring are related to that of parents. It occurs in asexually propagated species where progeny genotypes are identical to their parents and identical among themselves. In self-fertilizing homozygous species, progeny genotypes also are identical to their parents, but in this case there is no dominance because there are no heterozygotes.

3.2 COVARIANCE BETWEEN RELATIVES AS A LINEAR FUNCTION OF GENETIC VARIANCES

The most common cases of covariance between relatives are based mainly on the references of Cockerham (1954, 1963) and Kempthorne (1954, 1957). In general, these cases may be classified as follows: (1) Noninbred relatives--(a) from a noninbred reference population and (b) from an inbred reference population. (2) Inbred relatives--(a) gene frequency is one-half and (b) gene frequency is general.

3.2.1 Noninbred Relatives
Relatives from a Noninbred Reference Population. A very common situation in maize breeding involves noninbred relatives from a noninbred population. A general method for finding covariance between relatives when epistasis is present is given by Kempthorne (1954). According to his method the covariances depend on two coefficients, ϕ and ϕ', which vary according to the genetic resemblance between relatives. The coefficient ϕ is defined as the probability of two relatives receiving by descent the same genes from one or more common ancestors tracing back through one parent; ϕ' is defined in the same way for the parent of the opposite sex. The coefficients for the components of variance are found, using σ_x^2 as a general notation, where x contains r A subscripts (for additive effects) and s D subscripts (for dominance effects). Using the following formula for the covariance of relatives, Cov $(X, Y) = \Sigma[(\phi + \phi')/2]^r(\phi\phi')^s$, the coefficients C of σ_A^2 and σ_D^2 can be obtained.

For example, in the covariance between full-sibs, $\phi = 1/2$ and $\phi' = 1/2$, so

$$\sigma_A^2: \quad r = 1, \ s = 0 \qquad C = (1/2)^1 \cdot (1/4)^0 = 1/2$$

σ_D^2: $r = 0$, $s = 1$ $C = (1/2)^0 \cdot (1/4)^1 = 1/4$

σ_{AA}^2: $r = 2$, $s = 0$ $C = (1/2)^2 \cdot (1/4)^0 = 1/4$

σ_{DD}^2: $r = 0$, $s = 2$ $C = (1/2)^0 \cdot (1/4)^2 = 1/16$

σ_{AD}^2: $r = 1$, $s = 1$ $C = (1/2)^1 \cdot (1/4)^1 = 1/8$

The expectations of the genetic components of variance for the covariance of full-sibs becomes $(1/2)\sigma_A^2 + (1/4)\sigma_D^2 + (1/4)\sigma_{AA}^2 + (1/8)\sigma_{AD}^2 + (1/16)\sigma_{DD}^2$.

Relatives from an Inbred Reference Population. When an inbred reference population is under consideration, Cockerham's (1963) general formula can be used: $\text{Cov}(x, y) = \theta_1\sigma_A^2 + \theta_2\sigma_D^2 + \theta_1^2\sigma_{AA}^2 + \theta_2^2\sigma_{DD}^2 + \theta_1\theta_2\sigma_{AD}^2 + \theta_1^3\sigma_{AAA}^2 + \dots$, where θ_1 and θ_2 are expressed as a function of F. Obviously, it also includes the case of a noninbred reference population (F = 0).

Both Cockerham's (1963) and Kempthorne's (1954) coefficients are presented in Table 3.1 for the most common covariances of relatives; F is the coefficient of inbreeding in the reference population.

Table 3.1. Coefficients of additive and dominance components of variance in the covariances of noninbred relatives. (adapted from Cockerham 1963; Kempthorne 1954).

Relatives	From Cockerham		From Kempthorne	
	θ_1	θ_2	ϕ	ϕ'
Parent-offspring[†]	1/2	0	1	0
Half-sibs	(1+F)/4	0	1/2	0
Full-sibs	(1+F)/2	$[(1+F)/2]^2$	1/2	1/2
Uncle-nephew	(1+F)/4	0	1/2	0
Half-uncle nephew	(1+F)/8	0	1/4	0
First cousins	(1+F)/8	0	1/4	0

[†] For noninbred parent because it is one of the relatives.

3.2.2 Inbred Relatives

The most common procedure involving inbreeding in maize is continuous selfing starting from a noninbred reference population S_0. The S_0 plants are selfed, giving rise to S_1 families. Selfing is continued and pedigree maintained so that in each generation of inbreeding the relatives can be traced back to a common parent in any previous generation. Actually, this is only one specific situation. In general, we can start in any generation t of inbreeding and continue selfing to the t' and t" generations. Usually t' = t + 1 and t" = t' + 1, but not necessarily (Cockerham 1963), and two situations exist relative to the gene frequency and the genetic model.

Gene Frequency is One-half. When gene frequency is assumed to be one-half, the covariance between relatives in the same generation of inbreeding may be expressed as a linear function of the components of genetic variance from the noninbred population. The coefficients are

$$\theta_1 = (1 + F_t) \quad \text{and} \quad \theta_2 = [(1 + F_t)/(1 - F_t)](1 - F_g)^2$$

where F_t is the inbreeding coefficient of the last common parent in the self-ing chain and F_g is the inbreeding coefficient of progenies or relatives. The coefficient for additive effects θ_1 depends only on the inbreeding of the common parent and is not affected by the inbreeding of progenies or relatives. On the other hand, θ_2 depends on the inbreeding both of parents F_t and of progenies F_g. For example, let us consider continuous selfing starting from a noninbred population S_0.

	General notation	Generation	F
	i	S_0	0
	ij	S_1	1/2
	ijk	S_2	3/4
	ijkl	S_3	7/8

(tree diagram with nodes: 1 → 11, 12; 11 → 111, 112; 12 → 121, 122; 2 → 21, 22; 21 → 211, 212; 22 → 221, 222; 222 → 2221, 2222)

As an illustration we will find the genetic variance expressed among S_2 families. It has already been noted that variance among groups of relatives may be expressed as covariance between relatives within groups. The total ge-netic variance among S_2 families, therefore, includes the variance among S_2 families tracing back to the same S_0 plant and the variance among groups of S_2 families tracing back to different S_0 plants; i.e., $\sigma^2_{S_2}$ (total) $= \sigma^2_{S_1/S_0} + \sigma^2_{S_0}$, where S_0 and S_1 are taken in terms of S_2 data. This is similar to the nested mating design, which is described in Chap. 4.

We have $\sigma^2_{S_0}$ = Cov tgg = Cov 022, which is the covariance between individ-uals in the S_2 generation whose last common parent is one S_0 plant. In the selfing series given above this corresponds to the covariance between individ-uals that have the same i but different j in the general notation. The covar-iance is

$$\text{Cov } 022 = (1 + 0)\sigma^2_A + [(1 + 0)/(1 - 0)](1 - 3/4)^2\sigma^2_D = \sigma^2_A + (1/16)\sigma^2_D$$

The other term $\sigma^2_{S_1/S_0}$ = Cov t'gg - Cov tgg = Cov 122 - Cov 022, where Cov 122 is the covariance between individuals in the S_2 generation whose last common parent is in the S_1 generation. It can be seen that this is the covariance between individuals that contain the same ij but different k in the general notation. Thus we have

$$\text{Cov } 122 = (1 + 1/2)\sigma^2_A + [(1 + 1/2)/(1 - 1/2)](1 - 3/4)^2\sigma^2_D = (3/2)\sigma^2_A + (3/16)\sigma^2_D$$

and Cov 122 - Cov 022 $= (1/2)\sigma^2_A + (2/16)\sigma^2_D$. Hence the total genetic variance among S_2 families is

$$\sigma^2_{S_2} = \sigma^2_A + (1/16)\sigma^2_D + (1/2)\sigma^2_A + (2/16)\sigma^2_D = (3/2)\sigma^2_A + (3/16)\sigma^2_D$$

The total genetic variance in the S_2 generation when gene frequency is one-half is $(1 + F_g)\sigma_A^2 + (1 - F_g^2)\sigma_D^2 = (7/4)\sigma_A^2 + (7/16)\sigma_D^2$, so σ^2 (total) $- \sigma^2$ (among lines) $= (1/4)\sigma_A^2 + (1/4)\sigma_D^2$ is expected to be the genetic variance within S_2 families over the population of families.

When relatives are in different generations of inbreeding (g and g'), the coefficients are $\theta_1 = 1 + F_t$ and $\theta_2 = [(1 + F_t)/(1 - F_t)](1 - F_g)(1 - F_{g'})$, which includes the situation when inbreds are in the same generation (g = g'). However, when relatives are in different generations the covariance can only be translated into components of genetic variance when there are no additive by dominance epistasis and no dominance types of epistasis (Cockerham 1963). For example, the covariance between parent and offspring when parents are in the S_0 generation is Cov 001 $= \sigma_A^2 + (1/2)\sigma_D^2 + \sigma_{AA}^2 + \sigma_{AAA}^2 + \ldots$ if we assume no epistasis involving dominance effects.

Gene Frequency is General. In this case covariance between relatives can only be linearly expressed as a function of the components of variance under the restriction of the genetic model. The only adequate model is one including additive and additive types of epistatic effects. There is, therefore, only one coefficient, $\theta_1 = 1 + F_t$, and the components of variance are defined for the noninbred generation. Even under dominance effects we can have a good approximation to the additive model with low inbreeding of parents ($F_t \simeq 0$) or high inbreeding of progenies or relatives ($1 - F_g \simeq 0$) as shown by Cockerham (1963).

Cockerham presented the general limitations of expressing covariances of relatives in linear functions of components of genetic variance, which are summarized in Table 3.2. When F = 0 no limitation is imposed either on the genetic model or on gene frequencies. With partially inbred relatives (F < 1) in the same generation, the expression of covariances as linear functions of

Table 3.2. Situations for which the covariance of relatives can be expressed as a linear function of components of genetic variance (Cockerham 1963).

Relatives	Inbreeding coefficient	Gene frequencies	Genetic model
Noninbred[†]	F = 0	unspecified	Unlimited
In the same generation of inbreeding	0 < F < 1	unspecified	Additive and all additive types of epistasis
	F = 1	unspecified	Unlimited but including only additive and all additive types of epistasis
	0 ≤ F ≤ 1	one-half	Unlimited
In different generations of inbreeding		unspecified	Additive and all additive types of epistasis
		one-half	Additive, dominance, and all additive types of epistasis

[†]From either noninbred or inbred reference populations.

genetic variances depends on assumptions about either gene frequency or genetic model. For example, the covariance among S_1 families falls into this category because $F = 1/2$. Thus for gene frequency unspecified, Cov $S_1 = \sigma_A^2$ + all additive types of epistasis. When only additive effects are assumed (additive model), it is equal to the variance among S_1 families as already shown. If gene frequency is one-half, the covariance also can be linearly related to dominance variance and Cov $S_1 = \sigma_A^2 + (1/4)\sigma_D^2$ + all types of epistasis $[\sigma_{AA}^2 + (1/16)\sigma_{DD}^2 + \ldots]$.

When relatives are completely homozygous ($F = 1$), there is no limitation on either gene frequencies or genetic model. There is no dominance variance, however, because there are no heterozygotes; consequently only additive and additive types of epistasis effects are considered. Completely inbred relatives can only be obtained by selfing, so the last common parent must have $F_t = 1$ and then Cov tgg $= 2\sigma_A^2$. For the selfing series the approach to homozygosity is theoretically asymptotic; and we can say that the limit of F_g is unity when n, the number of selfing generation, increases toward infinity.

When relatives are in different generations of inbreeding, it is required that the components of genetic variance be translated from one generation to another. This is only valid, assuming either a completely additive model or unimportant dominance effects, because of high inbreeding level of the relatives. If gene frequencies are assumed to be one-half, restriction on the genetic model must be only in the absence of epistasis involving dominance effects.

REFERENCES

Cockerham, C. C. 1954. An extension of the concept of partitioning hereditary variance for analysis of covariances among relatives when epistasis is present. *Genetics* 39:859-82.

———. 1956. Analysis of quantitative gene action. Brookhaven Symp. Biol. 9:53-68.

———. 1963. Estimation of genetic variances. In *Genetic Statistics and Plant Breeding*, W. D. Hanson and H. F. Robinson, eds., pp. 53-94. NAS-NRC Publ. 982.

Falconer, D. S. 1960. *Introduction to Quantitative Genetics*. Ronald Press, New York.

Fisher, R. A. 1918. On the correlation between relatives on the supposition of Mendelian inheritance. *Trans. R. Soc. Edinburgh* 52:399-433.

Kempthorne, O. 1954. The correlation between relatives in a random mating population. *Proc. R. Soc. London* B143:103-13.

———. 1955. The theoretical values of correlations between relatives in random mating populations. *Genetics* 40:153-67.

———. 1957. *An Introduction to Genetic Statistics*. Wiley, New York.

Pearson, F. 1904. On a generalized theory of alternative inheritance with special reference to Mendel's laws. *Proc. Trans. R. Soc.* A203:53-86.

Schnell, F. W. 1963. The covariance between relatives in the presence of linkage. In *Genetic Statistics and Plant Breeding*, W. D. Hanson and H. F. Robinson, eds., pp. 468-83. NAS-NRC Publ. 982.

Weinberg, W. 1908. Uber Vererbungsgesetze beim Menschen. *Z. Indukt. Abstamm.-u. Vererblehre* 1:377-92.

———. 1910. Weitere Beitrage zur Theorie der Verebung. *Arch. Rass.-u Ges. Biol.* 7:35-49, 169-73.

Wright, S. 1921. Systems of mating. I. The biometric relation between par-
    ent and offspring. *Genetics* 6:111-23.
Yule, G. U. 1906. On the theory of inheritance of quantitative compound
    characters on the basis of Mendel's laws. Third Int. Conf. Genet. Pp.
    140-42.

4

Hereditary Variance: Mating Designs

Information of the total phenotypic variation that is conditioned by the joint action of genetic and environmental forces is very important for the breeder in making decisions for the allocation of resources and expected response to selection. Plant breeders observe and measure phenotypes that are the expression of genotypes in a particular set of environments; i.e., $p_i = u + g_i + e_i + (ge)_i$, where the phenotype p_i is the sum of an overall mean u, a genotypic effect g_i, an environmental effect e_i, and an interaction effect of genotypes with environments $(ge)_i$. This linear summation of effects shows that units of measure p_i include not only genotypic effects but effects of environments on genotypes, which may or may not be repeatable from one environment to another. The environmental "noise" is a very important parameter in maize breeding because it is often beyond the control of the breeder. In some instances some of the macroenvironmental factors can be controlled (e.g., supplemental moisture by irrigation; application of fertilizers, herbicides, and insecticides; and tillage methods). For some we have previous information (e.g., soil types, previous cropping patterns, and long-time weather records); some are uncontrollable because of the uncertainties of the weather (e.g., cloud cover, occurrence of heavy rains and high temperatures, and in temperate areas the duration of frost-free days) and ravages of insects and diseases, which also are influenced by environmental conditions. Hence we measure a phenotype that includes not only the genotypic effect but the influences of an unestimable number of environmental factors on the genotype that are regular and irregular and predictable and unpredictable. Additionally, microenvironmental factors are often more obscure than macroenvironmental factors; these variations occur because of minor variations of the treatments applied to the experimental area, unevenness of moisture flow and penetration, and husbandry practices that may cause minor plant damage. Because the phenotype is a joint expression of the genotypic and environmental effects, our main interest is to determine what proportion of the phenotypic expression is due to genotypic and environmental effects.

As Cockerham (1956a) has emphasized, the genotypic effect for a particular genotype is the difference between the mean of all the phenotypes with that genotype and the mean of all the phenotypes in the population. Hence the genotypic effect is defined only in contrast with other genotypes in the same environments. If the genotypes vary in expression from one environment to another, the relative contrasts among genotypes can either remain the same or change in relative magnitude and sign. Because the contrasts among genotypes may change, and evidence shows that they often do, it becomes necessary to evaluate genotypes in different environments to determine their general performance. The changes in order, ranking, and relative values among genotypes

over several environments is called the genotype-environment interaction,
which is due mostly to macroenvironmental effects. Properly designed experiments repeated over environments will permit estimation of the variability due
to environmental effects and determine its relative importance for the particular genotypes evaluated for the particular sets of environments tested. The
total variability for the linear model becomes $\sigma_p^2 = \sigma_g^2 + \sigma_e^2 + \sigma_{eg}^2$, assuming no
correlation of genotypes and environments. Proper experimental designs and
randomization procedures will tend to minimize the correlations of genotypes
and environments, and the additive model permits estimation of components of
variance.

We are basically concerned with variance components estimation in this
chapter. Estimation is based on the analysis of variance for balanced data
and on expected mean squares proper for the model. In estimating components
of genetic variance, a random model is assumed and a random sample of the material for analysis is always required. The estimation procedure involves two
steps (Searle 1971):

1. In the analysis of variance appropriate to the model, equate observed
mean squares to their expected values; the expected values are linear functions of the unknown variance components, so the resulting equations will be a
set of simultaneous linear equations in the variance components.

2. Solve the equations established above; the solutions are the estimators of the variance components.

The expected values of mean squares in the analysis of variance do not
need assumptions of normality because the variance component estimators obtained by the analysis of variance method of estimation do not, of themselves,
depend on normality assumptions. However, the resulting estimators have limited properties, listed by Searle (1971) as follows:

Unbiasedness. Estimators of variance components derived by the analysis
of variance method from balanced data are always unbiased, whether the model
is random or mixed.

Minimum variance. Variance component estimators obtained by the analysis
of variance method are minimum variance quadratic unbiased. This means that
among all estimators of σ^2 that are both quadratic functions of the observations and unbiased, those derived by the analysis of variance method have the
smallest variance.

Negative estimates. Variance components are, by definition, positive.
Despite this, estimates obtained by the analysis of variance method can be
negative. Searle (1971) gives some suggestions to overcome this situation.
In maize, it seems, negative estimates may be due to an inadequate model (genetic designs to estimate epistatic variance), inadequate sampling (small numbers), and inadequate experimental techniques (competition among progenies).

For estimation of components of variance, we will consider mating designs
that develop progenies for evaluation. All mating designs include progenies
that involve relationships among relatives having known genetic components of
variance (see Chap. 3). For purposes of estimating components of genetic and
environmental variance, the progenies developed from the mating design will be
evaluated over environments in an appropriate experimental design. From the
analysis of variance of experimental designs, expectations will be expressed
in terms of the appropriate components of variance; from the components of
variance, translations will be made to the appropriate relationships (covari-

ances) of relatives based on the mating design used; finally, translations will be made from the relationships of relatives to the theoretically determined functions of genetic components of variance for the covariances of relatives. Unless specified for particular situations, the following assumptions are necessary to give the exactness of the genetic composition of the covariances of relatives of the different mating designs: normal Mendelian diploid inheritance, no environmental correlations with relatives, no maternal effects, linkage equilibrium in the population sampled, noninbred relatives, and relatives as random members of some specified population. Maize generally exhibits regular diploid inheritance, and maternal effects usually are not important. Proper use of experimental design and randomization will ensure no correlation of environmental effects with relatives. Noninbred relatives are evaluated in most instances, but either inbred or noninbred parents can be used to produce the relatives as long as the parents are unrelated; in each instance, the parents must have originated from a common population. Because we wish to make inferences from our estimates of genetic components of variance, it is necessary that the relatives are random members of some reference population. The one remaining assumption of linkage equilibrium will probably cause more problems in maize than the others. For certain types of populations (e.g., F_2 and recently formed synthetics from inbred lines), linkages could be important in biasing the estimates of components of variance. The possible effects of linkages will be indicated for the different types of populations studied.

Some mating designs are used more extensively than others, but each has its advantages and disadvantages depending on the reference population under consideration and the information desired. Further information on mating designs in plants is given by Cockerham (1963).

4.1 BIPARENTAL PROGENIES

One of the simplest mating designs for estimation of genetic variance in a reference population is the one that is designated biparental progenies by Mather (1949). This mating design involves crossing pairs of individual plants taken at random in the population. For maize, individual pairs of plants can be crossed reciprocally to produce two ears, which can be bulked for evaluation. To permit interpretations relative to the reference population, it is desirable that one make as many crosses as facilities permit for growing and making measurements. If n plants are chosen, there will be n/2 crosses; the information estimable from the among- and within-cross sources of variation are shown in Table 4.1 for the degrees of freedom df, mean squares MS, and expected mean squares E(MS). An F-test of differences among crosses

Table 4.1. Analysis of variance among and within biparental crosses.

Source	df[†]	MS	E(MS)
Among crosses	(n/2) − 1	M_2	$\sigma_w^2 + k\sigma_c^2$
Within crosses	(n/2)(k − 1)	M_1	σ_w^2
Total	(nk/2) − 1		

[†]n and k refer to the number of parents sampled and plants within each cross, respectively.

Table 4.2. Analysis of variance of biparental progenies replicated within environments.

Source	df[†]	MS	E(MS) Components of variance	E(MS) Covariance of relatives
Replications	$r - 1$			
Among crosses	$(n/2) - 1$	M_3	$\sigma_w^2 + k\sigma_p^2 + rk\sigma_c^2$	$\sigma_w^2 + k\sigma_p^2 + rk \; \text{Cov FS}$
Error	$(r-1)[(n/2)-1]$	M_2	$\sigma_w^2 + k\sigma_p^2$	$\sigma_w^2 + k\sigma_p^2$
Total	$r(n/2)-1$			
Within crosses	$r(n/2)(k-1)$	M_1	σ_w^2	$\sigma_{we}^2 + (\sigma_G^2 - \text{CovFS})$

[†] r, n, and k refer to the number of replications, parents, and plants, respectively.

can be made to determine if they are greater than within-cross variations. It also is recognized that an intraclass correlation r_I can be calculated from the analysis of variance in Table 4.1: $r_I = \sigma_c^2/(\sigma_c^2 + \sigma_w^2)$.

This type of mating design will provide information needed to determine if significant genetic variation is present in a population but no information is available for the type of genetic variation. Since any among-group component of variance is equal to the covariance of individuals within the groups, σ_c^2 in Table 4.1 is equal to the covariance of full-sibs, Cov FS. Growing biparental progenies in replicated trials will give the analysis of variance shown in Table 4.2. Similar to Table 4.1, an F-test to determine if the variation among crosses is significantly different from zero, and an intraclass correlation can be computed. If one desires an alternative to the F-test for testing the variation among crosses, a chi-square can be calculated by dividing the among-cross sum of squares by the error mean square.

Mather (1949) has presented the genetic expectations for the among- and within-biparental progenies. The cross component of variance, σ_c^2, is the variance among biparental progeny means, which is $\sigma_c^2 = (1/2)\sigma_A^2 + (1/4)\sigma_D^2$ [i.e., $(1/4)D + (1/16)H$ in Mather's notation], or the Cov FS. The within-cross component, σ_w^2, is the mean variance of biparental progenies and includes genetic (σ_{wg}^2) and environmental (σ_{we}^2) sources of variation. The genetic variation within crosses is the total genetic variance, σ_G^2, minus the Cov FS: $\sigma_G^2 - \text{Cov FS} = (1/2)\sigma_A^2 + (3/4)\sigma_D^2$ [i.e., $(1/4)D + (3/16)H$ in Mather's notation]. Hence, we have four unknowns (σ_A^2, σ_D^2, σ_p^2, and σ_{we}^2) and only three mean squares. An estimate of σ_p^2 can be obtained from the experimental error mean square (M_2), but an estimate of σ_{we}^2 is not available from the analysis of variance. If we make the assumption that the dominance effects are zero, we can, however, obtain an estimate of heritability (h^2). The expressions reduce to $\sigma_c^2 = (1/2)\sigma_A^2$ and $\sigma_w^2 = (1/2)\sigma_A^2 + \sigma_{we}^2$. Both components of variance can be estimated from the expected mean squares listed in Table 4.2. Estimates of σ_A^2, σ_p^2, and σ_{we}^2 can be obtained as $2(M_3 - M_2)/(rk) = 2\sigma_c^2 = \sigma_A^2$, $\sigma_p^2 = (M_2 - M_1)/k$, and $\sigma_{we}^2 = (\sigma_w^2 - \sigma_c^2)$. From the estimates of σ_A^2, σ_p^2, and σ_{we}^2, one can determine what proportion of the total variation, for the assumptions used, in the population is under additive genetic control: $\hat{h}^2 = \hat{\sigma}_A^2/(\hat{\sigma}_A^2 + \sigma_p^2 + \sigma_{we}^2)$. The approximation becomes more accurate as σ_D^2 approaches zero. A variance of the estimate of h^2 can be calculated because the components were calculated from linear functions of independent mean squares. The variance of a ratio θ is approximately

$$V(\theta) = V(X)/Y^2 - 2[X \text{ Cov}(X, Y)]/Y^3 + (X^2/Y^4)V(Y)$$

where $\hat{\sigma}_A^2 = X$ and $\hat{\sigma}_A^2 + \sigma_p^2 + \sigma_{we}^2 = Y$ (Kempthorne 1957). Snedecor (1956, p. 262) has shown that if the variance component was computed from a linear function of independent mean squares, the approximate variance V of $\hat{\sigma}_i^2$ is determined as

$$V(\hat{\sigma}_i^2) = (2/f^2) \sum_i (\lambda_i^2 M_i^2)/(df_i + 2)$$

where $\lambda_i = \pm1$ and M_i are the mean squares used to determine the component of variance, df_i is the degrees of freedom of the respective mean squares, and f is the coefficient of the component of variance. Since $\hat{\sigma}_c^2$ was calculated as $(M_3 - M_2)/(rk)$, the variance of $\hat{\sigma}_c^2$ is

$$V(\hat{\sigma}_c^2) = \frac{2}{(rk)^2} \left[\frac{M_3^2}{(n/2) + 1} + \frac{M_2^2}{(r - 1)(n/2) + 1} \right]$$

A satisfactory but conservative approximation to the standard error (SE) of heritability that is easier to calculate was suggested by Dickerson (1969). If we estimate $\hat{h}^2 = 2\hat{\sigma}_c^2/(2\hat{\sigma}_c^2 + \sigma_p^2 + \sigma_{we}^2)$, then $SE(\hat{h}^2) = 2SE(\hat{\sigma}_c^2)/(2\hat{\sigma}_c^2 + \sigma^2 + \sigma_{we}^2)$ where $SE(\hat{\sigma}_c^2)$ is the square root of the variance of $_c^2$ and the denominator is the phenotypic variance. (Also, see Knapp et al. 1985 and Knapp 1986.)

The use of biparental progenies provides the breeder with limited information on the relative importance of additive genetic variance. It is a simple mating design to use, but the information may not be sufficient to formulate a long-term breeding program. The information provided by biparental progenies gives an indication if indeed sufficient genetic variability is present in the population to warrant a selection program.

4.2 PARENT-OFFSPRING REGRESSION

Assume a reference population and make individual plant measurements for the traits of interest (e.g., plant and ear height, days to maturity, yield, disease resistance). Harvest seed of the measured plants in the population and measure the same traits in the offspring of each parent. Our task is to determine the degree of association between the traits measured in the parents and in their respective offspring. Our reference population is that from which the parents were derived; in maize it could be either a broad genetic base variety or an F_2 population derived from a cross of two inbred lines. Because maize is cross-pollinated, the individual plants measured would need controlled pollinations for parental control.

Robinson et al. (1949) gave an example of the use of parent-offspring in maize. Randomly chosen S_0 plants used as males were crossed to another set of randomly chosen S_0 plants used as females. Measurements for quantitative traits were made for the plants used as males and females. Progenies from the crosses were evaluated in trials, and the same traits were measured in the replicated progeny trials. Crosses can be made in either open-pollinated varieties or F_2 populations. If the reference population is an F_2 population, a common procedure is to measure F_2 plants and their progeny means. There is some conflict in terminology, but in both instances S_0 parental plants and their progeny are evaluated. Because the F_2 population is the reference population, the individual F_2 plants are equivalent to S_0 parents in an open-pollinated variety. In both instances we want to minimize selection of parental S_0 plants because we want to estimate genetic parameters relative to the reference population--whether it is an F_2 of two inbred lines or an open-pollinated, a synthetic, or a composite variety--to provide an unbiased estimate of heritability.

Our analysis is the regression of the n progeny measurements (Y, the dependent variable) on the S_0 parental plant measurements (X, the independent variable). The standard regression model of $Y_i = a + bX_i + e_i$ is used, where Y_i is the mean measurement of the offspring, X_i is the measurement of the parental S_0 plant, b is the regression of Y_i on X_i, and e_i is the error associated with the Y_i. We want to find

$$b = \Sigma xy/\Sigma x^2 = \sum_i (X_i - \overline{X})(Y_i - \overline{Y})/\sum_i (X_i - \overline{X})^2 = \sigma_{xy}/\sigma_x^2$$

because n - 1 is common to the numerator and denominator. We know that σ_{xy} has the following genetic component (covariance of parent-offspring, Cov PO):

$$\text{Cov PO} = \sum_i \left(\frac{1 + F}{2}\right)^i \sigma_{A^i}^2$$

When F = 0, Cov PO $= (1/2)\sigma_A^2 + (1/4)\sigma_{AA}^2 + (1/8)\sigma_{AAA}^2 + \ldots$; the total variation of the parental measurements is σ_x^2. Then, assuming no epistasis, $b = (1/2)\sigma_A^2/\sigma_x^2 = \sigma_A^2/(2\sigma_x^2)$.

To determine the heritability estimate on an individual plant basis of the traits from parent-offspring regression, we can calculate $\hat{h}^2 = 2b = \sigma_A^2/\sigma_x^2$. A standard error can be calculated rather easily for the heritability estimate of parent-offspring regression. From standard statistical texts the standard error (SE) of the regression coefficient is determined as:

$$\sigma_b^2 = [\sum_i y_i^2 - (\sum_i x_i y_i)^2/\sum_i x_i^2]/[(n - 2)\sum_i x_i^2] = \text{average of deviations from regression squared}/\sum_i x_i^2$$

Hence, $\text{SE}(b) = (\sigma^2/\Sigma_i x_i^2)^{1/2}$. The standard error of the heritability estimate on an individual basis becomes $\text{SE}(\hat{h}^2) = 2\text{SE}(b)$.

Parent-offspring regressions also can be obtained by evaluating the progeny from the cross of two individuals; for this case we would have the regression of offspring on the mean of the parents. Mean parent-offspring regression also would be amenable for use in maize because of the ease in making controlled crosses between two individuals. Progeny would be produced in the same manner as described for biparental progenies. Instead of taking measurements only on offspring, as for biparental progenies, we also want to record measurements of traits in each parent used in crosses and regress offspring measurements Y on those of the mean of the pairs of parents, $\overline{X}$. If we use $\overline{X} = (X_1 + X_2)/2$ for the midparent, the midparent-offspring regression, Cov $\overline{PO}$ becomes $b = \sigma_{xy}/\sigma_x^2$. Assuming $\sigma_{X_1}^2 = \sigma_{X_2}^2$ and that X_1 and X_2 are uncorrelated,

$$\sigma_{\overline{X}}^2 = (1/4)\sigma_{X_1}^2 + (1/4)\sigma_{X_2}^2 = (1/2)\sigma_x^2.$$ Thus $b = (1/2)\sigma_A^2/[(1/2)\sigma_x^2] = \sigma_A^2/\sigma_x^2$ and $h^2 = b$.

Midparent-offspring regression also would have utility for estimation of heritability in dioecious plant species (e.g., hops, hemp, asparagus, date palms, and willow trees). The species need not be dioecious but 2n parents must be measured for midparent-offspring regression, whereas only n parents must be measured for parent-offspring regression. In all instances the reference population from which the parents are derived is random mated and noninbred. Sufficient sampling and no selection of the parents evaluated should be ensured for the estimates to have meaning relative to the reference population.

In maize it also is convenient to measure the S_0 plants in either a broad genetic base population or an F_2 population and self the S_0 plants to obtain S_1 progenies. Measurements can be taken on the S_0 plants and S_1 progenies in replicated trials. Because the S_0 plants were selfed, the measurements represent the male and female parents; hence the S_0 plant measurements would be for both parents. Regression of S_1 progenies on the S_0 plants provides an estimate of Cov PO = $\sigma_A^2 + (1/2)\sigma_D^2 + \sigma_{AA}^2$ (p=q=0.5, see Chap. 3). It is seen that Cov PO for S_1 progenies on S_0 plants includes dominance and dominance type epistasis in addition to additive and additive type epistasis. A broad sense heritability estimate can be determined as $h^2 = b = $ (Cov PO)$/\sigma_x^2$, where σ_x^2 is the phenotypic variability among the S_0 plants. This would be a conservative estimate of h^2 because the genetic variability among a population of S_0 plants would include all the genetic variability ($\sigma_A^2 + \sigma_D^2 + \sigma_{AA}^2 + ...$), whereas Cov PO includes only half the dominance variance and fractions of the epistatic sources of variation. If allelic (dominance) interaction effects are assumed minor, the estimates of heritabilities for Cov PO of S_1 progenies on S_0 plants would be similar to Cov PO of progenies on one parent and Cov $\overline{PO}$ of progenies on the mean of parents. The estimate of σ_A^2, however, is not exactly the same as in the previous formula unless either p = q = 0.5 or dominance is absent.

The two commonly used methods of estimating heritability by use of parent-offspring regression ($2b = h^2$ and $b = h^2$) are valid when the parents are noninbred, as in a random mating population. If the parents are inbred or related, Smith and Kinman (1965) have shown that the previous inbreeding will cause an upward bias in the estimate of heritability. The bias is not generally severe for bisexual populations, but it becomes more important in self-pollinated populations. Smith and Kinman (1965) have shown that the correct estimator for the general case is $b/(2r_{XY})$, where r_{XY} is a measure of the relationship between the parent Y and its offspring X (Kempthorne 1957). Malécot (1948) defined r_{XY} as the probability that a random gene at a specific locus in X is identical by descent to a random gene at the same locus in Y. Hence the estimation $b/(2r_{XY}) = h^2$ provides for an adjustment for the mating system (bisexual or self-pollination) and for the known level of inbreeding or relationship of the parents.

Instances of parent-offspring regressions that may be used to estimate heritability for continuous self-pollinations are as follows (Smith and Kinman 1965):

Parent-offspring generation	r_{XY}	$h^2 = b/(2r_{XY})$
F_1, F_2	1/2	b_{F_2,F_1}
F_2, F_3	3/4	$(2/3)b_{F_3,F_2}$
F_3, F_4	7/8	$(4/7)b_{F_4,F_3}$

$$F_4, \ F_5 \qquad\qquad 15/16 \qquad\qquad (8/15)b_{F_5, F_4}$$
$$F_5, \ F_6 \qquad\qquad 31/32 \qquad\qquad (16/32)b_{F_6, F_5}$$

It is seen that adjustments for level of inbreeding make the estimates more conservative when we have continuous self-fertilization. The same estimator is valid for bisexual populations except that r_{XY} also depends on level of inbreeding of the parents $[Y, (F_Y)]$ and degree of relationship between the two parents (r_{YZ}). If the two parents are noninbred and unrelated, $2r_{XY} = 1/2$; but if both parents are identical homozygotes (both F_Y and r_{YZ} are equal to one), $2r_{XY} = 2$, which is the same as for continuous self-pollination.

Ordinarily, the parents are noninbred and unrelated for most instances when parent-offspring regression is used for estimation of heritability in maize. But self-fertilization can be used to generate different inbred generations by controlled pollination. If some inbred generations are used to determine heritabilities, adjustments are needed to account for the level of inbreeding. Smith and Kinman (1965) have discussed the situations and shown the relations that are necessary to account for the inbreeding and/or relatedness of the parents when using parent-offspring regression for estimating heritabilities. In maize, therefore, we can determine estimates of parent-offspring regressions by (1) regression of selfed progeny on parents (selfing method), (2) regression of offspring on one parent (half-sib method), and (3) regression of offspring on the mean of two parents (full-sib method).

The progeny of the parent-offspring that are regressed on the mean of the parents also can be analyzed as shown in Table 4.2. We would have two equations that would permit estimation of the variance due to dominance effects because Cov FS $= (1/2)\sigma_A^2 + (1/4)\sigma_D^2$ and Cov $\overline{PO} = (1/2)\sigma_A^2$, assuming no epistasis. Hence $4(\text{Cov FS} - \text{Cov } \overline{PO})$ is an estimate of σ_D^2. If individual plant data are collected for the progenies that resulted from the cross of two parents, we have three equations that permit estimation of three parameters. For instance, additive × additive epistatic variance could be estimated because Cov FS $= (1/2)\sigma_A^2 + (1/4)\sigma_D^2 + (1/4)\sigma_{AA}^2$, Cov $\overline{PO} = (1/2)\sigma_A^2 + (1/4)\sigma_{AA}^2$, and $\sigma_G^2 - \text{Cov}$ FS $= (1/2)\sigma_A^2 + (3/4)\sigma_D^2 + (3/4)\sigma_{AA}^2$. If nonadditive effects are small relative to additive effects, each equation has similar expectations for σ_A^2 and the observed variances would be similar. If nonadditive effects are important, the observed component for the within-plot variance would deviate from the other two.

4.3 CROSS CLASSIFICATION
4.3.1 Diallel

The diallel mating design has been used and abused more extensively than any other in maize and other plant species. The diallel mating design can be very useful if properly analyzed and interpreted. Although extensive theoretical research and discussion have been presented, the main problem seems to arise from interpretations and inferences that can be made about estimates obtained from analysis of the diallel crosses. As the name implies, crosses are made in pairs for n number of parents. Diallel crossing schemes and analyses have been developed for parents that range from inbred lines to broad genetic base varieties. After the crosses are made, evaluated, and analyzed, inferences regarding the types of gene action can be made. It is important, however, that the assumptions and limitations of the diallel mating design are realized when one interprets the data.

First, let us consider how the crosses are made for the diallel mating design. If one has a complete diallel, it will include all possible crosses and parents. If n parents are included in the diallel and crossed in pairs, the number of permutations of n parents taken two at a time becomes $n!/(n-r)!$ or $n!/(n-2)!$ because $r = 2$, which reduces to $n(n-1)$. For this case the reciprocal crosses (e.g., 1×2 and 2×1 for parents 1 and 2) are made among the n parents. If reciprocal crosses are not included in the crosses among n parents, the possible arrangements of n parents taken r at a time become $n!/[(n-r)!r!]$ or $n!/[(n-2)!2!]$ because $r = 2$; this reduces to $n(n-1)/2$. Assume 10 parents are included for diallel crossing; we would have 90 crosses if reciprocals are made and 45 if reciprocals are ignored. If the 10 parents are included for evaluation we would have 100 entries for the first case and 55 for the second. Table 4.3 summarizes several possible examples of the entries available from the diallel crossing of different numbers of parents.

Obviously, as the number of parents increases, the number of possible crosses increases very rapidly. After one includes 10 to 15 parents, the number of crosses to make and to evaluate becomes somewhat unmanageable. Should one want to estimate the genetic variation present within a population, it would not be unreasonable to include 100 individuals to represent the range of genotypes within the population. The inclusion of only the combination of crosses (ignoring reciprocals) among 100 parents would require making and evaluating 4950 crosses for a complete diallel. It is readily apparent that the number of parents one can include to produce the crosses is an important factor in the diallel mating design.

The mechanical procedures for making the diallel crosses will vary among crop species (self- vs. cross-pollinators) and within crop species (inbred vs. noninbred parents). If the parents are relatively homozygous (inbred lines),

Table 4.3. Possible number of entries from the diallel mating design for different numbers of parents.

Number of parents	Combinations		Permutations	
	Crosses	Total[+]	Crosses	Total[‡]
5	10	15	20	25
6	15	21	30	36
7	21	28	42	49
8	28	37	56	64
9	36	45	72	81
10	45	55	90	100
15	105	120	210	225
20	190	210	380	400
50	1225	1275	2450	2500
100	4950	5050	9900	10000
n	$n(n-1)/2$	$n(n+1)/2$	$n(n-1)$	n^2

[+]Total number of entries refers to all cross combinations and parents; i.e., $n + n(n-1)/2 = n(n+1)/2$.

[‡]Total number of entries refers to all cross permutations and parents; i.e., $n + n(n-1) = n^2$.

the series of diallel crosses can be made by repeating each parent for each
combination of crosses and making paired-row crosses; the only limitation to
the number of plants included and cross-pollinated for each paired-row cross
is the quantity of seed needed for testing the crosses. By use of paired-row
crosses, seed produced on each parent can be bulked for each cross combination
or kept separate if each cross permutation is desired.

Diallel crosses among a set of maize inbred lines are usually not too
difficult, provided the proper timing is made for the flowering of the male
and female inflorescences. Diallel crosses among a set of maize populations
are handled similarly to inbred lines, but the sampling of the population genotypes increases the number of individual plants included in the population
crosses. Amount of seed usually is not a problem, but the number of crosses
between plants required to sample the populations increases the space and time
needed. Crosses between 10 plants of inbred lines may be sufficient for seed
needs, whereas 100 to 150 (or more) plants are necessary to adequately sample
the genotypes in a population. Diallel crosses among pure lines of *Avena*,
Hordeum, *Triticum*, and *Glycine*, for example, would be similar to inbred lines
of maize, but the degree of difficulty in making the crosses often is the limiting factor. For a multiflowered self-pollinated species such as *Nicotiana*,
crosses can be made quite easily with sufficient quantities of seed and small
numbers of plants. It is obvious, therefore, that the diallel mating design
can be used for most crop species, and the extent of its use depends on the
difficulty of crossing and the resultant seed supplies. Seed produced on each
set of crosses between parents either can be kept separate if one desires to
test for reciprocal and maternal effects or bulked if one is interested only
in performance of the crosses.

The diallel series of crosses are grown in replicated tests, with appropriate randomizations, to determine the relative merit of the parents in
crosses. If 10 to 12 parents are included in the diallel crosses, a randomized complete block design should be satisfactory in most instances. Incomplete block designs should be considered if the number of crosses is large and
the environmental variability among experimental units is great. If we assume
that only crosses among parents are tested in e environments, the initial
analysis of variance to determine if the variation among crosses is significantly different from zero is shown in Table 4.4. If no significant differences exist, there is no need to proceed further because apparently no detectable differences were contributed by the parents to their offspring.

Table 4.4. Analysis of variance of a diallel set of crosses among n
parents evaluated in e environments.

Source	df	MS	E(MS) Model I	Model II
Environments (E)	$e-1$			
Replications/E	$e(r-1)$			
Crosses (C)	$[n(n-1)/2]-1$	M_3	$\sigma^2+reK_c^2$	$\sigma^2+r\sigma_{ce}^2+re\sigma_c^2$
C x E	$(e-1)\{[n(n-1)/2]-1\}$	M_2	$\sigma^2+rK_{ce}^2$	$\sigma^2+r\sigma_{ce}^2$
Pooled error	$e(r-1)\{[n(n-1)/2]-1\}$	M_1	σ^2	σ^2
Total	$er[n(n-1)/2]-1$			

An orthogonal subdivision of the sums of squares for crosses is valid when the crosses mean square is significantly different from zero. Before the experiments were conducted, an important decision was made about the parents included to make the crosses: Are the parents the reference genotypes or are the parents random genotypes from some reference population? The answer to this question has great implications in the interpretations made from the analysis of the diallel mating design, and it usually has been the basic feature in arguments for and against the utility of that design to provide the information desired by the researcher. Usually, the assumption made about the parents to be included, not how the experiment was conducted and analyzed, causes difficulties in the interpretation of the estimated parameters. Griffing (1956) and Cockerham (1963) have discussed the diallel analysis in detail as well as the analysis of variance for fixed models (model I, where the parents are the genotypes under consideration) and random models (model II, where the parents are a sample of genotypes from a reference population). Model I estimates apply only to the genotypes included and cannot be extended to some hypothetical reference population. Model II estimates are interpreted relative to some reference population from which the genotypes included are an unselected sample.

The partition of the cross sums of squares is shown in Tables 4.5 and 4.6. For the present assume five parents are included in the diallel, which obviously does not apply to the model II analysis, and only the ten cross combinations are evaluated in five replications. Table 4.5 shows that each parent is included in a cross with each of the others--four specific crosses, each including a common parent. The average performance of each parent in the four crosses is determined from the marginal means. Hence total variation among crosses (variation among all cells in Table 4.5, or 9 df) can be partitioned into variation among margins (4 df for the five parents included) and variation among cells within margins or of individual cells about the margins, which have a common parent. There are 5 df for the variation among cells within margins because we have five independent observations among the cells in Table 4.5; i.e., if we know the marginal values, three cell values for parent 1, and two cell values for parent 2, the remaining cell values may be calculated. In other words, restrictions are imposed on the deviations of the cell means from the marginal means; $\sum_i c_{ij} = 0$ for each i and 1 df is lost for

Table 4.5. Example of the ten cross combinations possible from the diallel crossing of five parents.

| Parents | Parents | | | | | Margins |
	1	2	3	4	5	
1		X_{12}	X_{13}	X_{14}	X_{15}	$X_{1.}$
2			X_{23}	X_{24}	X_{25}	$X_{2.}$
3				X_{34}	X_{35}	$X_{3.}$
4					X_{45}	$X_{4.}$
5						$X_{5.}$

Table 4.6. Orthogonal partition of the crosses sum of squares in the analysis of variance of the diallel cross shown in Table 4.5.

Source	df[+] General	df[+] Example	MS	E(MS) Model I	E(MS) Model II
Replications	$r-1$	4			
Crosses	$[n(n-1)/2]-1$	9	M_2	σ^2+rK^2	$\sigma^2+r\sigma_c^2$
Among margins	$n-1$	4	M_{21}	$\sigma^2+[r(n-2)/(n-1)]K_g^2$	$\sigma^2+r\sigma_s^2+r(n-2)\sigma_g^2$
Among cells/margins	$n(n-3)/2$	5	M_{22}	$\sigma^2+\{2r/[n(n-3)]\}K_s^2$	$\sigma^2+r\sigma_s^2$
Error	$(r-1)\{[n(n-1)/2]-1\}$	36	M_1	σ^2	σ^2
Total	$r[n(n-1)/2]-1$	49			

[+] r and n refer to the number of replications and parents, respectively.

each restriction. Thus the degrees of freedom for variation among cells within margins become $10 - 5 = 5$. F-tests can be made to determine whether variation among margins and cells within margins are significantly different from zero.

We now add additional terminology that is commonly used in the diallel analysis of variance. Sprague and Tatum (1941) introduced the concepts of *general combining ability* (GCA) and *specific combining ability* (SCA) to distinguish between the average performance of parents in crosses or margins (GCA) and the deviation of individual crosses from the average of the margins (SCA). The concepts of GCA and SCA are extensively used in plant breeding and have particular significance to the diallel mating design. If we insert the GCA and SCA terms in Table 4.6, we have Table 4.7. F-tests for both models can be made to test for GCA and SCA mean squares. For model I, M_{21} and M_{22} are tested with M_1; whereas for model II, M_{22} is tested with M_1 and M_{21} with M_{22}. If M_{22}, however, is not different from zero, M_1 also is the appropriate mean square to test M_{21}.

Table 4.7. Diallel analysis of variance for a fixed (model I) and random set (model II) of n parents to produce the n(n-1)/2 crosses.

	df[+]	MS	E(MS) Model I	E(MS) Model II
Replications	$r-1$			
Crosses	$[n(n-1)/2]-1$	M_2	$\sigma^2+rK_c^2$	$\sigma^2+r\sigma_c^2$
GCA	$n-1$	M_{21}	$\sigma^2+[r(n-2)/(n-1)]K_{GCA}^2$	$\sigma^2+r\sigma_{SCA}^2+r(n-2)\sigma_{GCA}^2$
SCA	$n(n-3)/2$	M_{22}	$\sigma^2+\{2r/[n(n-3)]\}K_{SCA}^2$	$\sigma^2+r\sigma_{SCA}^2$
Error	$(r-1)\{[n(n-1)/2]-1\}$	M_1	σ^2	σ^2
Total	$r[n(n-1)/2]-1$			

[+] r and n refer to the number of replications and parents, respectively.

In model I the parents are the population, whereas in model II the parents are a sample from a population. The distinction between the two models is important not only for the analysis of variance but for the information derived from the analysis of variance. Because the parents are the population for model I, estimation of components of variance is not appropriate but estimation of the effects of each pair of parents for specific crosses (SCA) and for all crosses that include a common parent (GCA) is appropriate and valid. No apologies are needed from the experimenter for the estimation of effects rather than of components of variance; GCA and SCA effects are more informative than components of variance for the model I analysis. Also, estimated effects are applicable only for the parents included and would be different if the parents were tested with a different group of parents. F-tests in model I show that differences may exist; hence estimation of the GCA and SCA effects show the relative magnitude and sign of the effects for each parent and each cross.

The model for the analysis of variance is

$$X_{ijk} = u + r_k + g_i + g_j + s_{ij} + p_{ijk}$$

where u is the mean, r_k is the replication effect, g_i and g_j are the GCA effects, s_{ij} is the SCA effect, and p_{ijk} is the experimental error for the X_{ijk} observation (k = 1, 2,..., r; i = j = 1, 2,..., n). For model I, GCA and SCA effects are estimated, respectively, as

$$\hat{g}_i = \{1/[n(n - 2)]\}(nX_{i.} - 2X_{..})$$

$$\hat{s}_{ij} = X_{ij} - [1/(n - 2)](X_{i.} + X_{.j}) + \{2/[(n - 1)(n - 2)]\}X_{..}$$

From Table 4.6, the experimental error σ^2 is estimated from the mean square M_1. The variance of any specific X_{ij} cross is $\hat{\sigma}^2/r$, the variance of the difference between any two crosses is $2\hat{\sigma}^2/r$, and the variance for the mean (marginals) of the crosses where one parent is common is $\hat{\sigma}^2/[(n - 1)r]$. The variances of the GCA and SCA effects are

$$\hat{\sigma}^2(g_i) = \{(n - 1)/[n(n - 2)]\}\hat{\sigma}^2 \quad \text{and} \quad \hat{\sigma}^2(s_{ij}) = [(n - 3)/(n - 1)]\sigma^2$$

The model I analysis, therefore, yields considerable information about the fixed set of parents included--information that can be useful for the selection of parents that have good general combining ability in a series of crosses and good specific combining ability for specific pairs of parents. This type of information is quite useful to maize breeders, particularly if the selected set of parents represents an elite group of inbred lines that are possible candidates as parent seed stock for the production of single-cross hybrids.

For the model II analysis, estimation of the components of variance is of prime interest. To relate the variance components in Table 4.7 to types of gene action in the reference population, it is helpful to write the expected mean squares in terms of genetic relationships of relatives and translate from the covariances of relatives to the genetic components of variance. It then is necessary to determine what types of relatives are included in the diallel mating design. The variation among margins (GCA) is due to differences among parents (sires or dams--usually not distinguished as such in plants), which is σ^2_{GCA} in Table 4.7. This may be stated in another manner: Variance among parents is equal to covariance of members (or crosses having a common parent, Table 4.5) with a common parent. Variability among i parents can be shown to be

$\sigma^2_{y_i} = E_i(X_i - \overline{X})^2 = E_i(X_i^2) - \overline{X}^2$. If we have j individuals within families of the common i parent, covariance is $Cov(y_{ij}, y_{ij'}) = E[(X_{ij} - \overline{X})(X_{ij'} - \overline{X})] = E(X_{ij}X_{ij'} - \overline{X}^2) = E(X_i^2) - \overline{X}^2$ because $E(X_{ij}) = E(X_{ij'}) = X_i$ and $E(X_{ij}X_{ij'}) = X_i^2$, assuming symmetrical distribution of sibs around family mean. Hence σ^2_{GCA} is equal to covariance of half-sibs (Cov HS) because one parent is common for all the crosses included in the marginal mean of a particular parent. Next, let us consider sources of variation within the component σ^2, which is estimated from M_1 (Table 4.7). Experimental error (σ^2) includes a plot error variance (σ^2_p) and variance among individuals within plots (σ^2_w). Therefore, $\sigma^2 = \sigma^2_w/k + \sigma^2_p$, where k is the number of individuals measured within each plot. Variance within plots includes plant-to-plant environmental variance (σ^2_{we}) and genetic variance (σ^2_{wg}) among individuals if $F \neq 1$; hence $\sigma^2_w = \sigma^2_{wg} + \sigma^2_{we}$. Individuals having the same parents are full-sibs, and consequently variations among individuals having the same parents are covariances of full-sibs (Cov FS). Within-progeny genetic variance can be expressed as $\sigma^2_{wg} = \sigma^2_G - $ Cov FS, where $\sigma^2_G = \sigma^2_A + \sigma^2_D + \sigma^2_{AA} + \ldots$, total genetic variance. Then $\sigma^2_{wg} = \sigma^2_G - $ Cov FS $= (1/2)\sigma^2_A + (3/4)\sigma^2_D + (3/4)\sigma^2_{AA} + \ldots$ for $F = 0$.

Because Cov HS is $(1/4)\sigma^2_A + (1/16)\sigma^2_{AA} + \ldots$ and σ^2_{wg} is $(1/2)\sigma^2_A + (3/4)\sigma^2_D + (3/4)\sigma^2_{AA} + \ldots$, then $(3/4)\sigma^2_A + (3/4)\sigma^2_D + (13/16)\sigma^2_{AA} + \ldots$ of total genetic variance ($F = 0$) is accounted for. Remaining genetic variation is among crosses within each parent (sire or dam). The crosses within each parent are half-sibs but consist of full-sib individuals. We can determine variation among crosses for each parent, but again this does not tell us anything about genetic relationships. The crosses consist of full-sib individuals, and therefore the variance among cross means is equal to Cov FS - 2 Cov HS, where

Table 4.8. Analysis of variance of diallel crosses for $n(n-1)/2$ crosses of n parents where expected mean squares, E(MS), are expressed in terms of covariances of relatives.

Source	df[†]	MS	E(MS)[†]
Replications	r-1		
Crosses	[n(n-1)/2]-1	M_2	$\sigma^2 + r\sigma^2_c$
GCA	n-1	M_{21}	$\sigma^2 + r(\text{Cov FS} - 2\text{Cov HS}) + r(n-2)\text{Cov HS}$
SCA	n(n-3)/2	M_{22}	$\sigma^2 + r(\text{Cov FS} - 2\text{Cov HS})$
Error	(r-1){[n(n-1)/2]-1} M_1		σ^2[‡]
Total	r[n(n-1)/2]-1		

[†] r and n refer to the number of replications and parents, respectively.

[‡] If individual plant data are taken, σ^2 in Table 4.8 will be equal to $[(\sigma^2_G - \text{Cov FS}) + \sigma^2_{we}]/k + \sigma^2_p$, where k is the number of plants.

we subtract 2 Cov HS because each of the parents is included in each cross. From the relationship of relatives we can then write the model II analysis of variance as shown in Table 4.8.

We have two types of relatives in the diallel mating design for use in making translations to genetic variances, i.e., Cov HS and Cov FS. If we obtained within-plot data, we would have another source of information relative to genetic variance. Because we know Cov HS $= (1/4)\sigma_A^2 + (1/16)\sigma_{AA}^2 + \ldots$ and Cov FS $= (1/2)\sigma_A^2 + (1/4)\sigma_D^2 + (1/4)\sigma_{AA}^2 + \ldots$ for F = 0, we can use these to obtain $\hat\sigma_A^2$ and $\hat\sigma_D^2$ under the assumption of no epistasis. Assuming no epistasis, $\hat\sigma_A^2 = 4 \widehat{\text{Cov}} \text{ HS} = 4(M_{21} - M_{22})/[r(n - 2)]$, and $\hat\sigma_D^2 = 4[\widehat{\text{Cov}} \text{ FS} - 2 \widehat{\text{Cov}} \text{ HS}] = 4(M_{22} - M_1)/r$. If we collect individual plant data and have $\hat\sigma_{we}^2$, we can obtain $\hat\sigma_{wg}^2$, which is $(1/2)\sigma_A^2 + (3/4)\sigma_D^2$. A rough check on the assumption of no epistasis then can be tested as $\hat\sigma_{wg}^2 - 3 \text{ Cov FS} + 4 \text{ Cov HS}$. If epistasis is either absent or relatively small, this relation will be zero or some relatively small value. Otherwise, the presence of significant epistasis will cause this relation to be greater than zero because the difference of the relation theoretically includes the following relative proportions of the various types of epistasis: $(1/4)\sigma_{AA}^2 + (1/2)\sigma_{AD}^2 + (3/4)\sigma_{DD}^2 + \ldots$. This test, however, has two major disadvantages: (1) the great variability of individual plant data and (2) the large errors that are associated with such estimates obtained from complex linear functions.

We have assumed the inbreeding of the parents used in making the crosses was zero (F = 0). Quite often inbred parents (F = 1) are used, such as inbred lines of maize and cultivars of self-pollinated crop species. The model I analysis would be appropriate for selected, elite inbred lines, and no complications are encountered estimating the GCA and SCA effects. This information is often very useful to the maize breeder. The inclusion of inbred lines as parents for the model II analysis, however, does change the coefficients of the components of genetic variance of the covariances of relatives. For F = 1, Cov HS $= (1/2)\sigma_A^2 + (1/4)\sigma_{AA}^2 + \ldots$, and Cov FS $= \sigma_A^2 + \sigma_D^2 + \sigma_{AA}^2 + \ldots$. Hence $\hat\sigma_A^2$ and $\hat\sigma_D^2$, assuming no epistasis, are as follows: $\hat\sigma_A^2 = 2 \widehat{\text{Cov}} \text{ HS}$ and $\hat\sigma_D^2 = \widehat{\text{Cov}} \text{ FS} - 2 \widehat{\text{Cov}} \text{ HS}$. Individual plant data collected on crosses of inbred lines would include only σ_{we}^2, which is the within-plot environmental variance. The only restriction in estimating components of genetic variance from the model II analysis that uses inbred lines as parents is that inbred lines are an unselected sample from the reference population to which $\hat\sigma_A^2$ and $\hat\sigma_D^2$ are to be applied.

We have considered only one of the possible diallel mating systems. Other possible designs include the combination of crosses and parents and the permutations of crosses with or without parents. Griffing (1956) listed the four choices as methods 1, 2, 3, and 4. Combinations of crosses without parents is probably the most commonly used method in maize because parents are usually inbred lines and the vigor of parents (F = 1) and crosses among parents (F = 0) frequently cause complications in field designs used to evaluate parents and crosses. Border rows are necessary if mature plant traits are measured and differences in seedling vigor cause problems in measuring traits in the juvenile stage of development. Cockerham (1963) presented an analysis for partitioning the cross sums of squares for GCA, SCA, maternal, and reciprocal sources of variation when all possible permutations among parents are

Table 4.9.   Diallel analysis that includes n parents and their n(n-1)/2 crosses.

Source	df[†] General	df[†] r=3; n=10	Sums of squares	MS
Replications	r-1	2		
Entries	$[n(n+1)/2]-1$	54	S_2	M_2
Parents	n-1	9	S_{21}	M_{21}
Parents vs. crosses	1	1	S_{22}	M_{22}
Crosses	$[n(n-1)/2]-1$	44	S_{23}	M_{23}
GCA	n-1	9	S_{231}	M_{231}
SCA	$n(n-3)/2$	35	S_{232}	M_{232}
Error	$(r-1)\{[n(n+1)/2]-1\}$	108	S_1	M_1
Total	$[rn(n+1)/2]-1$	164		

[†]r and n refer to the number of replications and parents, respectively.

included. If parents are included for evaluation, the source of variation of parents vs. crosses provides another test (in addition to SCA) for the importance of nonadditive effects. If we assume the parents and all combinations of crosses (no reciprocals) are included for analysis, we can orthogonally partition variation among entries as shown in Table 4.9. For the model I analysis, tests can be made for variation among parents and parents vs. crosses, and among crosses. Partitioning of the crosses sums of squares would be the same as shown in Table 4.8. The analysis given in Table 4.8 is the same as Griffing's experimental method 2 except that SCA has $n(n-3)/2$ df rather than the $n(n-1)/2$ df given by Griffing. The difference of n df is for the separation of among parents and parents vs. crosses from the SCA source of variation. The parents vs. crosses comparison is a test for heterosis, which also is due to nonadditive genetic effects.

Diallel analyses that include the parents are frequently used for open-pollinated, synthetic, and composite varieties. For these instances we are interested in the variety performance itself as well as the variety crosses. Border rows on the plots may not be needed because the differences in vigor are usually small. Variety and variety cross evaluations are important in maize breeding for (1) determining the relative potential of varieties as breeding populations and (2) evaluating the response of varieties to different recurrent selection schemes. Because the varieties usually are a selected sample of the most promising available, the model I analysis is appropriate to determine the GCA effects of the varieties and SCA effects of the variety crosses.

4.3.2 Gardner–Eberhart Analysis II

Gardner and Eberhart's (1966) analysis II also is useful for evaluating varieties and their crosses. The Gardner-Eberhart analysis II model includes the n parent varieties and their $n(n - 1)/2$ variety crosses, but the partitioning of the entry sums of squares differs from the analysis given in Table

4.9. The following models are used to determine the sums of squares for the analysis shown in Table 4.10, following the notation of Gardner and Eberhart:

1. $X_{jj'} = u + (1/2)(v_j + v_{j'}) = (B'G)_1$
2. $X_{jj'} = u + (1/2)(v_j + v_{j'}) + v\bar{h} = (B'G)_2$
3. $X_{jj'} = u + (1/2)(v_j + v_{j'}) + v\bar{h} + v(h_j + h_{j'}) = (B'G)_3$
4. $X_{jj'} = u + (1/2)(v_j + v_{j'}) + v\bar{h} + v(h_j + h_{j'}) + vs_{jj'} = (B'G)_4$

In each of the models, u, v_j, $\bar{h}$, h_j, and $s_{jj'}$ indicate the mean and variety and heterosis effects. The coefficient v in these models is zero when $j = j'$ and one when $j \neq j'$. Because the phenomenon of heterosis is important, the analysis maximizes the information on variety performance and the expression of heterosis of their crosses. Estimates of the variety and heterosis effects can be determined for each of the constants in the models. Four of the mean squares of the diallel analysis (Table 4.9) and analysis II of the Gardner-Eberhart model (Table 4.10) are equivalent; i.e., entry and error mean squares are the same, average heterosis mean square is equal to the parents vs. crosses mean square, and specific heterosis mean square is equal to the SCA mean square. The variety mean square in Table 4.10 is not equivalent to the parent mean square in Table 4.9 because the variety mean square includes information of the performance of varieties themselves and in variety crosses. Also, the GCA mean square (Table 4.9) is not equivalent to the variety heterosis (Table 4.10) because $g_i = (1/2)v_j + h_j$ so that $M_{21} + M_{231} = M_{21}^1 + M_{222}^1$. The analysis in Table 4.10 is a nonorthogonal partition of the entry sums of squares, but the sums of squares can be obtained by sequentially fitting the four models.

Table 4.10. Analysis of variance of n parents and their n(n-1)/2 variety crosses for variety and heterosis effects (analysis II of Gardner-Eberhart model).

Source	df[†]	Sums of squares	Mean squares Gardner-Eberhart	Diallel
Replications	r-1			
Entries	$[n(n+1)/2]-1$	S_2^1	M_2^1	$= M_2$
Varieties (v_i)	n-1	$S_{21}^1 = (B'G)_1 - CF$	M_{21}^1	
Heterosis (h_{ij})	n(n-1)/2	$S_{22}^1 = (B'G)_4 - (B'G)_1$	M_{22}^1	
Average ($\bar{h}$)	1	$S_{221}^1 = (B'G)_2 - (B'G)_1$	M_{221}^1	$= M_{22}$
Variety (h_i)	n-1	$S_{222}^1 = (B'G)_3 - (B'G)_2$	M_{222}^1	
Specific (s_{ij})	n(n-3)/2	$S_{223}^1 = (B'G)_4 - (B'G)_3$	M_{223}^1	$= M_{232}$
Error	$(r-1)\{[n(n+1)/2]-1\}$	S_1^1	M_1	$= M_1$
Total	$[rn(n+1)/2]-1$			

[†] r and n refer to the number of replications and parents, respectively.

We have not considered diallel mating designs grown in several environments. The analyses are the same except for additional sources of variation
for interaction of main effects with environments. Analysis of the diallel
mating design repeated over environments was given by Matzinger et al. (1959).
Nothing changes in regard to the assumptions regarding parents relative to
model I and model II analyses except that estimates of interaction of effects
(model I) and variances (model II) with environments can be made. An example
of a diallel that is repeated over environments is shown in Table 4.11, and an
analysis that has partitioned environments into years and locations was given
by Matzinger et al. If we assume environmental effects are random, F-tests of
main effects use the same mean squares for models I and II. If environmental
effects are fixed, as are parents in model I, all main effects and their interactions with environments are tested with the pooled error for model I.

For the model II analysis, estimates of components of genetic variance
permit estimation of heritability (h^2). From Table 4.8 the source of variation due to GCA is the covariance of half-sibs, or $(1/4)\hat{\sigma}_A^2$, and the variation
due to SCA is the covariance of full-sibs minus two times the covariance of
half-sibs, or $(1/4)\hat{\sigma}_D^2$, assuming no inbreeding of the parents and no epistasis.
Using the components of variance shown in Table 4.7, heritability based on the
means of r plots (entry-mean basis) can be calculated as

Table 4.11. Diallel analysis of n parents and their $n(n-1)/2$ crosses repeated over environments.

| | df[†] | | Expected mean squares | |
	General	e=6; r=3; n=10	Model I	Model II
Environments (E)	e-1	5		
Replications/E	e(r-1)	12		
Entries	$[n(n+1)/2]-1$	54	$\sigma^2 + r\sigma_{en}^2 + erK_n^2$	$\sigma^2 + r\sigma_{en}^2 + er\sigma_n^2$
Parents	n-1	9		
Parents vs. crosses	1	1		
Crosses	$[n(n-1)/2]-1$	44		
GCA	n-1	9		
SCA	n(n-3)/2	35		
E × entries	$(e-1)\{[n(n+1)/2]-1\}$	270	$\sigma^2 + r\sigma_{en}^2$	$\sigma^2 + r\sigma_{en}^2$
E × parents	(e-1)(n-1)	45		
E × parents vs. crosses	e-1	5		
E × crosses	$(e-1)\{[n(n-1)/2]-1\}$	220		
E × GCA	(e-1)(n-1)	45		
E × SCA	(e-1)[n(n-3)/2]	175		
Pooled error	$e(r-1)\{[n(n+1)/2]-1\}$	648	σ^2	σ^2
Total	$[ern(n+1)/2]-1$	989		

[†] e, r, and n refer to the number of environments, replications, and parents, respectively.

$$h^2 = 4\hat{\sigma}^2_{GCA}/(\hat{\sigma}^2/r + 4\hat{\sigma}^2_{SCA} + 4\hat{\sigma}^2_{GCA})$$

which is commonly referred to as heritability in the narrow sense. Standard errors of the heritability estimate can be calculated from the variance of a ratio of two sets of variance components, as illustrated for biparental progenies. Dickerson (1969, pp. 36-79) presented a simplified general formula for calculating conservative approximations of the standard errors of heritabilities, based on the methods of Graybill et al. (1956) and Graybill and Robertson (1957). The simplified formula is $\sigma_{(\hat{X}/\hat{Y})} = (C/\hat{Y})[V(\hat{X})]^{1/2}$, which neglects the terms involving $V(\hat{Y})$ and $Cov(\hat{X}, \hat{Y})$. In the formula for the estimate of heritability, $\hat{\sigma}^2_{GCA}$ is equivalent to $\hat{X}$ and the total, or phenotypic, variance is equivalent to $\hat{Y}$. Hence the standard error (SE) of our estimate of heritability becomes

$$SE(h^2) = 4SE(\hat{\sigma}^2_{GCA})/(\hat{\sigma}^2/r + 4\hat{\sigma}^2_{SCA} + 4\hat{\sigma}^2_{GCA})$$

The variance of the estimate of σ^2_{GCA} is obtained in the usual manner from Table 4.7 as

$$\frac{2}{[r(n-2)]^2}\left[\frac{M^2_{21}}{n+1} + \frac{M^2_{22}}{n(n-3)/2+2}\right]$$

Estimates of components of variance for one environment include a genotype-environment bias, so that $(\sigma^2_A + \sigma^2_{AE})$ is estimated by σ^2_A. If diallel experiments are repeated over environments, an estimate of σ^2_A unbiased by environmental interactions can be obtained, and the phenotypic variance would include additional terms due to the interaction of σ^2_{GCA} and σ^2_{SCA} with environments. Estimates of heritability from the diallel mating design are only as good as estimates of components of variance. Because of the number of parents that can be included in the diallel, standard errors of the components of variance may be quite large, particularly for the GCA source of variation. Standard errors on estimates of heritability, consequently, may be large.

Sokol and Baker (1977) and Baker (1978) have recently reviewed the critical issues involved in the use of the diallel mating design. The critical issues concern the choice of model for analysis of data, i.e., model I (fixed genotypic effects) or model II (random genotypic effects). Baker (1978) emphasizes that two assumptions are critical for interpreting the results of the diallel analyses: (1) independent distribution of genes in the parents included and (2) no epistasis. Neither assumption seems valid for the small number of parents usually included in a diallel set of crosses. Independent distribution of genes at n loci cannot occur unless a minimum of 2^n parents are included for the diallel set of crosses. The assumption of no epistasis is commonly made, and estimation of the relative importance of epistasis usually has not been fruitful (Chap. 5). But epistatic effects unpredictably affect the GCA and SCA mean squares, variances, and effects (Baker 1978). He concluded that most diallel experiments are restricted to estimation of GCA and SCA mean squares and effects.

Use of diallel mating designs has been common in self-fertilizing species. To overcome the problem of limited hybrid (F_1) seed often experienced from crosses of pure line cultivars, F_1 generation seed may be advanced one generation to produce F_2 generation seed so as to have adequate seed for testing. The analyses of diallel mating designs discussed so far included noninbred progeny produced from either noninbred or inbred (or partially inbred) parents; i.e., the progeny evaluated were noninbred but parents may have some level of inbreeding or none. Use of F_2 generation (or any other advanced generation obtained by inbreeding) seed in the evaluation trials of crosses among parents means the progenies evaluated have some level of inbreeding. Stuber (1970) examined the situations that included evaluation of inbred progenies where they are generated by bulk selfing of each F_1 cross. The main effect of evaluating inbred progenies is the change in coefficient of the dominance variance. Although the evaluation of inbred progenies of a cross is applicable for both cross- and self-fertilizing species, its greatest usefulness would be for self-fertilizers. Stuber also discussed how the evaluation of progenies with different levels of inbreeding that are developed from parents with different levels of inbreeding may be used in the estimation of genetic components of variance.

4.3.3 Design II

The design II mating design or factorial design was described by Comstock and Robinson (1948). Basic features of the design II and diallel mating designs are quite different, but the genetic information obtained from the two designs is similar. For the diallel design the same parents are used as males and females, whereas different sets of parents are used as males and females (see Table 4.12) for design II. If four parents are included in the diallel design, we have 6 cross combinations and 12 cross permutations. For comparison, if we include a set of eight parents in design II, we have 16 crosses (vs. 12 for all cross permutations in the diallel design) but twice as many parents are included. With either design the number of crosses increases rapidly as the parents included increase, but the number of crosses is considerably less for design II, particularly when greater numbers of parents are used (Table 4.13). Approximately half as many crosses are produced when ten or more parents are used. For a fixed number of experimental units, therefore, approximately twice as many parents can be used in the experiment. This is an advantage of design II, particularly if one wishes to estimate the genetic parameters of a reference population.

From Table 4.12 it is obvious we have a cross-classification design for analysis. Consequently, we will have sources of variation for males, females, and the interaction of males with females. The form of the analysis of variance when m males are crossed with f females and evaluated in r replications is shown in Table 4.14. The expected mean squares expressed in terms of the covariance of relatives are similar to those for the diallel analysis in Table

Table 4.12. Comparison of the diallel and design II mating designs for the possible crosses among parents.

Parents (females)	Diallel Parents (males)				Parents (females)	Design II Parents (males)			
	1	2	3	4		1	2	3	4
1	---	X_{12}	X_{13}	X_{14}	5	X_{15}	X_{25}	X_{35}	X_{45}
2	X_{21}	---	X_{23}	X_{24}	6	X_{16}	X_{26}	X_{36}	X_{46}
3	X_{31}	X_{32}	---	X_{34}	7	X_{17}	X_{27}	X_{37}	X_{47}
4	X_{41}	X_{42}	X_{43}	---	8	X_{18}	X_{28}	X_{38}	X_{48}

Table 4.13. Comparisons of the number of crosses produced by the diallel and design II mating designs for n parents.

Number of parents	Number of crosses		Ratio of crosses design II/diallel
	Diallel	Design II	
4	6	4	0.67
6	15	9	0.60
8	28	16	0.57
10	45	25	0.56
16	120	64	0.53
20	190	100	0.53
30	435	225	0.52
50	1225	625	0.51
100	4950	2500	0.50
n	$n(n-1)/2$	$n^2/4$	$n/[2(n-1)]$

4.8. The expectations of males and females for design II are equivalent to GCA, and the male × female source is equivalent to SCA of the diallel analysis. Because we have two sets of parents in design II, we have two independent estimates of GCA. Appropriate F-tests can be made to test for the differences among males and among females and for the interactions of males and females. Similar to the diallel analysis, the model I analysis provides estimates of GCA effects for males and females and SCA effects for males × females.

The model II analysis gives estimates of components of genetic variance that are estimable from covariances of relatives. From Table 4.14, $\sigma_m^2 = \sigma_f^2$ = Cov HS = $(1/4)\sigma_A^2$ for F = 0 and $(1/2)\sigma_A^2$ for F = 1, and $\sigma_{mf}^2 =$ Cov FS − Cov HS$_m$ − Cov HS$_f$ = $(1/4)\sigma_D^2$ for F = 0 and σ_D^2 for F = 1; estimates are under the assumption of no epistasis in all instances. Two independent estimates of σ_A^2 are calculated for F = 0 as $\hat{\sigma}_{A_m}^2 = 4(M_5 - M_3)/(rf)$ and $\hat{\sigma}_{A_f}^2 = 4(M_4 - M_3)/(rm)$. The variance V of the estimates of σ_A^2 are

$$V(\hat{\sigma}_{A_m}^2) = \frac{16}{(rf)^2} \cdot 2 \left[\frac{M_5^2}{m+1} + \frac{M_3^2}{(m-1)(f-1)+2} \right]$$

$$V(\hat{\sigma}_{A_f}^2) = \frac{16}{(rm)^2} \cdot 2 \left[\frac{M_4^2}{f+1} + \frac{M_3^2}{(m-1)(f-1)+2} \right]$$

When inbreeding of the parents is zero, σ_D^2 is estimable as $\hat{\sigma}_D^2 = 4(M_3 - M_2)/r$, which has a variance of

$$V(\hat{\sigma}_D^2) = \frac{16}{r^2} \cdot 2 \left[\frac{M_3^2}{(m-1)(f-1)+2} + \frac{M_2^2}{(r-1)(mf-1)+2} \right]$$

Table 4.14. Analysis of variance of the design II mating design in one environment.

Source	df	Mean squares	Expected mean squares (model II)	
			Components of variance	Covariances of relatives
Replications	$r-1$			
Males	$m-1$	M_5	$\sigma^2 + r\sigma_{fm}^2 + rf\sigma_m^2$	$\sigma^2 + r(\text{Cov FS} - \text{Cov HS}_f - \text{Cov HS}_m) + rf\text{Cov HS}_m$
Females	$f-1$	M_4	$\sigma^2 + r\sigma_{fm}^2 + rm\sigma_f^2$	$\sigma^2 + r(\text{Cov FS} - \text{Cov HS}_f - \text{Cov HS}_m) + rm\text{Cov HS}_f$
Males × females	$(m-1)(f-1)$	M_3	$\sigma^2 + r\sigma_{fm}^2$	$\sigma^2 + r(\text{Cov FS} - \text{Cov HS}_f - \text{Cov HS}_m)$
Error	$(r-1)(mf-1)$	M_2	σ^2	σ^2
Total	$rmf-1$			
Within plot	$rmf(k-1)$	$M_1^{\dagger}$		

$^{\dagger}M_1$ is the within-plot mean square and includes the within-plot genetic variance (σ_{wg}^2) and environmental variance (σ_{we}^2); $\sigma_{wg}^2 = \sigma_G^2 - \text{Cov FS}$, thus $\sigma^2 = [\sigma_{we}^2 + (\sigma_G^2 - \text{Cov FS})]/k + \sigma_p^2$, where σ_G^2 is the total genetic variance, σ_p^2 is the plot error variance, and k is the number of plants measured in each plot.

Estimates of components of variance characterize the population from which the parents were a random sample.

The design II mating design has not been used nearly as extensively in maize as the diallel, but it seems to merit further consideration. The mechanics of making crosses when the parents are inbred lines are no different from those for diallel designs. For noninbred S_0 plants, however, multiple crosses on female plants are not possible in maize. Unselected S_1 progenies developed from a population can be used in crosses with unselected S_0 plants used as males; this requires making crosses on several plants (5 to 10) and bulking the seed for testing. Multiple pollinations from the S_0 male plants are possible in maize, but care must be taken not to break or damage the male inflorescence if multiple pollinations are needed for more than one day. The main problem of crossing S_0 plants as males onto a series of S_1 progenies is coordinating the time of flowering because of protandry of males and delayed flowering of S_1 progenies. It is suggested that the S_0 plants be delay-planted to increase opportunities of simultaneous flowering of S_1 progenies and S_0 plants. Hallauer (1970) effectively used design II for estimation of genetic components of variance in maize populations by use of noninbred parents. Care must be taken, however, to minimize selection in producing S_1 progenies, in choice of S_0 plants used as males, and in sufficient sampling of genotypes within each of the S_1 progenies.

It seems that design II has the following advantages over diallel designs if one is interested in estimating components of variance of a reference population: (1) more parents can be included for a given level of resources, (2) two independent estimates of σ_A^2 are available, (3) an estimate of σ_D^2 is determined directly from the mean squares, and (4) a greater number of parents can be included by subdividing parents into sets. Advantages (1) and (4) are related and can be used to increase the sampling of the reference population. Grouping of parents into sets permits pooling the sums of squares over sets. We are interested in obtaining estimates of components of variance rather than comparisons of means. If interest is primarily in estimation of genetic components of variance of a reference population, sets of diallel crosses also could be pooled so that advantages (1) and (4) relative to the diallel are not great. Generally, however, the diallel mating design assumes all possible crosses among a set of parents. For example, if 20 parents of a reference population are considered, we will have 190 crosses for the diallel and 100 crosses for design II (Table 4.15). But if we subdivide the 20 parents into two sets (n' = 10) of design II crosses, we will need to make 50 crosses. Similarly, if we subdivide 20 parents into four sets (n' = 5) of 5 parents for diallel crossing, we would have 40 diallel crosses. Sampling 200 parents from a population would not be unreasonable: 40 five-parent (n' = 5) diallels result in 400 crosses and 20 ten-parent (n' = 10) design IIs result in 500 crosses. Table 4.15 summarizes some of the options made possible by partitioning the parents into sets of n' parents, remembering that each design II set includes twice as many parents as each diallel.

The analysis of variance (Table 4.16) for parents grouped in sets includes a source due to sets, but otherwise the expectations of the mean squares of males, females, and males × females are the same for the components of variance and the covariances of relatives. An analysis is conducted on each set, and sums of squares and degrees of freedom are pooled over sets. Repeating the experiments over environments gives the analysis of variance shown in Table 4.17. For the analysis repeated over environments, direct F-tests can be made for all sources of variation except for males and females. Satterthwaite's (1946) approximate test procedure can be used to synthesize

Table 4.15. Number of crosses possible from the diallel and design II mating designs and possible options for different sample sizes of parents for partitioning into sets.

Number of parents	Number of crosses									
	Sets of diallel					Sets of design II				
	1	2	4	10	20	1	2	4	10	20
6	15	--	--	--	--	9	--	--	--	--
10	45	20	--	--	--	25	12[+]	--	--	--
20	190	90	40	--	--	100	50	24[+]	--	--
40	780	380	180	60	--	400	200	100	40	--
80	3,160	1,560	760	280	120	1,600	800	400	160	80
100	4,950	2,450	1,200	450	200	2,500	1,250	624[+]	250	120[+]
200	19,900	9,900	4,900	1,900	900	10,000	5,000	2,500	1,000	500
n	$n(n-1)/2$	$2[n'(n'-1)/2]$‡	$4[n'(n'-1)/2]$	$10[n'(n'-1)/2]$	$20[n'(n'-1)/2]$	$n^2/4$	$2(n'^2/4)$	$4(n'^2/4)$	$10(n'^2/4)$	$20(n'^2/4)$

[+] Number of males and females are not equal.

‡ n' depends on the grouping of parents in sets.

Table 4.16. Analysis of variance of design II sets pooled over sets for
one experiment for model II.

Source	df	Mean squares	Expected mean squares
Sets	$s-1$[†]		
Replications/sets	$s(r-1)$		
Males/sets	$s(m-1)$	M_4	$\sigma^2 + r\sigma_{fm}^2 + rf\sigma_m^2$
Females/sets	$s(f-1)$	M_3	$\sigma^2 + r\sigma_{fm}^2 + rm\sigma_f^2$
Males x Females/sets	$s(m-1)(f-1)$	M_2	$\sigma^2 + r\sigma_{fm}^2$
Pooled error	$s(r-1)(mf-1)$	M_1	σ^2[‡]
Total	$s(rmf-1)$		

[†]s, r, m, and f refer to the number of sets, replications, males, and
females, respectively.

[‡]If individual plant data are taken, σ^2 will be equal to $[(\sigma_G^2 - \text{Cov FS})$
$+ \sigma_{we}^2]/k + \sigma_p^2$, where k is the number of plants measured per plot and
σ_p^2 is the experimental plot error.

mean squares that have the same expected value except for the effect being
tested; e.g., for females, $M_6 + M_2$ can be tested with $M_4 + M_3$ with the following degrees of freedom:

$$n_1 = (M_6 + M_2)^2/\{M_6^2/[s(f-1)] + M_2^2/[s(m-1)(f-1)(e-1)]\}$$

$$n_2 = (M_4 + M_3)^2/\{M_4^2/[s(m-1)(e-1)] + M_3^2/[s(f-1)(e-1)]\}$$

Tables 4.16 and 4.17 assume random effects of parents and environments; hence
estimates of the components of genetic variance are interpreted relative to
the reference population and how they interact with environments.

A measure of the average dominance of genes in the expression of the
trait analyzed can be determined by the components of genetic variance estimated from the expected mean squares. If we assume a population in linkage
equilibrium at $p = q = 0.5$ and no epistasis, the components of genetic variance are estimated as

$$\sigma_A^2 = (1/2)\sum_i a_i^2 \quad \text{and} \quad \sigma_D^2 = (1/4)\sum_i d_i^2$$

When $F = 0$, $\sigma_m^2 = \sigma_f^2 = \text{Cov HS} = (1/4)\sigma_A^2$, and $\sigma_{mf}^2 = \text{Cov FS} - \text{Cov HS}_m - \text{Cov HS}_f$
$= (1/4)\sigma_D^2$. The average dominance of genes affecting the trait can be deter-
mined as

$$\bar{d} = (2\sigma_{mf}^2/\sigma_m^2)^{1/2} = (2\sigma_{mf}^2/\sigma_f^2)^{1/2}$$

From the calculated ratio we can determine which of the following levels of

Table 4.17. Analysis of variance of design II repeated over environments for model II.

Source	df	Mean squares	Expected mean squares
Environments (E)	e-1[†]		
Sets (S)	s-1		
S × E	(e-1)(s-1)		
Replications/S/E	es(r-1)		
Males/S	s(m-1)	M_7	$\sigma^2 + r\sigma^2_{fme} + rf\sigma^2_{me} + re\sigma^2_{mf} + ref\sigma^2_m$
Females/S	s(f-1)	M_6	$\sigma^2 + r\sigma^2_{fme} + rm\sigma^2_{fe} + re\sigma^2_{mf} + rem\sigma^2_f$
Males × females/S	s(m-1)(f-1)	M_5	$\sigma^2 + r\sigma^2_{fme} + re\sigma^2_{mf}$
Males/S × E	s(m-1)(e-1)	M_4	$\sigma^2 + r\sigma^2_{fme} + rf\sigma^2_{me}$
Females/S × E	s(f-1)(e-1)	M_3	$\sigma^2 + r\sigma^2_{fme} + rm\sigma^2_{fe}$
Males × females/S × E	s(m-1)(f-1)(e-1)	M_2	$\sigma^2 + r\sigma^2_{fme}$
Pooled error	es(r-1)(mf-1)	M_1	σ^2
Total	esrmf-1		

[†]e, s, r, m, and f refer to the number of environments, sets within an environment, replications, males, and females, respectively.

dominance of genes were operative: 0 is no dominance, 0 to 1 is partial dominance, 1 is complete dominance, and a value that exceeds 1 is termed overdominance. Examination of the assumptions for calculation of average dominance of genes shows that bias may be important. If digenic epistasis is present, the estimate of d will be biased upward because Cov FS - Cov HS_m - Cov HS_f shows that we have a contribution of $(1/8)\sigma^2_{AA} + (1/8)\sigma^2_{AD} + (1/16)\sigma^2_{DD}$ in the estimate of σ^2_{mf}. Effects of linkage bias depend to some extent on the reference population under consideration. If a large random mating population is considered, linkage bias probably is minimal; whereas in an F_2 population created from two inbred lines, linkage disequilibrium may be important. Coupling phase linkages would not be a source of bias in estimates of d because both $\hat\sigma^2_A$ and $\hat\sigma^2_D$ have a positive or upward bias if the population is in linkage disequilibrium; i.e., $\hat\sigma^2_A$ and $\hat\sigma^2_D$ are biased but d is not. If the genes are in repulsion phase linkages (which is more likely when crossing two inbred lines to correct weaknesses in both), the expression could be the same as for overdominance in the expression of independently segregating genes, although none of the linked genes was individually more than partially dominant to their alleles. Repulsion phase linkages cause an upward or positive bias in the estimate of σ^2_D (same as coupling phase linkages) but cause a downward or negative bias in estimates of σ^2_A. Hence σ^2_D will be overestimated and σ^2_A underestimated, which results in an overestimate of d. Gene frequencies will not be known but are expected to be approximately p = q = 0.5 for F_2 populations and may not diverge greatly from 0.5 for populations that have not been under intentional selection. Populations under long-term selection pressures may have a significant departure from 0.5, however.

Estimates of heritability can be calculated from use of estimates of σ_A^2 from male and female components of variance. Assuming noninbred parents and no epistasis, $\sigma_A^2 = 4\sigma_m^2$ (Table 4.16) and an estimate of h^2 based on the mean of r plots (an entry-mean basis) for one environment is

$$h^2 = \frac{4\hat{\sigma}_m^2}{\hat{\sigma}^2/r + 4\hat{\sigma}_{mf}^2 + 4\hat{\sigma}_m^2}$$

which has an approximate standard error of

$$SE(h^2) = \frac{4SE(\hat{\sigma}_m^2)}{\hat{\sigma}^2/r + 4\hat{\sigma}_{mf}^2 + 4\hat{\sigma}_m^2}$$

A similar estimate of heritability can be calculated from the female source of variation. Also, if the degrees of freedom are equal for male and female sources of variation, an estimate of σ_A^2 can be made from the mean square obtained by pooling the male and female degrees of freedom and sums of squares. Standard errors of components of variance are calculated in the usual manner, and the standard error of σ_A^2 obtained from the mean square by pooling the males and females would be reduced because twice as many degrees of freedom are included in the denominator. If individual plant data are collected, an estimate of heritability on an individual plant basis can be calculated as

$$h^2 = 4\hat{\sigma}_m^2/(\hat{\sigma}_w^2 + \hat{\sigma}_p^2 + \hat{\sigma}_{mf}^2 + \hat{\sigma}_f^2 + \hat{\sigma}_m^2), \text{ where } \hat{\sigma}_p^2 \text{ is plot error}$$

and $\hat{\sigma}_w^2 = \hat{\sigma}_{we}^2 + (\hat{\sigma}_G^2 - \text{Cov FS})$. If the parents are noninbred (F = 0), $\hat{\sigma}_w^2$ includes environmental variance among plants within a plot (σ_{we}^2) and the genetic variance within plots, $(\sigma_G^2 - \text{Cov FS}) = (1/2)\sigma_A^2 + (3/4)\sigma_D^2$, assuming no epistasis. If the parents are homozygous, σ_w^2 of course includes only σ_{we}^2. The standard error of the individual plant heritability estimate is

$$SE(h^2) = 4SE(\hat{\sigma}_m^2)/(\hat{\sigma}_w^2 + \hat{\sigma}_p^2 + \hat{\sigma}_{mf}^2 + \hat{\sigma}_f^2 + \hat{\sigma}_m^2).$$

Estimates of heritability unbiased by genotype-environment interaction and based on the mean of re plots can be calculated from Table 4.17 as

$$h^2 = \frac{4\hat{\sigma}_m^2}{\hat{\sigma}^2/(re) + 4\hat{\sigma}_{fme}^2/e + 4\hat{\sigma}_{me}^2/e + 4\hat{\sigma}_{mf}^2 + 4\hat{\sigma}_m^2}$$

with a standard error of

$$SE(h^2) = \frac{4SE(\hat{\sigma}_m^2)}{\hat{\sigma}^2/(re) + 4\hat{\sigma}_{fme}^2/e + 4\hat{\sigma}_{me}^2/e + 4\hat{\sigma}_{mf}^2 + 4\hat{\sigma}_m^2}$$

An estimate of heritability also is available from the female component of variance, but the best estimate would be obtained from pooling male and female sums of squares. It seems that design II is very useful for estimation of genetic variances in a population. In plants we usually are able to have balanced data, and pooling the male and female sums of squares provides an estimate of σ_A^2 with reasonable errors. If only a few selected parents are included, design II has no advantages over the diallel for estimating genetic effects of parents (GCA) and their crosses (SCA); the same information can be obtained from the diallel and design II. Design II also is useful when there is a grouping of male-sterile and male-fertile parents, such as the A and B lines of *Sorghum bicolor*.

4.3.4 Partial Diallel

The partial diallel design was developed by Kempthorne and Curnow (1961) to increase the number of parents that can be included in the diallel. The mechanical procedures for developing crosses and the principles of analyzing data are similar to those of the complete diallel. The major difference between partial and complete diallels is the number of crosses made among parents. For the diallel, $n(n - 1)/2$ combinations of crosses are made among n parents, whereas for the partial diallel only a portion of the possible crosses are made for each parent.

Assume one has the resources to grow $ns/2$ crosses, where n is the number of parents and s is a whole number greater than or equal to two. If we have n parents, the following crosses can be sampled:

1 X (k + 1)	2 X (k + 2)	...	n X (k + n)
1 X (k + 2)	2 X (k + 3)		n X (k + n + 1)
•	•		•
1 X (k + s)	2 X (k + 1 + s)	...	n X (k + n - 1 + s)

where $k = (n + 1 - s)/2$ is a whole number (not to be confused with the k used elsewhere in this chapter). All the k above n are reduced by multiples of n to be between 1 and n. For k to be a whole number, we do not want both n and s to be odd or both even. Each parent occurs in s crosses, and the number of crosses sampled is $ns/2$. If $s = n - 1$, it corresponds to the complete diallel of $n(n - 1)/2$ cross combinations.

As an example, assume we have the resources to make and grow 120 crosses. If $ns/2 = x$, we can determine the number of parents that can be included for number of crosses s; i.e., $n = 2x/s$. Table 4.18 summarizes relations among mating designs for the number of parents that can be included if 120 crosses are made and evaluated. We can include 2.5 (when s = 6) to 5 (when s = 3)

Table 4.18. Number of parents that can be included to produce 120 crosses for the diallel, partial diallel, and design II mating designs.

Number	Diallel	Partial diallel				Design II
		s=3	s=4[†]	s=5	s=6[†]	n=8, sets=8
Parents, n	16	80	60	48	40	64
Crosses	120	120	120	120	120	128

[†]Not valid because for k to be a whole number, n and s cannot both be even.

times more parents in the partial diallel relative to the diallel when we consider the number of crosses is fixed at 120. Design II is similar to s = 4 (a
nonvalid s value) when eight sets of eight parents are included. For estimation of genetic variances, the partial diallel is superior to the diallel because a greater number of parents can be included for a fixed limit on resources.

The particular crosses made and tested among the n parents is determined
from the calculated k value. If we assume n = 80 and s = 3, we will have (80
X 3)/2 = 120 crosses, and k = (80 + 1 - 3)/2 = 39. Hence, the cross combinations among the 80 parents will be:

1 X 40	2 X 41	...	39 X 78	40 X 79	41 X 80	42 X 1^{+}	...	80 X 39^{+}
1 X 41	2 X 42		39 X 79	40 X 80	41 X 1^{+}	42 X 2^{+}		80 X 40^{+}
1 X 42	2 X 43		39 X 80	40 X 1	41 X 2^{+}	42 X 3^{+}		80 X 41^{+}

$^{+}$A reciprocal cross – one or the other not to be grown.

Each parent, therefore, is equally represented in the crosses.

The experimental design model, the genetic model, the analysis of variance, and the covariances of relatives of the partial diallel are the same as
for the diallel. Degrees of freedom and coefficients of expected mean squares
will be different because of the sampling of crosses among parents (Table
4.19). In addition to including a greater number of parents, the partial diallel also has a more even distribution of degrees of freedom for GCA and SCA
(Table 4.20) because more parents are included for a given number of crosses.
In the diallel the degrees of freedom are usually small for the GCA mean
square relative to the SCA mean square. This may be a serious disadvantage in
the estimation of σ_g^2 unless σ_g^2 is small compared with $\sigma^2 + r\sigma_s^2$. For s = 3
nearly twice as many degrees of freedom are included for GCA compared with
SCA. For s = 4, the degrees of freedom are nearly equal, but 4 is not valid
because then k is not a whole number for even n. When s = 5, the ratio of GCA
to SCA df is 0.65. With a more even distribution of degrees of freedom, components of variance will be estimated with approximately the same precision.

Table 4.19. Analaysis of variance of the partial diallel evaluated in one environment, model II.

Source	df	Mean squares	Expected mean squares
Replications	$r-1^{+}$		
Crosses	(ns/2)-1	M_3	
GCA	n-1	M_{31}	$\sigma^2 + r\sigma_s^2 + [rs(n-2)/(n-1)]\sigma_g^2$
SCA	n(s/2-1)	M_{32}	$\sigma^2 + r\sigma_s^2$
Error	(r-1)[(ns/2)-1]	M_2	$\sigma^{2\,\ddagger}$
Total	(rns/2)-1		
Within	r(ns/2)(k-1)	M_1	

$^{+}$r, n, s, and k refer to the number of replications, parents, crosses per parent, and plants within
a plot, respectively.

$\ddagger$ If individual plant data are taken, σ^2 will be equal to $[(\sigma_g^2 - Cov\ FS) + \sigma_{we}^2]/k + \sigma_p^2$, where k is
the number of plants measured per plot and σ_p^2 is the experimental plot error.

Table 4.20. Distribution of the degrees of freedom for the
partial diallel and diallel for 120 crosses grown
in two replications in one environment.

| | | Degrees of freedom | | |
| | | Partial diallel | | |
Source	Diallel	s = 3	s = 4[†]	s = 5
Replications	1	1	1	1
Crosses	119	119	119	119
GCA	15	79	59	47
SCA	104	40	60	72
Error	119	119	119	119
Total	239	239	239	239

[†]s = 4 is not valid because for k to be a whole number, n and s
cannot both be even.

Kempthorne and Curnow (1961) also determined the relative efficiency of
the partial diallel to the complete diallel by comparing the relative yielding
capacities of the crosses through methods of estimation. Two methods of estimation were designated A and B: In A the unsampled crosses are estimated by $\hat{u}$
$+ \hat{g}_i + \hat{g}_j$ but sampled crosses are estimated by cross means y_{ij}; in B both sampled and unsampled crosses are estimated by $\hat{u} + \hat{g}_i + \hat{g}_j$. Method B is preferable to method A only if $r\sigma_s^2/\sigma^2$ is small. The methods are identical when s
= 2, but s has to be greater than 2 for σ_s^2 to be estimable. If s > 2, A is
preferred to B only if $\sigma^2/(r\sigma_s^2) < 1$. If r > 1, a decision between A and B is
determined by the estimate of $\sigma^2/(r\sigma_s^2)$ from the analysis of variance. A third
comparison C was made by estimating performances of the crosses by crossing
the parents with each of t common testers. When n = 5, the partial diallel
with s = 2 is preferred to use of the common tester only if $\sigma^2/(r\sigma_s^2) < 1/3$.
When $n \geq 7$, the use of one common tester is preferred to the partial diallel
with s = 2 for all values of $\sigma^2/(r\sigma_s^2)$.

Since its introduction the partial diallel has had limited testing (Jensen 1959). For a small number of selected parents the partial diallel will
not provide any more information than can be obtained from the complete diallel. For a larger number of parents design II is simpler to use than the partial diallel. The partial diallel provides another alternative, but the design does not seem to have potential for extensive use. Estimates of components of variance and heritabilities would be similar to those given for the
complete diallel.

4.4 NESTED DESIGN (DESIGN I)

The design I mating scheme also was introduced by Comstock and Robinson
(1948). Except for the diallel, design I probably has been used more frequently in maize than any of the other mating designs. The procedures for

producing progenies and the analysis of variance are quite different from
those described for the diallel, partial diallel, and design II mating schemes
because the mating design is a nested rather than a cross-classified design.
Design I is appropriate only for estimating genetic components of variance for
a reference population. Hence the model II analysis is applied because the
parents used in producing progenies for testing are an unselected sample from
the reference population.

Assume a random mating population in linkage equilibrium as the reference
population: S_0 plants chosen at random are designated as males that are mated
to a set of randomly chosen S_0 plants designated as females. Female plants
are different from male plants, and each male is mated to a different set of
females; i.e., m males are each mated to f females to produce mf progenies for
evaluation. The genetic structure of the progenies includes full-sibs that
have both parents in common and half-sibs that have a male parent in common.
If each male is mated to four females, we will have the following matings:

$$m_1 \times f_1 = p_{11} \quad m_2 \times f_5 = p_{25} \quad \cdots \quad m_i \times f_j = p_{ij}$$
$$f_2 = p_{12} \qquad f_6 = p_{26} \qquad f_k = p_{ik}$$
$$f_3 = p_{13} \qquad f_7 = p_{27} \qquad f_l = p_{il}$$
$$f_4 = p_{14} \qquad f_8 = p_{28} \qquad f_n = p_{in}$$

Individuals within each p_{ij} progeny are full-sibs. The p_{ij}, p_{ik}, p_{il}, p_{in}
progenies are half-sibs because they have a common i male parent. The model
for one environment is

$$Y_{ijk} = u + m_i + f_{ij} + r_k + e_{ijk}$$

where u is the mean, m_i is the effect of the ith male, f_{ij} is the effect of
the jth female mated to the ith male, r_k is the replication effect, and e_{ijk}
is the experimental error. Because the mating design is nested, expected mean
squares are obtained by a hierarchical type of design. Also, because of the
genetic structure of the mating design, expected mean squares can be expressed
in the more useful covariances of relatives (Table 4.21). Direct F-tests can
be made for males and females-within-males mean squares, and males and females-within-males components of variances can be estimated from the appropriate mean squares. Covariances of relatives, however, relate the genetic components of variance. The male component (Cov HS) is the same genetically as
the GCA of the diallel and partial diallel and among males and among females
of design II. The among-females-within-males component, however, has a different expectation than shown in previous mating designs. For design I each
male is mated to a different group of females. Therefore, only one half-sib-
ship is included, and the component, females within males, is Cov FS – Cov HS
= $(1/4)\sigma_A^2 + (1/4)\sigma_D^2$ for F = 0. We do not have a direct estimate of σ_D^2 from
the mean squares of design I, which can only be obtained by solving for expectations of components of variance:

$$\sigma_D^2 = 4\sigma_{f/m}^2 - 4\sigma_m^2 = 4[(\text{Cov FS} - \text{Cov HS}) - \text{Cov HS}]$$

$$= 4[(1/4)\sigma_A^2 + (1/4)\sigma_D^2 - (1/4)\sigma_A^2] = \sigma_D^2$$

The importance of not having a "clean" estimate of σ_D^2 is exemplified by the

Table 4.21. Analysis of variance of the design I mating design for one environment.

Source	df	Mean squares	Expected mean squares	
			Components of variance	Covariances of relatives
Replications	$r-1$[†]			
Males	$m-1$	M_4	$\sigma^2 + r\sigma^2_{f/m} + rf\sigma^2_m$	$\sigma^2 + r(\text{Cov FS} - \text{Cov HS}) + rf\text{Cov HS}$
Females/males	$m(f-1)$	M_3	$\sigma^2 + r\sigma^2_{f/m}$	$\sigma^2 + r(\text{Cov FS} - \text{Cov HS})$
Error	$(r-1)(mf-1)$	M_2	σ^2	σ^2
Total	$rmf-1$			
Within	$rmf(k-1)$	M_1[‡]		

[†]r, m, f, and k refer to number of replications, males, females within males, and plant within plots, respectively.

[‡]$M_1 = \sigma^2_w = (\sigma^2_{we} + \sigma^2_{wg}) = [\sigma^2_{we} + (\sigma^2_G - \text{Cov FS})]$.

Hence, $M_2 = \sigma^2 = [\sigma^2_{we} + (\sigma^2_G - \text{Cov FS})]/k + \sigma^2_p$, where σ^2_p is the experimental plot error. For $F = 0$ and no epistasis, $\sigma^2_G - \text{Cov FS} = (1/2)\sigma^2_A + (3/4)\sigma^2_D$.

calculation of standard errors of estimates of components of variance. From Table 4.21 the estimated variance of σ^2_A is

$$V(\hat{\sigma}^2_A) = \frac{16 \times 2}{r^2 f^2} \left[\frac{M_4^2}{m+1} + \frac{M_3^2}{m(f-1)+2} \right]$$

The estimate of σ^2_D was obtained as $4(\sigma^2_{f/m} - \sigma^2_m)$, where $\sigma^2_{fm} = (M_3 - M_2)/r$ and $\sigma^2_m = (M_4 - M_3)/(rf)$. Therefore, the estimate of the variance of σ^2_D is approximately

$$V(\hat{\sigma}^2_D) = \frac{16 \times 2}{r^2 f^2} \left[\frac{M_4^2}{m+1} + \frac{(f+1)^2 M_3^2}{m(f-1)+2} + \frac{f^2 M_2^2}{(r-1)(mf-1)+2} \right]$$

Because of the complicated function used to estimate σ^2_D, the variance of the estimate usually is quite large.

It is essential that the males and the females mated to each male are randomly chosen in the reference population. Any stratification of the individuals selected as males and females could reduce the estimates of σ^2_m and σ^2_{fm}. Randomly chosen males mated to a group of females that are similar in date of flowering to the male could reduce the variability among females within males (σ^2_{fm}). Assortative mating of males with females can be reduced by delayed planting of a portion of the population from which the males are chosen (Lindsey et al. 1962). Early flowering female silks can be held for later pollination by covering the ear shoot to prevent fertilization. Randomly chosen males from the delayed planting can be crossed with the females without regard to vigor, plant size, maturity, etc. It can be argued that this forced mating is not representative of a cross-pollinated population because of forced crossing among physiologically isolated segments of a random mating population, but it does satisfy the assumptions of the genetic model.

Because we are interested in obtaining valid estimates of genetic components of variance of our reference population, we want to include an adequate sample of genotypes from our reference population. If a large number of males are mated with females, the size of replications required to include all progenies could be quite large. To reduce replication size and attempt to increase the precision of our experiment, Comstock and Robinson (1948) suggested grouping progenies into sets by males, which is similar to the technique discussed for design II. If 100 male plants are each mated with 4 females, we have 400 full-sib progenies for testing. To reduce replication size, five sets of 20 males each or 80 full-sib progenies may be considered. Or assume that 10 males each crossed to 4 females are included in one set of 40 full-sib progenies. The final choice of number of progenies to include in a set depends on the experimenter's knowledge of soil variability and past experience with size of replication for local control of experimental error. After the number of progenies included in a set has been determined, the two alternatives for arranging the progenies in the field are: (1) replications within sets or (2) sets (as subblocks) within replications. Each set is analyzed as shown in Table 4.21 and then pooled across sets for degrees of freedom and sums of squares as shown in Table 4.22. There are only small differences in the distribution of degrees of freedom; therefore, the major difference between the two field arrangements may be in magnitude of error sums of squares. Intuitively, it seems that replications within sets would be a preferable arrangement for local control of experimental error.

The analysis of the design I experiment repeated over environments is shown in Table 4.23. Translation of components of variance to covariances of relatives permits estimation of components of genetic variances (σ_A^2 and σ_D^2) and their interactions with environments. F-tests and estimation of components of variance can be made directly for all sources of variation except males within sets, for which Satterthwaite's (1946) approximation can be used. Design I has been used frequently in maize, and it is good for extensive sampling of S_0 plants in a population. In comparison with mating designs previously discussed, design I is the easiest for producing a large number of progenies in maize. The nested structure of the progenies makes design I amenable for grouping them in sets, and the pooling across sets is straightforward.

Estimates of additive genetic variance (σ_A^2) and total genetic variance (σ_G^2), assuming no epistasis, are obtained directly from the mean squares of the analysis of variance. If an estimate of dominance variance (σ_D^2) is desired, it can be obtained as the difference between the females-within-males and the males components of variance, but the variance of σ_D^2, unfortunately, is usually quite large. Similar to diallel and design II analyses, design I provides GCA information for males. If male plants are self-pollinated, early test information is obtained and males with superior GCA can be included in breeding nurseries as S_1 progenies. Average dominance of the genes also can be determined from design I analyses. Measure of the dominance of genes has the same interpretation as that given for design II. For restrictions of no epistasis, linkage equilibrium, and $p = q = 0.5$, expectations of the components of variance are

$$\sigma_m^2 = \text{Cov HS} = (1/4)\sigma_A^2 \quad \text{where} \quad \sigma_A^2 = (1/2)\sum_i a_i^2$$

Table 4.22.　Analysis of variance of design I experiment pooled over sets in one environment.

| | df | | Mean | |
Source	General	Example	squares	Expected mean squares[†]
Replications within sets				
Sets	$s-1$[‡]	9		
Replications/sets	$s(r-1)$	10		
Males/sets	$s(m-1)$	90	M_4	$\sigma^2 + r\sigma^2_{f/m} + rf\sigma^2_m$
Females/males/sets	$sm(f-1)$	300	M_3	$\sigma^2 + r\sigma^2_{f/m}$
Pooled error	$s(mf-1)(r-1)$	390	M_2	σ^2
Total	$srmf-1$	799		
Within plots	$srmf(k-1)$		M_1[§]	
Sets within replications				
Replications	$r-1$	1		
Sets	$s-1$	9		
Sets × replications	$(r-1)(s-1)$	9		
Males/sets	$s(m-1)$	90	M_4	$\sigma^2 + r\sigma^2_{f/m} + rf\sigma^2_m$
Females/males/sets	$sm(f-1)$	300	M_3	$\sigma^2 + r\sigma^2_{f/m}$
Pooled error	$(r-1)(smf-1)$	390	M_2	σ^2
Total	$rsmf-1$	799		
Within plots	$rsmf(k-1)$		M_1[§]	

[†]Covariances of relatives are as shown in Table 4.21.

[‡]r, s, m, f, and k refer to the number of replications (2), sets (10), males (10), females (4), and number of plants within plots, respectively, in the example.

[§]See Table 4.21.

$$\sigma^2_{f/m} = \text{Cov FS} - \text{Cov HS} = (1/4)\sigma^2_A + (1/4)\sigma^2_D \quad \text{where} \quad \sigma^2_D = (1/4)\sum_i d^2_i$$

$$\overline{d} = \left[\frac{2(\sigma^2_{f/m} - \sigma^2_m)}{\sigma^2_m}\right]^{1/2} = \left(\frac{2\sigma^2_D}{\sigma^2_A}\right)^{1/2} = \left[\frac{(1/2)\sum_i d^2_i}{(1/2)\sum_i a^2_i}\right]^{1/2}$$

Heritability estimates based on the mean of r plots can be determined from components of variance given in Table 4.22 for one environment as follows:

$$h^2 = \frac{4\hat{\sigma}^2_m}{\hat{\sigma}^2/r + 4\hat{\sigma}^2_{f/m}}$$

Table 4.23. Analysis of variance of design I experiments pooled over sets and repeated over environments for sets within replications.

Source	df	Mean squares	Expected mean squares[+]
Environments (E)	$e - 1$[‡]		
Replications/E	$e(r-1)$		
Sets/replications/E	$er(s-1)$		
Males/sets	$s(m-1)$	M_6	$\sigma^2 + r\sigma^2_{ef/m} + rf\sigma^2_{em} + reo^2_{f/m} + refo^2_m$
Females/males/sets	$ms(f-1)$	M_5	$\sigma^2 + r\sigma^2_{ef/m} + reo^2_{f/m}$
E × males/sets	$(e-1)s(m-1)$	M_4	$\sigma^2 + r\sigma^2_{ef/m} + rf\sigma^2_{em}$
E × females/males/sets	$(e-1)ms(f-1)$	M_3	$\sigma^2 + r\sigma^2_{ef/m}$
Pooled error	$es(r-1)(mf-1)$	M_2	σ^2
Total	$esrmf-1$		
Within plots	$esrmf(k-1)$	M_1[§]	

[+] Covariances of relatives are as shown in Table 4.21.

[‡] e, r, s, m, f, and k refer to the number of environments, replications with sets, sets, males, females within males, and plants within plots, respectively.

[§] See Table 4.21.

For noninbred parents and no epistasis, $\hat{\sigma}^2_m = (1/4)\sigma^2_A$ and $\hat{\sigma}^2_{f/m} = (1/4)\sigma^2_A$ $+ (1/4)\sigma^2_D$; hence $4\sigma^2_{f/m}$ includes additive and dominance variance. The standard error of the estimate of heritability is approximately

$$SE(h^2) = \frac{4SE(\hat{\sigma}^2_m)}{\hat{\sigma}^2/r + 4\hat{\sigma}^2_{f/m}}$$

where $SE(\hat{\sigma}^2_m)$ is the square root of

$$\frac{2}{(rf)^2}\left[\frac{M_4^2}{s(m-1)+2} + \frac{M_3^2}{sm(f-1)+2}\right]$$

Heritability estimates for individual plant selection can be calculated as $h^2 = 4\hat{\sigma}^2_m/(\hat{\sigma}^2_w + \hat{\sigma}^2_p + \hat{\sigma}^2_{f/m} + \hat{\sigma}^2_m)$, where $\hat{\sigma}^2_w$ is the estimate of the within-plot variability and $\hat{\sigma}^2_p$ is estimate of plot error. Standard error of h^2 can be calculated as

$$SE(h^2) = \frac{4\ SE(\hat{\sigma}^2_m)}{\hat{\sigma}^2_w + \hat{\sigma}^2_p + \hat{\sigma}^2_{f/m} + \hat{\sigma}^2_m}$$

These estimates of heritability are for one environment and would include an unknown bias because of genotype-environment interactions; i.e., $4\sigma_m^2 = \sigma_A^2 + \sigma_{AE}^2$. From the analysis repeated over environments (Table 4.23) an estimate of heritability unbiased by genotype-environment interactions and based on the mean of re plots would be

$$h^2 = \frac{4\hat{\sigma}_m^2}{\hat{\sigma}^2/(re) + 4\hat{\sigma}_{ef/m}^2/e + 4\hat{\sigma}_{f/m}^2}$$

Noninbred parents usually have been used in making design I progenies. If, however, parents are either partially inbred or homozygous, the coefficients for the calculation of σ_A^2 would have to be changed accordingly; e.g., Cov HS = $[(1 + F)/4]\sigma_A^2$. Thus if F = 1 and $\sigma_m^2 = (1/2)\sigma_A^2$, all estimates of heritability given are in the narrow sense. A check on the relative importance of nonadditive epistatic effects can be determined by comparing $4\hat{\sigma}_m^2 = \sigma_A^2 + (1/4)\sigma_{AA}^2 + \ldots$; $4\hat{\sigma}_{f/m}^2 = \sigma_A^2 + \sigma_D^2 + (3/4)\sigma_{AA}^2 + \ldots$; and $2(\hat{\sigma}_m^2 + \hat{\sigma}_{f/m}^2) = \sigma_A^2 + (1/2)\sigma_D^2 + (1/2)\sigma_{AA}^2 + \ldots$. All relations include all the σ_A^2 but they have different proportions of nonadditive variance, with $4\hat{\sigma}_{f/m}^2$ being the greatest and $4\hat{\sigma}_m^2$ the least bias due to epistasis. Estimates of heritability determined by use of the male component of variance contain less bias due to epistasis than the other two components.

4.5 DESIGN III

The design III mating design also was developed by Comstock and Robinson (1948) to estimate the average level of dominance of genes affecting the traits evaluated. Genes for F_2 populations that were developed from crossing two inbred lines often were in the overdominant range when average level of dominance was estimated. Estimates of average level of dominance assumed linkage equilibrium of the populations. Because F_2 populations were sampled, linkage effects could have been a serious bias in estimating σ_A^2 and σ_D^2. If linkage effects are present, estimates of σ_D^2 are always biased positively, regardless of whether linkage is in the coupling or the repulsion phase. Linkage bias for estimates of σ_A^2, however, depends on the phase of linkage; σ_A^2 is underestimated for repulsion phase linkages and overestimated for coupling phase linkages. Because the measure of dominance of genes was in the overdominant range it was speculated that repulsion phase linkages had overestimated σ_D^2 and underestimated σ_A^2.

The design III mating design, therefore, has primarily been used in maize F_2 populations to determine the effects of linkages on the estimates of σ_A^2, σ_D^2, and average level of dominance. The initial reference population is the F_2 population. Progenies evaluated are developed by backcrossing individual S_0 plants of the F_2 population to both parents; the parents, of course, are the two inbred lines used to produce the F_2 population. Randomly chosen F_2 plants are used as males in crossing to each of the inbred line parents. Progenies for evaluation will be pairs of progenies for each F_2 male parent. Entries include m pairs of progenies, one pair for each F_2 male, and the parental lines. The analysis of variance replicated in one environment is

Table 4.24. Analysis of variance of design III progenies tested in one environment.

Source	df	Mean squares	Expected mean squares
Replications	$r-1$[†]		
Parents	1		
Males	$m-1$	M_3	$\sigma^2 + 2r\sigma_m^2$
Males × parents	$m-1$	M_2	$\sigma^2 + r\sigma_{mp}^2$
Error	$(r-1)(2m-1)$	M_1	σ^2
Total	$2mr-1$		

[†] r and m refer to number of replications and male plants, respectively.

shown in Table 4.24. Mean square expectations show a component of variance among males (σ_m^2) and one for the interaction of males and inbred parents (σ_{mp}^2). Direct F-tests for σ_m^2 and σ_{mp}^2 are possible with the error term, but they do not tell us anything about genetic structure of the progenies and how they relate to components of genetic variances. Assuming no linkage among segregating genes for the trait being studied and no epistasis in the action of segregating genes, Comstock and Robinson (1952) showed

$$(M_3 - M_1)/(2r) = \sigma_m^2 = (1/2)\sum_i p_i q_i a_i^2$$

$$(M_2 - M_1)/r = \sigma_{mp}^2 = \sum_i p_i q_i d_i^2$$

Because gene frequencies of segregating loci in an F_2 population are expected to be one-half, $\sigma_m^2 = (1/8)\sum_i a_i^2 = (1/4)\sigma_A^2 = $ Cov HS and $\sigma_{mp}^2 = (1/4)\sum_i d_i^2 = \sigma_D^2$. Thus $\sigma_A^2 = 4\sigma_m^2$ and $\sigma_{mp}^2 = \sigma_D^2$. From the expectations of the σ_m^2 and σ_{mp}^2 components of variance, a measure of the dominance of genes can be obtained as

$$\bar{d} = \left(\frac{\sigma_{mp}^2}{2\sigma_m^2}\right)^{1/2} \quad \text{or} \quad \left[\frac{(M_2 - M_1)/r}{2(M_3 - M_1)/(2r)}\right]^{1/2} = \left(\frac{M_2 - M_1}{M_3 - M_1}\right)^{1/2}$$

The interpretation of the ratio is the same as for design II.

Design III also provides exact F-tests of two hypotheses concerning the relative importance of dominance effects: (1) that dominance is not present (this can be tested by comparison of the M_1 and M_2 mean squares of Table 4.24; except for sampling errors the M_2 mean square will be greater than the M_1 mean square only when some level of dominance occurs at one or more loci), and (2) that dominance is complete (which is sometimes assumed for genes determining quantitative traits that exhibit heterosis). For the assumption of independent segregation of loci and independent action of genes (no epistasis), expected values of M_3 and M_2 are $M_3 = \sigma^2 + 2r\sigma_m^2 = \sigma^2 + (r/4)\sum_i a_i^2$ and $M_2 = \sigma^2$

$+ (r/4)\Sigma_i d_i^2$. Expected mean squares are similar when $\Sigma_i a_i^2 = \Sigma_i d_i^2$; and deviations of the ratio, M_3/M_2, from unity in either direction indicate that $\Sigma_i a_i^2$ $\neq \Sigma_i d_i^2$ and that dominance is not complete.

To sample the F_2 populations adequately, the number of pairs of progenies may be greater than we desire to include in one replication. The same procedures used for designs I and II for local control of experimental error can be used for design III by grouping pairs of progenies into sets. Each set, therefore, is analyzed as in Table 4.24 and then sums of squares and degrees of freedom across sets are pooled. If the experiment is repeated across environments, the analysis is as shown in Table 4.25. Direct F-tests are available for each mean square and components of variance can be calculated directly from the mean squares with their appropriate standard errors. In addition to providing a measure of the dominance of genes for the expression of a trait, design III also is an excellent mating design for estimation of σ_A^2 and σ_D^2 for an F_2 population, assuming no linkage and epistasis. The combined analysis provides estimates of the interaction of the additive and dominance effects with environments.

It is valid to estimate heritabilities of the traits measured in the design III analysis because the F_2 population is a valid reference population. Sufficient sampling is needed to obtain valid estimates of components of variance to determine average level of dominance. Hence, the components are useful for estimating heritability in the narrow sense in the F_2 population. From the components given in Table 4.24, an estimate of heritability based on the mean of r plots can be determined as follows: $h^2 = 4\hat{\sigma}_m^2/(\hat{\sigma}^2/r + \hat{\sigma}_{mp}^2 + 4\hat{\sigma}_m^2)$.

Table 4.25. Analysis of variance of design III progenies pooled over sets within environments and combined across environments.

Source	df	Mean squares	Expected mean squares
Environments (E)	$e-1$[+]		
Sets	$s-1$		
E X sets	$(e-1)(s-1)$		
Replications/sets/E	$se(r-1)$		
Parents/sets	s		
E X parents/sets	$s(e-1)$		
Males/sets	$s(m-1)$	M_5	$\sigma^2 + 2r\sigma_{me}^2 + 2re\sigma_m^2$
Males X parents/sets	$s(m-1)$	M_4	$\sigma^2 + r\sigma_{mpe}^2 + re\sigma_{mp}^2$
E X males/sets	$(e-1)s(m-1)$	M_3	$\sigma^2 + 2r\sigma_{me}^2$
E X males X parents/sets	$(e-1)s(m-1)$	M_2	$\sigma^2 + r\sigma_{mpe}^2$
Pooled error	$es(r-1)(2m-1)$	M_1	σ^2
Total	$es2mr-1$		

[+] e, s, r, and m refer to the number of environments, sets within an environment, replications, and pairs of progenies for each male parent, respectively.

Estimates of heritabilities in which the estimate of σ_A^2 is not biased by σ_{AE}^2 can be obtained if the experiment is repeated over environments (Table 4.25) as

$$h^2 = \frac{4\hat{\sigma}_m^2}{\hat{\sigma}^2/(re) + \hat{\sigma}_{mpe}^2/e + \hat{\sigma}_{mp}^2 + 4\hat{\sigma}_{me}^2/e + 4\hat{\sigma}_m^2}$$

which is on the mean basis of re plots. Approximate standard errors of heritability estimates can be obtained as previously shown for the other mating designs.

Design III is powerful in testing for presence of dominance effects, but linkage biases may be serious in the estimation of σ_A^2 and σ_D^2 in the F_2 populations where effects of linkage are expected to be at a maximum. Tests for effects of linkage biases can be made by advancing the F_2 and successive generations by random mating (Gardner et al. 1953; Gardner 1963). Random mating can be done by either isolation plantings or hand-pollination but with no selection of male and female plants in the matings. If the first random mating in the F_2 population is designated as synthetic 1, eight generations of random mating of plants, for example, permit recombination of linked loci and, at least for loose linkages, linkage equilibrium would be approached. The approach to linkage equilibrium obviously depends on the rate of recombination and may require many generations of random mating for tightly linked gene combinations. After the F_2 population has been advanced by random mating for, say, eight generations, the design III mating scheme is repeated for the F_2 and F_2 synthetic 8 population. Estimates of σ_A^2, σ_D^2, and $\bar{d}$ from the analysis of the two populations are compared to determine effects of linkage bias on the estimates. Gardner and Lonnquist (1959) have reported estimates testing effects of linkage and found that they were an important bias in the estimates of σ_A^2 and σ_D^2. Hence estimates of overdominance in the F_2 populations were pseudooverdominance because of the joint action of linked genes rather than overdominance at individual loci.

4.6 TRIALLEL AND QUADRALLEL

Cockerham (1961) determined the covariances, in terms of genetic components of variance, between all possible pairs of hybrid relatives among single-cross, three-way, and double-cross hybrids produced from a group of parents that originated from the same population. Analysis of single-cross hybrids was shown in the various diallel analyses in which only additive and dominance genetic variances were estimable, with the assumption of no epistatic effects. Rawlings and Cockerham (1962a, 1962b) presented the triallel (three-way crosses) and quadrallel (double-cross hybrids) analyses to clarify gene action involved in hybrids and to provide estimates of genetic components of variance and tests of genetic hypotheses. As in the model II analyses of the diallel it is important to emphasize that parents used to produce hybrids are randomly chosen from a reference population, which is necessary for estimation of genetic variance components and making tests of genetic hypotheses. Maize breeders usually classify their parent lines for use in hybrids and produce hybrids relative to unrelatedness of parent lines. This is valid, but the researcher should not attempt to estimate components of genetic variance and interpret estimates relative to some reference population. Hence for model II analyses of diallel, triallel, and quadrallel mating designs, the parents of the hybrids are not related because they originated from randomly chosen S_0 plants in the reference population. The parents per se can be inbred lines, but they are not related in the sense of having common parents in their ancestry. It is assumed that all lines have the same level of inbreeding.

Because results of triallel and quadrallel analyses are to be interpreted relative to a particular reference population, it is necessary to sample the reference population adequately. If we have n lines, $n(n - 1)(n - 2)/6$ three-way cross combinations are possible. The possible arrangements or permutations of the three crosses are $3[n(n - 1)(n - 2)/6]$, assuming no reciprocal crosses. Similarly, if we have n parents, there are $n(n - 1)(n - 2)(n - 3)/24$ possible double crosses; and considering the three possible arrangements of four parents, we will have $3[n(n - 1)(n - 2)(n - 3)/24]$ possible double crosses. Table 4.26 illustrates relative proportions of the three types of hybrids for n parents. Even for n = 10 the number of three-way and double crosses is large for testing. Fifty parents from a population is not unreasonable, but the number of crosses to test is unmanageable. The reduction in the number of crosses for testing (but including a reasonable number of parents to sample reference population genotypes) can be obtained by partitioning the parents into sets and pooling sums of squares and degrees of freedom across sets, as illustrated for previous designs.

Basic models and analyses of variance of triallel and quadrallel designs repeated over environments are the same as those given for the diallel in Table 4.4 for model II. Direct F-tests of crosses and crosses with environment mean squares are available to determine if further partitioning of the crosses sums of squares is informative. If the crosses mean square is nonsignificant, further analyses will be fruitless. For diallel analyses the crosses sums of squares were orthogonally partitioned into average performance of a parent (GCA) and interaction of parents in specific crosses (SCA). Orthogonal partitioning of the triallel and the quadrallel also depends on the number of lines common among crosses and the arrangement of lines within crosses. For example, the $12[(4 \times 3 \times 2)/2]$ three-way crosses possible from four parent lines (A, B, C, and D) are as follows:

Table 4.26. Number of possible hybrids available for different numbers of parents, ignoring reciprocal crosses.

Number of parents	Number of hybrids		
	Single crosses	Three-way crosses	Double crosses
5	10	30	15
6	15	60	45
7	21	105	105
8	28	168	210
9	36	252	378
10	45	360	630
20	190	3,420	14,535
50	1,225	58,800	690,900
n	$n(n-1)/2$	$n(n-1)(n-2)/2$	$n(n-1)(n-2)(n-3)/8$

1.	(A × B)C	5.	(A × D)B	9.	(B × D)A
2.	(A × B)D	6.	(A × D)C	10.	(B × D)C
3.	(A × C)B	7.	(B × C)A	11.	(C × D)A
4.	(A × C)D	8.	(B × C)D	12.	(C × D)B

From the basic model of the triallel, $Y_{i(jk)l} = u + r_i + C_{i(jk)} + e_{i(jk)l}$, we can partition the $C_{i(jk)}$ cross sums of squares as shown by Rawlings and Cockerham (1962a) (Table 4.27), where $C_{i(jk)}$ is defined as a linear function of uncorrelated effects as $C_{i(jk)} = (g_i + g_j + g_k) + (s_{2ij} + s_{2ik} + s_{2jk}) + s_{3ijk} + o_{1i} + o_{1(j)} + o_{i(k)} + (o_{2aij} + o_{2aik} + o_{2ajk}) + (o_{2bi(j)} + o_{2bi(k)}) + o_{3ijk}$.
From the linear model, expected mean squares in Table 4.27 are expressed in terms of components of variance. The components of variance have the following interpretations:

σ_g^2 = average effect of lines averaged over all orders, e.g., 1, 2, 3, 4, 5, 6, 7, 9, and 11 for A

σ_{s2}^2 = two-line interaction effect of lines appearing together averaged over all orders, e.g., 1, 2, 3, 5, 7, and 9 for A × B

σ_{s3}^2 = three-line interaction effect of lines appearing together averaged over all orders, e.g., 1 for (A × B)C

σ_{o1}^2 = one-line order effect of lines as a parent, e.g., 7, 9, and 11 for A

Table 4.27.   Sources of variation for the triallel analysis of variance (adapted from Rawlings and Cockerham 1962a).

Source	df	Mean squares	Expected mean squares
Three-way crosses	$3pC_3-1$	C*	
One-line general	P_1	G*	$\sigma^2 + 3r\sigma_{s_3}^2 + 6rp_3\sigma_{s_2}^2 + (3rp_2p_3/2)\sigma_g^2$
Two-line specific	$pp_3/2$	S_2^*	$\sigma^2 + 3r\sigma_{s_3}^2 + 3rp_4\sigma_{s_2}^2$
Three-line specific	$pp_1p_5/6$	S_3^*	$\sigma^2 + 3r\sigma_{s_3}^2$
One-line order	P_1	O_1^*	$\sigma^2 + r\sigma_{o_3}^2 + 3rp_2\sigma_{o_{2b}}^2 + (rp/3)\sigma_{o_{2a}}^2 + (rpp_2/3)\sigma_{o_1}^2$
Two-line order (a)	$pp_3/2$	O_{2a}^*	$\sigma^2 + r\sigma_{o_3}^2 + (2rp_1/3)\sigma_{o_{2a}}^2$
Two-line order (b)	$P_1P_2/2$	O_{2b}^*	$\sigma^2 + r\sigma_{o_3}^2 + 2rp_3\sigma_{o_{2b}}^2$
Three-line order	$pp_2p_4/3$	O_3^*	$\sigma^2 + r\sigma_{o_3}^2$
Error	$(r-1)(3pC_3-1)$	E*	σ^2

Table 4.28. Coefficients of components of variance from the triallel analysis of variance expressed as functions of the genetic components of variance translated from the covariances of relatives.

Component of variance	Coefficients of genetic components of variance for F = 1								
	σ^2_A	σ^2_D	σ^2_{AA}	σ^2_{AD}	σ^2_{DD}	σ^2_{AAA}	σ^2_{AAD}	σ^2_{ADD}	σ^2_{DDD}
σ^2_g	2/9	0	1/16	0	0	25/1152	0	0	0
$\sigma^2_{s_2}$	0	1/9	25/288	1/16	1/36	49/768	41/1152	1/64	1/144
$\sigma^2_{s_3}$	0	0	0	1/24	1/24	3/64	5/96	1/24	1/32
σ^2_{01}	1/8	0	9/64	0	0	49/512	0	0	0
σ^2_{02a}	0	1/4	1/32	9/64	1/16	3/64	41/512	9/256	1/64
σ^2_{02b}	0	0	0	1/64	0	3/256	9/512	1/256	0
σ^2_{03}	0	0	0	1/32	1/8	0	13/256	13/128	3/32

σ^2_{o2a} = two-line order interaction effects of lines averaged over orders, e.g.,
3, 5, 7, and 9 for A and B

σ^2_{o2b} = two-line order interaction effects of parent and grandparent lines due
to particular order, e.g., 7 and 9 for A and B

σ^2_{o3} = three-line order interaction effects of parent and grandparent lines
due to particular order, e.g., 1 for (A × B)C

General (g) and specific (s2, s3) effects are analogous in meaning to effects of diallel analyses, but we have not been confronted with order effects
in prior analyses. The order effects (o1, o2a, o2b, o3) occur because of the
arrangement of parents in three-way crosses and because of the ancestry of the
lines, parental and grandparental, in crosses. F-tests to determine which effects are significantly different from zero can be made directly for all except one-line order effects; similar to previous mating designs, comparisons
of mean squares determine if differences exist but do not provide any genetic
information. Again, it becomes necessary to express expected mean squares in
terms of covariances of relatives because their composition can be expressed
in terms of genetic components of variance.

Rawlings and Cockerham (1962a) showed that there were nine covariances of
relatives of three-way cross relatives that could be translated into genetic
components of variance, whereas in previous mating designs we usually had only
two covariances of relatives. The triallel analysis has seven mean squares
for estimation of components of variance. Translating from covariances of
relatives, genetic components for components of variance in Table 4.27 are
shown in Table 4.28. From Table 4.28, components of variance σ^2_g and σ^2_{o1} include only additive effects and additive × additive epistatic effects. Variance components σ^2_{s2} and σ^2_{o2a} include dominance and all types of epistasis or
deviations from an all-additive model; σ^2_{s3} includes all types of epistatic effects except additive × additive. If normality of genetic effects in the model is assumed, F-tests can be used to test some genetic hypotheses; e.g., the
null hypothesis that σ^2_g = 0 = G/S$_2$ will be a test for the relative importance
of additive genetic effects. If tests of hypotheses indicate significant genetic effects, it would be logical to proceed to obtain estimates of σ^2_A, σ^2_D,
σ^2_{AA}, etc.

The quadrallel model and analysis are similar in form to the triallel,
but covariances of relatives and coefficients of genetic components of variance are different. There are eight covariances of relatives and seven orthogonal partitions of double crosses sums of squares. Tests of hypothesis of
appropriate mean squares and their genetic interpretations are similar to
those of the triallel. Those interested in the quadrallel analysis should refer to Rawlings and Cockerham (1962b).

Triallel and quadrallel mating designs and analyses are of interest because of the number of independent mean squares for estimation of genetic components of variance. With most of the mating designs the assumption of no
epistasis was imposed to provide estimates of σ^2_A and σ^2_D. Because we have more
mean squares available, it is possible to estimate some of the lower order
epistatic variances. If one writes the genetic expectations of mean squares
(and there are eight mean squares, including one for experimental error, in
the triallel and the quadrallel) in terms of genetic components of variance,
the equations may be solved by least squares, weighted least squares, or maxi-

mum likelihood to obtain estimates of the genetic model tested. The first
models tested may exclude epistasis to determine the degree of fit. If deviations are significant, a model that includes lower order epistatic terms
(e.g., σ_{AA}^2) may be fit. Because the mean squares have different variances,
Hayman (1958) suggested the method of maximum likelihood to solve the equations. Because one iteration of the maximum likelihood method is often sufficient, the method of weighted least squares may be preferable for estimation.

Triallel and quadrallel mating designs have not been used extensively because they (1) are relatively new in development, (2) are complex mating designs with complex analyses, (3) require a large number of crosses to sample
the population adequately, and (4) require two or more growing seasons to produce crosses before they can be tested. Because both analyses are for a specific reference population, most breeding programs do not have an unselected
group of parent lines from one population. The large number of crosses to
test can be reduced by grouping parent lines into sets. Table 4.26 shows that
the number of crosses can be drastically reduced for a group of 60 parent
lines, for the triallel crosses are reduced from 102,660 for one set of 60
parent lines to 600 for 10 sets of 6 lines in each set. Wright et al. (1971)
used diallel and triallel mating designs to estimate genetic variances in a
synthetic, Krug Hi I Synthetic 3. Mean squares from the diallel and the triallel were expressed as functions of genetic components of variance and used
to test genetic models that included σ_A^2, σ_D^2, and digenic epistasis. Epistasis
was not estimable from this analysis, even though 11 equations were available
to fit σ_A^2, σ_D^2, σ_{AA}^2, and experimental error. It seems the potential of the
triallel and the quadrallel for estimation of genetic variances in a reference
population is limited because of complexity in obtaining the parents and
crosses. Generally, such selected lines are of diverse origins, and the estimates of genetic variances would not be valid. As for diallel analyses, effects may be estimated for selected lines, but these estimations apply only to
the lines with which they were crossed; the effects could and probably will
have different effects when crossed to another group of lines.

Formulation of the composition of genetic variances among single crosses,
three-way crosses, and double crosses from a common group of parent lines,
however, has an important significance in maize breeding programs that are
committed to inbred line and hybrid development. Cockerham (1961) showed that
variation among single crosses is always greater than among three-way crosses,
and variation among three-way crosses is always greater than among double
crosses. If we assume all parents have an inbreeding coefficient of F = 1,
the genetic variance content of the crosses components of variance (Table
4.26) is shown in Table 4.29. If only additive genetic effects are assumed,

Table 4.29. Coefficients of components of genetic variance of σ_C^2
among unrelated single, three-way, and double crosses.

Type of crosses	Components of genetic variance (F = 1)					
	σ_A^2	σ_D^2	σ_{AA}^2	σ_{AD}^2	σ_{DD}^2	σ_{AAA}^2
Single	1	1	1	1	1	1
Three-way	3/4	1/2	9/16	3/8	1/4	27/64
Double	1/2	1/4	1/4	1/8	1/16	1/8

the relative advantage of single, three-way, and double crosses is 1:3/4:1/2; i.e., variation among single crosses will be twice that of double crosses if we have only additive effects. If nonadditive variance is important, the relative advantage increases for single crosses over three-way and double crosses.

4.7 INBRED LINES

Use of inbred lines as parents of the crosses for diallel, design II, triallel, and quadrallel designs has been discussed. In all instances the crosses tested were noninbred, but the parents could be noninbred (F = 0), partially inbred (0 < F < 1), or completely inbred (F = 1). The effect of using inbred parents was to increase coefficients of components of genetic variance for covariances of relatives; e.g., for F = 0, Cov HS = $(1/4)\sigma_A^2 + (1/16) \cdot \sigma_{AA}^2$, and for F = 1, Cov HS = $(1/2)\sigma_A^2 + (1/4)\sigma_{AA}^2$. Another method for estimation of genetic variances in a population is to test the unselected inbred lines themselves. Although no mating design is used, variability among inbred lines can be used as an estimate of genetic variability of a reference population. Inbred lines will refer to all lines that have some measure of inbreeding, such as S_1 (F = 0.5), S_2 (F = 0.75), and lines assumed to be completely inbred (F = 1). Analysis and translations are facilitated, however, if level of inbreeding is the same for all lines included for testing (see Chap. 3).

Use of inbred lines for estimation of genetic variances requires the same assumption that was imposed when inbred lines were used as parents in crosses; i.e., the inbred lines are an unselected sample of genotypes from the reference population. Adequate sampling of the reference population is necessary, and no selection is imposed on the S_0 plants included to produce the inbred lines. If S_1 lines are used, adequate sampling with a minimum of natural and artificial selection is not too serious. As the inbreeding is continued, however, deleterious recessives are uncovered and it becomes more difficult to obtain a representative sample of the original S_0 plants of the reference population. The degree of difficulty in maintaining a line from each S_0 plant will depend on the genetic load of the reference population. The frequency of deleterious recessives in an open-pollinated variety is probably greater than for a synthetic variety formed from elite inbred lines because many of the deleterious recessives probably were purged in the development of elite lines used to form the synthetic. Therefore, except for mutation the frequency of deleterious recessives would be expected to be less in the synthetic populations. Results from research in developing inbred lines from open-pollinated varieties (Eberhart et al. 1966) and from a synthetic variety (Hallauer and Sears 1973) seem to support this hypothesis. Single-seed descent seems to be the logical method for developing inbred lines beyond the S_1 generation (Brim 1966). The ideal situation would be to have an inbred line for each S_0 plant sampled in the reference population.

The inbred lines developed are themselves tested to determine variability among lines. If an adequate sample, say 196 lines, is available, the lines may be grown in one large replication or partitioned into sets, as illustrated for the mating designs. A 14 × 14 lattice design also may be an appropriate field design to use for increasing the local control of experimental error. Measurements are made on inbred plots in each environment, and analyses on plot means is the ordinary analysis of variance for entries tested in different environments (Table 4.30). Direct F-tests and estimates of components of variance can be made from mean squares shown in Table 4.30.

Because the inbred lines are assumed to be a random sample of genotypes

Table 4.30. Analysis of variance of inbred lines repeated over
            environments.

Source	df	Mean squares	Expected mean squares
Environments (E)	$e-1^{\dagger}$		
Replications/E	$e(r-1)$		
Inbred lines	$n-1$	M_4	$\sigma^2 + r\sigma_{ge}^2 + re\sigma_g^2$
E × inbred lines	$(e-1)(n-1)$	M_3	$\sigma^2 + r\sigma_{ge}^2$
Pooled error	$e(r-1)(n-1)$	M_2	$\sigma^{2\,\ddagger}$
Total	$ern-1$		
Within	$ern(k-1)$	M_1	

† e, r, n, and k refer to the number of environments, replications
within environments, inbred lines, and individual plants measured
within each plot, respectively.

‡ $\sigma^2 = (\sigma_{we}^2 + \sigma_{wg}^2)/k + \sigma_p^2$, where σ_p^2 is experimental plot error.

from the reference population, model II analyses are appropriate. To obtain
estimates of genetic variance in the reference population, it is necessary to
translate (as shown for previous mating designs) components of variance to genetic components of variance. The genetic composition of the variance component due to inbred lines, for different levels of inbreeding, is shown in
Chap. 2. Dominance effects dissipate rapidly with inbreeding. Because we
have only one equation, the assumption of no dominance can be made, and the
inbred line component of variance provides an estimate of the additive genetic
variance. For instance, variance among S_1 lines provides an estimate of σ_A^2.
It also is shown that when F = 1 for the inbred lines, the variation is doubled among those with no genetic variation within the lines. For the case of
F = 1 the inbred line variance component gives an estimate of $2\sigma_A^2$. Depending
on the level of inbreeding, therefore, the estimate of σ_A^2 from the inbred
line component of variance can range from σ_A^2 (F = 0.5) to $2\sigma_A^2$ (F = 1.0). The
seriousness of the dominance bias in the estimate of σ_A^2 is shown in Table
2.11. If p = q = 0.5 (no dominance effects), variance among lines will be due
only to additive effects.

Because estimates of σ_A^2 can be obtained from the inbred line tests, estimates of heritability on a progeny mean basis can be obtained. If, for instance, S_1 lines are evaluated in experiments repeated over environments (Table 4.30), an estimate of heritability can be obtained as

$$h^2 = \frac{\hat{\sigma}_g^2}{\hat{\sigma}^2/(re) + \hat{\sigma}_{ge}^2/e + \hat{\sigma}_g^2}$$

Because S_1 lines are used, the bias from dominance effects is $(1/4)\sigma_D^2$ if p ≠ q (dominance effects are present). The estimate of heritability is not biased by genotype-environment interaction effects because an estimate of σ^2_{AE} is provided by the combined analysis. Approximate standard error of the heritability estimate is

$$SE(h^2) = \frac{SE(\hat{\sigma}^2_g)}{\hat{\sigma}^2/(re) + \hat{\sigma}^2_{ge}/e + \hat{\sigma}^2_g}$$

where the standard error of σ^2_g is obtained from Table 4.30 as the square root of

$$\frac{2}{(re)^2}\left[\frac{M^2_4}{n+1} + \frac{M^2_3}{(e-1)(n-1)+2}\right]$$

Because the component of variance $\hat{\sigma}^2_g$ estimates σ^2_A, we do not have a coefficient for the standard error of σ^2_A, as in previous mating designs.

If lines that are near homozygosity are evaluated, we do not have any bias due to dominance included in our estimate of σ^2_A from the component σ^2_g. For $F = 1$, $\sigma^2_g = 2\sigma^2_A$, and an estimate of heritability based only on additive effects can be obtained in the usual way.

Use of inbred lines for estimation of σ^2_A of a reference population seems enticing, but use of inbred lines has at least two serious handicaps. First, one needs to develop a set of unselected inbred lines that are representative of genotypes of the reference population; this becomes increasingly difficult as the level of inbreeding increases. Second, the time required to develop inbred lines, particularly for greater levels of inbreeding, would be greater than for mating designs using noninbred parents. Neither handicap is present if S_1 inbred lines are used. There has been some speculation that information obtained from inbred line tests would not be as good as information obtained from use of noninbred material because of larger experimental errors associated with inbred line tests. Present evidence, however, does not indicate that experimental errors of inbred tests are any greater than those for noninbred tests. Use of S_1 lines seems to be a good method for estimation of σ^2_A in maize populations if departures from $p = q = 0.5$ and no dominance are not serious.

4.8 SELECTION EXPERIMENTS

All mating designs discussed require that progenies be developed and their evaluation in experiments repeated over environments to provide estimates of genetic components of variance. To ensure proper sampling of the reference population, large experiments are needed to estimate genetic parameters with reasonable sampling errors. Also, experiments need to be repeated over environments to obtain estimates of genetic parameters unbiased by environmental effects. Consequently, conducting experiments to obtain genetic information can be very expensive in time and resources. The primary purpose for obtaining estimates of genetic parameters is to provide guidelines in developing breeding programs and to predict future gain from selection. Often the breeder has populations under selection for different traits but has not determined relative proportions of differences among progenies due to genetic and to environmental forces. Selection experiments, which include adequate sampling and testing in properly conducted field experiments repeated over environments, also provide information necessary for predicting future gain from selection. One case of a selection experiment not providing information is mass selection, where we do not have progenies for testing in replicated tests.

Any selection methods that use progeny information for selecting the best individuals to recombine for population improvement will provide estimates of genetic variance (under the assumption of no nonadditive effects), genetic-environmental interaction variance, and experimental error. If we use Table 4.30 as an example, direct estimates of these components can be determined from linear functions of the mean squares. Genetic composition of the estimate of σ_g^2 depends on the type of progeny evaluation. Table 1.2 lists types of progenies that may be evaluated. For example, if half-sib progenies are evaluated, σ_g^2 contains $(1/4)\sigma_A^2$ and, assuming F = 0 and no epistatic variance, $\sigma_A^2 = 4\sigma_g^2$. If S_1 progenies are evaluated, $\sigma_g^2 = \sigma_A^2$, assuming p = q = 0.5 or no dominance effects. Since σ_g^2 is estimated as $M_4 - M_3$, the variance of our estimate is $[2/(r^2e^2)]\{M_4^2/[(n + 1) + M_3^2/[(e - 1)(n - 1)] + 2\}$. Similarly, estimates of the interaction of genetic effects with environments and its variance can be calculated from M_3 and M_2 mean squares. A coefficient of genetic variation can be calculated as $(\sigma_g/\bar{x}) \times 100$, where $\bar{x}$ is the mean of all progenies. Estimates of heritability can be calculated as shown for the case of inbred lines.

For the initial sampling of a population undergoing selection, estimates of σ_A^2 should be valid if experiments are repeated over environments, assuming no epistasis. Although estimates of σ_A^2 may be biased upward by dominance and epistatic effects, the bias will not be any more serious than for the assumption of no epistasis imposed on the analysis for most of the mating designs. Additive genetic variance is the component of genetic variance useful to the breeder in a selection program because it is fixable; nonadditive intraloci effects (dominance) and interloci effects involving dominance (epistasis) are not fixable. Estimates of σ_A^2 are dependent on gene frequency. If selection is effective, which is the primary objective of selection--i.e., changing gene frequency of the trait(s)--estimates of σ_A^2 will change in cyclical selection programs. An estimate of σ_A^2, however, is available from each cycle of selection. Estimates of σ_A^2 may change with cycles of selection but should be equally valid if sampling was sufficient (which is necessary for effective selection) for each cycle. Estimates of σ_A^2 will vary among cycles (because of sampling) and introduce an unknown source of error. Standard errors for estimates of σ_A^2 can be calculated to determine confidence limits in successive cycles of selection. Changes in gene frequency for most quantitative traits are not expected to be great for several cycles of selection. Thus estimates of σ_A^2 for three to five cycles may be similar, and in most instances probably will be, except for sampling. A pooled estimate of σ_A^2 can be obtained from the pooled analysis that will probably be a better estimate than can be obtained from a mating design evaluated in fewer environments. The pooled estimate also should be an excellent parameter for predicting selection for the next three to five cycles. Variation in estimates of σ_A^2 for different cycles of selection also can be influenced by the extent of recombination of selected parents to form the next cycle population. Effects of recombination could be particularly evident if selected parents are inbred (e.g., S_2) because of retention of large linkage blocks. Environmental effects for different cycles could be important in expansion or contraction of variability σ_g^2 among progenies tested; e.g., the range among progenies may be expanded in a favorable environment and compressed in a very unfavorable one. Pooled estimates of σ_A^2 would tend to dilute the extremes and give a more valid estimate when pooled or combined across environments within cycles and across cycles. The genetic coefficient of variation calculated for each cycle also would provide a measure of the range among progenies relative to the overall mean for progenies.

Estimates of genetic variance from selection experiments that use progeny evaluation is predicated on the thesis that the researcher is initiating a long-term cyclical selection program. If the decision is made to initiate such a program, estimates of σ_A^2 calculated from σ_g^2 to predict expected progress are better than those obtained from special mating designs. Also, research activity can be given to the selection study without diverting resources to a special study. If the experimenter, however, desires to have estimates for a specific population to answer a specific question, the only recourse is to use a specific mating design. It must be emphasized that adequate sampling and evaluation are as important for valid estimates of genetic parameters as for a long-term selection experiment that can involve the span of the researcher's career (Lamkey and Hallauer 1987).

4.9 SPECIAL CASE OF p = q = 0.5

Gene frequencies of our populations generally are not known. It is shown in Chap. 2 that means and genetic variances of populations depend on gene frequencies and change as the frequencies change. In certain types of populations, however, expected average gene frequency is known; they are F_2 populations produced from inbred lines (equivalent to pure lines or cultivars in self-pollinated species) that are assumed to be homozygous and homogeneous (F = 1). If we cross two inbred lines to form the F_1 hybrid and then self-pollinate the heterozygous but homogeneous F_1 plants, we derive the F_2 population that will segregate at each heterozygous locus in the well-known Mendelian ratio of 1/4 homozygous dominant, 1/2 heterozygous, and 1/4 homozygous recessive (Table 4.31). If the two parents have the same allele at some loci, there will be no segregation in the F_2 population. Because we cross two inbred lines, average gene frequency at segregating loci will be 0.5; hence expected frequencies of genotypes are known. The constant C for genotypic values in Table 4.31 includes actions of all genes not under consideration as well as

Table 4.31. Distribution of genotypes and genotypic values of an F_2 population formed by crossing two inbred lines, where A is the desirable allele and a is the undesirable allele.

Genotypic	Frequency of genotype	Genotypic value	Coded genotypic value	
			Falconer (1960)	Mather (1949)
AA	$p^2 = 0.25$	C + a	a	d
Aa	$2pq = 0.50$	C + d	d	h
aa	$q^2 = 0.25$	C − a	−a	−d

nonheritable forces. Mather (1949) and Mather and Jinks (1971) have used the model given in Table 4.31 to develop expected means and genetic components of variance for the different types of generations that can be produced from crossing two inbred lines. The average gene frequency of 0.5 can be viewed from two standpoints: (1) average gene frequency of desirable alleles and undesirable alleles will be 0.5 at each locus or (2) average gene frequency of desirable alleles and undesirable alleles will be 0.5 over all loci.

For estimation of genetic components of variance, linkages of genes and interloci genetic effects (which in most if not all instances are not valid) are assumed absent. Linkage effects on estimates of genetic components of variance can be reduced by random mating of the F_2 population until linkage equilibrium of the genes is approached. We will always have linkages, but an equilibrium of linkages can be approached by random mating. Epistatic effects probably are present also, but Mather (1949) has devised tests to determine seriousness of epistatic effects. Tests for nonadditivity of genetic effects among loci include comparing different generations and how much they deviate from the additive model. If nonadditive effects are detected among loci, Mather has suggested that a transformation of the data be made before analysis. Transformation to an additive scale among loci would satisfy the assumption of no epistasis for estimation of main effect--additive and dominance variance (nonadditivity within loci), which is total genetic variance of the two alleles at each locus. From Table 4.31 we have at each locus three genotypes or 2 df that can be partitioned as (1) additive genetic variance, which is the sum of squares due to regression; and (2) dominance variance, due to deviations from regression.

Procedures used by Mather (1949) and Mather and Jinks (1971) to arrive at the means and the proportions of total genetic variance due to additive and dominance effects will be indicated only briefly; the interested reader should refer to the references for details. For the special case of $p = q = 0.5$, F_2 is our reference population and is equivalent to the open-pollinated, synthetic, and composite variety populations used as reference populations for other mating designs for interpreting genetic components of variance. Hence the S_0 plants of a variety are equivalent to F_2 plants, and the populations are similar in genetic structure (see Table 2.2 vs. Table 4.31) for developing the relations. The two inbred parents and the F_1 hybrid of the two parents contain no genetic variability and are merely building blocks for constructing the F_2 population. From Table 4.31 and procedures outlined in Chap. 2 the mean of the F_2 population is $(1/4)a + (1/2)d + (1/4)(-a) = (1/2)d$ for one locus. Similarly, total genetic variance in the F_2 population is $(1/4)a^2 + (1/2)d^2 + (1/4)(-a)^2 - [(1/2)d]^2 = (1/2)a^2 + (1/4)d^2$ for one locus. Summing over all independent loci segregating for the trait in question, the mean is $(1/2)\Sigma_i d_i$ and total genetic variance is $(1/2)\Sigma_i a_i^2 + (1/4)\Sigma_i d_i^2$. For brevity we will designate the mean as $(1/2)d$ and total genetic variance as $(1/2)A + (1/4)D$. (Note: For consistency we will use A for Mather's D and D for Mather's H to conform with the usage of σ_A^2 for additive genetic variance and σ_D^2 for dominance variance.) Relations shown in Chap. 2 and the ones developed for the F_2 population can be shown to be equivalent if we refer to the basic definitions of the mean and total genetic variance of a population. It is shown in Chap. 2 that the mean is $2pqd$, which on substitution of $p = q = 1/2$ is $(1/2)d$. Total genetic variance is shown to be $2pq[a + (q - p)d]^2 + 4p^2q^2d^2$, which on substitution is $(1/2)a^2 + (1/4)d^2$. Hence the relations of the total genetic variance of a reference population $[\sigma_A^2 + \sigma_D^2$ and $(1/2)A + (1/4)D]$ are equivalent except that the expected gene frequency has been substituted in the definitions of σ_A^2 and σ_D^2 in the latter.

Total genetic variance of the F_2 population can be estimated by use of the parents, the F_1 population (to provide an estimate of the environmental effects), and the F_2 population. If we desire to determine relative proportions of total F_2 genetic variance due to additive and dominance effects, we need to develop additional generations. One common procedure is to self-pollinate individual S_0 plants in the F_2 generation to advance to the S_1 level, commonly referred to as the F_3 generation. For maize, controlled self-pollinations are needed. If individual plants in the F_2 generation are measured, parent-offspring regressions also can be determined by regressing the F_3 progeny means (values of S_1) on the F_2 parent plants (values of S_0), which is σ_A^2 + $(1/2)\sigma_D^2$ or $(1/2)D + (1/8)H$ in Mather's (1949) notation. Again, it is important to adequately sample the F_2 population for estimates of genetic components of variance to be representative of the population variability. The genotypic array of the F_3 generation is $(1/4)AA + (1/2)[(1/4)AA + (1/2)Aa + (1/4)aa] + (1/4)aa$, which, by collecting frequencies of different classes of genotypes, becomes $(3/8)AA + (1/4)Aa + (3/8)aa$. The F_3 generation mean is determined as $(3/8)a + (1/4)d + (3/8)(-a)$ or $(1/4)d$. Two sources of genetic variation in the F_3 generation are (1) variation among F_3 progeny means and (2) mean variation of F_3 progenies. Means of the F_3 progeny array are $(1/4)a$ for AA, $(1/2)d$ for Aa, and $(1/4)(-a)$ for aa. Hence the F_3 progeny variance $\sigma_{gF_3}^2$ is determined as

$$\sigma_{gF_3}^2 = \{(1/4)a^2 + (1/2)[(1/2)d]^2 + (1/4)(-a)^2\} - [(1/4)d]^2$$

$$= (1/2)a^2 + (1/8)d^2 - (1/16)d^2 = (1/2)a^2 + (1/16)d^2$$

By summing over loci we have $\sigma_{gF_3}^2$ = $(1/2)A + (1/16)D$, which is equal to σ_A^2 + $(1/4)\sigma_D^2$ from the definitions in Chap. 2. The mean variance of F_3 progenies is the variation within the progenies. From the genotypic arrays, the AA and aa types will not contain any genetic variation and need not be considered further. The F_3 progenies originating from the Aa F_2 plants will have segregation ratios similar to the F_2 population. The frequency of this progeny in the F_3 generation is 0.5 and thus equals $(1/2)[(1/4)AA + (1/2)Aa + (1/4)aa]$ with a mean of $(1/2)d$. The mean F_3 variance $(\overline{\sigma}_{gF_3}^2)$ becomes

$$\overline{\sigma}_{gF_3}^2 = (1/2)\{(1/4)a^2 + (1/2)d^2 + (1/4)(-a)^2 - [(1/2)d]^2\}$$

$$= (1/2)[(1/2)a^2 + (1/4)d^2] = (1/4)a^2 + (1/8)d^2$$

Summing over all loci, $\overline{\sigma}_{gF_3}^2$ = $(1/4)A + (1/8)D$, which is equivalent to $(1/2)\sigma_A^2$ + $(1/2)\sigma_D^2$. If we want to sum the effects for all loci, effects for individual loci can be added (since we assumed effects for each locus are additive or made so by transformation); and we have the expressions for the phenotypic variances given by Mather (1949) and Mather and Jinks (1971):

$$\sigma_{F_2}^2 = (1/2)A + (1/4)D + E_1$$

$$\sigma_{\overline{F}_3}^2 = (1/2)A + (1/16)D + E_2$$

$$\overline{\sigma}_{F_3}^2 = (1/4)A + (1/8)D + E_1$$

where the E components are the nonheritable variances associated with the respective components of variance.

The same techniques can be used to derive the mean and variance of subsequent generations (F_4, F_5, etc.) derived from the F_2 population. Mather (1949) has listed the more common ones in Table 9 of *Biometrical Genetics*. Backcross populations derived from the inbred lines and their F_1 populations are similar to F_2 populations because each S_0 plant is a unique entity and cannot be repeated. Each S_0 plant can be measured, but no estimate of error is available. (However, measurements of individual S_0 backcross plants can be used to compute parent-offspring regressions.)

Backcross generations can be advanced by selfing individual S_0 backcross plants, as was done by advancing F_2 plants to the F_3 generation. If the F_1 hybrid Aa is backcrossed to the parent having the desirable allele, the backcrossed genotypic array (BC1) becomes $(1/2)AA + (1/2)Aa$. By substituting genotypic values for respective genotypes, the BC1 mean becomes $(1/2)a + (1/2)d$; the BC1 variance is $(1/2)a^2 + (1/2)d^2 - [(1/2)a + (1/2)d]^2$ or $(1/4)A + (1/4)D - (1/2)AD + E_1$, by summing over all loci and including the nonheritable variance E_1. If we self individual backcross plants, the genotypic array of the backcrossed-selfed (BS1) generation progenies becomes $(1/2)AA + (1/2)[(1/4)AA + (1/2)Aa + (1/4)aa]$. Substituting genotypic values for genotypes, we find that the BS1 mean is $(1/2)a + (1/2)[(1/4)a + (1/2)d + (1/4)(-a)]$ or $(1/2)a + (1/4)d$. The variance among BS1 progeny means is $(1/2)a^2 + (1/2)[(1/2)d]^2 - [(1/2)a + (1/4)d]^2$, which becomes $(1/4)A + (1/16)D - (1/4)AD + E_2$ after collecting terms and summing over all loci. Mean variance of BS1 progenies ($\overline{\sigma}^2_{gBS1}$) would involve only the progenies that arose from selfing the Aa plants and is similar to the mean variance in the F_3; hence $\overline{\sigma}^2_{BS1}$ is $(1/4)a^2 + (1/8)D^2$ or $(1/4)A + (1/8)D + E_1$. The same calculations would be needed to determine the variance of the population formed by crossing the F_1 hybrid to the parent having the opposite allele, which is designated as the BC2 population.

Plants of the BC2 are selfed to produce the BS2 progenies. The BC2 population of $(1/2)Aa + (1/2)aa$ has a mean of $(1/2)d - (1/2)a$ and a variance of $(1/4)A + (1/4)D + (1/2)AD + E_1$. Selfed progenies of the BC2 population also contain two sources of variation: (1) variance among BS2 progenies or $(1/4) + (1/16)D + (1/4)AD + E_2$ and (2) mean variance of BS2 progenies or $(1/4)A + (1/8)D + E_1$. Expectations of the variances of the two backcross populations can be summarized as follows:

$$\sigma^2_{BC1} = (1/4)A + (1/4)D - (1/2)AD + E_1 \qquad \sigma^2_{\overline{BS2}} = (1/4)A + (1/16)D + (1/4)AD + E_2$$

$$\sigma^2_{BC2} = (1/4)A + (1/4)D + (1/2)AD + E_1 \qquad \overline{\sigma}^2_{BS1} = (1/4)A + (1/8)D + E_1$$

$$\sigma^2_{\overline{BS1}} = (1/4)A + (1/16)D - (1/4)AD + E_2 \qquad \overline{\sigma}^2_{BS2} = (1/4)A + (1/8)D + E_1$$

In the BC1, BC2, BS1, and BS2 generations the additive and dominance effects are not separable; if they are of similar magnitude, one can combine the equations for estimates of D and H. For instance, $\sigma^2_{BC1} + \sigma^2_{BC2} = (1/2)A + (1/2)D + 2E_1$ and $\sigma^2_{\overline{BS1}} + \sigma^2_{\overline{BS2}} = (1/2)A + (1/8)D + 2E_2$. As for F_2 and F_3 populations, E_1 and E_2 terms refer to the error variance among individual plants and progeny means, respectively.

Expressions for the different generations can be used to estimate additive and dominance variances. As noted for the expressions, we have two error

terms for which we need estimates. An estimate of the error variance for the
F_2, BC1, and BC2 populations can be obtained as the within variance for environmental effects of the inbred parent lines and their F_1 hybrid. Use of environmental effects estimated from inbred parent lines and their F_1 hybrids of
maize has been subject to criticism because variation among inbred individuals
may be greater than expected and that among F_1 hybrid individuals less than
expected among segregating individuals in an F_2 or backcross population. Estimates of E_2 obtained by pooling sums of squares and degrees of freedom of
inbred parents and F_1 hybrids, however, usually are the only estimates available. A technique suggested by Warner (1952) eliminates the necessity of estimating environmental effects on individual plant measurements of inbred parents and their F_1 hybrid. Warner's method provides an estimate of $(1/2)A$,
which then can be used to calculate an estimate of heritability h^2 on an individual plant basis; the method requires measurement of the F_2 and both backcross populations:

Population	Expected components of variance
F_2	$(1/2)A + (1/4)D + E_1$
$2(F_2)$	$A + (1/2)D + 2E_1$
BC1 + BC2	$(1/2)A + (1/2)D + 2E_1$
	$\overline{(1/2)A}$
$2F_2 - (BC1 + BC2)$	

$$h^2 = \frac{(1/2)A}{\sigma^2_{F_2}} = \frac{(1/2)A}{(1/2)A + (1/4)D + E_1}$$

The method of estimating $(1/2)A$ (or σ^2_A) does not require a direct estimate of
E_2 but it requires the assumption that nonheritable components of variance are
comparable for the F_2 and backcross populations. This assumption seems to be
logical because the populations (F_2 and combined backcrosses) include the same
genotypes, but the combined backcrosses have twice the frequency of the genotypes. A breeder working with a plant species or trait with little information can obtain some preliminary information by this procedure. A relatively
high heritability estimate on an individual plant basis would indicate that a
simple selection procedure, such as mass selection, would be effective.

Another example of how expected components of variance for different generations can provide genetic information is the evaluation of F_3 progenies,
which are equivalent to S_1 lines. Two inbred lines are crossed, their F_1 hybrid is selfed, and an adequate sample of F_2 plants is self-pollinated to produce F_3 progenies, i.e., F_3 seed produced on F_2 plants. If proper pollination
techniques are used, adequate F_3 seed should be available for testing in replicated trials. Assume that the experimental entries include the parents (P_1
and P_2), their F_1 hybrid, and 100 F_3 progenies. One possible form of the
analysis of variance of entries included in replicated tests is shown in Table
4.32. A direct F-test can be made to determine if the differences among F_3
progenies are significant; if they are, $\sigma^2_{gF_3} = (M_{32} - M_2)/r = (1/2)A + (1/16)D$,
which is equivalent to the variation among S_1 progenies, $\sigma^2_A + (1/4)\sigma^2_D$.
Mean variance of F_3 progenies is estimated from the M_{12} mean square,
which has an expectation of $\sigma^2_{F_3} = (1/4)A + (1/8)D + E_1$. Because the parents
and the F_1 hybrid are homogeneous, M_{11} provides an estimate of individual

Table 4.32. Analysis of variance of two inbred parents, their F_1 hybrid, and 100 F_3 entries tested in one environment.

Source	General	Example	Mean square	Expected mean squares
Replications	r-1	3		
Entries	n-1	102	M_3	
Among generations	g-1	3	M_{31}	$\sigma^2 + r\sigma_g^2$
Among F_3 progenies	p-1	99	M_{32}	$\sigma^2 + r\sigma_{g\bar{F}_3}^2$
Error	(r-1)(n-1)	306	M_2	σ^2
Total	rn-1			
Within	rn(k-1)	3708	M_1	
F_3 entries	rp(k-1)	3600	M_{12}	$\sigma_{wg}^2 + \sigma_{we}^2$
Homogeneous entries	rh(k-1)	108	M_{11}	σ_{we}^2

plant environmental effects. An estimate of E_2 is provided by the M_2 mean square and is an estimate of experimental error on a plot basis for F_3 progeny means; i.e., $E_2 = M_2/r$, the variance of a mean. Hence we have four mean squares and four unknowns, A, D, E_1, and E_2:

$$\sigma_{g\bar{F}_3}^2 = (1/2)A + (1/16)D = (M_{32} - M_2)/r$$

$$\sigma_{gF_3}^2 = (1/4)A + (1/8)D = M_{12} - M_{11}$$

$$M_2/r = E_2 \text{ and } M_{11} = E_1$$

Estimates of E_1 and E_2 are made directly from the mean squares in Table 4.32. To estimate A and D, we have two equations: $(1/2)A + (1/16)D = (M_{32} - M_2)/r$ and $(1/4)A + (1/8)D = M_{12} - M_{11}$. After solving for A and D, we find $A = (4/3) \cdot \{[2(M_{32} - M_2)/r] - (M_{12} - M_{11})\}$ and $D = (16/3)[2(M_{12} - M_{11}) - (M_{32} - M_2)/r]$. Variance V of the estimates can be obtained rather easily for E_1, E_2, and A, but as shown in the mating designs, the estimate of D was obtained from a complex function of mean squares.

$$V_{E_1} = \frac{2M_{11}^2}{rn(k-1)+2} \qquad V_{E_2} = \frac{2M_2^2}{(r-1)(n-1)+2}$$

$$V_A = \frac{128}{9r^2}\left[\frac{M_{32}^2}{p+1} + \frac{M_2^2}{(r-1)(n-1)+2}\right] + \frac{32}{9}\left[\frac{M_{12}^2}{rp(k-1)+2} + \frac{M_{11}^2}{rh(k-1)+2}\right]$$

$$V_D = \frac{512}{9}\left\{\frac{4M_{12}^2}{rp(k-1)+2} + \frac{4M_{11}^2}{rn(k-1)+2} + \frac{1}{r^2}\left[\frac{M_{32}^2}{p+1} + \frac{M_2^2}{(r-1)(n-1)+2}\right]\right\}$$

This example includes the minimum equations necessary to estimate four parameters. We have a perfect fit in this example because the number of equations equals the number of parameters we wished to estimate.

If we had included the backcross (BC1 and BC2) and backcross-selfed populations, three additional equations would be available for the estimation of A and D:

$$\sigma^2_{\overline{BS1}} + \sigma^2_{\overline{BS2}} = (1/2)A + (1/8)D + 2E_2$$

$$\overline{\sigma}^2_{BS1} = (1/4)A + (1/8)D + E_1$$

$$\overline{\sigma}^2_{BS2} = (1/4)A + (1/8)D + E_1$$

An analysis of variance that includes the parents and their F_1 (to provide an estimate of environmental effects for individual plants) and the F_3, BS1, and BS2 progenies is shown in Table 4.33. Homogeneous errors were assumed among the F_3, BS1, and BS2 progeny means, but the error mean squares can

Table 4.33. Analysis of variance of an experiment grown in one environment that includes the parents, their F_1 hybrid; and F_3, BS1, and BS2 progenies.

Source	df	Mean squares	Expected mean squares
Replications	r-1		
Entries	n-1	M_3	
Among generations	g-1	M_{31}	
Entries/groups			
Among F_3	p-1	M_{32}	$\sigma^2 + r\sigma^2_{g\overline{F}_3}$
Among BS1	s-1	M_{33}	$\sigma^2 + r\sigma^2_{g\overline{BS1}}$
Among BS2	t-1	M_{34}	$\sigma^2 + r\sigma^2_{g\overline{BS2}}$
Error	(r-1)(n-1)	M_2	σ^2
Total	rn-1		
Within	rn(k-1)	M_1	
Within F_3 progenies	rp(k-1)	M_{11}	$\sigma^2_{we} + \sigma^2_{wg}$
Within BS1 progenies	rs(k-1)	M_{12}	$\sigma^2_{we} + \sigma^2_{wg}$
Within BS2 progenies	rt(k-1)	M_{13}	$\sigma^2_{we} + \sigma^2_{wg}$
Within P_1, P_2, F_1	rh(k-1)	M_{14}	σ^2_{we}

Table 4.34. Summary of the means, variance among progenies, and mean progeny variance with continuous inbreeding and the comparison of the relations using Mather's (1949) notation.

Generation		Mean		Variance among progenies†		Mean progeny variance†	
F_i	S_i	Mather	general	Mather	general	Mather	general
F_1		h	d	0	0	0	0
F_2	S_0	$(1/2)h$	$(1/2)d$	$(1/2)D+(1/4)H$	$\sigma_A^2+\sigma_D^2$	0	0
F_3	S_1	$(1/4)h$	$(1/4)d$	$(1/2)D+(1/16)H$	$\sigma_A^2+(1/4)\sigma_D^2$	$1/4D+1/8H$	$(1/2)\sigma_A^2+(1/2)\sigma_D^2$
F_4	S_2	$(1/8)h$	$(1/8)d$	$(3/4)D+(3/64)H$	$(3/2)\sigma_A^2+(3/16)\sigma_D^2$	$1/8D+1/16H$	$(1/4)\sigma_A^2+(1/4)\sigma_D^2$
F_5	S_3	$(1/16)h$	$(1/16)d$	$(7/8)D+(7/256)H$	$(7/4)\sigma_A^2+(7/64)\sigma_D^2$	$1/16D+1/32H$	$(1/8)\sigma_A^2+(1/8)\sigma_D^2$
$\cdots$	$\cdots$	$\cdots$	$\cdots$	$\cdots$	$\cdots$	$1/32D+1/64H$	$(1/16)\sigma_A^2+(1/16)\sigma_D^2$
						$\cdots$	$\cdots$
F_{10}	S_8	$(1/512)h$	$(1/512)d$	$\sim D$	$\sim 2\sigma_A^2$	0	0

† See Table 3.2 for situations in which estimates of σ_D^2 are valid.

be partitioned in like manner as entries. A complete listing of genetic and error components of variance for each source in Table 4.33 is as follows:

$$\sigma^2_{F_3} = (1/2)A + (1/16)D + E_2$$

$$\bar{\sigma}^2_{F_3} = (1/4)A + (1/8)D + E_1$$

$$\sigma^2_{\overline{BS1}} + \sigma^2_{\overline{BS2}} = (1/2)A + (1/8)D + 2E_2$$

$$\bar{\sigma}^2_{BS1} = (1/4)A + (1/8)D + E_1$$

$$\bar{\sigma}^2_{BS2} = (1/4)A + (1/8)D + E_1$$

$$\sigma^2/r = E_2$$

$$M_{14} = E_1$$

The estimates of A, D, E_1, and E_2 are not as easily determined as when the number of equations is equal to the number of unknowns. In this example we have seven equations and four unknowns for which we desire to obtain estimates. The best procedure is to use the least squares analysis or, more appropriately, a weighted least squares analysis.

For maize, and particularly self-pollinated species, additional generations that increase the number of equations available for estimating components of variance can be obtained by additional generations of inbreeding, each of which provides estimates of variance among progenies and mean variance of progenies. If, for example, the F_3 and F_4 generations are included, we will have six equations and four parameters that we wish to estimate. A summary of expectations of variance among progenies and mean variance of progenies is shown in Table 4.34. For comparison, the expressions used by Mather (1949) and Mather and Jinks (1971) are shown with the equivalent expressions for σ^2_A and σ^2_D. Also, the equivalency of inbreeding generations is shown for the F and S designations commonly used with inbreeding. Because the F_2 generation is the reference population, the S designation is included to show the relations commonly used for broad genetic base populations. It is seen that the variance among F_2 plants is equivalent to the variance among S_0 plants. Albeit the F_2 population was obtained by selfing the F_1 hybrid, this selfing was merely to develop the F_2 population with an expected gene frequency of 0.5. Also, the variance among F_3 progenies is equivalent to variation among S_1 progenies. Genetic relations, as expected, are the same, but the differences in terminology are often confusing. The F_2 is our reference population for the special case of $p = q = 0.5$; and although it is obtained by inbreeding (selfing), its genotypic structure is equivalent to a random mating noninbred population with the same gene frequency.

Estimates of genetic and environmental parameters estimated from F_2 populations are applicable only for the populations developed from the specific pair of inbred lines. Estimates are expected to be different if other pairs of inbred lines are used. Estimates obtained for F_2 populations indicate relative types of variation present and future prospects of selection. But estimates usually are for only short-term objectives in comparison with populations that have a broad genetic base. Composition of components of variance in terms of additive genetic and dominance components of variance of F_2 and broad genetic base populations are equivalent if one assumes that the gene frequency of segregating loci in broad genetic base populations is 0.5.

Linkage effects on estimates of additive genetic and dominance variances obtained from populations developed from inbred lines are the same as those discussed for design III; dominance variance will be biased upward regardless of linkage phase, and additive genetic variance will be biased downward for repulsion and upward for coupling phase linkages. As for design III, equilibrium of linkages can be approached by random mating of the F_2 population for four or more generations (Hanson 1959). The F_2 and backcross populations are commonly used as base populations in applied maize breeding programs by either crossing two elite inbred lines or crossing an elite line to another line that has a specific trait to incorporate in the elite line. For F_2 populations, linkage groups may be highly desirable; whereas in backcross populations the breaking of tight linkages may be highly desirable and may be difficult, particularly if they have an undesirable pleiotropic effect.

In many instances, estimates of genetic variances are not important objectives of the breeder; inbred line development has the highest immediate priority, and the breeder wants to attain inbred lines as quickly as possible. Estimates of components of variance for F_2 populations, therefore, are important for (1) obtaining some preliminary information on the inheritance of a trait and (2) predicting gain for cyclical selection. In the first instance, the two inbred parents often represent the extremes in the expression of the trait under study and effects of linkage bias could be important in estimates of genetic variation (due to additive and dominance effects) that are needed to estimate the heritability of a trait. In the second instance, long-term cyclical selection experiments usually are not conducted in F_2 populations because of the limited range of genetic variability. A few cyclical selection programs, however, have been conducted to answer specific questions relative to selection theory, in which valid estimates of components of variance are necessary to predict future genetic gain and to show how predicted gain compares with observed gain (Robinson et al. 1949). Continued progress from selection in F_2 populations may result because of genetic variability released from breakup of linkages in recombination generations of each cycle.

Epistasis was assumed to be absent, or the data were transformed to an additive scale before analysis. The assumption of no epistasis is no more serious than for other mating designs. It seems that epistasis is relatively minor compared to additive and dominance effects; most attempts to estimate epistatic variance, however, have been for broad genetic base populations. For F_2 and backcross populations developed from inbred lines, epistatic effects may be more important than for broad genetic base populations. It is important, therefore, that the possible bias due to epistasis be checked by comparing the means of different generations, as given by Mather (1949). Estimates of components of variance and heritability are valid for the model of $p = q = 0.5$, but one must not extend the results to other populations. This restriction on interpretation of results is the same for broad genetic base populations, but the range of genotypes is more restrictive for populations developed from inbred lines than for broad genetic base populations such as open-pollinated, synthetic, and composite varieties.

4.10 EPISTASIS

Epistatic effects were assumed absent in estimation of additive genetic and dominance variances in all the mating designs. It seems logical, however, that whatever the magnitude of epistatic effects, they must be present in the functioning of a genotype. Epistatic effects have been demonstrated in the expression of traits involving two loci, and thus it does not seem reasonable that epistatic effects are not operative in the expression of a complex quantitative trait such as yield. Although they most certainly are operative, the important aspect is what proportion of the total genetic variance can be attributed to epistatic variance. If the proportion of epistatic variance to total genetic variance is relatively small, the bias in our estimates caused by assuming no epistasis will not seriously hinder our selection progress. Of the total epistatic variance, only additive types of epistasis are fixable or usable by the breeder.

Epistasis, as defined for quantitatively inherited traits, is purely a statistical description and does not define a physiological function or expression. Epistatic variance is an orthogonal partitioning of the interaction among loci (interallelic interaction); whereas the variability of the three phases within a locus (e.g., AA, Aa, and aa) is partitioned into 1 df for sums of squares due to regression (σ_A^2) and 1 df for the deviations from regression (dominance, σ_D^2, or interaction within a locus). Hence epistasis, in a statistical sense, is the nonadditivity of effects among loci as contrasted with dominance effects, which are due to nonadditivity within a locus.

A simple operational model to illustrate how epistasis is defined can be shown for an agronomic experiment that includes two fertilizer elements (say nitrogen and phosphorus), each at three levels. We will have a 3^2 factorial experiment with nine combinations of nitrogen and phosphorus fertilizer treatments. Hence we have 8 df that can be partitioned into additive and nonadditive effects for nitrogen and phosphorus and for the interactions of their linear and quadratic effects. The linear effects of the two fertilizer treatments are analogous to additive genetic effects; quadratic effects of the three levels of each fertilizer treatment are analogous to dominance deviations; and the linear × linear (additive × additive), linear × quadratic (additive × dominance), and quadratic × quadratic (dominance × dominance) interactions of the two fertilizer treatments are analogous to epistatic effects for interaction among the three genotypes at each of two loci. As shown in Chap. 2, however, this simple agronomic example is not exactly analogous to a genetic situation because we are dealing with the frequency of genes and, consequently, with the frequency of genotypes.

Cockerham (1954) developed the relations for partitioning total epistatic variance into the three types of epistasis--additive × additive, additive × dominance, and dominance × dominance for a two-locus model. Higher order epistasis follows by including additional loci. For the simple fertilizer experiment the linear sums of squares for nitrogen and phosphorus would be analogous to estimation of additive genetic variance pooled over two loci. Similarly, dominance variance is estimated from nonadditivity within a locus (or nonadditivity for the three levels of each fertilizer treatment partitioned as the quadratic) and epistasis from nonadditivity among loci (or interaction between different levels of the two fertilizer treatments).

Cockerham (1956a) illustrated different genetic models for the expression of different types of epistasis for a two-locus situation. He also suggested how epistasis may be estimated by use of different mating designs and different inbreeding levels of parents included in these designs. Combinations of mating designs and levels of inbreeding provide additional equations and differences in coefficients of genetic components of variance for covariances of relatives. Two attempts to estimate epistatic variance in maize populations were reported by Eberhart et al. (1966) and Silva and Hallauer (1975). In both instances, progenies developed by designs I and II were evaluated, where the parents were noninbred (F = 0) for design I matings and inbred (F = 1) for design II matings. The inbred parents were unselected inbred lines developed from the same populations from which the noninbred parents were sampled. Hence two samplings of S_0 plants were made from the population: (1) S_0 plants used as noninbred parents of design I matings and (2) S_0 plants that were the progenitors of the unselected inbred lines. Because expectations of covariances of relatives are different for the two levels of inbreeding, additional equations are available for estimation.

Tables 4.35 and 4.36 give examples of sources of variation and coefficients of components of variance for an experiment that includes design I and II progenies evaluated in one environment and including within-plot variability. Coefficients of components of variance in Table 4.35 were obtained from expected mean squares for each source of variation. Nine equations are available, and a genetic model that includes digenic epistatic components could be fitted by least squares analysis. Although we have sufficient equations for estimation of epistatic variance, the results so far have not been encouraging because Eberhart et al. (1966) and Silva and Hallauer (1975) were unable to obtain realistic estimates of digenic epistatic components. Silva and Hallauer, for example, found that σ_A^2 accounted for 93.2% of total genetic variance for yield and that inclusion of σ_A^2 and σ_D^2 in the model accounted for 99% of total genetic variation, with no improvement in the fit when σ_{AA}^2 was included. Because the mean squares have unequal variances, it was considered that weighted least squares or maximum likelihood methods of estimation would be an improvement. But when the models included more than one digenic epi-

Table 4.35. Expected components of genetic variance for the components of variance of the expected mean squares for design I and II mating designs.

Components of variance	Genetic components of variance						
	σ_A^2	σ_D^2	σ_{AA}^2	σ_{AD}^2	σ_{DD}^2	σ^2	σ_{we}^2
Design I (F = 0)							
σ_m^2	1/4		1/16				
$\sigma_{f/m}^2$	1/4	1/4	3/16	1/8	1/16		
σ^2						1	
σ_w^2	1/2	3/4	3/4	7/8	15/16	1	
Design II (F = 1)							
σ_m^2	1/2		1/4				
σ_f^2	1/2		1/4				
σ_{fm}^2		1	1/2	1	1		
σ^2						1	
σ_w^2							1

Table 4.36. Expectation of mean squares in terms of genetic and environmental variance components pooled over sets when within plot
variances are available for design I and II mating designs.

Source	Components of variance						
	σ^2_A	σ^2_D	σ^2_{AA}	σ^2_{AD}	σ^2_{DD}	σ^2_{we}	σ^2
Design I[†]							
Males (M)/sets (S)	3.554	0.582	1.207	0.345	0.227	0.109	1
Females/M/S	0.554	0.582	0.457	0.345	0.227	0.109	1
Error	0.054	0.082	0.082	0.095	0.102	0.109	1
Within plot	0.500	0.750	0.750	0.875	0.938	1.000	0
Design II[‡]							
Males (M)/sets (S)	4.000	2.000	3.000	2.000	2.000	0.110	1
Females (F)/S	4.000	2.000	3.000	2.000	2.000	0.110	1
(M x F)/S	0.000	2.000	1.000	2.000	2.000	0.110	1
Error	0.000	0.000	0.000	0.000	0.000	0.110	1
Within plot[§]	0.000	0.000	0.000	0.000	0.000	1.000	0

[†]$r = 2$, $f = 6$; [‡]$r = 2$, $m = 4$, $f = 4$; [§]harmonic means of the number of plants
per plot (k) were $k = 9.2$ for design I and $k = 9.1$ for design II.

static component they usually were negative and unrealistic, with much greater
standard errors. A model that included as many terms as permitted by the number of independent equations made the X-matrix nearly singular. Repeating the
experiments over environments also provides additional equations that can be
used to estimate interaction of genetic effects with environments.

It was previously shown that a series of F-tests can be made for triallel
and quadrallel analyses of variance to test hypotheses that include epistasis.
If the F-tests indicate significant variation for different sources of variation, one can proceed to estimate genetic components of variance, including
digenic epistasis. For instance, in the triallel analysis there are nine covariances of relatives that have different genetic expectations. Solutions of
the set of simultaneous linear equations of the mean squares (or components of
variance) will then permit the fitting of genetic models that include epistasis. Wright et al. (1971) used triallel and diallel analyses to estimate
epistasis by use of 60 unselected inbred lines developed from a selected
strain of Krug Yellow Dent, which is an open-pollinated variety. From pooled
analyses of the triallel and diallel, nine linear equations were available to
fit genetic models that included trigenic additive epistatic variance. Unweighted least squares and maximum likelihood estimation procedures were compared for estimation of genetic components of variance. Regardless of genetic
model and estimation procedure used, σ^2_A accounted for the largest proportion
of total genetic variance. Fitting the error and a six-parameter genetic mod-
el (σ^2_A, σ^2_D, σ^2_{AA}, σ^2_{AD}, σ^2_{DD}, σ^2_{AAA}) showed that it was not possible to obtain realistic estimates of epistatic components of variance, although significant
epistatic effects were detected in triallel analyses of variance. Negative

estimates of epistatic variance components were frequent in the unweighted
analysis, but standard errors of estimates indicated the estimates were within
the range of zero. Negative estimates of epistatic components also were frequent by the maximum likelihood procedure, but they were smaller and usually
had smaller standard errors. Only 0.27% and 0.24% of total genetic variation
of yield for unweighted and weighted estimation procedures, respectively, were
attributed to the four epistatic components of variance; whereas fitting the

error and σ_A^2 accounted for 99% and 97% of total variation for yield.

Chi et al. (1969) used a complex mating design in the Reid Yellow Dent
open-pollinated variety that included 66 covariances within and among 11
branches of a family, which represented five broad categories of relatives:
full-sibs, half-sibs, cousins, uncle-nephew, and no relationship. Seven genetic models were fitted to observed mean squares that included up to trigenic
additive epistasis. They used unweighted least squares analyses and found no
evidence of epistasis in the Reid Yellow Dent variety. Estimates from a model

that included σ_A^2 and σ_D^2 indicated that the major part of the total genetic
variance could be attributed to these two parameters.

Although large studies have been conducted, including a sufficient number
of mean squares for the fitting of genetic models that included epistasis,
they all failed to obtain realistic estimates of epistatic components of vari-

ance. In all instances, the reduced models that included error, σ_A^2, and σ_D^2
accounted for most of the total variation. It seems that it may be either impossible to estimate components of epistatic variance with the present genetic
models or the relative proportion of total genetic variance due to epistatic
variance is quite small; both factors probably are related to the sequential
fitting of genetic models.

Chi et al. (1969) have indicated that one of the main problems involves
coefficients of digenic and trigenic epistatic components of variance because
they are highly correlated with those of additive and dominance variance components. The correlation occurs because the coefficients of epistatic components of variance are generated by either squaring or multiplying the coefficients of additive and dominance variance components. This is evident from
the general relations for covariances of relatives shown in Chap. 2; e.g.,

Cov HS = $[(1 + F)/2]^{io}\sigma_{io}^2$. Hence this inherent property of the model for covariances of relatives reduces the sensitivity of the model for detecting
epistasis. For the model that included six components of genetic variance
used by Chi et al., the correlation matrix of coefficients of genetic components of variance is shown in Table 4.37. The correlations of coefficients

Table 4.37. Correlation matrix of coefficients of genetic components of variance.

	σ_A^2	σ_D^2	σ_{AA}^2	σ_{AD}^2	σ_{DD}^2	σ_{AAA}^2
σ_A^2	1.00					
σ_D^2	0.75	1.00				
σ_{AA}^2	0.92	0.93	1.00			
σ_{AD}^2	0.71	0.98	0.92	1.00		
σ_{DD}^2	0.67	0.95	0.89	0.99	1.00	
σ_{AAA}^2	0.81	0.96	0.97	0.98	0.97	1.00

among epistatic components of variance and of σ_D^2 with the epistatic components
are very high in all instances, e.g., σ_{AA}^2 has a correlation of 0.92 with σ_A^2.
Consequently, high correlations of the coefficients will give greater standard
errors of estimates of genetic components of variance.

Several assumptions were used in obtaining estimates of components of genetic variance: (1) randomness of parents, (2) diploid inheritance, (3) no
linkage effects or linkage equilibrium, (4) no maternal effects, and (5) additive environmental and genotypic effects. The studies of Eberhart et al.
(1966), Wright et al. (1971), and Silva and Hallauer (1975) used unselected
inbred lines developed from populations. In all instances selection was minimized, but it is also obvious that it was impossible to maintain the progeny
of each original S_0 plant during the inbreeding process. Eberhart et al. and
Wright et al. had a relatively small sample of inbred lines developed from the
original S_0 plants sampled. Silva and Hallauer had a sample of 160 inbred
lines, but the estimation of epistatic variance was no more successful than
those of Eberhart et al. and Wright et al. The effects of linkage could be a
disturbing element, and Cockerham (1956b) and Schnell (1963) have shown that
linkage effects increase the coefficients of epistatic terms of covariances of
relatives. In all studies except Wright et al., the parents were derived from
populations that should approach linkage equilibrium. Complete linkage of all
factors controlling a trait on each chromosome does not seem probable in
maize. Linkage bias on epistatic components would not be of much importance
unless the average recombination frequency was 0.1 or less. Hence effects of
linkage should not have been an important source of bias in all studies. Assumptions of diploid inheritance, no maternal effects, and additivity of genetic and environmental effects have not been found to be invalid in maize.
The small coefficients of epistatic terms relative to the coefficients of additive and dominance terms are an important factor in estimation of epistatic
components of variance. The smallness of coefficients also contributes to
greater standard errors of epistatic estimates.

Qualitative evidence of epistatic effects, however, has been reported
from comparison of means of different types of hybrids. In most instances the
hybrids were produced from inbred lines and desirable epistatic combinations
would be fixed in those lines. The use of the same set of inbred lines to
produce a balanced set of single, three-way, and double-cross hybrids would
indicate effects of recombination for epistatic effects. Recombination would
be present in the one single cross used to produce the three-way crosses and
in both single crosses for production of the double-cross hybrids. Comparisons of different permutations of crosses for the same set of inbred lines
give qualitative evidence of the presence of epistatic effects for specific
combinations of inbred lines. This type of evidence for epistatic effects
cannot be quantified to determine its relative importance to other types of
genetic variance, but it demonstrates that epistatic effects may be important
for specific combinations of inbred lines.

4.11 GENERATION MEANS

All the methods of estimation discussed so far have been concerned with
estimation of genetic variances. It also has been shown that genetic variances were determined from the summation of squared effects for each locus.
We will briefly consider a different type of genetic analysis of gene action
because we will be concerned with the relative importance of genetic effects,
not genetic variances. This type of genetic analysis does not involve development of progenies that have a family structure of sibships, but it includes
genetic populations (or generations) that are similar to those discussed for

the special case of $p = q = 0.5$. Instead of estimating genetic variation within generations, we will concern ourselves with relative genetic effects estimated from the means of different generations. Mather (1949) presented several generation comparisons to test for additiveness of genetic effects for estimation of σ_A^2 and σ_D^2. If the scale of measurement deviated from additivity, he suggested a transformation to make the effects additive. The generation models were extended to include estimation of epistatic effects.

Several models have been developed for analysis of generation means (Anderson and Kempthorne 1954; Hayman 1958, 1960; Van der Veen 1959; and Gardner and Eberhart 1966). Because of similarity in nomenclature we will use the Hayman (1958, 1960) model to illustrate the type of genetic information obtained from generation mean analyses. As an example, consider the generations produced from the cross of two inbred lines that were given in the discussion of the special case of $p = q = 0.5$. For estimation of genetic effects we will use the means of each generation rather than develop progenies within the segregating generations.

Hayman (1958) defined his base population as the F_2 population resulting from a cross of two inbred lines. If they differ by any number of unlinked loci, expectations of parents and their descendant generations in terms of genetic effects relative to the F_2 generation are as follows:

$$P_1 = m + a - (1/2)d + aa - ad + (1/4)dd$$
$$P_2 = m - a - (1/2)d + aa + ad + (1/4)dd$$
$$F_1 = m \quad\quad + (1/2)d \quad\quad\quad\quad + (1/4)dd$$
$$F_2 = m$$
$$F_3 = m \quad\quad - (1/4)d \quad\quad\quad\quad + (1/16)dd$$
$$F_4 = m \quad\quad - (3/8)d \quad\quad\quad\quad + (9/64)dd$$
$$F_5 = m \quad\quad - (7/16)d \quad\quad\quad\quad + (49/256)dd$$
$$F_6 = m \quad\quad - (15/32)d \quad\quad\quad + (225/1024)dd$$
$$BC1 = m + (1/2)a \quad\quad\quad + (1/4)aa$$
$$BC2 = m - (1/2)a \quad\quad\quad + (1/4)aa$$
$$BC1^2 = m + (3/4)a \quad\quad\quad + (9/16)aa$$
$$BC2^2 = m - (3/4)a \quad\quad\quad + (9/16)aa$$
$$BS1 = m + (1/2)a - (1/4)d + (1/4)aa - (1/4)ad + (1/16)dd$$
$$BS2 = m - (1/2)a - (1/4)d + (1/4)aa + (1/4)ad + (1/16)dd$$

or in general the observed mean = $m + \alpha a + \beta d + \alpha^2 aa + 2\alpha\beta ad + \beta^2 dd$, where α and β are the coefficients of a and d. Because the F_2 mean is $(1/2)d$ and the F_1 mean is equal to d, the F_1 mean relative to the F_2 mean has an added increment of $(1/2)d$. Terms a and d are the same as those illustrated in the special case of $p = q = 0.5$, where a indicates additive effects and d indicates dominance effects. Hayman (1958) used lowercase letters to indicate summation over all loci by which the two inbred lines differ. Thus a measures the pooled additive effects; d the pooled dominance effects; and aa, ad, and dd the pooled digenic epistatic effects.

The different generations listed can be produced rather easily for cross- and self-pollinated species. Hand pollinations are necessary for all generations in maize, but selfing generations can be obtained naturally in self-pollinated species. Bulks of progenies of each generation are evaluated in replicated experiments repeated over environments, and generation means can be determined for traits under study. In growing different generations, one should be cognizant of two important considerations in order to have valid estimates of the generation means: (1) Sufficient sampling of segregating generations is necessary to have a representative sample of genotypes. In parental and F_1 generations no sampling is involved, but F_2, F_3, F_4,..., and backcross generations will be segregating and sample size has to be considered. (2) In maize it is necessary to consider the level of inbreeding of each generation, and it becomes necessary to have sufficient border rows in experimental plots to minimize competition effects of adjacent plots.

Several different possibilities exist for the type and number of generations that can be included in a generation mean experiment. If the two parents and the F_1, F_2, and F_3 generations are evaluated, we have five means for comparison. Expectations of each generation can be determined and a least squares analysis made to estimate m, a, and d with a fair degree of precision. For this simple experiment we can also make a goodness-of-fit test (observed means compared with predicted means) to determine the sufficiency of the model for m, a, and d to explain the differences among the generation means.

Letting m = general mean, a = sum of signed additive effects, and d = signed dominance effects, we have the following expressions with F_2 as the base population:

$$P_1 = m + a \qquad F_1 = m + (1/2)d$$
$$P_2 = m - a \qquad F_2 = m$$
$$F_3 = m - (1/4)d$$

By use of the technique suggested by Mather (1949), the five equations can be reduced to the following normal equations:

$$5m + (1/4)d = Q_1(P_1 + P_2 + F_1 + F_2 + F_3)$$
$$2a = Q_2(P_1 - P_2)$$
$$(1/4)m + (5/16)d = Q_3(F_1 + F_3)$$

In matrix form the set of equations is

$$\begin{bmatrix} 5 & 0 & 1/4 \\ 0 & 2 & 0 \\ 1/4 & 0 & 5/16 \end{bmatrix} \begin{bmatrix} m \\ a \\ d \end{bmatrix} = \begin{bmatrix} Q_1 \\ Q_2 \\ Q_3 \end{bmatrix}$$

Solving for parameters m, a, and d, we get the following estimates:

$$\hat{m} = (5/24)Q_1 - (1/6)Q_3$$

$$\hat{a} = (1/2)Q_2$$

$$\hat{d} = -(1/6)Q_1 + (10/3)Q_3$$

The estimates of m, a, and d can be inserted in the predicted values and can be compared with observed values for each generation. If the squares of deviations of the expected from the observed are significant, the three estimated parameters are not sufficient to explain differences among the generation means; this is a goodness-of-fit test for epistasis and/or linkage to determine if the three parameters included are sufficient or if more are needed.

The best procedure would be to sequentially fit the successive models starting with the mean and add one term with each successive fit. Tests of residual mean squares can be made for each model to determine how much of the total variation among generations is explained by different parameters in the model. High-speed computers facilitate such computations, and a weighed least squares analysis can be done rather easily.

The similarity of genetic populations included for this simple generation mean experiment and genetic populations used for estimation of genetic variances for the special case of p = q = 0.5 is obvious. If necessary measurements have been made for the different genetic populations, we also can obtain the following sets of equations:

Variance among F_2 individuals = $\sigma_A^2 + \sigma_D^2 + E_1$

Variance among F_3 progeny means = $\sigma_A^2 + (1/4)\sigma_D^2 + E_2$

Variance within F_3 progenies = $(1/2)\sigma_A^2 + (1/2)\sigma_D^2 + E_1$

Covariance between F_2 individuals and F_3 progeny means = $\sigma_A^2 + (1/2)\sigma_D^2$

Variance among parents and F_1 individuals = E_1

Experimental error = E_2

Six equations are available for estimation of two heritable and two nonheritable sources of variation. Direct estimates of E_1 and E_2 are available, but an unweighted (or preferably a weighted) least squares analysis can be used to estimate the four parameters (σ_A^2, σ_D^2, E_1, and E_2) from the six equations. If one estimates the genetic effects a and d (by use of generation mean analysis) and the genetic variances σ_A^2 and σ_D^2, there probably will be little relation in the magnitude of the two sets of estimates. This should be expected because in the first instance we are estimating the sum of the signed genetic effects; whereas in the second instance we are estimating variances that are the squares of the genetic effects. For maize it seems that estimates of the d effects are usually greater, especially if the F_1 generation is included. On the other hand, estimation of genetic variances in maize usually shows that estimates of σ_A^2 are similar to or greater than estimates of σ_D^2. The expression of heterosis in F_1 crosses of two inbred lines of maize probably has a much greater effect on the estimate of d for maize than for many other crop species.

Limitations of the generation mean analysis if the model includes epistatic effects were discussed by Hayman (1960). Briefly, if the residuals are not significantly different from zero after m, a, and d are fitted, we have unique estimates for a and d. However, if it is necessary to include epistatic effects in the model, estimates of digenic epistatic effects are unique but estimates of a and d are confounded with some of the epistatic effects. Hence if epistatic effects are not present, estimates of a and d effects are meaningful and unbiased by linkage disequilibrium; if epistatic effects are present, estimates of a and d effects are biased by epistatic effects and linkage disequilibrium (if present). Estimation of digenic epistatic effects is unbiased if linkage of interacting loci and higher order epistatic effects are absent. Because of the bias in the estimates of a and d effects when a model that includes epistatic effects is used, the relative importance of a and d effects vs. epistatic effects cannot be directly assessed. Some indication of their relative importance may be gained by comparing residual sums of squares after fitting the three-parameter (m, a, and d) and the six-parameter (m, a, d, aa, ad, and dd) models.

It seems that the primary function of generation mean analysis is to obtain some specific information about a specific pair of lines. How useful the information obtained from generation mean analysis is to the maize breeder is not obvious. For quantitative traits the estimates of genetic effects would be quite different for different pairs of lines, depending on the relative frequency of opposing and reinforcing effects for the specific pair of lines studied. The cancellation of opposing effects may confound interpretations, but a complete diallel of interested lines could be used to determine which have opposing and reinforcing effects. Generation mean analysis is amenable for use in self-pollinated species because limited hand pollinations are required to produce the different generations; hence generation mean analysis may provide some information on the relative importance of nonadditive genetic effects for the justification of a hybrid breeding program. The relative importance of dominance effects could be determined by comparing different generations derived from the F_1 generation, which would involve only the cross of the two parents to produce the F_1 generation. For maize, however, controlled pollinations would be necessary for all generations.

Generation mean analysis has some advantages and disadvantages in comparison with mating designs used for estimation of genetic components of variance. Some advantages are: Because we are working with means (first order statistics) rather than variances (second order statistics), the errors are inherently smaller. We can rather easily extend generation mean analysis to more complex models that include epistasis, but the main effects (a and d) are not unique when epistatic effects are present. Generation mean analysis is equally applicable to cross- and self-pollinating species. Smaller experiments are required for generation mean analysis to obtain the same degree of precision. Generation mean analysis also has some serious disadvantages, one of which is that an estimate of heritability cannot be obtained; a corollary disadvantage is that one cannot predict genetic advance because estimates of genetic variances are not available. Cancellation of effects may be a significant disadvantage because, say, dominance effects may be present but opposing at various loci in the two parents and cancel each other. Generation mean analysis does not reveal opposing effects, but this may be overcome to some extent by a balanced set of diallel crosses.

This discussion for generation mean analysis is restricted to the use of parents being either inbred or pure lines, i.e., relatively homozygous and homogeneous. Generation mean analysis has been extended to populations generated from parents that are not homozygous. Robinson and Cockerham (1961) presented an analysis for two varieties and n alleles at each locus. Gardner and Eberhart (1966) included n varieties with only two alleles at a locus. The analysis by Robinson and Cockerham (1961) is orthogonal in the partitioning of sums of squares, and tests can be made to determine the presence of nonadditive effects. All the generation mean analyses provide information on the relative importance of genetic effects, but the information in most instances may not be useful to applied breeders, particularly those that are conducting long-term selection programs.

REFERENCES

Anderson, V. L., and O. Kempthorne. 1954. A model for the study of quantitative inheritance. *Genetics* 39:883-98.

Baker, R. J. 1978. Issues in diallel analysis. *Crop Sci.* 18:533–36.

Brim, C. A. 1966. A modified pedigree method of selection in soybeans. *Crop Sci.* 6:220.

Chi, R. K.; S. A. Eberhart; and L. H. Penny. 1969. Covariances among relatives in a maize variety (*Zea mays* L.). *Genetics* 63:511–20.

Cockerham, C. C. 1954. An extension of the concept of partitioning hereditary variance for analysis of covariance among relatives when epistasis is present. *Genetics* 39:859–82.

———. 1956a. Analysis of quantitative gene action. Brookhaven Symp. Biol. 9:53–68.

———. 1956b. Effects of linkage on the covariances between relatives. *Genetics* 41:138–41.

———. 1961. Implications of genetic variances in a hybrid breeding program. *Crop Sci.* 1:47–52.

———. 1963. Estimation of genetic variances. In *Statistical Genetics and Plant Breeding*, W. D. Hanson and H. F. Robinson, eds., pp. 53–94. NAS-NRC Publ. 982.

Comstock, R. E., and H. F. Robinson. 1948. The components of genetic variance in populations of biparental progenies and their use in estimating the average degree of dominance. *Biometrics* 4:254–66.

———. 1952. Estimation of average dominance of genes. In *Heterosis*, J. W. Gowen, ed., p. 494–516. Iowa State Univ. Press, Ames.

Dickerson, G. E. 1969. Techniques for research in quantitative animal genetics. In *Techniques and Procedures in Animal Science Research*. Am. Soc. Anim. Sci., Albany, New York.

Eberhart, S. A.; R. H. Moll; H. F. Robinson; and C. C. Cockerham. 1966. Epistatic and other genetic variances in two varieties of maize. *Crop Sci.* 6:275–80.

Falconer, D. C. 1960. *Introduction to Quantitative Genetics*. Ronald Press, New York.

Gardner, C. O. 1963. Estimates of genetic parameters in cross-fertilizing plants and their implications in plant breeding. In *Statistical Genetics and Plant Breeding*, W. D. Hanson and H. F. Robinson, eds., pp. 225–52. NAS-NRC Publ. 982.

Gardner, C. O., and S. A. Eberhart. 1966. Analysis and interpretation of the variety cross diallel and related populations. *Biometrics* 22:439–52.

Gardner, C. O., and J. H. Lonnquist. 1959. Linkage and the degree of dominance of genes controlling quantitative characters in maize. *Agron. J.* 51:524–28.

Gardner, C. O.; P. H. Harvey; R. C. Comstock; and H. F. Robinson. 1953. Dominance of genes controlling quantitative characters in maize. *Agron. J.* 45:186–91.

Graybill, F. A., and W. H. Robertson. 1957. Calculating confidence intervals for genetic heritability. *Poult. Sci.* 36:261–65.

Graybill, F. A.; F. Martin; and G. Godfrey. 1956. Confidence intervals for variance ratios specifying genetic heritability. *Biometrics* 12:99–109.

Griffing, B. 1956. Concept of general and specific combining ability in relation to diallel crossing systems. *Australian J. Biol. Sci.* 9:463–93.

Hallauer, A. R. 1970. Genetic variability for yield after four cycles of reciprocal recurrent selections in maize. *Crop Sci.* 10:482–85.

Hallauer, A. R., and J. H. Sears. 1973. Changes in quantitative traits associated with inbreeding in a synthetic variety of maize. *Crop Sci.* 13:327–30.

Hanson, W. D. 1959. The breakup of initial linkage blocks under selected mating systems. *Genetics* 44:857–68.

Hayman, B. I. 1958. The separation of epistatic from additive and dominance variation in generation means. *Heredity* 12:371–90.

———. 1960. The separation of epistatic from additive and dominance variation in generation means. II. *Genetica* 31:133–46.

Jensen, S. D. 1959. Combining ability of unselected inbred lines of corn from incomplete diallel and topcross tests. Ph.D. dissertation, Iowa State Univ., Ames.

Kempthorne, O. 1957. *An Introduction to Genetic Statistics*. Wiley, New York.

Kempthorne, O., and R. N. Curnow. 1961. The partial diallel cross. *Biometrics* 17:229–50.

Knapp, S. J. 1986. Confidence intervals for heritability for two-factor mating design single environment linear models. *Theoret. Appl. Genet.* 72:857–91.

Knapp, S. J., W. W. Stroup, and W. M. Ross. 1985. Exact confidence intervals for heritability on a progeny mean basis. *Crop Sci.* 25:192–94.

Lamkey, K. R., and A. R. Hallauer. 1987. Heritability estimated from recurrent selection experiments in maize. *Maydica* 32:61–78.

Lindsey, M. F.; J. H. Lonnquist; and C. O. Gardner. 1962. Estimates of genetic variance in open-pollinated varieties of Corn Belt corn. *Crop Sci.* 2:105–8.

Malécot, G. 1948. *Les Mathématiques de l'Hérédité*. Masson et Cie, Paris.

Mather, K. 1949. *Biometrical Genetics*. Methuen, London.

Mather, K., and J. L. Jinks. 1971. *Biometrical Genetics*. Cornell Univ. Press, Ithaca.

Matzinger, D. F.; G. F. Sprague; and C. C. Cockerham. 1959. Diallel crosses of maize in experiments repeated over locations and years. *Agron. J.* 51:346–50.

Rawlings, J., and C. C. Cockerham. 1962a. Triallel analysis. *Crop Sci.* 2:228–31.

———. 1962b. Analysis of double cross hybrid populations. *Biometrics* 18:229–44.

Robinson, H. F., and C. C. Cockerham. 1961. Heterosis and inbreeding depression in populations involving two open-pollinated varieties of maize. *Crop Sci.* 1:68–71.

Robinson, H. F.; R. E. Comstock; and P. H. Harvey. 1949. Estimates of heritability and the degree of dominance in corn. *Agron. J.* 41:353–59.

Satterthwaite, F. E. 1946. An approximate distribution of estimates of variance components. *Biom. Bull.* 2:110–14.

Schnell, F. W. 1963. The covariance between relatives in the presence of linkage. In *Statistical Genetics and Plant Breeding*, W. D. Hanson and H. F. Robinson, eds., pp. 468–83. NAS–NRC Publ. 982.

Searle, S. R. 1971. Topics in variance component estimation. *Biometrics* 27:1–74.

Silva, J. C., and A. R. Hallauer. 1975. Estimation of epistatic variance in Iowa Stiff Stalk Synthetic maize. *J. Hered.* 66:290–96.

Smith, J. D., and M. L. Kinman. 1965. The use of parent-offspring regression as an estimator of heritability. *Crop Sci.* 5:595–96.

Snedecor, G. W. 1956. *Statistical Methods*. Iowa State Univ. Press, Ames.

Sokol, M. J., and R. J. Baker. 1977. Evaluation of the assumptions required for the genetic interpretation of diallel experiments in self-pollinated crops. *Canadian J. Plant Sci.* 57:1185–91.

Sprague, G. F., and L. A. Tatum. 1942. General vs. specific combining ability in single crosses of corn. *J. Am. Soc. Agron.* 34:923–32.

Stuber, C. W. 1970. Estimation of genetic variances using inbred relatives. *Crop Sci.* 10:129–35.

Van der Veen, J. H. 1959. Tests of non-allelic interaction and linkage for quantitative characters in generations derived from two diploid pure lines. *Genetica* 30:201-32.

Warner, J. N. 1952. A method for estimating heritability. *Agron. J.* 44:427-30.

Wright, J. A.; A. R. Hallauer; L. H. Penny; and S. A. Eberhart. 1971. Estimating genetic variance in maize by use of single and three-way crosses among unselected inbred lines. *Crop Sci.* 11:690-95.

5

Hereditary Variance:
Experimental Estimates

The mating designs described in Chap. 4 have been used extensively in maize to determine relative proportions of total variation that are governed by genetic and environmental forces and to characterize genetic variation due to additive and nonadditive effects. Maize is amenable to study by the different mating designs because of the ease in obtaining sufficient quantities of seed for testing by cross- and self-fertilization. Because maize is a naturally cross-fertilized crop species, variability within maize populations was obvious to researchers. Until the late 1940s maize breeders and researchers emphasized the development of procedures for increasing effectiveness and efficiency of inbred line and hybrid development based on the principles given by East (1908), Shull (1908, 1909), and Jones (1918). Because of the variability within populations maize breeders, unlike animal breeders, did not concern themselves with attempting to characterize the types of genetic variability present and how this could influence effectiveness of selection of lines and their expression in hybrids until the publications of Comstock and Robinson (1948) and Mather (1949). Similarly, improvement of the populations was generally ignored. The papers by Jenkins (1940), Hull (1945), and Comstock et al. (1949) integrated possible effects of types of gene action on efficiency of selection and stimulated interest in maize populations and their improvement by breeders.

Since the 1940s, researchers have been very active in estimation of genetic and environmental components of variance for different types of maize populations. Additionally, they have attempted to determine the relative proportions of total genetic variance that are attributable to additive and nonadditive effects, both intra- and interallelic. Voluminous literature has developed, presenting the empirical results of these studies and how they compare with theory and different selection procedures. Several different population types have been sampled because of the interest in possible differences of genetic variability among populations. Because of commitment to hybrid use the ratio of dominance variance to additive genetic variance has received considerable attention and has been estimated in several studies. The F_2 populations developed from a cross of two inbred lines have been studied in some instances to estimate specific parameters; e.g., gene frequency expected to be 0.5 for all segregating loci permits estimation of the average level of dominance of genes affecting the trait under study. Also, F_2 populations and their advanced generations have been studied to estimate effects of linkages on estimates of additive and dominance variances.

5.1 EXPERIMENTAL RESULTS FROM THE LITERATURE

Table 5.1 summarizes the estimates of additive variance σ_A^2 and dominance variance σ_D^2 available from many studies abstracted from the literature for 19 different traits of maize. Most estimates were obtained by use of mating designs I, II, and III, but a few estimates for F_2 populations were obtained by techniques given by Mather (1949). No estimates from diallel analyses were included. The last column of Table 5.1 shows the number of estimates included for each trait because the precision of estimates is determined by the number of estimates included in the average. Estimates of components of variance shown in Table 5.1 are averages for each trait that were reported in the literature. The greatest number of estimates was reported for yield, and averages for each parameter show that the ratio of dominance to additive variance was quite large for yield; dominance variance seems important in the expression of yield.

If estimates of the ratio σ_D^2/σ_A^2 for each study are totaled and divided by the total number of estimates, we obtain 0.9377 (Table 5.1). If, however, we divide the average estimate of the dominance variance by the average estimate

Table 5.1. Summary for 19 traits for the average estimates of additive ($\hat{\sigma}_A^2$) and dominance ($\hat{\sigma}_D^2$) components of variance and their standard errors (SE), ratio of dominance to additive genetic variance ($\hat{\sigma}_D^2/\hat{\sigma}_A^2$), and heritability ($\hat{h}^2$) on a plot basis.

Trait	$\hat{\sigma}_A^2$	SE($\hat{\sigma}_A^2$)	$\hat{\sigma}_D^2$	SE($\hat{\sigma}_D^2$)	$\hat{\sigma}_D^2/\hat{\sigma}_A^2$	$\hat{h}^2$(%)	No. of estimates
Yield, g[†]	469.1	174.3	286.8	210.1	0.9377 (0.6113)[‡]	18.7	99
Plant height, cm	212.9	51.6	36.2	46.5	0.5338 (0.1700)	56.9	45
Ear height, cm	152.7	35.5	11.1	36.5	0.3743 (0.2324)	66.2	52
Number of ears ($\times 10^3$)	45.9	13.2	11.8	9.9	0.4366 (0.2875)	39.0	39
Ear length, cm ($\times 10^2$)	152.4	37.8	50.4	47.3	0.3746 (0.2480)	38.1	36
Ear diameter, cm ($\times 10^2$)	4.6	1.1	0.9	1.1	0.3269 (0.2391)	36.1	35
Kernel-row number ($\times 10^2$)	189.0	45.5	14.5	77.8	0.1774 (0.2407)	57.0	18
Kernel weight, g	34.9	8.5	9.5	9.4	0.5544 (0.2435)	41.8	11
Days to flower	4.0	0.9	-0.1	0.9	0.6598 (-----)	57.9	48
Grain moisture, %	7.2	1.7	0.5	2.5	0.4801 (0.2361)	62.0	4
Oil, % ($\times 10^2$)	82.2	15.6	8.7	8.8	0.1808 (0.1897)	76.7	4
Lodging, % ($\times 10^3$)	126.1	33.6	-30.2	24.2	0.0265 (----)	----	5
Number of tillers ($\times 10^2$)	26.9	6.0	-1.6	----	0.1850 (----)	71.9	5
Kernel depth ($\times 10^3$)	18.7	4.2	5.0	3.6	0.5114 (0.2673)	29.2	7
Cob diameter ($\times 10^3$)	16.6	2.8	3.4	3.0	0.2131 (0.2048)	37.0	6
Husk extension ($\times 10^2$)	54.8	10.4	25.2	14.7	0.4598	49.5	3
Husk score ($\times 10^2$)	65.2	1.0	20.4	12.9	0.3128	35.9	3
Flag leaf number ($\times 10^2$)	67.8	----	18.0	----	0.2654	----	1
Flag leaf length ($\times 10^2$)	154.0	----	58.6	----	0.3805	----	1

[†]Units on per plant basis. [‡]Ratio is for average estimates of σ_A^2 and σ_D^2.

of the additive variance, we obtain 0.6113, shown in parentheses in Table 5.1. The difference between the two estimates of the ratio σ_D^2/σ_A^2 arises because of the frequent occurrence of negative estimates of σ_D^2. When the estimate of σ_D^2 was reported as a negative value, it was consequently not possible to determine the ratio of σ_D^2/σ_A^2. An unbiased average of σ_D^2 was obtained by including negative estimates in the algebraic summation of estimates of σ_D^2 from the reported studies. If estimates of σ_D^2 are in reality either very small positive values or zero, negative experimental estimates are not unexpected. Hence the average of the ratio σ_D^2/σ_A^2 would be biased upward because the ratios in individual studies that had negative estimates of σ_D^2 were not available; if the true estimates were very small positive values, the ratio σ_D^2/σ_A^2 would be small. The true value of a parameter for a particular trait for a particular population will be approached by repeated experimentation; thus the best estimates of σ_A^2 and σ_D^2 are those averaged across all the reported studies, provided we make the broad generalization that maize is our population. It seems, therefore, that the best estimate of the ratio σ_D^2/σ_A^2 is the one determined from the average estimates of σ_A^2 and σ_D^2. Although the ratio σ_D^2/σ_A^2 calculated from the averages of σ_A^2 and σ_D^2 is lower than the average of ratios, considerable dominance variance is expressed for yield. Assuming no epistasis and linkage effects, σ_A^2 on the average accounted for 61.2% and σ_D^2 accounted for 38.8% of the total genetic variation for yield.

The standard errors of σ_A^2 and σ_D^2 also are included in Table 5.1. Relative to σ_A^2 and σ_D^2, standard errors for yield were much larger for σ_D^2. Many of the estimates for σ_D^2 were obtained from studies that used the design I mating scheme, and it is known that the standard errors for these estimates are large because of the complex function used to make the estimate. For designs II and III, σ_D^2 can be estimated directly from the expected mean squares and σ_A^2 can be estimated directly from the expected mean squares from all the mating designs. Consequently the average standard error of σ_A^2 is only 37.2% as large as the average estimate of σ_A^2, whereas the average standard error of σ_D^2 is 73.2% as large as the average estimate of σ_D^2. For individual experiments the standard error of σ_D^2 was often greater than the estimate of σ_D^2.

The ratio σ_D^2/σ_A^2 was considerably lower for other traits than for yield. In most instances the ratio when calculated from the average of estimates σ_A^2 and σ_D^2 was lower than the average of the reported ratios σ_D^2/σ_A^2. It seems that the greatest proportion of total genetic variance can be attributed to additive effects for most traits. A relatively large average ratio was obtained for days to silk, but the average of estimates shows that the average estimate for σ_D^2 is a small negative value. Similar to yield, frequent negative estimates of σ_D^2 were obtained, and the ratio σ_D^2/σ_A^2 was not estimable. Hence the average of the ratios is biased upward because only the ratios that had positive estimates of σ_D^2 were included. Similarly, the best estimate of the ratio seems to be the one from the unbiased average of estimates of σ_A^2 and σ_D^2, which for days to silk must be near zero. The average ratio σ_D^2/σ_A^2 for kernel weight and kernel depth indicated similar estimates of σ_A^2 and σ_D^2. In both instances, however, the ratio was considerably lower when unbiased averages of σ_A^2 and σ_D^2 were compared; average estimates of σ_A^2 were about four times greater than average estimates of σ_D^2.

Estimates of heritabilities were calculated on a per plot basis, where the necessary components of variance were given in the reported studies. As Hanson (1963) has emphasized, the unit used for reporting heritabilities is very important in plant research. Because several of the studies were conducted in only one environment and did not include genotype-environment components of variance, estimates of heritability were calculated as $h^2 = \sigma_A^2/(\sigma^2 + \sigma_D^2 + \sigma_A^2)$, where σ^2 is the plot experimental error. Estimates of heritability for the 16 traits were distributed according to the following ranges:

Heritability estimates, %	Traits
$h^2 > 70$	Percent oil, number of tillers
$50 < h^2 < 70$	Plant height, ear height, kernel-row number, days to flower, grain moisture
$30 < h^2 < 50$	Number of ears, ear length, ear diameter, kernel weight, husk extension, husk score, cob diameter
$h^2 < 30$	Yield and kernel depth

The magnitude of average heritability estimates reflects number of estimates reported in the literature and the complexity of the traits. Yield is the most economically important trait in maize, and its heritability is the lowest of all traits. Yield also had a relatively large proportion of the total variance accounted for by σ_D^2. Kernel depth, number of ears, ear length, and ear diameter are components of yield, and their estimates of heritability were about twice as large as that for yield. Yield results from the total expression of the genotype from the time seed is planted until harvest; consequently, yield itself is the combined expression of genotype and environment throughout the duration of the growing season. Yield components, however, are determined during certain stages of the ontogeny of the genotype, so their expression depends on just a portion of the growing season. Number of ears per plant, for example, is determined by the combination of genetic and environmental forces from six weeks before flowering to flowering time. If environmental conditions and genetic composition of the genotype are favorable for more than one ear per plant before and at flowering time, the plants will have more than one ear. If the combination of conditions is unfavorable for more than one ear before and at flowering, more than one ear will not result even though optimum conditions may occur for ear development and grain fill after flowering. Yield components of kernel depth, ear length, and ear diameter are influenced to a large extent by environmental conditions after flowering.

Kernel-row number is another yield component, but it has a relatively high average heritability estimate (57.0%). Kernel-row number also is determined in relatively early stages of plant ontogeny and is less affected by environmental conditions at the time of flowering and from flowering to maturity; i.e., environmental forces can affect depth of kernels, kernel weight, and size of ear development after flowering, but the number of kernel rows is not altered. Days to flowering and grain moisture are measures of maturity of maize, and they both have relatively large average heritability estimates. The relatively large heritability values for maturity are in concert with experiences of applied breeding programs because it has been relatively easy to change the maturity of maize, as evidenced by range of adaptation of maize for latitude and elevation.

Average estimates of the parameters shown in Table 5.1 also were calculated for five types of maize populations: (1) F_2 populations developed from a cross of two inbred lines; (2) synthetics developed from recombination of elite inbred lines (usually 10 to 24 lines); (3) open-pollinated varieties; (4) variety crosses produced by crossing either open-pollinated varieties or synthetic varieties; and (5) composites developed by intermating material of diverse origin, such as open-pollinated and synthetic varieties, hybrids, and inbred lines. There are large differences for average estimates for the five types of maize populations (Table 5.2); and again, the best estimates for each type of population depend on the number of experimental estimates included. (Table 5.2 summarizes information from many studies abstracted from the literature.) The last column and last row of Table 5.2 show the number of estimates available for the five types of populations and for each parameter, respectively. For example, 37 estimates of σ_A^2 were reported for open-pollinated varieties vs. only 10 estimates for composites. The reason for the seeming discrepancy in the number of estimates available for each parameter is that (1) standard errors were calculated for σ_A^2 and σ_D^2, (2) estimates of σ_D^2 were not available from the use of inbred lines, (3) negative estimates did not permit the estimation of the ratio σ_D^2/σ_A^2, and (4) experimental error was not included in the literature to permit calculation of heritability on a plot basis. Average estimates of σ_A^2 and σ_D^2 were largest for composite and F_2 populations and smallest for synthetic varieties. Standard errors of σ_A^2 and σ_D^2 also

Table 5.2. Summary of estimates of additive dominance and genetic components of variance, ratio of dominance variance to additive variance, and heritabilities for yield for five types of maize populations.

Type of populations	$\hat{\sigma}_A^2$	$SE(\hat{\sigma}_A^2)$	$\hat{\sigma}_D^2$	$SE(\hat{\sigma}_D^2)$	$\hat{\sigma}_D^2/\hat{\sigma}_A^2$	$\hat{h}^2$	$\hat{\sigma}_A^2/\hat{\sigma}_D^2$	No. of reports
F_2	585.1	338.5	451.0	593.0	1.0022 (0.7708)[+]	24.4	1.30	24
Synthetics	225.9	59.3	128.6	83.4	0.8255 (0.5690)	22.9	1.76	15
Open-pollinated	503.8	178.9	245.8	320.8	0.7619 (0.4879)	18.9	2.16	37
Variety crosses	306.2	139.2	292.2	32.0	1.3854 (0.9540)	13.4	1.05	13
Composites	721.9	432.0	281.8	----	1.3335 (0.3902)	13.8	2.56	10
Average	468.6	229.6	279.9	257.3	0.9377 (0.6343)	18.7	1.76	
No. of estimates	99	55	82	41	72	43	82	99

[+]Ratio is for the average estimates of $\hat{\sigma}_A^2$ and $\hat{\sigma}_D^2$.

were largest for composite and F_2 populations. The average estimate of σ_A^2 for composites was greater than for F_2 populations, but the average estimate of σ_D^2 was greater for F_2 populations. The average ratio σ_D^2/σ_A^2 was greater than one for F_2, variety-cross, and composite populations and less than one for synthetic and open-pollinated variety populations. In all instances, however, the ratios estimated from unbiased averages of σ_A^2 and σ_D^2 were less than one but still relatively large for F_2 and variety-cross populations.

For each of the five types of populations, the average σ_A^2 value was greater than the average σ_D^2 value, ranging from 2.56 times greater for composite populations to 1.05 times greater for variety-cross populations. Additive genetic variance σ_A^2 is considerably greater than dominance variance σ_D^2 for synthetic, open-pollinated, and composite populations, indicating that σ_A^2 is more important for yield than σ_D^2 in these populations. Average standard errors were particularly large for average estimates of σ_A^2 and σ_D^2 for F_2 populations and for σ_D^2 for open-pollinated varieties. Nearly all estimates for F_2 populations and many for open-pollinated varieties were reported prior to 1960. Growing conditions for maize during the 1950s often were less than optimum, which may have contributed to larger errors in estimation. The design I mating scheme usually was used, and this mating design has inherently large standard errors for σ_D^2. Also, improvements in experimental technique, experimental design, and analysis probably were important in contributing to smaller errors in experiments conducted in more recent years.

Distributions of estimates of σ_A^2 for yield for each experiment for the five types of populations are shown in Fig. 5.1, which comprises information from many studies abstracted from the literature. Before plotting each estimate of σ_A^2, the arrangement of the five types of populations was made on the basis of expected relative magnitude of estimates of σ_A^2; i.e., F_2 populations have the least variability and composites the greatest. But Fig. 5.1 shows that distributions of individual estimates of σ_A^2 are as great for F_2 populations as for open-pollinated and composite variety populations. Averages of σ_A^2 (solid lines) and of standard errors of σ_A^2 (dashed lines) also are included in Fig. 5.1. Average standard errors of σ_A^2 overlap for all types of populations except for average estimates of synthetic and open-pollinated varieties. Except for the synthetic variety and variety crosses, few estimates are considerably different from the majority. Study of individual experiments from which extremely large estimates of σ_A^2 were reported also show very large estimates of standard errors of σ_A^2. One negative estimate (-144) of σ_A^2 also was reported in an open-pollinated variety.

The most surprising feature of the distributions is the difference between distributions of F_2 and synthetic variety populations. The F_2 populations seem to have greater variability on the average than synthetic varieties. Because synthetic varieties were developed by recombining elite inbred lines, it would seem that variability within synthetic varieties should be at least equivalent to variability of F_2 populations. Synthetic varieties would be equivalent to advanced generations of multiple F_2 populations. The stand-

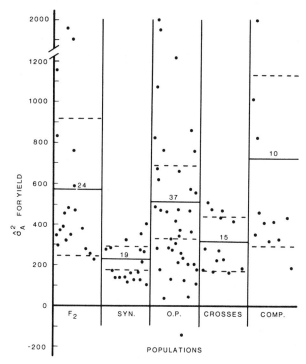

Fig. 5.1.   Distribution of estimates of additive genetic variance for five
            types of populations (see Table 5.2).

ard errors of estimates of σ_A^2 for synthetic varieties, however, were much
smaller (5.7 times) than those for F_2 populations. Most of the estimates for
synthetic varieties also were obtained since 1960 by use of mating designs
other than design I.   The greater estimates of σ_A^2 for F_2 populations compared
to synthetic varieties may be a function of gene frequency. Assuming a genetic model of partial to complete dominance, estimates of σ_A^2 for F_2 populations
whose gene frequencies at the segregating loci would be 0.5 would be greater
than σ_A^2 in synthetic varieties which may have gene frequencies greater than
0.5. Synthetic varieties usually were developed by intermating elite inbred
lines. Also, gene frequencies in open-pollinated varieties may be near 0.5
or lower and, with partial to complete dominance, σ_A^2 might be near the
maximum. Estimates of σ_A^2 for variety cross populations have definitions
slightly different from those for intrapopulations and may not be equivalent.

Because estimates of σ_A^2 are gene-frequency dependent, differences among
the five types of populations may be influenced as much by frequency of segregating loci as by number of segregating loci. From Fig. 5.1 it seems that
sufficient variability is present in any type of population to expect significant progress from selection. The maize breeder must decide whether to work
with populations to obtain improved versions of elite lines (F_2 populations),
with sample populations having considerable variation for development of new
lines (open-pollinated and composite varieties), or with sample populations
that seemingly have less genetic variation (hopefully at a higher frequency)
for new lines than other types (synthetic varieties).

Some of the first estimates of genetic components of variance in maize were obtained from F_2 populations, which were chosen because their expected gene frequencies were 0.5 to provide estimates of average level of dominance of genes affecting the traits studied. One of the first questions raised was the effect of linkage bias on estimates of σ_A^2 and σ_D^2. If linkages were primarily in the repulsion phase, linkage bias would underestimate σ_A^2 and overestimate σ_D^2. Estimates of average level of dominance for yield often were in the overdominance range, suggesting that at least some loci were giving an expression of overdominance or that pseudooverdominance was expressed because of linkage bias in estimates of σ_A^2 and σ_D^2. The next logical step was to test for effects of linkage bias on estimates of σ_A^2 and σ_D^2 by randomly mating the F_2 populations for several generations to permit genetic recombination and an approach to linkage equilibrium. The F_2 and advance random mated generations (F_n) of F_2 populations were resampled and estimates of σ_A^2 and σ_D^2 compared for possible linkage bias. Results from such studies are given in Table 5.3 for four traits.

Estimates of average level of dominance for F_n populations are lower than those for F_2 populations in all instances. For yield, the average of estimates for average level of dominance was slightly greater than 1 for the F_2 populations, whereas the average of estimates was 0.5969 for F_n populations. Average estimates of average level of dominance were about 0.5 for plant and ear height and ear number in F_2 populations, but they were about 0.3 in F_n populations. Although estimates were smaller, average levels of dominance in F_2 and F_n populations followed the same trend when unbiased averages of σ_A^2 and σ_D^2 were used to estimate average levels of dominance. It seems, therefore, that partial dominance rather than complete or overdominance was the predominant expression of gene action for yield. Estimates of overdominance obtained in F_2 populations apparently were pseudooverdominance because of repulsion phase linkages. Levels of dominance were not as great for the other three traits, but linkages also must have influenced estimates of σ_A^2 and σ_D^2 because

Table 5.3. Estimates of additive ($\hat{\sigma}_A^2$) and dominance ($\hat{\sigma}_D^2$) variances and ratio of $\hat{\sigma}_D^2/\hat{\sigma}_A^2$, and average degree of dominance ("a") for four traits estimated in the F_2 populations and F_2 populations advanced by random mating.

Trait	F_2				F_n				Ratio
	$\hat{\sigma}_A^2$	$\hat{\sigma}_D^2$	$\hat{\sigma}_D^2/\hat{\sigma}_A^2$	"a_2"	$\hat{\sigma}_A^2$	$\hat{\sigma}_D^2$	$\hat{\sigma}_D^2/\hat{\sigma}_A^2$	"a_n"	"a_n"/"a_2"
Yield, g	554.5	537.5	1.1021	1.0498	623.4	169.8	0.3564	0.5969	0.5685
	(13)[†]	(13)	(0.9693)	(0.9845)	(4)	(4)	(0.2723)	(0.5218)	
Plant height, cm	189.4	42.0	0.2599	0.5098	255.5	20.5	0.0801	0.2830	0.5551
	(7)	(7)	(0.2217)	(0.4708)	(2)	(2)	(0.0802)	(0.2831)	
Ear height, cm	84.3	14.4	0.2571	0.5070	134.5	8.0	0.0594	0.2437	0.4806
	(7)	(7)	(0.1708)	(0.4132)	(2)	(2)	(0.0594)	(0.2437)	
Ear number ($\times 10^3$)	43.5	10.2	0.2906	0.5390	87.0	6.7	0.1135	0.3368	0.6248
	(8)	(8)	(0.2344)	(0.4841)	(2)	(2)	(0.0770)	(0.2774)	

[†] Numbers in parentheses indicate the number of estimates included in the average.

levels of dominance were reduced in all instances. Comparisons of average estimates of σ_A^2 and σ_D^2 show that σ_A^2 increased and σ_D^2 decreased in all instances from F_2 to F_n populations. The last column of Table 5.3 shows the relative change of average level of dominance, which ranges from about 40% to 50% from F_2 to F_n populations.

Compilation of experimental estimates of the parameters shown in Tables 5.1 and 5.2 for 19 different agronomic traits of maize point to one general conclusion: Genetic variability is present in the five arbitrary types of maize populations for all traits and a major portion of the genetic variability is additive genetic variance. For selection purposes additive genetic variance is of primary importance, and thus selection should be effective in most populations. The most surprising feature of experimental estimates is lack of any significant trend in relative magnitude of additive genetic variance among five types of populations for yield (see Fig. 5.1). Distributions of estimates of σ_A^2 for the populations themselves do not show any major advantage for one type of population; a priori it seemed that F_2 populations would have the least and composites the greatest. The summary of estimates was discussed ignoring effects of epistasis and genotype-environment interactions, which will be considered later.

5.2 IOWA STIFF STALK SYNTHETIC

Iowa Stiff Stalk Synthetic (BSSS) is a synthetic variety developed in 1933-34 by G. F. Sprague (1946) from the following 16 lines: Ia.I159, Ia.I224, Ia.Os420, Ia.WD456, Ind.461-3, Ill.12E, CI617, CI540, Ill.Hy, Oh.3167B, Ind.AH83, Ind.Tr9-1-1-6, F1B1-7-1, A3G-3-1-3, CI187-2, and LE23. They were developed by various breeders and were chosen for their resistance to stalk breakage. BSSS is considered above average as a source population of inbred lines that are above average for combining ability with other elite lines. Surveys by the American Seed Trade Association show that lines originating from BSSS are used extensively in hybrids in the U.S. Corn Belt (Sprague 1971; Zuber 1975). BSSS is an important breeding population of the cooperative federal-state maize breeding program located at Ames, Iowa, and it is used extensively in selection programs for yield improvement and resistance to maize pests. Relative to other maize populations BSSS is characterized as having above average stalk quality, vigorous plant type, dark green leaf coloration, good ear size, no distinctive features for pest resistance, and full-season maturity for the central U.S. Corn Belt. Yield of BSSS as a variety itself is only average, but it has above average combining ability in crosses with other synthetic varieties (Hallauer and Malithano 1976). For the arbitrary A and B grouping of lines in the North Central region of the United States, lines from BSSS are classified in the B grouping. Inbred lines extracted from BSSS usually give excellent hybrid yield performance in crosses with Lancaster-type lines (Zuber and Darrah 1980; Darrah and Zuber 1986). BSSS seems more uniform in phenotype than other maize populations, but crosses among elite lines extracted from BSSS often express high-yields, suggesting genetic variability for yield within BSSS (Lamkey and Hallauer 1986).

Because of the importance of BSSS in the maize research program at the Iowa station, quantitative genetic studies have been conducted to characterize genetic variability present in the BSSS population. Estimates of genetic parameters for BSSS have been obtained by use of design I and II mating schemes, unselected inbred lines, and recurrent selection programs involving half-sib and selfed progeny evaluations. Detailed estimates of genetic components of variance for BSSS are given as an example of how estimates of a particular population either agree or disagree with those given in Tables 5.1 and 5.2, which, however, include estimates for BSSS. The methods used for obtaining estimates for BSSS also are indicated briefly.

Table 5.4 shows estimates for eight traits for four sets of experiments. Our purpose is to summarize the estimates of genetic components of variance and to show how they compare by reference to data of studies that used different types of progenies and methods of estimation. Details of specific experiments were reported previously (Hallauer 1970, 1971; Obilana and Hallauer 1974; Silva and Hallauer 1975; and Bartual and Hallauer 1976). In Table 5.4, experiment 1 used progenies developed by design II mating scheme; experiment 3 used progenies developed from design I and II mating schemes; and experiments 2 and 4 included unselected inbred lines developed by single seed descent by selfing and full-sibbing methods of inbreeding, respectively. Hence estimates of genetic parameters were obtained from evaluation of noninbred progenies for experiments 1 and 3 and of inbred progenies for experiments 2 and 4. In all instances, however, estimates of genetic components of variance were adjusted for inbreeding and made equivalent for the base population, BSSS.

Comparisons of estimates of components of variance among the four experiments show some large differences, but in most instances the estimates show striking similarities. If we use the standard errors of estimates of components of variance to compare the independent estimates, most are within range of the sampling error. Three of the estimates of σ_A^2 for yield were very similar (experiments 1, 2, and 3, Table 5.4); whereas the estimates of experiment 4 were considerably greater. Experiment 4 estimates were obtained in two similar environments, and consequently the estimate of σ_{AE}^2 was small; the experiment 4 estimate may be overestimated because of bias due to genotype-environment interaction. Estimates of components of genetic variance and their interactions with environments were amazingly similar for experiments 1 and 3 that included the evaluation of noninbred full-sib progenies; all estimates were similar in all instances. Both estimates of σ_A^2 for plant height obtained from experiments 2 and 4, which included inbred progeny evaluation, were greater than those for experiments 1 and 3, which included noninbred progeny evaluation. Greater discrepancies existed among estimates for days to flower than for other traits, but the estimates were obtained in fewer environments; if genotype-environment interaction were important, the bias on the estimates would be greater, particularly for inbred progenies. Estimates of heritability were similar for most traits, but they tended to be greater for experiments that evaluated inbred progenies. For yield, estimates of heritability from inbred data were about double those from noninbred data. Because only one source of variation was present for inbred progeny evaluation trials, it was not possible to estimate σ_D^2, which would be negligible at the level of inbreeding of the inbred lines used (F = 99%).

Comparison of estimates given in Table 5.1 with those given in Table 5.4 for BSSS show that estimates for BSSS tended to be smaller than those summarized from the literature for all populations. For example, the average estimate of σ_A^2 in Table 5.1 is 469.1 ± 174.3 vs. 188 ± 24 for BSSS in Table 5.4. The estimate of σ_D^2 (287 ± 210 in Table 5.1) also is much greater than the estimate of σ_D^2 for BSSS (179 ± 29 in Table 5.4). Although average estimates of σ_A^2 and σ_D^2 for BSSS are 60% and 38% smaller than those in Table 5.1, standard errors of estimates of σ_A^2 and σ_D^2 also are 84% smaller for BSSS. If the preci-

Table 5.4. Summary of the component of variance estimates for BSSS obtained from four experiments.

Trait	Experi-ment[†]	$\hat{\sigma}^2_A$	$\hat{\sigma}^2_{AE}$	$\hat{\sigma}^2_D$	$\hat{\sigma}^2_{DE}$	$\hat{\sigma}^2_D/\hat{\sigma}^2_A$	$\hat{\sigma}^2$	$\hat{h}^2$ (%) Plot	$\hat{h}^2$ (%) Progeny mean[‡]
Yield	1	156±29	83±22	174±37	74±13	1.12	387±13	17.8	34.9
	2	147±16	44±5	----[§]	----	----	129±7	45.9	80.2
	3	166±24	91±10	184±21	7.2±12	1.10	364±8	18.9	59.1
	4	283±29	22±6	----	----	----	157±10	61.2	89.4
	X̄	188±24	60±11	179±29	73±12	1.11	259±10	24.8#	41.4#
Plant height	1	143±15	22±4	25±5	5±7	0.17	55±2	57.2	76.8
	2	191±18	15±2	----	----	----	58±3	72.3	92.9
	3	141±10	12±1	15±2	7±1	0.11	46±1	63.8	82.9
	4	191±18	4±2	----	----	----	47±3	78.9	95.4
	X̄	166±15	13±2	20±4	6±4	0.14	52±2	64.6	81.7
Ear height	1	126±13	10±2	13±3	9±4	0.10	30±1	67.0	83.8
	2	102±10	3±1	----	----	----	35±2	72.8	93.7
	3	107±7	9±1	9±1	6±1	0.08	31±1	66.0	84.8
	4	111±11	1±1	----	----	----	26±2	80.4	96.0
	X̄	112±10	6±1	11±2	8±2	0.09	30±2	67.1	84.4
Ear length ($\times 10^2$)	1	104±14	22±9	24±12	−10±23[¶]	0.23	185±6	31.0	62.6
	2	129±15	58±26	----	----	----	242±14	30.1	68.4
	3	121±11	22±3	41±6	23±4	0.34	121±3	28.9	61.4
	4	215±20	13±5	----	----	----	120±10	61.8	89.8
	X̄	142±15	29±11	32±9	6±14	0.28	167±8	37.8	66.5
Ear diameter ($\times 10^2$)	1	2.6±0.4	1.1±0.3	0.4±0.4	0.8±0.8	0.15	5.8±0.2	24.3	56.5
	2	4.1±2.0	3.9±1.5	----	----	----	11.5±1.0	21.0	56.0
	3	2.9±0.3	0.5±0.1	0.8±0.1	0.7±0.1	0.28	4.7±0.1	30.2	59.4
	4	6.0±0.5	0.2±0.1	----	----	----	4.0±0.2	58.8	89.0
	X̄	3.9±0.8	1.4±0.5	0.6±0.2	0.8±0.4	0.22	6.5±0.4	29.5	52.0
Cob diameter ($\times 10^2$)	1	1.9±0.2	0.1±0.1	0.1±0.2	0.1±0.4	0.05	2.8±0.1	38.0	75.0
	2	1.8±1.0	1.4±0.0	----	----	----	6.0±0.0	19.6	55.1
	3	1.4±0.1	0.2±0.1	0.2±0.1	0.2±0.1	0.14	2.4±0.1	31.8	65.6
	4	1.9±0.2	0.1±0.1	----	----	----	2.0±0.1	47.5	73.7
	X̄	1.8±0.4	0.4±0.1	0.2±0.2	0.2±0.2	0.10	3.3±0.1	30.5	65.4
Kernel depth ($\times 10^2$)	1	1.5±0.3	1.0±0.2	0.5±0.3	−0.1±0.6[¶]	0.33	4.8±0.2	19.2	47.9
	2	1.4±1.0	1.0±0.0	----	----	----	5.6±0.1	17.5	52.5
	3	1.5±0.2	0.2±0.1	0.5±0.1	0.3±0.1	0.33	3.9±0.1	23.4	43.3
	4	3.2±0.4	0.3±0.1	----	----	----	5.6±0.1	35.2	75.6
	X̄	1.9±0.5	0.6±0.1	0.5±0.2	0.1±0.4	0.33	5.0±0.1	23.4	54.8
Days to flower	1	30.5±5.8	----	8.5±6.7	----	0.28	2.2±0.2	74.0	77.5
	2	8.6±0.6	0.4±0.2	----	----	----	4.8±0.2	62.3	90.2
	3	4.2±0.3	0.3±0.1	0.5±0.1	0.2±0.1	0.12	1.3±0.1	64.6	82.6
	4	16.8±1.6	8.6±0.6	----	----	----	3.0±0.3	59.2	83.3
	X̄	15.0±2.1	0.4±0.2	4.5±3.4	0.2±0.1	0.20	2.8±0.2	65.5	74.4

[†]The mating designs used to obtain the component of variance estimates were (1) design II with 320 full-sib progenies evaluated in three environments (Hallauer 1971); (2) 247 unselected inbred lines developed by selfing evaluated in three environments (Obilana and Hallauer 1974); (3) 800 full-sib progenies developed by use of designs I and II evaluated in six environments (Silva 1974); and (4) 231 unselected inbred lines developed by full-sibbing evaluated in two experiments (Bartual and Hallauer 1976).

[‡]Progeny mean heritabilities were estimated assuming three environments and two replications in each environment; i.e., $\sigma^2_A/(\sigma^2/re + \sigma^2_{ge}/e + \sigma^2_A)$.

[§]No estimates were available.

[¶]Negative estimates were assumed to be zero in the calculation of heritabilities.

#Estimates of heritabilities were computed from the average estimates of the components of variance.

sion of estimates is considered and we use two standard errors of the respective estimates to determine significance, estimates of BSSS are not significantly different from those in Table 5.1 for yield. Estimates of plant height, ear height, ear length, and ear diameter for BSSS are similar to those given in Table 5.1 and are within the range of sampling error. However, variation of cob diameter and kernel depth in BSSS is less than that for other populations summarized in Table 5.1; and the estimate of σ_A^2 for days to flower was about four times greater than the average estimate of all populations in Table 5.1. Estimates of σ_A^2 for days to flower for BSSS are probably biased upward because of the genotype-environment interaction. Experiment 3 in Table 5.4 had an estimate of σ_A^2 for days to flower that was similar to that given in Table 5.1.

Heritability estimates on a plot basis were very similar for BSSS and the average of all populations for yield, plant height, ear height, ear length, ear diameter, cob diameter, kernel depth, and days to flower. Considering sampling errors associated with estimates of components of variance, the similarities of heritability estimates in Tables 5.1 and 5.4 are remarkable. Relative heritability estimates for the eight traits measured in BSSS are the same as those for the average of all populations in Table 5.1.

The most distinctive difference between the BSSS estimates and the average for all varieties was in the estimate of σ_D^2 for yield. The average estimate of σ_D^2 in Table 5.1 is 39% smaller than the average estimate of σ_A^2, whereas average estimates of σ_A^2 and σ_D^2 for BSSS are nearly equal (Table 5.4). The averages of the ratio $\hat{\sigma}_D^2/\hat{\sigma}_A^2$ are about one in both instances (0.94 vs. 1.11), but ratios of average estimates of σ_A^2 and σ_D^2 are 0.61 (Table 5.1) vs. 0.95 (Table 5.4). If, for the BSSS population, we consider only experiments 1 and 3 for which estimates of both σ_A^2 and σ_D^2 were obtained, the average estimate of σ_A^2 (161) is less than the average estimate of σ_D^2 (179); hence $\hat{\sigma}_D^2/\hat{\sigma}_A^2$ is 1.18.

Comparison of estimates of σ_A^2 of BSSS with σ_A^2 in Table 5.2 for the average of 15 studies of synthetic varieties shows that the estimates are similar, 188 vs. 226. The average estimate of σ_D^2 for BSSS, however, is greater than the average estimate for all synthetic varieties, albeit the standard errors show that the difference between the two average estimates of σ_D^2 is probably not significant. It seems that the BSSS population relative to other synthetic varieties is average for $\hat{\sigma}_A^2$ and above average for $\hat{\sigma}_D^2$ for yield. For the other traits, BSSS has genetic variability similar to other synthetic varieties for plant height, ear height, ear length, and ear diameter but below average for cob diameter and kernel depth. Except perhaps for $\hat{\sigma}_D^2$, no evidence distinguishes BSSS from other synthetic variety populations; i.e., the estimates of σ_A^2 and h^2 were similar to the average of all synthetic variety populations. Compared to the mean of all populations estimates of σ_A^2 and σ_D^2 were smaller, but standard errors of estimates also were smaller. Resistance to stalk breakage is a very important trait in the production and growing of hybrids in mechanized maize culture. Perhaps the fortuitous choice of lines with good stalk strength and strong dominance of genes for yield were the contributing factors for BSSS being a good source population of inbred lines having good combining ability (Hallauer et al. 1983).

5.3 SELECTION EXPERIMENTS

Long-term selection programs for the cyclical improvement of breeding populations as source materials for applied breeding programs also provide evidence of the genetic variability present within populations. Estimates of genetic variability are only as good as the sampling techniques used in obtaining progenies for evaluation. Effects of sampling are equally important whether one is sampling populations to develop progenies by some mating design for estimation of variance components or sampling populations to test progenies in a cyclical improvement program. In either instance, the validity of estimates of variance components and progeny evaluations for populations is only as good as the sampling used. In most instances the sample sizes were greater when mating designs (e.g., designs I and II) were used than when testing progenies included in evaluation trials of cyclical selection programs. In cyclical selection programs, however, estimates of genetic variation among progenies tested in each cycle of selection are available. As shown in Chap. 2, components of genetic variance are dependent on gene frequencies of the population under study; changes in gene frequency of the population would affect estimates of components of variance. The primary objective of all selection trials is to change gene frequency, i.e., to increase the frequency of desirable alleles for the trait under selection. Use of variation among progenies after several cycles of selection as an estimate of the component of variance for original unselected populations, therefore, may be invalid. If, as expected, changes in gene frequency are relatively small from cycle to cycle of selection for a complex trait such as yield, the small change would not have a large effect on estimates of components of variance. Sampling errors may in fact be greater than absolute changes in gene frequency in the precision of estimates of components of variance. Estimates of components of variance obtained by combining data from long-term cyclical selection programs may provide better estimates of the population under selection because of repeated sampling of the population in each cycle of selection, assuming changes in gene frequency are not great in successive cycles of selection. Sufficient sampling and change in gene frequency, therefore, are antagonistic in regard to estimates from selection experiments compared with a study by use of a mating design.

Estimates of progeny components of variance and their interactions with environments will be summarized for nine recurrent selection studies for grain yield being conducted at the Iowa station (Table 5.5). The nine selection studies include nine populations: (1) Iowa Stiff Stalk Synthetic with seven cycles of half-sib, BSSS13, and S_2 selection, BS13, and (2) reciprocal recurrent selection, BSSS(R), with Iowa Corn Borer Synthetic No. 1 (BSCB1) as the tester and BSCB1(R) undergoing reciprocal recurrent selection with BSSS(R) as the tester; Krug Hi Synthetic 3 (BSK), a strain of Krug Yellow Dent (Lonnquist, 1949), that has undergone half-sib selection with (3) inbred line testers, BSK(HI), and (4) S_1 progeny selection, BSK(S); (5) Alph, an open-pollinated variety, that has undergone half-sib selection with an inbred line (B14) tester; (6) BS2, (7) BS16, and (8) BSTL undergoing S_2 recurrent selection; and (9) BS10 and BS11 undergoing reciprocal full-sib selection. Grain yield was emphasized in all instances. Populations, cycles of selection completed, and testers used are listed in Table 5.5. The purposes of providing estimates of genetic variation for recurrent selection studies are twofold: (1) to compare estimates obtained from selection trials with those shown in Tables 5.1 to 5.4 obtained via different mating designs; and (2) to show how estimates of genetic variance have changed in successive cycles of recurrent selection. Some studies have been in progress for nearly 40 years, others for only 20. Experimental and testing procedures have changed somewhat during the selection studies and these will be indicated for each. All estimates of σ_A^2 were

Table 5.5. Listing of the populations undergoing recurrent selection for
            yield improvement at Ames, Iowa.

Selection population	Code name	Tester	Progenies evaluated	Cycles of selection
(1) Iowa Stiff Stalk Synthetic	BSSS13	Ia13	half-sibs	7
	BS13(S)	----	S_2 progenies	4
(2) Iowa Stiff Stalk Synthetic	BSSS(R)	BSCB1	half-sibs	11
(2) Iowa Corn Borer Synthetic No. 1	BSCB1(R)	BSSS	half-sibs	11
(3) Krug Hi I, Synthetic 3	BSK(HI)	Inbred[+]	half-sibs	8
(4) Krug Hi I, Synthetic 3	BSK(S)	----	S_1 progenies	8
(5) Alph open-pollinated variety	BS12(HI)	B14	half-sibs	7
(6) Eto Composite x Early Lines	BS2(S)	----	S_2 progenies	5
(7) Eto Composite	BS16(S)	----	S_2 progenies	4
(8) (Lancaster x Tuxpeno) Lancaster	BSTL(S)	----	S_2 progenies	5
(9) Iowa Two-ear Synthetic	BS10(FR)	BS11	full-sibs	9
(9) Pioneer Two-ear Composite	BS11(FR)	BS10	full-sibs	9

[+]Different testers were used in different cycles of selection.

converted to grams per plant to make them equivalent to those in Tables 5.1 to
5.4. In making the conversions, we used an average plant density of 39,042
plants/ha.

Detailed data of the experiments that have undergone recurrent selection
for seven or more cycles are summarized in Tables 5.6 to 5.11. Estimates of
variance components, heritabilities, and genetic coefficients of variation for
each cycle of selection are included. A similar trend is noted for each population for successive cycles of selection: Estimates of the progeny component
of variance ($\hat{\sigma}_g^2$) are greatest in the initial two cycles of selection, are
smallest for cycles 2 to 4, and increase in the following cycles. This trend
is not as great for the BSK(S) selection population as for the other five selection programs. It is believed that the following items can partially explain the decrease and then increase of the genetic variance component estimate: (1) Progenies were tested in fewer environments (one or two in most instances) in the first two cycles of selection, and the genotype-environment
interaction biased estimates of genetic variance upward. (2) Cycles 2 to 4
were often tested when environmental conditions were unfavorable because of a
moisture stress. (3) Before 1970 all test plots were hand harvested with
gleaning for dropped ears and ears on broken stalks; since 1970 all plots have
been mechanically harvested with no retrieval of any ears not collected by the
harvester.

The effect of genotype-environment interaction is found in the summary of
recurrent selection programs given in Table 5.12. After conversion of component of variance estimates from the selection programs, they were averaged to
obtain the values at the bottom of Table 5.12. Average estimates of σ_{AE}^2
(166.8 ± 69.2) and σ_A^2 (311.2 ± 72.2) show that the estimate of σ_{AE}^2 is 53.6% as

Table 5.6. Summary of the estimates of variance components for BSSS13 with Ia13 as the tester for seven cycles of half-sib selection and two cycles of S_2 recurrent selection.

| Selection population | Year | No. of trials | Reps/ trial | Variance component estimates[†] | | | | $\hat{h}^2$ | Genetic CV, % | Average yield q/ha |
				$\hat{\sigma}^2$	$\hat{\sigma}^2_{ge}$	$\hat{\sigma}^2_{g}$	$\hat{\sigma}^2_A$[‡]			
BSSS13C0	1940	1	3	41.9±3.2	----	12.9±3.3	338.5±86.6	48.0	10.2	35.1
BSSS13C1	1948	1	6	15.3±0.8	----	4.2±0.8	110.2±21.0	62.2	3.8	53.8
BSSS13C2	1952	2	3	29.3±2.1	5.9±2.2	9.9±2.6	259.8±68.2	58.9	4.8	65.1
BSSS13C3	1955	2	3	63.2±5.1	13.5±5.6	-1.6±3.6	-42.0±94.5	----	----	44.9
BSSS13C4	1958-59	4	3	37.7±2.0	7.0±1.8	3.4±1.3	89.2±34.1	58.6	2.6	69.6
BSSS13C5	1962	4	2	32.5±2.5	3.0±2.1	10.3±2.3	270.3±60.4	65.2	3.8	85.0
BSSS13C6	1965	4	2	30.4±2.4	6.6±2.2	9.1±2.2	238.8±57.7	56.2	4.5	67.4
Pooled (C0-C6)		1.9¶	2.8¶	38.0±1.2	6.6±1.0	7.9±1.1	207.3±28.9	42.7	5.0	----
BS13(S)C0	1972	3	2	37.2±2.4	23.4±4.5 (3.9±0.8)§	58.9±10.4 (9.8±1.7)§	257.6±45.5	76.7	14.6	52.5
BS13(S)C1	1975	3	2	125.0±6.1	54.0±12.4 (13.5±3.1)§	80.2±17.2 (20.0±4.3)	350.5±75.2	62.6	22.3	40.2
BS13(S)C2	1978	3	2	78.4±6.6	21.5±6.9 (5.4±1.7)§	45.6±22.8 (11.4±5.7)§	199.3±99.6	69.2	12.9	52.2
BS13(S)C3	1981	3	2	65.8±6.7	48.4±9.6 (12.1±2.4)§	27.8±8.4 (7.0±2.1)§	121.5±36.7	50.7	11.3	46.5
BS13(S)C4	1984	3	2	42.8±3.6	10.5±3.7 (2.6±0.6)§	28.4±5.7 (7.1±1.4)§	124.1±24.9	72.8	16.4	39.8
Pooled (C0-C4)		3	2	65.4±2.5	32.0±3.2	53.7±12.0	234.7±52.4	71.3	15.9	46.2

[†]Variance component estimates in quintals per hectare. [‡]Estimates of σ^2_A for half-sib and S_2 selection converted to grams per plant. §Values adjusted to $(1/4)\sigma^2_A$ relative to half-sib selection. ¶Values are harmonic means.

Table 5.7. Summary of the estimates of variance components for yield for BSSS(R) with BSCB1(R) as the tester in a reciprocal recurrent selection experiment.

Selection population	Year	No. of trials	Reps/ trial	Variance component estimates [+]				$\hat{h}^2$	Genetic CV, %	Average yield, q/ha
				$\hat{\sigma}^2$	$\hat{\sigma}^2_{ge}$	$\hat{\sigma}^2_g$	$\hat{\sigma}^2_A$ [‡]			
BSSS(R)C0	1950	1	3	27.9±3.0	---	13.3±3.4	349.0±89.2	58.8	7.9	46.2
BSSS(R)C1	1953	2	3	33.1±2.3	0.1±1.7	11.1±2.5	291.3±65.6	66.7	6.2	53.4
BSSS(R)C2	1956–57	4	2.7[§]	41.4±2.2	3.3±1.7	3.9±1.2	102.3±31.5	43.5	3.7	53.2
BSSS(R)C3	1960	2	3	24.5±1.7	4.0±1.8	4.2±1.7	110.2±44.6	43.7	2.8	74.5
BSSS(R)C4	1964	4	2	30.2±2.4	3.0±2.0	2.0±1.0	52.5±26.2	27.5	2.1	66.9
BSSS(R)C5	1970	4	2	64.6±4.1	4.6±4.7	13.2±3.3	346.4±86.6	56.0	5.3	68.0
BSSS(R)C6	1973	3	2	70.6±4.5	14.0±5.9	15.6±5.5	409.3±144.3	45.4	5.5	71.3
BSSS(R)C9[§]	1982	4	2	104.0±5.7	8.0±5.6	17.2±3.8	357.4±79.0	53.4	4.7	88.2
BSSS(R)C10[§]	1985	4	2	45.4±3.1	5.8±2.7	5.7±1.8	118.4±37.4	45.0	3.3	73.5
Mean		2.6[¶]	2.3[¶]	57.8±3.8	7.8±4.1	14.1±3.6	359.3±92.4	52.8	5.4	69.1

[+]Variance component estimates in quintals per hectare. [‡]Estimates of σ^2_A from half-sib and full-sib selection in grams per plant. [§]Estimates of σ^2_A for reciprocal full-sib selection with BSCB1 (see Table 5.8).
[¶]Values are harmonic means.

Table 5.8. Summary of estimates of variance components for yield for BSCB1(R) with BSSS(R) as the tester in reciprocal recurrent selection experiment.

Selection population	Year	No. of trials	Reps/ trial	Variance component estimates[+]			$\hat{\sigma}^2_A$[‡]	$\hat{h}^2$	Genetic CV, %	Average yield, q/ha
				$\hat{\sigma}^2$	$\hat{\sigma}^2_{ge}$	$\hat{\sigma}^2_g$				
BSCB1(R)C0	1950	1	3	14.9±1.6	---	20.8±3.7	545.8±97.1	80.7	10.8	42.0
BSCB1(R)C1	1953	2	3	34.1±2.8	2.3±2.3	16.9±3.8	443.4±99.7	71.2	7.7	53.4
BSCB1(R)C2	1956-57	4	3	43.1±2.2	4.6±1.7	2.8±1.1	73.5±28.9	37.1	3.2	51.8
BSCB1(R)C3	1960	2	3	26.0±1.9	7.5±2.4	6.8±2.4	178.4±63.0	45.7	3.5	74.1
BSCB1(R)C4	1964	4	2	23.5±1.8	5.8±1.8	5.7±1.5	149.6±39.4	56.7	3.6	66.5
BSCB1(R)C5	1970	4	2	87.6±5.6	12.1±6.8	11.0±3.8	288.6±99.7	44.0	4.9	67.5
BSCB1(R)C6	1973	3	2	81.3±5.2	11.6±6.4	32.9±8.8	863.3±230.9	65.3	8.0	72.0
BSCB1(R)C9[§]	1982	4	2	104.0±5.7	8.0±5.6	17.2±3.8	357.4±79.0	53.4	4.7	88.2
BSCB1(R)C10[§]	1985	4	2	45.4±3.1	5.8±2.7	5.7±1.8	118.4±37.4	45.0	3.3	73.5
Mean		2.6[¶]	2.3[¶]	59.1±3.9	8.2±4.4	13.7±3.7	349.1±93.6	53.8	5.4	68.5

[+]Variance component estimates in quintals per hectare. [‡]Estimates of σ^2_A from half-sib and full-sib selection in grams per plant. [§]Estimates for reciprocal full-sib selection with BSSS (see Table 5.7).
[¶]Values are harmonic means.

Table 5.9. Summary of estimates of variance components for BS12(HI) for seven cycles of half-sib recurrent selection by use of the inbred-line tester, B14.

Selection population	Year	No. of trials	Reps/ trial	Variance component estimates[+]				$\hat{h}^2$	Genetic CV, %	Average yield, q/ha
				$\hat{\sigma}^2$	$\hat{\sigma}^2_{ge}$	$\hat{\sigma}^2_g$	$\hat{\sigma}^2_A$[#]			
BS12(HI)C0	1950	1	3	4.8±0.6	---	18.8±3.2	493.3±84.7	92.2	11.6	37.2
BS12(HI)C1	1954	2	3	26.8±2.0	16.4±3.7	15.9±4.5	417.2±118.1	55.7	7.8	51.2
BS12(HI)C2	1959	4	3	36.8±1.8	8.8±1.9	4.4±1.5	115.4±39.4	45.3	2.8	74.8
BS12(HI)C3	1963	4	2	28.6±2.2	10.1±2.4	16.0±3.3	419.8±86.6	72.4	4.5	89.1
BS12(HI)C4	1967	3	2	46.2±4.2	1.2±3.3	3.8±1.9	99.7±49.8	31.7	2.4	82.4
BS12(HI)C5	1972	4	2	85.5±6.7	24.4±6.4	19.7±5.3	516.9±139.1	54.0	5.0	88.6
BS12(HI)C6	1975	3	2	211.9±17.3	11.7±13.4	38.0±11.4	997.1±299.1	49.2	9.7	63.7
BS12(HI)C7	1979	3	2	141.7±12.9	30.6±13.6	121.9±8.6	574.6±225.7	39.0	6.1	76.3
Average		2.4§	2.3§	72.8±6.0	14.7±6.4	17.3±5.0	454.2±130.3	54.9	6.2	70.4

[+]Variance component estimates from the selection experiments in quintals per hectare.

[#]Estimates of σ^2_A calculated from half-sib selection, converted to grams per plant.

§Values are harmonic means.

Table 5.10. Summary of estimates of variance components for yield for BSK(HI) for eight cycles of half-sib recurrent selection by use of a tester.

Selection population	Year	No. of trials	Reps/ trial	Variance component estimates[†]			$\hat{\sigma}^2_A$[‡]	$\hat{h}^2$	Genetic CV, %	Average yield, q/ha
				$\hat{\sigma}^2$	$\hat{\sigma}^2_{ge}$	$\hat{\sigma}^2_g$				
BSK(HI)C0	1954	2	3	25.0±2.8	5.1±2.1	6.4±2.1	167.9±55.1	48.8	4.0	63.1
BSK(HI)C1	1958-59	4	2	23.1±1.7	6.1±1.6	13.0±2.4	341.1±63.0	74.7	5.2	69.4
BSK(HI)C2	1962	4	2	26.6±2.1	12.6±2.5	3.0±1.5	78.7±39.4	31.6	2.4	71.6
BSK(HI)C3	1965	4	2	32.9±2.6	3.8±2.2	3.6±1.4	94.5±36.7	41.5	3.4	55.7
BSK(HI)C4	1968	4	2	18.5±1.4	6.3±1.5	5.6±1.4	146.9±36.7	58.9	3.4	69.2
BSK(HI)C5	1971	4	2	76.4±6.0	6.0±4.7	9.9±3.1	259.8±81.3	57.2	5.0	62.4
BSK(HI)C6	1974	3	2	115.4±10.4	-2.6±7.6	17.7±5.4	464.5±141.7	49.1	7.7	54.9
BSK(HI)C7	1977	3	2	109.0±12.1	30.6±12.2	17.3±8.4	454.0±220.4	28.9	8.5	48.8
Average		3.3[§]	2.1[§]	40.0±4.9	8.5±4.3	9.6±3.2	250.9±84.3	48.8	5.0	61.9

[†]Estimates of components of variance from selection trials in quintals per hectare.

[‡]Estimates of σ^2_A calculated from half-sib progenies and expressed as grams per plant.

[§]Values are harmonic means.

Table 5.11. Summary of estimates of variance components for yield for BSK(S) for eight cycles of S_1 progeny recurrent selection.

Selection population	Year	No. of trials	Reps/ trial	Variance component estimates[+]			$\hat{\sigma}^2_A$[‡]	$\hat{h}^2$	Genetic CV, %	Average yield, q/ha
				$\hat{\sigma}^2$	$\hat{\sigma}^2_{ge}$	$\hat{\sigma}^2_g$				
BSK(S)C0	1954	1	2	13.9±2.2	----	35.4±6.1	232.2±40.0	83.6	16.2	36.6
BSK(S)C1	1958-59	4	2	19.3±1.4	8.6±1.6	38.9±5.9	255.2±38.7	89.5	15.7	39.8
BSK(S)C2	1962	3	2	24.5±2.2	10.7±2.7	31.7±5.9	208.0±38.7	80.6	10.8	52.1
BSK(S)C3	1965	3	2	22.2±2.1	13.2±2.8	39.5±7.1	259.1±46.6	83.0	15.4	40.7
BSK(S)C4	1968	4	2	17.9±1.4	10.6±1.8	18.4±3.3	120.7±21.6	79.0	8.8	48.8
BSK(S)C5	1971	4	2	44.8±3.5	13.0±3.4	32.1±5.8	210.6±38.0	78.4	13.4	42.1
BSK(S)C6	1974	3	2	65.9±6.0	30.1±7.0	59.3±11.5	389.0±75.4	73.8	20.6	37.4
BSK(S)C7	1977	3	2	21.5±2.2	6.6±2.7	38.9±6.8	255.2±44.6	81.7	39.7	15.7
Average		2.6[s]	2	28.8±2.6	13.2±3.1	30.5±6.6	241.2±43.0	81.2	17.6	39.2

[+]Estimates of components of variance from selection trials in quintals per hectare.

[‡]Estimates of σ^2_A calculated from S_2 progenies and expressed as grams per plant.

[s]Value is harmonic mean.

Table 5.12. Summary of the components of variance estimates, heritabilities, and genetic components of variation for the selection experiments (g/plant).

Selection population	No. cycles of selection	Variance component estimates			h^2, %	Genetic CV, %
		σ^2	σ^2_{AE}	σ^2_A		
BS13[†]	11	641.4±37.0	156.5±20.1	214.5±64.0	68.1	10.7
BSSS(R)	11	1516.7±99.7	204.7±107.6	359.3±92.4	52.8	5.4
BSCB1(R)	11	1550.8±102.3	215.2±115.4	349.1±93.6	53.8	5.4
BS12(HI)	8	1910.3±157.4	385.7±167.9	454.2±130.3	54.9	6.2
BSK(HI)	8	1048.9±128.2	223.0±112.8	250.9±84.3	48.8	5.0
BSK(S)	8	755.7±68.2	104.4±24.5	241.2±43.0	81.2	17.6
BS2(S)	5	227.2±20.5	99.2±25.8	321.6±54.2	80.2	39.5
BSTL(S)	5	209.8±18.8	68.6±20.5	253.9±42.8	78.6	29.0
BS16(S)	4	157.3±12.7	78.7±15.7	223.3±34.5	80.2	29.9
BS10 X BS11(FR)	9	1246.8±87.8	132.4±81.3	444.5±82.6	61.3	9.1
Average	8.0	926.5±73.3	166.8±69.2	311.2±72.2	66.0	14.9

[†]BS13 includes average of estimates for BSSS13 and BS13 (see Table 5.6).

great as that of σ^2_A. This comparison therefore indicates that estimates of σ^2_A obtained in one environment would have on the average a 50% upward bias from genotype-environment interaction. In some instances [e.g., see BSSS13 and BSSS(R) in Table 5.12] the bias from genotype-environment interaction would be greater than 50%; $\hat{\sigma}^2_{AE}$ is 84.2% and 61.8% as large as $\hat{\sigma}^2_A$ for BSSS13 and BSSS(R), respectively. The magnitude of $\hat{\sigma}^2_{AE}$ relative to $\hat{\sigma}^2_A$ in Table 5.12 shows that estimates of σ^2_{AE} are greater for half-sib progenies than for selfed progenies. The five half-sib selection programs (although BSSS13 includes two cycles of S_2 progeny selection) have an average estimate of σ^2_{AE} that is 66.6% as large as the estimate of σ^2_A; whereas the four populations undergoing selfed progeny selection (either S_1 or S_2) have an average estimate of σ^2_{AE} that is only 32.8% as large as the average estimate of σ^2_A. A more direct comparison of estimates of σ^2_{AE} relative to σ^2_A can be made for BSK, which is undergoing half-sib and selfed progeny recurrent selection. Although the two selection programs were grown in separate experiments, progenies have been tested in the same years at the same locations each year. Average estimates of σ^2_A are very similar [250.9 for BSK(HI) vs. 241.2 for BSK(S)], but average estimates of σ^2_{AE} are smaller for BSK(S) [223.0 for BSK(HI) vs. 104.4 for BSK(S)]. Although most average estimates of σ^2_{AE} are within one standard error of each other, estimates of σ^2_{AE} tend to be smaller for the selfed progeny selection studies. The evidence shows, however, that genotype-environment interaction is an important factor in estimation of genetic variability. Estimates of σ^2_g obtained in cycles 0 and 1 probably were biased upward.

   The evidence is not as clear for reduction of estimates of σ_g^2 in Tables
5.6 to 5.10 for the middle cycles of selection. Each population showed a reduction in the estimates of σ_g^2 for cycles 2 to 4 and they were all obtained
from data collected during 1954-63. Moisture stress in evaluation trials was
common during this time, which suggests that the stress environment caused a
reduction in genetic variability expressed among the half-sib progenies tested
but had a small effect on the selfed progenies [BSK(S) in Table 5.11]. Evidence from other studies evaluating types of environments for maximizing expression of genetic variability is not extensive (Stevenson 1965; Arboleda and
Compton 1974). Data from selection trials, however, suggest that genetic variability was reduced in stress environments. Effects of environment seem more
plausible than a reduction in variability because of effective selection. In
most instances only two or three cycles of selection were completed, and
changes in gene frequency were probably not great enough to significantly
change the estimates of genetic variance. For BSSS13, estimates of σ_g^2 obtained in cycles 3 (-1.6 ± 3.6) and 4 (3.4 ± 1.3) in 1955 and 1958-59, respectively, are significantly smaller than those obtained for cycles 2 and 5 in
1952 and 1962, respectively (Table 5.6). Information in the selection experiments supports the hypothesis that genetic variability is compressed in stress
environments. It seems that reduction of genetic variability in the middle
cycles of selection was a function of the environments sampled rather than a
reduction from changes in gene frequency, particularly for a complex trait
such as yield.

   A dramatic increase in estimates of σ_g^2 in later cycles of selection can
be accounted for by the harvesting methods used to collect yield data. Before
1970 all yield test plots were hand harvested, and all ears were included;
since 1970, mechanical harvesters have been used to determine grain yields,
and, consequently, many ears were not included in yield measurements. Mechanical harvesting therefore measures only harvestable yield and does not measure
total yield potential of genotypes. The change in harvesting methods for
evaluation of progenies in selection studies has opposing views: (1) Mechanical harvesting will not permit the total yield of genotypes unless ears are
gleaned from plots by hand. (2) If populations undergoing selection are for
use in applied breeding programs, grain yield from standing plants is the important criterion. The increased genetic variability for yield from mechanically harvested cycles of selection probably was caused indirectly by the influence of ear shank attachment, root lodging, and stalk lodging, none of
which would have influenced relative yields among progenies by use of hand
harvesting procedures.
   Genetic coefficients of variation $(\sigma_g/\overline{x}) \times 100$ were calculated for each
cycle to express the genetic standard deviation as a fraction of the cycle
mean. Although the cycle means in Tables 5.6 to 5.11 are confounded by environmental effects, they exhibit an increasing trend from the original (C0) to
the last cycle of selection in all instances. Although comparisons are not
valid because of year effects, yields of BSSS13 increased from 35.1 q/ha for
cycle C0 to 67.4 q/ha in cycle C6 (Table 5.6). Seven cycles of half-sib selection nearly doubled yields of half-sib progenies; interestingly, when progeny evaluation was changed from half-sib to S_2, average yields of the C0 and
C1 cycles of S_2 selection were approximately the same as half-sib yield in
1940. Although it is assumed that numerically larger quantities tend to vary
more than numerically smaller quantities, the genetic coefficients of variation in Tables 5.6 to 5.11 do not show a trend with greater yields of later
cycles of selection. It does not seem that a trend is evident in the size of

the genetic coefficients of variation associated with average yields of successive cycles of selection; they are greater, however, when selfed progeny evaluation is used (Table 5.6, last four cycles, and Table 5.11). Genetic coefficients of variation follow the same trend as estimates of σ_g^2 rather than average yields of different cycles.

Although there are instances that suggest some reduction in genetic variance after the first cycles of selection, it seems that the genetic variance estimates in populations undergoing recurrent selection fluctuate among selection cycles (estimates in different environments for each cycle) without evidence of a consistent reducing trend. Consider, for example, estimates of genetic components of variance σ_A^2 and genetic coefficients of variation in Tables 5.7 and 5.8 for BSSS(R) and BSCB1(R) populations that are undergoing reciprocal recurrent selection. Ten cycles have been completed and estimates of σ_g^2 are similar in cycles C0 and C10 for both populations [BSSS(R): 13.3 ± 3.4 (C0) vs. 14.1 ± 3.6 (C10); and BSCB1(R): 20.8 ± 3.7 (C0) vs. 13.7 ± 3.7 (C10)]. Also, genetic coefficients of variation did not change greatly from C0 to C10, although average yields of C10 testcrosses of BSSS(R) and BSCB1(R) are 59% and 75% greater than C0 populations. Data show that selection programs are satisfying the two criteria of successful application of cyclical selection programs for maize improvement: (1) maintenance of genetic variability for future selection and (2) improvement of overall performance of population crosses in successive cycles of selection.

Estimates of σ_A^2 are given in Tables 5.6 to 5.11 for each cycle of selection, with adjustments made for half-sib and selfed progenies used in selection programs. Variation of estimates of σ_A^2 among cycles of selection follows the same trend as estimates of components of genetic variance σ_g^2. Because there is no evidence that additive genetic variance has changed among cycles of selection, estimates of σ_A^2 averaged for each population are summarized in Table 5.12. Additionally, estimates for three populations that have had only two and three cycles of selection are included for comparison. BSSS13 and BSSS(R) involve the same population but two different recurrent selection procedures, and estimates of σ_A^2 in Table 5.12 are very similar to those in Table 5.4 for BSSS. Estimates of σ_A^2 obtained from selection experiments (Table 5.12) are within one standard error of the average of estimates obtained from mating designs in Table 5.4. Hence it seems that valid estimates of σ_A^2 can be obtained from long-term selection experiments, provided sufficient sampling, testing, and cycles of selection are available. Experimental errors are inherently large for variance component estimates, and estimates obtained from a particular cycle of selection may deviate considerably from the average [see cycle C3 for BSSS13 in Table 5.6 and cycle C4 for BSSS(R) in Table 5.7]. Sampling errors and bias from genotype-environment interaction could cause erroneous conclusions from only one cycle. Only when sufficient cycles of selection, sampling, and testing have been completed will valid estimates be obtained. Usually, only 100 progenies were tested in each cycle of selection for BS13 and BSSS(R), and errors of estimation were about twice those obtained from mating designs that included 231 to 480 progenies for testing. Estimates of heritability, on a progeny mean basis, from the selection studies for BSSS13 (40.7%) and BSSS(R) (52.8%) also were similar to average heritability of the four estimates included in Table 5.4 (41.4%).

The BSSS [BS13 and BSSS(R)] and BSK [BSK(HI) and BSK(S)] populations seem to have less genetic variability than other populations (Table 5.2). For the two selection studies involving BSSS and BSK, average estimates of σ_A^2 are 286.9 and 246.0, respectively; whereas the average estimate of σ_A^2 for the other six populations undergoing selection is 341.1. The two estimates for BSSS, however, are in good agreement with the average of all estimates of σ_A^2 for synthetic populations in Table 5.2. The two estimates of BSK [250.9 for BSK(HI) and 241.2 for BSK(S)] also are in excellent agreement with that (σ_A^2 = 202) reported by Wright et al. (1971) by use of weighted least squares analysis of progenies developed from the diallel and triallel mating designs. Both BSSS and BSK are synthetic varieties developed by recombination of 16 and 8 lines, respectively. The 16 lines used to form BSSS were characterized as having strong stalks and thus originated from several sources. The 8 lines used to form BSK, however, originated from the Krug open-pollinated variety and were selected on the basis of topcross yields with the parental variety Krug used as the tester (Lonnquist 1949). Genetic variability in BSSS and BSK may have been limited by either the restricted sample of lines included in their synthesis or a greater gene frequency. BS12 is an open-pollinated variety of unknown origin that has about twice as much genetic variability, based on average estimates from selection studies (Table 5.12). BS16, BSTL, and BS2 have had fewer cycles of selection completed, and estimates of σ_A^2 were similar to those for BSSS and BSK. BSTL was developed from a variety cross, Lancaster X Tuxpeno (a Mexican variety), and backcrossed to Lancaster; its estimate of σ_A^2 was similar to that for BSSS. BS16 has the smallest estimate of σ_A^2. BS16 was developed by six cycles of mass selection for early flowering in ETO Composite, obtained from Colombia, South America (Hallauer and Sears, 1972). Limited evidence from BS16 suggests that selection within exotic germplasm reduced genetic variability available to the breeder.

Comparison of estimates of σ_A^2 from selection experiments with those obtained from use of mating designs shows that they are equally valid. If estimates of σ_A^2 from selection experiments were included in Table 5.2 and Fig. 5.1, they would not deviate from estimates obtained from use of mating designs. Figure 5.1 shows that all types of populations are expected to have sufficient additive genetic variance to show response from selection--some more than others. Estimates from selection experiments show variation among cycles of selection, but the range of estimates is not any greater than those obtained from individual studies by use of mating designs. Sampling and genotype-environment interaction probably account for variation of estimates among cycles of selection; these factors, however, are just as important if insufficient sampling and testing are used to obtain estimates from mating designs.

Summaries of estimates given in Tables 5.1, 5.2, and 5.12 were obtained from data collected during the same period, i.e., selection experiments conducted cyclically since the 1940s and use of mating designs to estimate genetic parameters that were stimulated by the publication of Comstock and Robinson in 1948. Summaries of both methods show similar results. Use of mating designs will provide estimates of genetic parameters more quickly if sufficient sampling and testing are used. Silva and Hallauer (1975) obtained the estimate of σ_A^2 = 166 ± 24 for yield in BSSS in four years (experiment 3 in Table 5.4); on the other hand, the estimate for yield of σ_A^2 = 214.5 ± 64.0 was obtained from 11 cycles of recurrent selection in BSSS, which required 27 years. Although the estimates are similar, they are not exactly comparable because Silva and Hallauer used intrapopulation progenies, whereas the BSCB1 population was used as tester for the BSSS population in the reciprocal recurrent selection program. About 800 progenies were evaluated in both instances; 800 full-sib progenies were tested that were developed from designs I and II mating designs and about 800 half-sib progenies (8 cycles × 100 half-sib progenies per cycle) were tested from the selection experiment. Estimates of σ_A^2 from the selection experiment were obtained, however, in 23 environments vs. only 6 environments for the mating designs. If we use standard errors of estimates of σ_{AE}^2 and σ_A^2 to determine significance, estimates of σ_{AE}^2 and σ_A^2 are not significantly different by the two methods of estimation.

Estimates of genetic parameters in maize can be obtained more quickly from use of mating designs than from selection experiments. If the experimenter wants to determine the genetic variability of a population to predict future progress from selection, the use of adequate mating designs becomes necessary. If the experimenter is patient and is willing to collect data from several cycles of selection before predicting progress from selection it is not necessary to develop progenies from use of mating designs and conduct special studies to obtain this information, which can be quite costly if adequate sampling and testing are included. To answer specific questions about a particular population for its potential use in a breeding program, it also is necessary to use some type of mating design.

5.4 EPISTASIS

Estimates of additive genetic variance and variance due to dominance deviations for maize populations summarized in Table 5.1 were obtained under the assumption of no epistasis. In most instances only one or two equations were available for estimation of genetic components of variance. Estimation, therefore, frequently was limited by the mating design used: (1) if only one source of variation was available (e.g., variability among full-sib families), all nonadditive sources (both dominance and epistasis) of variation were assumed absent; (2) if two sources of variation were available (e.g., variability among half-sibs and full-sibs within half-sibs), epistasis was assumed absent. The assumptions were necessary because of limitations of mating designs used to develop progenies for test. The simpler the mating design, the greater the restrictions needed for estimation. Fortunately, in maize it does not seem that restrictions imposed for the assumption of no epistasis seriously biased estimates of additive genetic and dominance variance components.

Development of more complex mating designs permitted estimation of additional components of genetic variance. In most instances the more complex mating designs permitted estimation of all types of digenic epistasis and in a few instances trigenic epistasis, e.g., additive by additive by additive epistasis. The primary objective of the more complex mating designs was to develop additional covariances of relatives to permit estimation of additional components of genetic variation. One of the first suggestions for estimation of epistatic variances was by Cockerham (1956); designs I and II mating designs were used with parents at two different levels of inbreeding, but the progenies evaluated were noninbred in both instances. The procedure suggested by Cockerham has been used by Eberhart (1961) and Silva (1974). Rawlings and Cockerham (1962a, 1962b) developed the triallel and quadrallel analyses that provide up to nine covariances of relatives; these analyses permitted F-tests for the presence of epistasis in the analyses of variance and estimation of epistatic components of variance. Wright (1966) used diallel and triallel analyses, which provided nine mean squares, for estimation of epistasis in Krug Hi I Synthetic 3. Chi (1965) used a complex mating design suggested by Kempthorne (1957, pp. 425-26) that included 11 variances and 55 covariances among relatives to estimate epistasis in an open-pollinated variety Reid Yellow Dent.

Estimation of epistatic components of variance, however, has not been generally satisfying. Most of the studies included adequate sampling and testing, but the results of estimation have been disappointing. It seems that epistasis for a complex trait, such as yield, must exist. Additive types of epistasis would be useful to the breeder in selection programs, but realistic estimates of additive by additive epistasis have not been obtainable. Hence either the genetic models used are inadequate or epistatic variance is small relative to total genetic variance of maize populations; both are probably involved. As shown in Chap. 2, it seems that a major problem is the inherent correlation of coefficients of epistatic components of variance with those of additive genetic and dominance variance components (see Chap. 4).

Estimates of epistatic variance for five maize populations from four independent studies are included in Table 5.13. Three populations (Jarvis, Indian Chief, and Reid Yellow Dent) are open-pollinated varieties, whereas the other two [BSK (a strain of Krug Yellow Dent) and BSSS] are synthetic varieties. Each of the studies used different methods, but they all arrived at the same conclusion: Realistic estimates of epistatic components of variance were not obtainable and they all reverted to simpler genetic models for estimation of additive genetic and dominance components of genetic variance. In some instances negative estimates of epistatic variance components were two times greater than their standard errors.

Examples are estimates of σ^2_{Ig} and σ^2_{Id} for Jarvis (Eberhart et al. 1966), σ^2_{DD} for BSK (Wright et al. 1971), and σ^2_{AD} and σ^2_{DD} for BSSS (Silva 1974). Eighteen estimates of epistatic variance components are included in Table 5.13 and 11 of the estimates are negative. None of the positive estimates is greater than two times its standard error and most are within one standard error. Eberhart et al. (1966) concluded, "Additive variance appeared to account for the largest proportion of the total genetic variance for all characters in both varieties." Chi et al. (1969), from the complex mating design used in Reid Yellow Dent, concluded, "The results indicated that epistatic variances were negligible in relation to the additive and dominance variance components for the seven characters studied. The high correlations among the coefficients of the genetic parameters inevitably reduced the sensitivity for detecting epistasis." Wright et al. (1971) used the maximum likelihood estimation method on the mean squares from diallel and triallel analyses to fit the error and a six-parameter genetic model; and they concluded, "It was not possible to obtain realistic estimates of the epistatic components of variance, although significant effects were detected in the analysis of variance. For the two-parameter genetic model, the largest proportion of the total genetic variance was additive for all traits." Silva and Hallauer (1975), from an extensive study of BSSS, concluded, "Epistatic variance was not an important component of the genetic variance for yield in BSSS. Additive genetic variance accounted for 93.2% of the total variability. Including the variance due to dominance deviations in the models accounted for more than 99% of the variation, with no improvement when additive by additive epistatic variance was included. Estimates of epistatic variances obtained from models that included more than one digenic epistatic component usually were negative and unrealistic with much greater standard errors. The use of a complete model (as many epistatic terms as permitted by the number of independent equations) made the X-matrix nearly singular."

Table 5.13. Estimates of epistatic components of variance for yield by use of different methods of estimation for five maize populations.

Population	Epistatic variance components			$\hat\sigma^2_A$	$\hat\sigma^2_D$
	$\hat\sigma^2_{Ig}$	$\hat\sigma^2_{Ia}$	$\hat\sigma^2_{Id}$		
Jarvis (Eberhart et al. 1966)[+]	-300±136	128±119	-428±187	640±113	407±156
Indian Chief (Eberhart et al. 1966)[+]	175±132	78±107	97±183	313±80	247±138
	$\hat\sigma^2_{AA}$	$\hat\sigma^2_{AD}$	Deviations		
Reid Yellow Dent (Chi et al. 1969)[‡]	8.12	5.40	7.04	646.70	37.73
	$\hat\sigma^2_{AA}$	$\hat\sigma^2_{AD}$	$\hat\sigma^2_{DD}$	$\hat\sigma^2_{AAA}$		
BSK (Wright et al. 1971)[§]						
Unweighted	-40±225	182±236	-225±96	-164±257	221±54	74±108
Weighted	132±286	291±270	-102±120	-100±335	265±40	160±26
BSSS (Silva 1974)[¶]	-94±165	-305±122	----	----	271±73	544±94
	-94±165	----	-204±81	----	271±73	424±71

[+] Estimates obtained by mathematical arrangements of genetic expectations of the components of variance, where σ^2_{Ig} includes all epistatic components of variance, σ^2_{Ia} includes only additive types of epistatic variance, and σ^2_{Id} includes all types of epistatic variance except additive by additive.

[‡] Analysis of variance estimates of genetic components of variance averaged over two years $(\times 10^{-2})$.

[§] Estimates obtained by ordinary least squares and weighted least squares methods of estimation.

[¶] Estimates obtained by the maximum likelihood method of estimation.

    After attempts to estimate epistatic components of variance in each of
the populations were found to be fruitless, each study resorted to the two-parameter genetic model for estimation of additive genetic and dominance components of variance. Estimates and their interactions with environments for the
simpler models are shown in Table 5.14. In most instances estimates of genetic parameters in Table 5.14 are smaller and have smaller standard errors than
those obtained from genetic models that included epistasis (Table 5.13). All
estimates of additive genetic variance and their interactions with environments exceed twice their standard errors. In most instances estimates of dominance also exceed twice their standard errors. If we consider estimates that
exceed twice their standard errors as being significantly different from zero,
the experiments seemed to have used adequate sampling and testing procedures
for estimation of σ_A^2 and σ_D^2 genetic parameters. Models and estimation procedures apparently were not adequate. Estimates of σ_A^2 in Table 5.14 are within sampling error of those given in Tables 5.1 and 5.2.

    Estimation of epistatic components of variance are briefly indicated for
the combined analysis reported by Silva and Hallauer (1975). Designs I and II
mating designs were imposed on BSSS to develop 800 full-sib progenies (480 design I and 320 design II full-sib progenies) that were evaluated in six environments. From the genetic expectations of mean squares for the designs I and
II combined analysis of variance, the following estimates were calculated.

Design I (F = 0)	Design II (F = 1)

$$\hat{\sigma}_m^2 = \widehat{Cov}\ HS = 34.8 \pm 11.4$$

$$4\hat{\sigma}_m^2 = \hat{\sigma}_A^2 = 139.3 \pm 45.5$$

$$\hat{\sigma}_{f/m}^2 = \widehat{Cov}\ FS - \widehat{Cov}\ HS$$
$$= 101.9 \pm 10.0$$

$$4(\hat{\sigma}_{f/m}^2 - \hat{\sigma}_m^2) = \hat{\sigma}_D^2 = 268.2 \pm 60.5$$

$$\hat{\sigma}_w^2 = \hat{\sigma}_{we}^2 + (\hat{\sigma}_G^2 - \widehat{Cov}\ FS)$$
$$= 1654.1 \pm 23.3$$

$$\hat{\sigma}_p^2 = 354.3 \pm 9.5$$

$$\hat{\sigma}_m^2 = \widehat{Cov}\ HS = 37.1 \pm 18.0$$

$$2\hat{\sigma}_m^2 = \hat{\sigma}_A^2 = 74.2 \pm 35.9$$

$$\hat{\sigma}_f^2 = \widehat{Cov}\ HS = 84.5 \pm 26.3$$

$$2\hat{\sigma}_f^2 = \hat{\sigma}_A^2 = 169.0 \pm 52.6$$

$$\hat{\sigma}_{fm}^2 = \widehat{Cov}\ FS - \widehat{Cov}\ HS_f - \widehat{Cov}\ HS_m$$
$$= 167.2 \pm 21.8$$

$$\hat{\sigma}_w^2 = \hat{\sigma}_{we}^2 = 1280.6 \pm 29.6$$

$$\hat{\sigma}_p^2 = 332.3 \pm 11.0$$

Additional calculations include estimates of σ_{wg}^2 from design I and σ_{HS}^2 from
design II. Because $\hat{\sigma}_w^2$ in design II full-sib progenies includes only the measurement and plant-to-plant environmental errors, an estimate of the within-
plot genetic variance can be obtained as 1654.1 - 1280.6, which becomes 343.4
$\pm$ 32.1, the estimate of σ_{wg}^2 or $\hat{\sigma}_G^2 - \widehat{Cov}\ FS$. The value of $\hat{\sigma}_{HS}^2$ was calculated
from the design II analysis by pooling degrees of freedom and sums of squares
for male and female sources of variation; $\hat{\sigma}_{HS}^2$ was obtained from expected mean
squares as 60.8 $\pm$ 16.2, which is the average of estimates of σ_m^2 and σ_f^2.
Hence, an estimate of σ_A^2 from design II is 121.6 $\pm$ 32.4.

    If we assume for the present no epistasis, the two independent estimates
of σ_A^2 are 139.3 $\pm$ 45.5 (design I) and 121.6 $\pm$ 32.4 (design II) and the esti-

Table 5.14. Estimates of additive genetic and dominance variances for yield with the assumption of no epistasis for five maize populations.

Populations		$\hat{\sigma}^2_A$	$\hat{\sigma}^2_{AE}$	$\hat{\sigma}^2_D$	$\hat{\sigma}^2_{DE}$	$\hat{\sigma}^2$
			Variance component estimates			
Jarvis, original (Eberhart et al. 1966)		247±105	234±95	574±183	202±193	1045±35
Jarvis, reconstituted (Eberhart et al. 1966)		374±58	214±45	127±33	150±43	1018±33
Indian Chief, original (Eberhart et al. 1966)		181±97	134±105	43±179	407±138	1108±41
Indian Chief, reconstituted (Eberhart et al. 1966)		259±45	189±39	140±31	25±41	1199±37
Reid Yellow Dent (Chi et al. 1969)		239±41	----	329±102	----	339±8
BSK, unweighted (Wright et al. 1971)		181±16	102±27	85±47	-29±90	538±78
BSK, weighted (Wright et al. 1971)		202±88	105±38	67±35	56±37	393±21
BSSS (Silva and Hallauer 1975)	ML[+]	169±24	92±10	193±21	75±12	185±7
	LS	138±34	149±63	73±147	-6±303	355±324
	WLS	150±34	91±16	189±30	76±17	185±11

[+]ML, LS, and WLS refer to the maximum likelihood, least squares, and weighted least squares methods of estimation, respectively.

mates of σ_D^2 are 268.2 ± 60.5 (design I) and 167.2 ± 21.8 (design II). The estimates of σ_A^2 and σ_D^2 are within two standard errors of each other but the estimates from design I have greater standard errors, particularly the estimate of σ_D^2. Before pooling male and female half-sibs in design II, the standard errors of the estimates of σ_A^2 were similar for both mating designs. Because the coefficient of inbreeding is different for the parents of design I (F = 0) and design II (F = 1), the covariances of half-sibs have different genetic expectations. If we include additive by additive epistasis σ_{AA}^2 in the genetic expectations, Cov HS for design I is $(1/4)\sigma_A^2 + (1/16)\sigma_{AA}^2$ and for design II is $(1/2)\sigma_A^2 + (1/4)\sigma_{AA}^2$. Coefficients of genetic expectations of the covariances of half-sibs are different; hence we can use expectations and observed values of the two covariances of half-sibs to estimate σ_A^2 and σ_{AA}^2: $\sigma_{HS}^2 = (1/2)\sigma_A^2 + (1/4)\sigma_{AA}^2 = 60.8$ (design II) and $\sigma_m^2 = (1/4)\sigma_A^2 + (1/16)\sigma_{AA}^2 = 34.8$ (design I). By solving the two equations, we find the estimate of σ_A^2 is 156.8 and the estimate of σ_{AA}^2 is -70.8. The negative estimate of σ_{AA}^2, however, is not significant because the large standard error (223.54) indicates it is not different from zero. This estimate of σ_{AA}^2 was obtained by use of two estimates of Cov HS from different mating designs and shows no evidence of epistatic variance. Also, no bias from dominance types of epistasis would be included.

Another estimate of σ_{AA}^2 can be calculated from the three sources of genetic variation in the design I analysis. Because within-plot variation was available from both mating designs, an estimate of the within-plot genetic variation can be obtained; $\hat{\sigma}_{wg}^2$ is 373.5 ± 32.1, which is the total genetic variance minus Cov FS. If we include σ_{AA}^2 in the expectations for the design I analysis, we have:

$$\hat{\sigma}_m^2 = (1/4)\hat{\sigma}_A^2 + (1/16)\hat{\sigma}_{AA}^2 = 34.8$$

$$\hat{\sigma}_{f/m}^2 = (1/4)\hat{\sigma}_A^2 + (1/4)\hat{\sigma}_D^2 + (3/16)\hat{\sigma}_{AA}^2 = 101.9$$

$$\hat{\sigma}_{wg}^2 = (1/2)\hat{\sigma}_A^2 + (3/4)\hat{\sigma}_D^2 + (3/4)\hat{\sigma}_{AA}^2 = 373.5$$

The solution of these three equations gives the following estimates of the three components of genetic variance: $\hat{\sigma}_A^2 = 36.6$, $\hat{\sigma}_D^2 = 63.2$, and $\hat{\sigma}_{AA}^2 = 410.4$. The value of $\hat{\sigma}_{AA}^2$ is very large and $\hat{\sigma}_A^2$ and $\hat{\sigma}_D^2$ are smaller than any of the previous estimates. By use of the same three equations, σ_{AA}^2 was deleted and σ_{DD}^2 included in the genetic expectations; the estimates were determined to be $\hat{\sigma}_A^2 = 139.2$, $\hat{\sigma}_D^2 = 234.2$, and $\hat{\sigma}_{DD}^2 = 136.8$. The inclusion of dominance by dominance epistasis increased the estimates of σ_A^2 and σ_D^2 to the levels at which epistasis was assumed absent. Additive by additive epistasis seemed to have a greater effect on σ_A^2 and σ_D^2 variance components than dominance by dominance epistasis for equations from the design I analysis. Both estimates of σ_A^2 given above are biased by epistatic variances, but it seems that the bias is

greater when σ^2_{AA} is not included in the model. The estimate of σ^2_{AA} from use
of covariances of half-sibs from designs I and II analyses, however, showed no
evidence of additive by additive epistasis.

Discussion of estimates of epistatic variance has been limited to estimates reported for yield. Estimates reported for different plant and ear
traits, however, show the same types of results.

Although quantitative estimates of epistatic variance in maize populations have not been convincing, reports have indicated that epistatic effects
are present in quantitative traits. Most of the evidence was obtained by use
of mean comparisons, which included observed and predicted performance of different types of hybrids (single, three-way, and double-cross), generations
having different levels of percent homozygosity, and comparisons of different
generations of inbreeding with theoretical levels of homozygosity. In most
instances the generations were derived by crossing two inbred lines. Genetic
models proposed by Anderson and Kempthorne (1954), Cockerham (1954), Hayman
and Mather (1955), and Hayman (1958, 1960) permit estimation of additive, dominance, and epistatic gene effects that are based on the factorial model used
in the design of experiments. In all instances, qualitative evidence rather
than quantitative evidence of epistasis is available.

Gamble (1962a, 1962b) obtained estimates of six genetic parameters (m, a,
d, aa, ad, and dd) from six generations (P1, P2, F1, F2, P1F1, P2F1) generated
from crosses among six inbred lines of maize. All lines had good general combining ability. Estimates of the six genetic parameters for the 15 crosses
obtained from generation means of four experiments were obtained for six
traits. Frequency of significant effects is summarized in Table 5.15 for each
trait. Dominance effects were significant in all crosses for all traits except kernel-row number. Occurrence of significant additive effects also was
high for all except yield, which was significant in 47% of the 15 crosses. It
seems that additive and dominance effects made a significant contribution to
inheritance of these traits for this particular set of crosses. Although not
as frequent as additive and dominance effects, significant epistatic effects
were frequent for all traits. Plant height had the greatest frequency of significant digenic epistatic effects among traits and yield the least. Of the
three digenic epistatic effects, a × d had the greatest occurrence and a × a
the least. Gamble concluded that all gene effects contributed to the inheritance of traits studied. For yield, estimates of dominance gene effects were
quite important and additive effects were low in magnitude and often nonsignificant. All dominance effect estimates for yield were positive, whereas
nine of the d × d estimates were negative. All except three of the a × a estimates were positive. Darrah and Hallauer (1972) and Sprague and Suwantaradon (1975) found similar results for yield and other traits. Hayman (1960),
however, pointed out that the presence of epistatic effects would bias estimates of additive and dominance effects.

Moll et al. (1963) used generation mean analysis to study inheritance of
resistance to brown spot (*Physoderma maydis*) in six crosses among four inbred
lines. Significant epistatic effects were detected in two crosses, additive
effects in three crosses, and nonsignificant dominance effects in all crosses.
Because of the genetic structure of populations studied, estimates of additive
and dominance genetic variances also were obtained. The estimates of σ^2_A exceeded twice their standard errors in four of the six crosses, but no estimates of σ^2_D exceeded twice their standard errors. Moll et al. (1963) concluded that much genetic variation in brown spot reaction is additive, but heritability seems low.

Table 5.15. Number of significant (0.05 and 0.01 probability levels) estimates of six genetic effects for 15 crosses among six inbred lines (Gamble 1962a, 1962b). Relative frequency of the total significant effects for all traits shown in last column.

| Genetic effect | Trait | | | | | | | Relative frequency, % |
	Yield	Kernel row number	Ear length	Ear diameter	Seed weight	Plant height	
Mean	15	15	15	15	15	15	100
Additive (a)	7	12	13	12	11	11	73
Dominance (d)	15	9	15	15	15	15	93
a × a	5	5	6	3	2	6	30
a × d	6	6	6	9	8	8	48
d × d	2	5	5	7	3	8	33

Hughes and Hooker (1971) reported similar results in a study of four crosses for the nature of gene action conditioning resistance to northern leaf blight (*Helminthosporium turcicum* Pass.). Significant epistatic effects were noted in all crosses by generation mean analysis, but additive genetic effects were of major importance in three of the four crosses. Assuming no epistasis, estimates of σ_A^2 were relatively large and significant whereas estimates of σ_D^2 were within the realm of zero. They concluded that northern leaf blight resistance was conditioned by a relatively low number of genes with primarily additive effects.

Hallauer and Russell (1962) also included genetic populations developed from a cross of two inbred lines that permitted estimation of genetic variances and effects for days from planting to flowering and grain moisture and seed weight at physiological maturity. Significant epistatic effects were noted in six of the nine instances with estimates of dominance effects greater than additive effects. Assuming no epistasis, estimates of σ_D^2 exceeded their standard errors but estimates of σ_A^2 were zero for seed weight and grain moisture at physiological maturity. The estimate of σ_A^2 for days to flowering was significant but the estimate of σ_D^2 was zero (Hallauer 1965); this is in contrast to the generation mean analysis that showed a positive and significant estimate of dominance effects and a small nonsignificant estimate of additive effects.

Epistatic effects can be detected by examining the relation between level of heterozygosity and performance of the quantitative trait. The basis for this relation stems from Wright's (1922) studies of hybrid vigor and inbreeding depression: If the change in performance is proportional to the change in heterozygosity, epistasis is either negligible or nondetectable and implies that the change is dependent on some level of dominance. This relation has been examined in maize by Kiesselbach (1933), Neal (1935), Stringfield (1950), Sentz et al. (1954), Robinson and Cockerham (1961), and Martin and Hallauer (1976); the first four reports generally supported the linear relation, whereas the last two reports found instances of a curvilinear (or epistasis) relation between percent heterozygosity and performance. A linear response does not mean an absence of epistatic effects, but it shows no net epistatic effects because cancellation of positive and negative epistatic effects could produce a linear relation. Although some conflicting results have been reported from use of the heterozygosity-performance relation to obtain evidence for the presence of epistasis, it seems reasonable that this should be expected because of the sample of parents included in the studies, environments used for testing, and traits measured. For example, Martin and Hallauer (1976) studied 21 crosses in each of four groups (first cycle, second cycle, good lines, and poor lines) of inbred lines. Each group included seven inbred lines and all possible F1 crosses were made among the seven inbred lines. For each of the 21 crosses within each group, F2, F3, backcross, and backcross-selfed generations were included to form four levels of heterozygosity (0%, 25%, 50%, and 100%). Epistasis was detected most frequently for ear diameter and kernel-row number and least frequently for yield for the four groups of lines, with poor lines having greatest frequency of epistasis for all traits. Only 6 instances out of 84 possible were significant (0.05% probability level) for yield for all groups, which would be expected to occur by chance. Figure 5.2 shows a nearly linear relation of yield with percent heterozygosity, whereas ear diameter has a slightly curvilinear relation. Linear sums of squares accounted for 98.2% and 92.8% of total variation for yield and ear di-

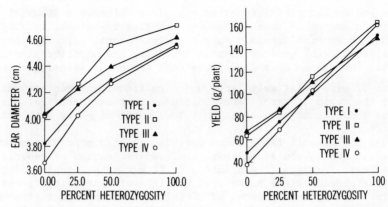

Fig. 5.2.   Relation of yield and ear diameter with level of heterozygosity
            generations produced from four types of inbred lines (Martin and
            Hallauer 1976).

ameter, respectively. Cumulative totals for occurrences of significant epistasis for the four groups were 6 for yield and 43 for ear diameter. Martin
and Hallauer's (1976) results also agreed with those of Sentz et al. (1954)
for effects of environments: Combining across environments made the response
to added heterozygosity more linear and decreased the number of crosses exhibiting significant epistatic effects.

     A curvilinear response to increased homozygosity by inbreeding also is
evidence for the presence of epistasis, where inbreeding is compared for different generations of unselected lines developed by single-seed descent (Hallauer and Sears 1973; Cornelius and Dudley 1974). Linear regressions were
highly significant in all instances, indicating additive gene action with some
level of dominance. Hallauer and Sears (1973) measured ten traits and linear
regression accounted for 92.9% (ear-row number) to over 99% (plant and ear
height and yield) of total variation. Hence the linear regression model accounted for most of the variation among generations, with less than 2% of total variation accounted for by quadratic and deviation sums of squares for six
of the ten traits. This also indicates that epistatic effects were of minor
importance.

     Sing et al. (1967) also used unselected lines developed from Jarvis and
Indian Chief open-pollinated varieties to develop hybrid progenies representing seven levels of inbreeding. Linear mean squares were significant in all
instances and the quadratic mean square in no instance. For yield, for example, linear sums of squares accounted for 95.6% and 96.0% of total variation
for Indian Chief and Jarvis, respectively. Sing et al. concluded that "if
epistatic effects are contributing to the variability in these two populations
they must be of the type which lead to a linear relationship between average
performance and inbreeding."

     Estimates of relative importance of epistatic effects have been obtained
by comparing observed and predicted means of single, three-way, and double-
cross hybrids produced usually from elite inbred lines. Unique epistatic complexes of genes may be important in single-cross performance, particularly if
elite inbred lines are used. Production of three-way and double-cross hybrids
affords an opportunity for genetic recombination in parental single crosses
used to produce the more complex hybrids. Recombination in parental single
crosses would disrupt unique gene complexes and cause differences between pre-

dicted and observed performances.   Comparisons of three-way and double-cross
performances, therefore, would provide a qualitative assessment of relative
importance of epistatic effects of parental single-cross performance.

Bauman (1959), Gorsline (1961), Sprague et al. (1962), Sprague and Thomas
(1967), Eberhart and Hallauer (1968), Stuber and Moll (1970), Otsuka et al.
(1972), and Stuber et al. (1973) have examined the relations among different
types of hybrids.   In most instances prediction methods based on single-cross
performance were used and epistasis was indicated if predicted performance
differed significantly from average performance of parental single crosses.
In nearly all instances in all studies, significant epistatic effects were detected in specific instances.   Two general conclusions are evident from all
the studies: (1) Although bias from epistasis may be substantial in some
unique combinations of lines, use of more complex procedures for predicting
three-way and double crosses is not warranted.   (2) Bias from genotype-environment interaction in prediction methods may be greater than bias from epistasis; hence adequate testing is probably more important than inclusion of
epistasis in prediction procedures.   Results of these studies indicate the
presence of epistasis in crosses among elite inbred lines, but epistasis seems
small in comparison with additive and dominance effects.

Russell and Eberhart (1970) and Russell (1970, 1976) used a method described by Fasoulas and Allard (1962) to estimate genetic effects of individual loci or short chromosome segments.   Their studies included backcross derived isogenic sublines of inbred lines B14 and Hy.   Three gene loci that permitted identification of genotypes were used in their studies.   Twenty-seven
genotypes were included in each study and total variation among genotypes
could be orthogonally partitioned into additive and dominance effects of each
locus and epistatic effects among loci, which include all digenic and trigenic
epistatic combinations.   Analysis was on genotype means, permitting qualitative estimates of epistatic effects.   Existence of epistatic effects was noted
for nine traits in each study.   In B14 for all traits except yield, additive,
dominance, and epistatic effects accounted for 52.5%, 6.2%, and 41.2%, respectively, of total variability among genotypes.   For yield per plant of B14, additive (31%) and dominance (28%) were nearly equal, with additive by additive
(16%) and dominance by dominance (14%) also prominent.   Additive effects (69%)
were greater in the expression of yield of Hy; dominance effects (23%) were of
secondary importance with all types of epistatic effects accounting for 8% of
total yield variability among the 27 genotypes.   These studies showed that
epistatic effects were present for the three sets of loci considered in each
inbred line.   B14 and Hy were elite inbred lines that were used extensively
after their release.   Obviously, yield is affected by more than three loci
(which may not be typical of the real situation), and epistasis should be considered when designing long-term breeding procedures.   Continuous improvement
of economically important traits of maize will require identifying unique complexes of genes, which undoubtedly will include favorable epistatic combinations.

Although estimates of components of epistatic variances generally have
not been successful for maize populations, mean comparisons of crosses produced among elite inbred lines invariably showed epistatic effects.   In the
first instance we are estimating variances from complex functions of covariances of relatives, often using either complex mating designs or combinations
of mating designs.   Large errors of estimates are inevitable.   Mean comparisons are first order statistics that are easier to estimate with smaller errors.   In the first instance we are working with a population that includes a
large collection of genotypes; and in the second we are comparing a few geno-

types among elite inbred lines. It may seem that we have an untenable situation, but differences probably exist because of types of genetic materials
studied and parameters estimated.

5.5 CORRELATIONS AMONG TRAITS

Usually more than one trait is measured on progenies evaluated either for
a specific trait in cyclical selection programs or in applied breeding programs that require a combination of traits to satisfy growers. Although yield
is usually the primary trait of interest, maturity, standability, grain quality, stalk quality, and resistance to pests are all corollary traits that the
modern maize breeder must consider for eventual usefulness of genotypes evaluated for yield. For instance, a high yielding genotype will have limited use
in the northern part of the U.S. Corn Belt if it is very late flowering, has
high grain moisture content at physiological maturity, and has a slow rate of
moisture loss between physiological maturity and harvest maturity. Similarly,
a high yielding genotype that has very poor stalk quality is not acceptable in
situations that use mechanical harvesting. It is only natural, therefore,
that attention is given to associations among traits during selection and
testing of genotypes.

It was established in classical genetics that many genes have manifold
effects; i.e., some genes seem to affect traits that are unrelated. Genes
that have manifold effects are pleiotropic. In pleiotropy the same gene affects different traits in a complementary way; whereas in epistasis different
genes affect the same trait. The existence of pleiotropic effects of genes in
classical genetic analysis would logically imply the existence of pleiotropic
effects for quantitative traits. Then it is possible that selection may be
exerted on secondary traits that have greater heritabilities than the primary
trait. For example, if genes with pleiotropic effects for kernel-row number
and yield are present, selection on the basis of kernel-row number rather than
yield itself may be more effective because kernel-row number has greater heritability (57.0% for kernel-row number vs. 18.7% for yield, Table 5.1). Success of selection, however, also depends on association between the traits--
kernel-row number and yield; if association is not large the effect of indirect selection for yield by use of kernel-row number will not be successful.
Indirect selection will be effective if (1) heritability of the secondary
trait is greater than that of the primary trait and (2) the genetic correlation between them is substantial.

Mode and Robinson (1959) investigated the concept of genetic correlations
for traits of maize under the assumption that genes exhibit pleiotropic effects. Linkage is another important cause of a correlation between traits,
but a random mating population in linkage equilibrium was assumed. Genetic
covariance of two traits could be partitioned in the same manner as genetic
variance. It was possible, therefore, to conduct an analysis of covariance
for two traits in the same fashion as an analysis of variance for each trait.
Because expected mean cross products had expectations similar to expected mean
squares, it was possible to determine genetic and phenotypic correlation coefficients from analysis of mating designs proposed by Comstock and Robinson
(1948).

Many reports in the literature giving estimates summarized in Table 5.1
also include estimates of genetic, additive genetic, and phenotypic correlations among traits. We will summarize only estimates of genetic correlations.
A few reports used the estimates to construct selection indexes, but most
merely reported associations among traits. Table 5.16 summarizes available
estimates of genetic correlations among 13 traits of maize. All correlations

Table. 5.16. Summary of genetic correlations among plant and ear traits with yield
             obtained by averaging the values reported in the literature.

Trait	Yield	Height		Ear per plant	Ear		Kernel row no.	Kernels per row
		Plant	Ear		Length	Diameter		
Plant height	0.26 (23)†	---	---	---	---	---	---	---
Ear height	0.31 (23)	0.81 (22)	---	---	---	---	---	---
Ears per plant	0.43 (16)	0.12 (11)	0.14 (11)	---	---	---	---	---
Ear length	0.38 (13)	0.22 (11)	0.08 (11)	0.03 (1)	---	---	---	---
Ear diameter	0.41 (13)	0.03 (11)	0.08 (11)	-0.08 (1)	-0.01 (13)	---	---	---
Cob diameter	0.10 (9)	0.14 (9)	0.12 (9)	---	0.03 (9)	0.67 (9)	---	---
Kernel depth	0.51 (9)	-0.11 (8)	-0.05 (9)	---	-0.18 (9)	0.72 (9)	0.61 (1)	0.66 (2)
Kernel row no.	0.24 (9)	0.00 (5)	0.25 (5)	-0.23 (2)	-0.16 (5)	0.57 (5)	---	---
Kernels per row	0.45 (6)	0.25 (3)	0.22 (3)	0.15 (2)	---	0.57 (1)	---	---
Kernel weight	0.25 (8)	0.05 (4)	0.05 (4)	0.05 (2)	-0.03 (3)	0.21 (3)	-0.33 (5)	0.27 (2)
Days to flower	0.14 (13)	0.32 (9)	0.42 (8)	-0.02 (6)	-0.15 (2)	---	---	---
Tiller number	0.06 (3)	---	---	---	---	---	---	---

†Number of estimates available for each pair of traits.

were calculated from components of variances and covariances for the different
mating designs. Numerous estimates of genetic correlations between the yield
and the plant and ear traits were available, and averages of estimates show
low association of yield with any trait.

In the example quoted previously for kernel-row number and yield, average
genetic correlation is 0.24 (Table 5.16). Hence indirect selection for yield
improvement by use of kernel-row number would be less effective than direct
selection for yield. assuming the same selection differentials for both
traits; i.e., 0.24 X $\sqrt{57.0}/\sqrt{18.7}$, which is 0.42. Although the heritability of
kernel-row number was 57.0% vs. 18.7% for yield, the genetic correlation
(0.24) between kernel-row number and yield was small enough that indirect selection was 58% less effective than direct selection. Kernel depth, ears per
plant, ear length, and ear diameter had greater genetic correlations with
yield than kernel-row number, but they also had smaller heritabilities than
kernel-row number. Indirect selection for yield improvement by these traits
would be 36% to 46% less effective than selection for yield itself, which had
an average heritability 50% smaller than the heritabilities for ear traits
(Table 5.1). Genetic correlations of traits with yield were, in all instances, too small to compensate for greater heritabilities.

Average genetic correlations with yield were larger for ear traits than
for plant and ear height, days to flower, and tiller number. Kernel depth,
for instance, had the highest correlation (0.51) with yield, but the coefficient of determination--the proportion of total sums of squares for yield that
can be explained by linear regression on kernel depth--is only 26%. In all
other instances, coefficients of determination with yield are 20% or less; and
with a few exceptions, average correlations among plant and ear traits, in-

cluding yield, are relatively small. Plant height and ear height had the
greatest correlation (r = 0.81), and some of the ear traits showed moderate
associations. Some of the ear traits would be expected to have some association because of the type of measurements used. For example, cob diameter and
kernel depth would be expected to show some association because both traits
are included in diameter measurements of the ear. Kernel depth was positively
correlated with kernel-row number (agreeing with the general observation that
kernel depth increases as kernel-row number increases) but at the expense of
kernel size (as evidenced by the negative correlation -0.33 between kernel-row
number and kernel weight). Except for correlations of plant and ear height
and kernel depth and ear diameter, coefficients of determination were less
than 50%.

Average correlations given in Table 5.16 were obtained for several different populations. As for average estimates of components of variance, correlations were determined among traits of Iowa Stiff Stalk Synthetic for comparison with those in Table 5.16. Estimates of correlations for Iowa Stiff
Stalk Synthetic are given in Table 5.17 for ten traits. All correlations giv-
en in Table 5.17 do not have the same number of estimates included, and fewer
estimates are included than for those in Table 5.16. Trends, however, are
similar for the two sets of correlations: Plant and ear height are highly
correlated and correlations among ear traits are relatively high. Correlation
of kernels per row with yield is greater in Table 5.17, but only one estimate
is available. Days to flower has a negative correlation with yield in Iowa
Stiff Stalk Synthetic whereas the average of 13 estimates in Table 5.16 is a

Table 5.17. Summary of correlations among plant and ear traits with yield for
             Iowa Stiff Stalk Synthetic.

| | Height | | | Ear | | Cob | | Kernel | |
	Yield	Plant	Ear	Length	Diameter	diameter	Row no.	Per row	Depth
Plant height	.05 (5) +								
Ear height	.05 (5)	.77 (4)							
Ear length	.45 (5)	.32 (3)	.0 (3)						
Ear diameter	.54 (5)	-.01 (3)	.08 (3)	.03 (5)					
Cob diameter	.13 (5)	.19 (3)	.23 (3)	.15 (5)	.67 (5)				
Kernel row no.	.45 (2)	---	---	.19 (2)	.70 (2)	.42 (2)	---		.60 (2)
Kernels per row	.84 (1)	---	---	---	---	---	---	---	---
Kernel depth	.65 (5)	.32 (2)	-.13 (3)	-.06 (5)	.71 (5)	.01 (5)	---	---	---
Days to flower	-.52 (2)	.27 (1)	.33 (1)	-.14 (2)	-.16 (2)	.15 (2)	-.21 (2)	---	.46 (2)

+Number of estimates available for each pair of traits.

small, positive value. The 2 estimates for Iowa Stiff Stalk Synthetic were obtained from two sets of unselected inbred progenies that tend to show a negative relation because of the effects of inbreeding that delay flowering and consequently seed set because of the sparsity of viable pollen.

Estimates of heritability for BSSS are given in Table 5.4 on a plot and progeny mean basis. The average of five estimates of the correlations between kernel depth and yield was 0.65, which was calculated from the analysis of variance and the covariance of progeny means. If we use average estimates of heritability for yield (41.4%) and kernel depth (54.8%) and assume the same selection differentials for each trait, we find that indirect selection for yield by use of kernel depth is only 74.8% as efficient as direct selection for yield. Only one estimate of the correlation of kernels per row with yield was available (r = 0.84, Table 5.17). If we use estimates of heritability reported by Bartual and Hallauer (1976), we find indirect selection for yield by use of kernels per row to be 15% less efficient than direct selection for yield. It would seem in this instance that indirect selection would be effective because of the high correlation of kernel depth with yield and the high heritability of kernels per row (91%); however, the estimate of heritability for yield also was large (90%). If we use the average estimate of heritability (41.4%) for BSSS from Table 5.4, indirect selection would be more effective than direct selection for yield but it would be biased because of the different precision of the parameters used to calculate predicted response.

Genetic correlations inherently have large errors. All the estimates given in Tables 5.16 and 5.17 were obtained using components of variance and covariance calculated from the analysis of variance and the covariances. No simple correlations between, say, inbred lines and their F_1 hybrids are included; these correlations are found in Chap. 8. Precision of average estimates of genetic correlations among the traits in Table 5.16 is quite different because of the different number of estimates available. For example, 23 genetic correlations between plant and ear height were reported, whereas only 3 were available for tiller number and yield.

Genetic correlations are of interest to determine degree of association between traits and how they may enhance selection. Genetic correlations are useful if indirect selection gives greater response to selection for a trait than direct selection for the same trait. This depends on estimates of heritabilities for each trait and genetic correlation between them. In the examples given, there were no instances where indirect selection for yield improvement by one of the ear components was more effective than direct selection for yield. Another example was reported by Cortez-Mendoza (1977), who analyzed ten generations of divergent mass selection for ear length. Direct selection for ear length was significant for both longer and shorter ears. Yield was not improved by selection for longer ears, but yield significantly decreased with selection for shorter ears. It was found that selection for increased ear length decreased kernel depth, which in Table 5.16 had a greater correlation with yield than ear length. It seems that indirect selection for a complex trait, such as yield, is not plausible. Yield is an expression of fitness and drastic changes in one component of yield are accompanied by adjustments in other component(s), implying the existence of correlated changes of gene frequencies. It seems that the most effective method for yield improvement is direct selection for yield itself; there may be correlated changes among yield components, but these correlated changes will be in concert with development of the most physiologically efficient genotype for expression of yield.

REFERENCES

Anderson, V. L., and O. Kempthorne. 1954. A model for the study of quantitative inheritance. *Genetics* 39:883-98.

Arboleda, R. F., and W. A. Compton. 1974. Differential response of maize to mass selection in diverse selection environments. *Theor. Appl. Genet.* 44:77-81.

Bartual, R., and A. R. Hallauer. 1976. Variability among unselected maize inbred lines developed by full-sibbing. *Maydica* 21:49-60.

Bauman, L. F. 1959. Evidence of non-allelic gene interaction in determining yield, ear height, and kernel row number in corn. *Agron. J.* 51:531-34.

Chi, K. R. 1965. Covariances among relatives in a random mating population of maize. Ph.D. dissertation, Iowa State Univ., Ames.

Chi, K. R.; S. A. Eberhart; and L. H. Penny. 1969. Covariances among relatives in a maize variety (*Zea mays* L.). *Genetics* 63:511-20.

Cockerham, C. C. 1954. An extension of the concept of partitioning hereditary variance for analysis of covariance among relatives when epistasis is present. *Genetics* 39:859-82.

————. 1956. Analysis of quantitative gene action. Brookhaven Symp. Biol. 9:53-68.

Compton, W. A.; C. O. Gardner; and J. H. Lonnquist. 1965. Genetic variability in two open-pollinated varieties of corn (*Zea mays* L.) and their F_1 progenies. *Crop Sci.* 5:505-8.

Comstock, R. E., and H. F. Robinson. 1948. The components of genetic variance in populations of biparental progenies and their use in estimating the average degree of dominance. *Biometrics* 4:254-66.

Comstock, R. E.; H. F. Robinson; and P. H. Harvey. 1949. A breeding procedure designed to make maximum use of both general and specific combining ability. *Agron. J.* 41:360-67.

Cornelius, P. L., and J. W. Dudley. 1974. Effects of inbreeding by selfing and full-sib mating in a maize population. *Crop Sci.* 14:815-19.

Cortez-Mendoza, H. 1977. Evaluation of ten generations of divergent mass selection for ear length in Iowa Long Ear Synthetic. Ph.D. dissertation, Iowa State Univ., Ames.

Darrah, L. L., and A. R. Hallauer. 1972. Genetic effects estimated from generation means in four diallel sets of maize inbreds. *Crop Sci.* 12:615-21.

Darrah, L. L., and M. S. Zuber. 1986. 1985 United States farm maize germplasm base and commercial breeding strategies. *Crop Sci.* 26:1109-13.

Da Silva, W. H., and J. H. Lonnquist. 1968. Genetic variances in populations developed from full-sib and S_1 testcross progeny selection in an open-pollinated variety of maize. *Crop Sci.* 8:201-4.

Dudley, J. W.; R. J. Lambert; and D. E. Alexander. 1971. Variability and relationships among characters in *Zea mays* L. synthetics with improved protein quality. *Crop Sci.* 11:512-14.

East, E. M. 1908. Inbreeding in corn. Connecticut Agric. Exp. Stn. Rep. 1907. Pp. 419-28.

Eberhart, S. A. 1961. Epistatic and other genetic variances in two varieties of corn (*Zea mays* L.). Ph.D. dissertation, North Carolina State Univ., Raleigh.

Eberhart, S. A., and A. R. Hallauer. 1968. Genetic effects for yield in single, three-way, and double-cross maize hybrids. *Crop Sci.* 8:377-79.

Eberhart, S. A.; R. H. Moll; H. F. Robinson; and C. C. Cockerham. 1966. Epistatic and other genetic variances in two varieties of maize. *Crop Sci.* 6:275-80.

El-Rouby, M. M., and L. H. Penny. 1967. Variation and covariation in a high oil population of corn (*Zea mays* L.) and their implications in selection. *Crop Sci.* 7:216-19.

El-Rouby, M. M.; Y. S. Koraiem; and A. A. Nawar. 1973. Estimation of genetic variance and its components in maize under stress and nonstress environments. I. Plant date. *Egyptian J. Genet. Cytol.* 2:10-19.

Fasoulas, A. C., and R. W. Allard. 1962. Nonallelic gene interactions in the inheritance of quantitative characters in barley. *Genetics* 47:899-907.

Gamble, E. E. 1962a. Gene effects in corn (*Zea mays* L.). I. Selection and relative importance of gene effects for yield. *Canadian J. Plant Sci.* 42:339-48.

————. 1962b. Gene effects in corn (*Zea mays* L.). II. Relative importance of gene effects for plant height and certain component attributes of yield. *Canadian J. Plant Sci.* 42:349-58.

Gardner, C. O., and J. H. Lonnquist. 1959. Linkage and the degree of dominance of genes controlling quantitative characters in maize. *Agron. J.* 51:524-28.

Gardner, C. O.; P. H. Harvey; R. E. Comstock; and H. F. Robinson. 1953. Dominance of genes controlling quantitative characters in maize. *Agron. J.* 45:186-91.

Goodman, M. M. 1965. Estimates of genetic variances in adapted and exotic populations of maize. *Crop Sci.* 5:87-90.

Gorsline, G. W. 1961. Phenotypic epistasis for ten quantitative characters in maize. *Crop Sci.* 1:55-58.

Gwynn, G. R. 1959. Relation between means and components of genotypic variance in biparental progenies of a variety of maize. Ph.D. dissertation, Iowa State Univ., Ames.

Hallauer, A. R. 1965. Inheritance of flowering in maize. *Genetics* 52:129-37.

————. 1968. Estimates of genetic variances in Iowa Long Ear Synthetic, *Zea Mays* L. *Adv. Front. Plant Sci.* 22:147-62.

————. 1970. Genetic variability for yield after four cycles of reciprocal recurrent selections in maize. *Crop Sci.* 10:482-85.

————. 1971. Change in genetic variance for seven plant and ear traits after four cycles of reciprocal recurrent selection for yield in maize. *Iowa State J. Sci.* 45:575-93.

Hallauer, A. R., and D. Malithano. 1976. Evaluation of maize varieties for their potential as breeding populations. *Euphytica* 25:117-27.

Hallauer, A. R., and W. A. Russell. 1962. Estimates of maturity and its inheritance in maize. *Crop Sci.* 2:289-94.

Hallauer, A. R., and J. H. Sears. 1972. Integrating exotic germplasm into Corn Belt maize breeding programs. *Crop Sci.* 12:203-6.

————. 1973. Changes in quantitative traits associated with inbreeding in a synthetic variety of maize. *Crop Sci.* 13:327-30.

Hallauer, A. R., and J. A. Wright. 1967. Genetic variances in the open-pollinated variety of maize, Iowa Ideal. *Züchter* 37:178-85.

Hanson, W. D. 1963. Heritability. In *Statistical Genetics and Plant Breeding,* W. D. Hanson and H. F. Robinson, eds., pp. 125-40. NAS-NCR Publ. 982.

Hayman, B. I. 1958. The separation of epistatic from additive and dominance variation in generation means. *Heredity* 12:371-90.

————. 1960. The separation of epistatic from additive and dominance variation in generation means. II. *Genetica* 31:133-46.

Hayman, B. I., and K. Mather. 1955. The description of genetic interaction in continuous variation. *Biometrics* 11:69-82.

Hughes, G. R., and A. L. Hooker. 1971. Gene action conditioning resistance to northern leaf blight in maize. *Crop Sci.* 11:180-84.

Hull, F. H. 1945. Recurrent selection and specific combining ability in corn. *J. Am. Soc. Agron.* 37:134-45.

Hyer, A. H. 1960. Non-allelic interactions in a population of maize derived from a cross of two inbred lines. Ph.D. dissertation, Iowa State Univ., Ames.

Jenkins, M. T. 1940. The segregation of genes affecting yield of grain in maize. *J. Am. Soc. Agron.* 32:55-63.

Jones, D. F. 1918. The effects of inbreeding and crossbreeding upon development. Connecticut Agric. Exp. Stn. Bull. 207:5-100.

Kempthorne, O. 1957. *An Introduction to Genetic Statistics.* Wiley, New York.

Kiesselbach, T. A. 1933. The possibilities of modern corn breeding. *Proc. World Grain Exhib. Conf.* (Canada) 2:92-112.

Laible, C. A., and V. A. Dirks. 1968. Genetic variance and selective value of ear number in corn (*Zea mays* L.). *Crop Sci.* 8:540-43.

Lamkey, K. R., and A. R. Hallauer. 1986. Performance of high x high, high x low, and low x low crosses of lines from BSSS maize Synthetic. *Crop Sci.* 26:1114-18.

Levings, C. S., III; J. W. Dudley; and D. E. Alexander. 1971. Genetic variance in autotetraploid maize. *Crop Sci.* 11:680-81.

Lindsey, M. F.; J. H. Lonnquist; and C. O. Gardner. 1962. Estimates of genetic variance in open-pollinated varieties of Corn Belt corn. *Crop Sci.* 2:105-8.

Lonnquist, J. H. 1949. The development and performance of synthetic varieties of corn. *Agron. J.* 41:153-56.

Lonnquist, J. H.; O. Cota A.; and C. O. Gardner. 1966. Effect of mass selection and thermal neutron irradiation on genetic variances in a variety of corn (*Zea mays* L.). *Crop Sci.* 6:330-32.

Marquez-Sanchez, F., and A. R. Hallauer. 1970a. Influence of sample size on the estimation of genetic variances in a synthetic variety of maize. I. Grain yield. *Crop Sci.* 10:357-61.

———. 1970b. Influence of sample size on the estimation of genetic variances in a synthetic variety of maize. II. Plant and ear characters. *Iowa State J. Sci.* 44:423-36.

Martin, J. M., and A. R. Hallauer. 1976. Relation between heterozygosis and yield for four types of maize inbred lines. *Egyptian J. Genet. Cytol.* 5:119-35.

Mather, K. 1949. *Biometrical Genetics.* Methuen, London.

Mode, C. J., and H. F. Robinson. 1959. Pleiotropism and the genetic variance and covariance. *Biometrics* 15:518-37.

Moll, R. H., and H. F. Robinson. 1966. Observed and expected response in four selection experiments. *Crop Sci.* 6:319-24.

Moll, R. H.; H. F. Robinson; and C. C. Cockerham. 1960. Genetic variability in an advanced generation of a cross of two open-pollinated varieties of corn. *Agron. J.* 52:171-73.

Moll, R. H.; D. L. Thompson; and P. H. Harvey. 1963. A quantitative genetic study of the inheritance of resistance to brown spot (*Physoderma maydis*) of corn. *Crop Sci.* 3:389-91.

Moreno-Gonzalez, M.; J. W. Dudley; and R. J. Lambert. 1975. A design III study of linkage disequilibrium for percent oil in maize. *Crop Sci.* 15:840-43.

Neal, N. P. 1935. The decrease in yielding capacity in advanced generations of hybrid corn. *J. Am. Soc. Agron.* 27:666-70.

Obilana, A. T., and A. R. Hallauer. 1974. Estimation of variability of quantitative traits in BSSS by using unselected maize inbred lines. *Crop Sci.* 14:99-103.

Otsuka, Y.; S. A. Eberhart; and W. A. Russell. 1972. Comparisons of prediction formulas for maize hybrids. *Crop Sci.* 12:325-31.

Rawlings, J., and C. C. Cockerham. 1962a. Triallel analysis. *Crop Sci.* 2: 228-31.

———. 1962b. Analysis of double cross hybrid populations. *Biometrics* 18: 229-44.

Robinson, H. F., and C. C. Cockerham. 1961. Heterosis and inbreeding depression in populations involving two open-pollinated varieties of maize. *Crop Sci.* 1:68-71.

Robinson, H. F.; C. C. Cockerham; and R. H. Moll. 1958. Studies on the estimation of dominance variance and effects of linkage bias. In *Biometrical Genetics*, O. Kempthorne, ed., pp. 171-77. Pergamon Press, New York.

Robinson, H. F.; R. E. Comstock; and P. H. Harvey. 1949. Estimates of heritability and the degree of dominance in corn. *Agron. J.* 41:353-59.

———. 1955. Genetic variances in open pollinated varieties of corn. *Genetics* 40:45-60.

Russell, W. A. 1971. Types of gene action at three gene loci in sublines of a maize inbred line. *Canadian J. Genet. Cytol.* 13:322-34.

———. 1976. Genetic effects and genetic effect × year interactions at three gene loci in sublines of a maize inbred line. *Canadian J. Genet. Cytol.* 18:23-33.

Russell, W. A., and S. A. Eberhart. 1970. Effects of three gene loci in the inheritance of quantitative characters in maize. *Crop Sci.* 10:165-69.

Sentz, J. C. 1971. Genetic variances in a synthetic variety of maize estimated by two mating designs. *Crop Sci.* 11:234-38.

Sentz, J. C.; H. F. Robinson; and R. E. Comstock. 1954. Relation between heterozygosis and performance in maize. *Agron. J.* 46:514-20.

Shull, G. F. 1908. The composition of a field of maize. Am. Breeders' Assoc. Rep. 4:296-301.

———. 1909. A pure-line method in corn breeding. Am. Breeders' Assoc. Rep. 5:51-59.

Silva, J. C. 1974. Genetic and environmental variances and covariances estimated in the maize (*Zea mays* L.) variety, Iowa Stiff Stalk Synthetic. Ph.D. dissertation, Iowa State Univ., Ames.

Silva, J. C., and A. R. Hallauer. 1975. Estimation of epistatic variance in Iowa Stiff Stalk Synthetic maize. *J. Hered.* 66:290-96.

Sing, C. F.; R. H. Moll; and W. D. Hanson. 1967. Inbreeding in two populations of *Zea mays* L. *Crop Sci.* 7:631-36.

Sprague, G. F. 1946. Early testing of inbred lines of corn. *J. Am. Soc. Agron.* 38:108-17.

———. 1964. Estimates of genetic variations in two open-pollinated varieties of maize and their reciprocal F_1 hybrids. *Crop Sci.* 4:332-34.

———. 1966. Quantitative genetics in plant improvement. In *Plant Breeding*, K. J. Frey, ed., pp. 315-54. Iowa State Univ. Press, Ames.

———. 1971. Genetic vulnerability to disease and insects in corn and sorghum. *Proc. Annu. Corn Sorghum Res. Conf.* 26:96-104.

Sprague, G. F., and K. Suwantaradon. 1975. A generation mean analysis of mutants derived from a doubled-monoploid family in maize. *Acta Biol.* 7: 325-32.

Sprague, G. F., and W. I. Thomas. 1967. Further evidence of epistasis in single and three-way cross yields in maize. *Crop Sci.* 4:355-57.

Sprague, G. F.; W. A. Russell; L. H. Penny; T. W. Horner; and W. D. Hanson. 1962. Effect of epistasis on grain yield in maize. *Crop Sci.* 2:205-8.

Stevenson, G. C. 1965. *Genetics and Breeding of Sugarcane* (Tropical Series). Longmans, Green, London.

Stringfield, G. H. 1950. Heterozygosis and hybrid vigor in maize. *Agron. J.*

42:145-52.

Stuber, C. W., and R. H. Moll. 1970. Epistasis in maize (*Zea mays* L.). II. Comparison of selected with unselected populations. *Genetics* 67:137-49.

Stuber, C. W.; R. H. Moll; and W. D. Hanson. 1966. Genetic variances and interrelationships of six traits in a hybrid population of *Zea mays* L. *Crop Sci.* 6:541-44.

Stuber, C. W.; W. P. Williams; and R. H. Moll. 1973. Epistasis in maize (*Zea mays* L.). III. Significance in predictions of hybrid performances. *Crop Sci.* 13:195-200.

Subandi, W., and W. A. Compton. 1974. Genetic studies in an exotic population of corn (*Zea mays* L.) grown under two plant densities. II. Choice of a density environment for selection. *Theor. Appl. Genet.* 44:193-98.

Webel, O. D., and J. H. Lonnquist. 1967. An evaluation of modified ear-to-row selection in a population of corn (*Zea mays* L.). *Crop Sci.* 7:651-55.

Williams, J. C.; L. H. Penny; and G. F. Sprague. 1965. Full-sib and half-sib estimates of genetic variance in an open-pollinated variety of corn, *Zea mays* L. *Crop Sci.* 5:125-29.

Wright, J. A. 1966. Estimation of components of genetic variances in an open-pollinated variety of maize by using single and three-way crosses among random inbred lines. Ph.D. dissertation, Iowa State Univ., Ames.

Wright, J. A.; A. R. Hallauer; L. H. Penny; and S. A. Eberhart. 1971. Estimating genetic variance in maize by use of single and three-way crosses among unselected inbred lines. *Crop Sci.* 11:690-95.

Wright, S. 1922. The effects of inbreeding and crossbreeding on guinea pigs. III. Crosses between highly inbred families. USDA Bull. 1121.

Zuber, M. S. 1975. Corn germplasm base in the United States: Is it narrowing, widening, or static? *Proc. Annu. Corn Sorghum Res. Conf.* 30:277-86.

Zuber, M. S., and L. L. Darrah. 1980. 1979 U.S. corn germplasm base. *Proc. Annu. Corn Sorghum Res. Conf.* 35:234-49.

6

Selection: Theory

Selection is the essence of plant and animal breeding and has played an important role in the history of living beings. Evolution (via natural selection) and domestication (via artificial selection) created and improved the crop plant species that are so important for human survival. Ever since the potential of certain plant species as food sources was recognized, selection has been practiced for more productive plant types. Particularly in maize, in addition to great advances achieved by domestication and early empirical breeding, significant improvements have been made by changes in breeding methods that have occurred mainly during this century. At the present, new selection methods for the genetic improvement of maize still are important in increasing food production.

Selection is ultimately the differential reproduction of genotypes. The purpose and the critical feature of artificial selection is to choose from a group of individuals those that will be allowed to reproduce so as to make selection as effective as possible for a given selection intensity. Several techniques have been devised to help the breeder make accurate decisions. Such techniques involve procedures based on genetic concepts (such as progeny test or family evaluation) and experimental procedures that attempt to minimize the effect of environments on the expression of genotypes. In practical maize breeding, a selection experiment can be divided into two distinct phases: (1) selection among populations and (2) selection of genotypes within a population.

6.1 SELECTION AMONG POPULATIONS

In the first phase of the selection procedure breeders must decide which populations are more suitable for their purposes. Selected populations can be used either directly for production or for breeding purposes. Eventually several selected populations can be combined to develop a unique base population for breeding purposes or two selected populations can be crossed and the first generation of hybrid population can be used directly for grain production.

The genetic potential of populations for breeding purposes may be evaluated simply by observation of their performance or by analysis of their pedigree, origin, and past selection records. Intrinsic genetic properties of populations can only be evaluated through genetic designs. Parameters of primary importance are the mean, the additive genetic variance and its relative magnitude, the coefficient of heritability, and additive genetic correlations among the most important traits. The mean of the trait that will be changed by selection is important because selection started at a lower level would require several cycles in a population with lower performance to attain a higher

159

level. The presence of genetic variance is a requirement to permit progress
from selection. Other genetic information, whenever possible to obtain, may
be useful as criteria for selection among populations. The cross between populations also provides valuable information relative to heterosis and combining ability. When a set of varieties or populations is available, information
can be obtained from use of the diallel cross analysis. The diallel cross
analysis proposed by Gardner and Eberhart (1966) has been widely used for the
evaluation of varieties because information about the performance of populations per se and in crosses is provided on the basis of variety effects v_i;
total heterosis effects $h_{jj'}$; and the components of heterosis $\bar{h}$ (average heterosis), h_j (variety heterosis), and $s_{jj'}$ (specific heterosis), as shown in
Chap. 4.

The intervarietal diallel cross also provides information that permits
prediction of composite means allowing selection among populations (composites) that could be synthesized from a fixed set of varieties. Thus selection
among populations is possible not only within a given set of varieties but also among all possible populations that could be synthesized from them by making use of prediction procedures (see Chap. 10 for details about prediction).
Genetic variability within newly formed populations is difficult to predict
but can be assured to a certain extent by combining genetically divergent varieties (see Fig. 5.1). The magnitude of heterosis is the most direct indication of amount of divergence among populations. Combining populations that
give high heterosis when crossed would assure greater genetic variability
within the newly formed population. However, Moll et al. (1965) pointed out
that heterosis increases up to a certain level of divergence between populations, beyond which it tends to decrease. Lack of heterosis, however, is not
always an indication of lack of genetic divergence between populations (Cress
1966).

6.2 SELECTION OF GENOTYPES WITHIN POPULATIONS

In this case one can consider a population either as finite in size and
consisting of a fixed set of defined genotypes or as a unique gene pool that
for practical purposes is a random sample of a population theoretically infinite in size. In the first instance we may include all types of hybrids from
inbred lines.

6.2.1 Selection among Hybrids

Lines drawn from a population may be considered a sample from a large
population of genotypes. But once the lines are drawn and crossed with lines
from other sources for hybrid combination tests, we have a fixed set of genotypes to consider. In this case selection would be among all possible hybrids
involving a fixed set of parental lines. For example, with n inbred lines,
selection could be done among $n(n - 1)/2$ single crosses and subsequently the
best hybrids could be used directly for grain production. Hybrid development,
however, is a somewhat static process because the objective is to obtain a
genotype or a group of genotypes that, once identified, may be reproduced in-
definitely without need for continuous selection. Actually the exact reproduction of hybrid genotypes is valid only for single crosses from completely
homozygous lines, but for practical purposes all other types of hybrids are
assumed to be reproducible in the same manner. Selection during the inbreeding process for line development has a continuous phase that ends at the time
when the lines are approximately homozygous. Exceptional hybrids are not considered static entities when the parent lines are continuously improved
through some method like backcross or convergent improvement.

Cockerham (1961) formalized the differences of selection among single-, three-way, and double-cross hybrids. He considered unselected lines developed from a population for the purpose of having a reference population for making selections among hybrids and predicting gain. Inbred lines can have the same or different levels of inbreeding. In most maize breeding programs, inbred lines used to produce hybrids are from different source populations; the object is usually to have the source populations as different as is feasible to enhance heterosis among crosses of lines. Cockerham's developments are informative, however, for showing the relative advantages of selecting among different types of hybrids. The expected mean squares for the three types of relatives were expressed in terms of covariances of relatives and translations made to genetic components of variance. Briefly, components of genetic variance show that selection among single-cross hybrids will always be twice as effective as selection among double-cross hybrids, with selection among three-way crosses intermediate to single and double crosses. The ratio of the additive genetic variance component is 1:3/4:1/2 for single, three-way, and double crosses, respectively. If nonadditive effects (dominance and epistasis) are important in differences expressed among the three types of hybrids, relative advantages of selection among single crosses increase over those among three-ways and double crosses. Hence selection effectiveness among single crosses will be even greater than among double crosses if nonadditive effects are important. Although Cockerham's results are only for hybrids produced from lines developed from one population, it seems the results also should be applicable for groups of hybrids developed from lines originating from different source populations.

Theoretical approaches for the understanding of the inbred-hybrid breeding system were also presented by Comstock (1964). Assume a random sample of inbred lines from a noninbred source population is crossed at random to produce single-cross hybrids. Any hybrid produced in this manner will have a genotype that could have occurred in the source population, and the probability of such a genotype in the hybrid will be the same as the probability of that genotype in the source population. Thus inbreeding followed by the crossing of inbred lines is not a method for creating new genotypes and selection among hybrids is ultimately selection from the same array of genotypes possible in the source population. In the same manner, when hybrids are developed by crossing inbred lines from two different source populations, the genotype of any such hybrid is one of the possible genotypes in the cross between the two populations and occurs with the same probability.

Advantages of the inbred-hybrid system to select genotypes are: (1) the selected genotypes can be reproduced and then submitted to extensive tests, allowing a greater precision of evaluation; (2) commercial utilization of hybrids requires that they are reproducible, which is only possible through the inbred-hybrid system; and (3) the process used for inbred line development involves sequential selection in that inbreds are eliminated at successive stages before the inbreeding is complete and thus, under effective selection, the probability of obtaining better genotypes is increased.

An expression to calculate theoretical gain by selection among hybrids obtained by crossing inbred lines from two source populations is given by Comstock (1964) as follows:

$$\Delta_G = k(\sigma_y^2/\sigma_{ph}) + v$$

where k is a function of effective selection intensity; σ_y^2 is the total genetic variance among hybrids because genotypes are reproducible, so the offspring have the same genotypes as the selected hybrids; σ_{ph} is the phenotypic standard deviation of the performance means on which selection is based, so precision of tests increases the gain from selection; and v is the hybrid vigor in variety crosses.

Although the theory presented above is important in understanding the inbred-hybrid breeding system, it is difficult to adapt in practical breeding for the following reasons: (1) inbred-line development involves a sequential process of selection so that some lines are eliminated by their poor performance in the first generations of inbreeding, while others are eliminated later on the basis of general combining ability tests; (2) in one of the populations a vigorous line is desired to be used as seed parent and the most vigorous lines are not always those that give the best hybrids; and (3) success of an inbred line development program is not always measured by hybrids developed from that specific program because some lines have been shown to produce very good hybrids in combination with some preexisting lines or lines developed from other programs using other populations; and (4) several attributes other than yield are considered in the selection program so that the effect of selection on particular traits is difficult to predict.

Another important question about inbred-hybrid breeding pertains to the effectiveness of selection within new samples from the same population. Theoretically, there is no advantage in resampling the same population because hybrid genotypes obtained by crossing lines in the second sample will be expected to be the same as from the first sample. The key factor is that given successive samples of equal size drawn in the same way from the same population, the probability of containing the most extreme individual is the same for all samples. On the other hand, given any number of samples, there is a finite though small probability that if one more sample is taken it will be found to contain a more extreme individual than the most extreme in any preceding sample. This is only a principle in sampling theory but from the breeding standpoint one can conclude that after a reasonable number of hybrids have been screened the probable return per unit effort in screening more from the same population is sharply decreased (Comstock 1964).

The fact that hybrid genotypes occur with the same probability as the same genotypes in the source population brings about an obvious conclusion: To increase the probability of obtaining better hybrids from a given population the most direct way is to improve the population itself, which can be done through some recurrent selection method. Also, new sampling from the same population is recommended only if the source population is being continuously improved in the interval from one sample to another.

6.2.2 Recurrent Selection within Populations

Selection within populations that are unique gene pools is directed toward the improvement of the gene pool itself, which is done by increasing the frequency of favorable alleles within the population. This is a dynamic process, since gene frequency is changed gradually following a recurrent selection procedure. Simultaneous selection in two populations via a reciprocal recurrent selection scheme also is included in this category. Recurrent selection procedures available to increase gene frequencies of favorable alleles may be divided into two main categories (Moll and Stuber 1974): (1) intrapopulation improvement and (2) interpopulation improvement. Discussion about these two systems of selection is presented throughout this chapter. Improvement of a population as a unique entity may be done for qualitative traits (usually controlled by one or few genes) or for quantitative traits (usually controlled by many genes), although the effect of selection is basically the same (i.e., change in gene frequencies). A complete theory about selection is extensive, and a more complete presentation can be found in some books of quantitative genetics or population genetics (Kempthorne 1957; Lerner 1958; Falconer 1960).

6.3 INTRAPOPULATION IMPROVEMENT: QUALITATIVE TRAITS

It is shown in Chap. 2 that gene frequency remains constant from generation to generation in a panmictic population in the absence of selection and
other disturbing factors. Let us consider one locus with two alleles with
frequency p(A) and q(a) in a maize population in Hardy-Weinberg equilibrium.
The genotypic array is p^2(AA):2pq(Aa):q^2(aa) and one can readily see that selection favoring either the dominant allele A or the recessive allele a will
result in a change of gene and genotypic frequencies. Selection favoring recessives is the easiest one and can be performed fully in just one cycle of
selection if penetrance or expressivity do not cause problems in the identification of recessive genotypes. For example, selection for sugary kernels is
included in this category. From a population segregating for sugary and normal kernels the breeder can easily obtain a completely sugary kernel population in only one cycle of selection.

Selection favoring recessives is common in maize breeding for several
traits, such as sweetness, opacity, brachysm, lack of ligules, etc. Most of
them are completely recessive but even in these cases selection efficiency depends on how extensively the trait is influenced by environmental factors and
genotypic background. For example, selection for brachytic plants in a heterogeneous population cannot usually be completed in only one generation if the
breeder has interest in other traits and wants to keep effective population
size as large as possible. The difficulty in selecting brachytic (br_2br_2)
plants is caused by the action of modifier genes that are polygenic in inheritance, resulting in a continuous distribution of phenotypes. So plants may
vary from extremely small to nearly normal in height within the dwarf class.
In the same way, other recessive genes may be affected by modifiers and then
selection favoring recessives may be difficult.

On the other hand, if selection is against recessives another difficulty
arises because recessive genes are retained in the heterozygous genotypes
within the population. Although one can recognize homozygous recessives, they
cannot be eliminated completely in only one generation of selection. If selection does not involve a progeny test, decrease in recessive homozygotes is
gradual and asymptotic and rate of decrease depends on initial gene frequency.
For the genotypic array p^2(AA):2pq(Aa):q^2(aa) and selection against the recessives, the reproductive rates are 1(AA):1(Aa):0(aa). After selection the new
genotypic proportion is $[p^2/(p^2 + 2pq)]$AA:$[2pq/(p^2 + 2pq)]$Aa and the new
gene frequencies are $p_1(A) = 1/(p + 2q)$ and $q_1(a) = q/(p + 2q)$ so that $p_1(A)$
+ $q_1(a) = 1.0$. Decrease in gene frequency is then $\Delta q = q_0 - q_1 = q_0^2/(1 + q_0)$,
where q_0 is the initial gene frequency. For the complete elimination of re-
recessive homozygotes the greatest selection responses occur when q_0 is high;
i.e., the greatest response when selecting favorable alleles is expected when
q_0 is high. If selection against recessives is continued, the new frequency
in any generation can be related to the original gene frequency as

$$q_n = q_0/(1 + nq_0)$$

which gives an asymptotic pattern for change in gene frequency. Theoretically
the limit of q_n is zero only after an infinite number of generations. However, one can alternatively calculate the number n of generations necessary to
reduce the frequency of recessives from an initial frequency q_0 to a low frequency, say q_n, using

$$n = (q_0 - q_n)/(q_0q_n)$$

If selection does not completely eliminate the recessive homozygotes in
each generation, the rate of change in gene frequency is lower than in the

preceding case. If genotypes have a survival rate of 1(AA):1(Aa):(1 - s)(aa),
where s is the selection intensity for recessive homozygotes, the change in
gene frequency is

$$\Delta_q = -[sq^2(1 - q)]/(1 - sq^2)$$

Figure 6.1 shows the change in gene frequency Δq for several levels of selection intensity and initial gene frequencies. Formulas for several other cases
are given in population genetics textbooks.

     Selection can be quite effective for characters if the heterozygotes have
an intermediate expression between the two homozygotes, as in the case of selection for yellow kernels. Most types of maize have colorless aleurone so
that the kernel color is due to endosperm color. Endosperm is a triploid (3n)
tissue and its genotype for color genes may be as follows:

Endosperm genotype (3n)	Embryo genotype (2n)	Kernel color
YYY	YY	orange
YYy	Yy	yellow
Yyy	Yy	light yellow
yyy	yy	white

Careful selection for dark colored kernels increases the probability of elimi-
nating heterozygous types and consequently leads to a rapid decrease in the
frequency of the undesirable allele y. On the other hand, selection for white
kernels may be more efficient because they are easier to recognize among the
four phenotypes for kernel color.

     In practical maize breeding, selection at the one gene level (as in the
case of selection for yellow endosperm, aleurone color, brachysm, lack of ligules, etc.) may be done concurrently with recurrent selection programs where
the greater emphasis is on quantitative traits. Some qualitative traits (such
as opacity and sweetness) require special programs and some type of progeny
test such as testcross or selfing is commonly used. Also selection in a backcross scheme is a very common procedure. Usually selection for qualitative
traits is not completely dissociated from important quantitative characters
such as yield, ear height, and lodging resistance.

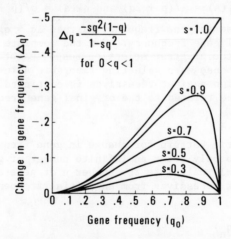

Fig. 6.1.   Change in gene frequency Δq for several selection intensities (s)
            and initial gene frequency q_0.

6.4 INTRAPOPULATION IMPROVEMENT: QUANTITATIVE TRAITS

Selection for a quantitative trait involves a quite different methodology from that used for a qualitative trait because the breeder is concerned with a large number of genes and genotypes that cannot be individually classified. Quantitative traits involve some metrical measure and not the direct knowledge of genotypic or gene frequencies and the exact genetic constitution of individuals. Therefore the tools used to study and measure selection effects must include statistical parameters such as means, variances, and covariances. Note however that change in gene and genotypic frequencies is the basic consequence of selection and changes in means and variances are the ultimate effect of those changes.

The breeder's first interest in results of selection is its effect on a population mean. Theoretical and practical inferences about selection efficiency are drawn from manipulation of genetic and environmental parameters of genetic and environment reference populations. Discussion about this topic is the subject of the remainder of this chapter.

6.5 RESPONSE TO SELECTION

Estimation of progress from selection has been one of the most important contributions of quantitative genetics to plant and animal breeding. One of its direct applications is concerned with the extent to which a given population is suitable for breeding purposes for either a given environment or a set of environments. Another important application is concerned with comparison of different selection methods. There are different approaches to determine theoretically the expected progress from selection. In any case a regression problem is involved.

For any breeding method, selection is based on a measurable entity (individual or family), which is the selection unit X that is related to some individual W in the improved population. Actually, W is taken as an individual representative of all those related to X. In some cases, W descends directly from X but in most instances there is a recombination unit R so that W is related to X through R; i.e., the improved population does not descend directly from selected individuals.

Expected progress from selection is related to the following basic question: If the best selection units are identified and recombined (directly or indirectly), what is the expected change in the population mean? When superior phenotypes are selected, it is assumed that superior genotypes are selected; i.e., phenotype and genotype are correlated to some extent. Otherwise, progress from selection would be impossible. The degree to which genotypic values of superior parents are transmitted to offspring depends on the heritability of the trait being selected. Considering only the linear relation between X and W for each unitary deviation in X, a response of b_{WX} is expected in W, where b_{WX} is the linear regression coefficient of W on X or $b_{WX} = Cov$ (W, X)$/\sigma_X^2$. The denominator σ_X^2 is the phenotypic variance of selection units. Denoting the mean of the selected group by $\overline{X}_S$, the deviation $(\overline{X}_S - \overline{X})$ is known as the selection differential s. So the first formula for the expected progress can be written as $\Delta_G = s \; Cov(W, X)/\sigma_X^2$. The term Cov(W,X) is a type of covariance between relatives; in some cases it can be expressed as a linear function of components of genetic variance of the reference population (see Chap. 3). This feature simplifies the prediction procedure.

The selection differential is a measure of the difference between the mean of a subset S containing the selected units and the mean of a set B containing all units so that S is contained in B, as represented in Fig. 6.2a. If the selection units have an approximately normal distribution Fig. 6.2a can

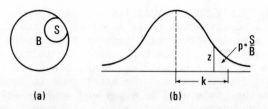

(a)   (b)

Fig. 6.2.   Schematic presentation of (a) a subset S selected from a basic set
            B of entities and (b) its proportion p in a normal distribution.

be represented in terms of frequency distribution approximately as in Fig.
6.2b, where p = S/B. If truncation selection is used (Fig. 6.2b), the selection differential can be expressed in terms of phenotypic standard deviation
units (i.e., k = s/σ_X) and then can be determined directly from properties of
the normal distribution whenever the sample size (number of selection units)
is greater than 50. Theory demonstrates that k = z/p, where p is the proportion of selected units and z is the height of the ordinate at the lower limit
of the selected group. The value of z can be obtained from Tables I and II of
Fisher and Yates (1948) as follows: In Table I enter with P where P = 2p because the normal curve is symmetrical. Table I gives a relative deviate x so
that a proportion 1 – P is within the range from –x to +x and our value of p
is for only one-half the curve. The value of x obtained in Table I is entered
in Table II and z is obtained directly. For example, if the top 10% is to be
selected from a large sample, then p = 0.10 and P = 0.20. From Table I, we
obtain x = 1.28, which we use to enter Table II, obtaining z = 0.1758. Thus
the selection differential in standard units is k = 0.1758/0.10 = 1.758.

Alternatively, k can be calculated with better approximation by obtaining
z directly from

$$z = [1/(2\pi)]e^{-x^2/2}$$

Table 6.1.   Selection differential (k) in standard deviation units for several selection
             intensities (p) for sample size greater than 50.

p	k	p	k	p	k	p	k
0.01	2.6652	0.11	1.7094	0.21	1.3724	0.31	1.1380
.02	2.4209	.12	1.6670	.22	1.3459	.32	1.1175
.03	2.2681	.13	1.6273	.23	1.3202	.33	1.0974
.04	2.1543	.14	1.5898	.24	1.2953	.34	1.0777
.05	2.0627	.15	1.5544	.25	1.2711	.35	1.0583
.06	1.9854	.16	1.5207	.26	1.2476	.36	1.0392
.07	1.9181	.17	1.4886	.27	1.2246	.37	1.0205
.08	1.8583	.18	1.4578	.28	1.2022	.38	1.0020
.09	1.8043	.19	1.4282	.29	1.1804	.39	0.9838
0.10	1.7550	0.20	1.3998	0.30	1.1590	0.40	0.9659

Table 6.2. Expected values of k (selection differential in standard units)
            for samples of size 50 or less.

n	Sample Size (N)				
	10	20	30	40	50
1	1.540	1.870	2.040	2.160	2.250
2	1.270	1.640	1.830	1.955	2.050
3	1.067	1.470	1.673	1.810	1.910
4	0.895	1.333	1.550	1.693	1.798
5	0.740	1.216	1.446	1.594	1.704
10	---	0.768	1.061	1.241	1.371
15	---	---	0.777	0.991	1.139
20	---	---	---	0.782	0.950

where e = 2.718282 and π = 3.141593. The value for x is obtained from Table I
of Fisher and Yates (1948). Values of k for some of the more common selection
intensities are given in Table 6.1.

If the group from which selection is to be made has size $N \leq 50$, then k
must be obtained from Table XX (Fisher and Yates 1948), where N varies from 2
to 50. If n (in the range from 1 to 25) is selected from a sample of N, the
expected value of k is the average of the first n values in the column corresponding to N. For example, if 5 individuals are selected from a sample of
20, the expected k value is the average of 1.87, 1.41, 1.13, 0.92, and 0.75;
i.e., k = 1.216. Table 6.2 gives some values of k for the most common cases.

In the procedure so far only one selection unit is considered. Some
breeding methods involve more than one selection unit, as for example in the
case of among and within family selection. In such cases there is one expected gain for each selection unit; expected gains must be summed to give total
gain. Also selection may be for only one or for both sexes. If selection is
for both sexes sum the expected gain for each sex to give total gain. At this
point a general formula to predict the gain from selection can be written:

$$\Delta_G = \sum_i \Delta_{G_i} = \sum_i k_i c_i [\sigma_A^2 (1 + \beta_i)/\sigma_{y_i}]$$

which comes from multiple linear regression theory. The subscript i denotes
either different selection units or different sexes; β_i is a deviation from
the additive genetic variance. When selection is not truncated, k_i must be
replaced by the selection differential ds in observed units and σ_y by σ_y^2,
where σ_y^2 is the phenotypic variance. The coefficient c_i, which varies according to the breeding method, is the coefficient of transformation of covariance
between relatives in components of genetic variance. When selection is for
both sexes at the same selection intensity, summation results in c being replaced by 2c.

6.6 INTRAPOPULATION SELECTION METHODS

The most common selection methods for maize intrapopulation improvement may be classified as follows:

1. Mass (phenotypic) selection
2. Family selection
   (a) Half-sib family selection
   (b) Full-sib family selection
   (c) Selfed (S_1 or S_2) family selection
   (d) Combined selection

6.6.1 Mass (phenotypic) Selection

For mass selection, individual plants are evaluated and selected phenotypically; i.e., no information other than their own phenotypes is used as a criterion for selection. Hence, the selection unit X is the individual phenotype, which is also the recombination unit; and the improved population W is obtained directly from X as follows:

$$X \longrightarrow W$$

(selection unit)   (improved population)

Usually the selected plants are pollinated by a random sample of pollen from the whole population (open pollination) so that selection is for only the female source of gametes, and there is no control for male gametes (pollen). The resemblance between X and W is due to a parent-offspring relationship and $Cov(X, W) = Cov(\text{parent-offspring}) = (1/2)\sigma_A^2$ (see Chap. 3).

Phenotypic variance among individual plants includes genetic variance among plants within the population and all types of environmental variance within the selection block. If epistasis is negligible, genetic variance is $\sigma_A^2 + \sigma_D^2$. Environmental variability can be expressed in terms of variance components if some experimental design is taken as reference, because mass selection is not based on replicated trials. For example, suppose the selection block is designed in a fashion similar to one replication of a randomized complete block design. Environmental variability within the block would be due to plot-to-plot environmental variance σ_p^2 and within-plot environmental variance σ_{we}^2. These components in the selection block are not expected to be exactly the same as those estimated over replicated trials. If we disregard this possible bias, phenotypic variance among individual plants is $\sigma_y^2 = \sigma_A^2 + \sigma_D^2 + \sigma_p^2 + \sigma_{we}^2$. Expected progress from selection for this form of mass selection can be written as

$$\Delta_G = \frac{k(1/2)\sigma_A^2}{(\sigma_A^2 + \sigma_D^2 + \sigma_p^2 + \sigma_{we}^2)^{1/2}}$$

Additive genetic variance can be estimated from mating designs given in Chap. 4. Phenotypic variance among individuals also can be estimated from a mating design where some type of family or families are evaluated over replicated trials. For example, if half-sib families are taken as reference, phenotypic variance in the expected progress from selection for individual plants is $\sigma_P^2 = \sigma_f^2 + \sigma_p^2 + (\sigma_G^2 - CovHS) + \sigma_{we}^2 = \sigma_A^2 + \sigma_D^2 + \sigma_p^2 + \sigma_{we}^2$ (see Chap. 2 for meanings of components).

    Assume in the above example that the trait under mass selection is plant
height and selected female parents are pollinated by the whole population
(nonselected male gametes). Also assume that we conduct mass selection for
the inverse situation, i.e., selection of male gametes but no control of female gametes. This is not a practical procedure but it could be done by covering female inflorescences (shoots) of all plants before flowering. At flowering time a sample of pollen is collected only from plants with the desired
height and all plants are randomly pollinated with the bulk sample of pollen.
The expected progress from selection would be, theoretically, the same as that
given for female selection if the same selection intensity is used.
    Selection for male gametes is feasible for traits that can be evaluated
before flowering. For this situation, undesirable plants would be detasseled
or eliminated so that selected plants would be pollinated by a sample of selected male gametes (provided by the same selected plants). An alternative
procedure to select for both male and female gametes is by selfing the selected plants. In this case the selfed progeny must be recombined in an isolated
block in the following year. If recombination can be done in a winter nursery
an additional year per cycle is not necessary. Selection for both sexes at
the same selection intensity leads to the following expected progress:

$$\Delta_G = \frac{k\sigma_A^2}{(\sigma_A^2 + \sigma_D^2 + \sigma_p^2 + \sigma_{we}^2)^{1/2}}$$

    Stratified mass selection was suggested by Gardner (1961), and the crite-
rion for selection is the deviation of each individual phenotype from the mean
phenotypic value of all plants within the same stratum. Hence

$$\text{Cov}(X, W) = \text{Cov}[(Y_{ij} - \overline{Y}_{i.}), W] = \text{Cov}(Y_{ij}, W) - \text{Cov}(\overline{Y}_{i.}, W)$$

where Y_{ij} is the jth (j = 1, 2,..., n) individual plant within the ith (i = 1,
2,..., s) stratum.
    Taking the ijth plant as the female plant from which W will descend,

$$\text{Cov}(Y_{ij}, W) = \text{Cov}(\text{parent-offspring}) = (1/2)\sigma_A^2$$

$$\text{Cov}(\overline{Y}_{i.}, W) = (1/n)\text{Cov}(\text{parent-offspring}) = [1/(2n)]\sigma_A^2$$

Thus,

$$\text{Cov}(X, W) = (1/2)\sigma_A^2 - [1/(2n)]\sigma_A^2 = (1/2)[(n - 1)/n]\sigma_A^2 \simeq (1/2)\sigma_A^2$$

if we consider n large enough that $(n - 1)/n \simeq 1$.
    The phenotypic variance within strata over the whole population is σ_y^2
$= \sigma_A^2 + \sigma_D^2 + \sigma_{we_s}^2$, where $\sigma_{we_s}^2$ is the environmental variability within strata.
So the expected progress would be

$$\Delta_G = \frac{k(1/2)\sigma_A^2}{(\sigma_A^2 + \sigma_D^2 + \sigma_{we_s}^2)^{1/2}}$$

which is doubled if selection is for both sexes at the same selection intensi-
ty. In the above formula, $\sigma_{we_s}^2$ should equal σ_{we}^2 (previously defined) if strata and plots from replicated trials are of the same size and under the same

environmental effects. If no stratification was used but the selection block
was assumed to be hypothetically divided into strata of any size, the phenotypic variance includes the environmental variance among different strata σ_s^2,
or $\sigma_y^2 = \sigma_A^2 + \sigma_D^2 + \sigma_s^2 + \sigma_{we_s}^2$. Efficiency of stratification can be found by
comparison of expected gain in each case; i.e., efficiency =

$$[(\sigma_A^2 + \sigma_D^2 + \sigma_s^2 + \sigma_{we_s}^2)/(\sigma_A^2 + \sigma_D^2 + \sigma_{we_s}^2)]^{1/2} = \{1 + [\sigma_s^2/(\sigma_A^2 + \sigma_D^2 + \sigma_{we_s}^2)]\}^{1/2}$$

The only purpose of this comparison is to show that the advantage of stratification increases for higher environmental variability among strata and stratification would show no advantage for $\sigma_s^2 = 0$.

In all cases considered so far, selection is assumed for only one environment and all variance components are estimated for one environment. If selection is directed to a population of environments, appropriate estimates to
predict genetic gain come from the population of environments. Gain is expected to be smaller than that for one environment if genotype-environment
interaction is large. Mass selection is conducted in specific environments
(e.g., locations), and genotype by environment interaction can only be estimated for a sample of environments. The phenotypic variance of the selection
units is defined for one environment as $\sigma_y^2 = \sigma_A^2 + \sigma_D^2 + \sigma_p^2 + \sigma_{we}^2$ or $\sigma_y^2 = \sigma_{A'}^2 +$
$\sigma_{D'}^2 + \sigma_{AE}^2 + \sigma_{DE}^2 + \sigma_p^2 + \sigma_{we}^2$, where σ_A^2 and σ_D^2 are biased estimates for one environment and $\sigma_{A'}^2$ and $\sigma_{D'}^2$ are for a population of environments. Expected gain
for a random sample of environments is a direct function of $\sigma_{A'}^2/\sigma_y^2$.

Gardner (1961) estimated components of variance by use of the design I
mating design in two locations; phenotypic variance was estimated by $\hat{\sigma}_y^2 = \hat{\sigma}_m^2$
$+ \hat{\sigma}_f^2 + \hat{\sigma}_{me}^2 + \hat{\sigma}_{fe}^2 + \hat{\sigma}_p^2 + \hat{\sigma}_w^2$, where σ_m^2 and σ_f^2 are the variances among males and
females, respectively; σ_{me}^2 and σ_{fe}^2 are variances due to interaction of male
and female effects by locations; σ_p^2 is the plot-to-plot environmental variance;
and σ_w^2 is phenotypic variance within plots. The average realized gain in four
generations was about 3.93% and the expected gain was 4.5%. Selection, as
well as yield tests of selected materials, was conducted in one location. On
the other hand, Hallauer and Sears (1969) obtained very little progress in two
maize populations when selection was in one location but yield tests for evaluation of progress were conducted at three different locations.

6.2.2 Family Selection

The primary difference between mass selection and family selection is
that family selection is based on some type of progeny test; i.e., plant genotypes are evaluated on the basis of average performance of their progeny. The
general scheme for family selection is as follows:

$$O \longrightarrow X$$
$$\searrow R \longrightarrow W$$

where O represents parental plants in the reference population, X is the selection unit, R is the recombination unit, and W represents an individual in
the improved population that is genetically related to X through R and O.

The important feature of family selection is that selection is based on family means (selection unit), which are obtained from replicated trials usually conducted over a set of environments.   Family means, therefore, are expected to show a phenotypic variance smaller than that for individual plants. Also, genotype-environment interaction has a less pronounced effect on results from selection, as can be seen in the phenotypic variance expression shown below.   The model for a family value is

$$Y_{ijk\ell} = m + f_i + b_j + E_k + (fE)_{ik} + e_{ij} + s_{ijk\ell}$$

where m = mean, f_i = families (i = 1, 2,..., p), b_j = replications/environments (j = 1, 2,..., r), E_k = environments (k = 1, 2,..., e), e_{ij} = experimental error associated with ith plot in jth replication, and $s_{ijk\ell}$ = plants/plots (ℓ = 1, 2,..., n).

Phenotypic variances among family means are

$$\sigma^2_{\bar{y}} = \sigma^2_f + \sigma^2_p/r + \sigma^2_w/nr \qquad \text{for one environment, and}$$

$$\sigma^2_{\bar{y}} = \sigma^2_{f'} + \sigma^2_{fE}/e + \sigma^2_p/(er) + \sigma^2_w/(ern) \qquad \text{for more than one environment}$$

where σ^2_f = variance (genetic) among family means for one environment, $\sigma^2_{f'}$ = variance (genetic) among family means over a series of environments, σ^2_{fE} = family by environment interaction variance, σ^2_p = plot-to-plot environmental variance (error variance), and σ^2_w = phenotypic variance within families.   In either case, the expected progress from selection is

$$\Delta_G = kc(\sigma^2_A + \beta_i)/\sigma_{\bar{y}}$$

To facilitate the description of predicted progress we consider families grown in a randomized complete block design in one environment.   Genotype-environment interaction can be easily introduced when more than one environment is considered (see Chap. 4).   Thus our discussion is based on the analysis of variance given in Table 6.3, in which estimates of all parameters are obtained on an individual plant basis.   Often the analysis of variance is performed either on plot totals or on plot means, and to express variance components on an individual plant basis the expectation of mean squares must be changed.

Table 6.3.   Structure of the analysis of variance for families evaluated in a randomized complete block design in one environment.

Source of variation	Degrees of freedom	Mean squares	Expected mean squares
Replications	r-1		
Families	f-1	M_1	$\sigma^2_w + n\sigma^2_p + nr\sigma^2_f$
Error	(r-1)(f-1)	M_2	$\sigma^2_w + n\sigma^2_p$
Within	rf(n-1)	M_3	σ^2_w

Half-sib Family Selection. Half-sib families can be used in intrapopulation improvement in several ways. One commonly used half-sib procedure is that suggested by Lonnquist (1964) and known as "modified ear-to-row" selection. This method involves selection among and within half-sib families. Selection among families is based on half-sib family means that are compared with the population mean or the mean of all families; i.e., the selection unit is X = $\overline{Y}_{i..} - \overline{Y}_{...}$, where Y_{ijk} is the phenotype of the kth plant of the ith family in the jth replication. A relationship of parent-offspring exists between R and W and R also is related as a half-sib with each of the nr plants of the ith family, so W is related as half-uncle-nephew with each plant in X. Thus

$$Cov(X, W) = Cov[(\overline{Y}_{i..} - \overline{Y}_{...}), W] = [1/(nr)]\sum_{jk} Cov(Y_{ijk}, W) = (1/8)\sigma^2_A$$

because $Cov(W, \overline{Y}_{...}) = 0$. In this case, $Cov(X, W)$ is half the genetic variance among half-sibs. The gain from selection among families is thus

$$\Delta_{G_1} = k_1(1/8)\sigma^2_A/\sigma_{\overline{P}}$$

In this case, $\sigma^2_f = \sigma^2_{HS}$ is the genetic variance among half-sib families; it is also the covariance between half-sibs and can be used to estimate the additive genetic variance, $\hat{\sigma}^2_f = Cov\ HS = (1/4)\hat{\sigma}^2_A$. However, selection experiments often are not large enough to provide good precision in genetic variance estimates.

The phenotypic variance $\sigma^2_{\overline{P}}$ among half-sib family means is $\sigma^2_{HS} = \sigma^2_f + \sigma^2/r + \sigma^2_w/(nr)$ and can be estimated directly from the corresponding mean square by $\hat{\sigma}^2_{\overline{HS}} = M_1/(nr)$. When half-sib families are evaluated over a series of environments the phenotypic variance is $\hat{\sigma}^2_{\overline{HS}} = \sigma^2_f + \sigma^2_{fe}/e + \sigma^2/(er) + \sigma^2_w/(enr)$.

In the case of e = 1 (one environment), σ^2_f is overestimated because it actually estimates $\sigma^2_f + \sigma^2_{fe}$, which causes an unknown bias in the expected gain. If the additive genetic variance is estimated as $\hat{\sigma}^2_A = 4\sigma^2_f$, then $\hat{\sigma}^2_A$ also is overestimated and the expected progress will be greater than the observed progress.

Within-family selection is conducted in the recombination block where all families are planted ear-to-row as females (which are detasseled) and pollinated by a sample of pollen produced by the interplanted male rows (a bulk of all families). Selection is conducted only within the best families, but selected plants are pollinated by the whole population so that among-family selection is for only one sex. The within-family selection is essentially phenotypic and usually is also for only one sex. It can be represented by R → W, where R is one individual within a half-sib family (which is itself the recombination unit for among-family selection). The selection unit is the difference between each individual phenotype and the family mean to which it belongs. However, the family mean is observed in just one block, so the selection unit is $Y_{ijk} - \overline{Y}_{ij.}$. A parent-offspring relationship exists between W and only one plant in the family; the remaining plants in the family are related to W as half-uncle-nephew. Hence

$$Cov(W, Y_{ijk} - \overline{Y}_{ij.}) = Cov(W, Y_{ijk}) - Cov(W, \overline{Y}_{ij.})$$

but

$$Cov(W, Y_{ijk}) = Cov(parent\text{-}offspring) = (1/2)\sigma^2_A$$

$$\text{Cov}(W, \overline{Y}_{ij.}) = (1/n)\text{Cov}(\text{parent-offspring}) + [(n-1)/n]\text{Cov}(\text{half-uncle-nephew})$$

$$= (1/n)\{(1/2)\sigma_A^2 + [(n-1)/8]\sigma_A^2\} \approx (1/8)\sigma_A^2$$

when n is large. Thus $\text{Cov}(W, Y_{ijk} - \overline{Y}_{ij.}) = (1/2)\sigma_A^2 - (1/8)\sigma_A^2 = (3/8)\sigma_A^2$.
Gain from selection within families is given by

$$\Delta_{G_2} = k_2 \frac{(3/8)\sigma_A^2}{\sigma_w}$$

where σ_w^2 is the phenotypic variance within half-sib families. Total expected
gain from selection among and within half-sib families is

$$\Delta_G = \Delta_{G_1} + \Delta_{G_2} = k_1 \frac{(1/8)\sigma_A^2}{\sigma_{\overline{HS}}} + k_2 \frac{(3/8)\sigma_A^2}{\sigma_w}$$

under the assumption that genetic variance within the selected families is the
same as that for the complete set of families.

An alternative procedure of modified ear-to-row selection is the recombination of only the best families after evaluation in replicated trials (Compton and Comstock 1976). In this case, remnant seed must be kept in good storage conditions for planting the next year. In the recombination block both
female and male gametes come from selected families so that selection among
families results in selection for both sexes. Within-family selection is also
possible in this case, but it usually is done only on the basis of female phenotypes so that the expected gain is the same as in the preceding case. The
total expected gain (Vencovsky 1969) is thus

$$\Delta_G = \Delta_{G_1} + \Delta_{G_2} = k_1 \frac{(1/4)\sigma_A^2}{\sigma_{\overline{HS}}} + k_2 \frac{(3/8)\sigma_A^2}{\sigma_w}$$

An additional year per cycle is not necessary if recombination is done in
off-season nurseries. Selection within families, however, is not recommended
in this case. This scheme, involving only among-family selection, usually is
referred to as "half-sib selection."

The effect of genotype-environment interaction is the same as that discussed in the preceding case; i.e., if interaction is of great magnitude and
families are evaluated over a series of environments, the additive genetic
variance estimate tends to be smaller than that obtained for one environment.
On the other hand, evaluation over environments has the advantage of decreasing the genotype-environment effect on phenotypic variance among family means.

For example, in a study reported by Paterniani et al. (1973), the effect
of genotype-environment interaction was well illustrated. Half-sib families
were developed from an irradiated (with gamma rays) population and a control
population and evaluated in two years. Additive genetic variance estimates
were practically unchanged from one year to another for each subpopulation and
were greater in magnitude for the irradiated one. However, combined analysis
(both years) revealed a significant family-years interaction in the irradiated
population leading to smaller estimates of σ_A^2 and expected progress. No interaction was detected for the nonirradiated population. The results are summarized in Table 6.4.

Gain from selection among half-sib families also can be increased by making use of prolific plants so that one ear is open pollinated and a second ear

Table 6.4. Additive genetic variance $(\hat{\sigma}_A^2)$ and mean $(\bar{x})$ estimates and expected progress (Δ_G) in
the population Centralmex (irradiated and control), using a half-sib family
performance in two years (Paterniani et al. 1973).

Treatment	$\hat{\sigma}_A^2$ $(x10^{-4})$			$\bar{x}$ (t/ha)			Δ_G (%)		
	1971	1972	Combined	1971	1972	Combined	1971	1972	Combined
Irradiated	1.2880	1.0624	0.1260	7.045	6.278	6.662	7.7	6.6	1.1
Control	0.7020	0.7184	0.7076	7.116	6.173	6.645	4.7	5.9	6.5

is selfed. Families are evaluated in replicated trials using seeds from open-
pollinated ears (half-sib families) and recombination is performed using
selfed seeds (S_1 families). If only the best S_1 families are recombined, selection is for both sexes and the expected gain is

$$\Delta_G = \frac{k(1/2)\sigma_A^2}{\sigma_{\overline{HS}}}$$

where $\sigma_{\overline{HS}}$ is the square root of the phenotypic variance among half-sib family
means (Empig et al. 1972).

Under this selection scheme an additional year is necessary because recombination units are inbred families (S_1) and selfing results in a greater
level of inbreeding. In the season following recombination the population is
again noninbred and can be used as source for new half-sib and S_1 families.
Recombination can be performed in off-season nurseries because there is no selection within families, and this procedure reduces to two years per cycle instead of three. Expected progress does not take into account selection for
prolificacy, but this may contribute to an additional gain from selection
since prolificacy seems to be a moderately heritable trait (Lonnquist 1967a;
Hallauer 1974).

The procedure reported by Lonnquist (1949) for the development of synthetic varieties also is a type of half-sib family selection. According to
this procedure a series of plants is selfed and at the same time each male
plant is crossed to a random sample of other plants of the population. The
crosses are made in the same way as in design I described in Chap. 4. Each
male plant, therefore, is evaluated in a number of families because individual
plants are related as full-sibs within families and as half-sibs among families. Selection among pollen parents is based on performance of the best
progenies in replicated trials, and the new population is produced by intercrossing the inbred progenies of the selfed male parents. If the crossed families are grown according to design I structure, several components of variance can be estimated and the expected improvement is

$$\Delta_G = \frac{k(1/2)\sigma_A^2}{\sigma_m^2 + \sigma_{me}^2 + [\sigma_f^2 + \sigma_{(f/m)e}^2]/n + \sigma_p^2/r + \sigma_w^2/(rk)}$$

where n is the number of plants to which each pollen parent was outcrossed, r
is the number of replications, k is the number of plants per plot, σ_m^2 and $\sigma_{f/m}^2$
are variances due to males and females/males effects, and σ_{me}^2 and $\sigma_{(f/m)e}^2$ are
variances due to interaction of males and females/males effects by environ-
ments (Robinson et al. 1955; Table 4.23).

Half-sib family selection also can be used in a selection scheme known as the testcross procedure. In this case the base population is crossed with a tester that can be either narrow base such as inbred lines or single crosses or broad base (including the parent population) such as varieties or populations. Families are evaluated in replicated trials, and selfed seeds from the best parents are used for recombination. Variances among tested half-sib families, as well as covariances between relatives, cannot be expressed as linear functions of genetic variances from the base population unless restrictions are imposed either on the gene frequencies or on the genetic model. For this reason a different approach for the interpretation of genetic variances, covariances, and expected gain must be introduced. The genetic variance among crosses (σ_f^2 from Table 6.3) at the one-locus level (see Chap. 2) is

$$\sigma_f^2 = (1/2)pq[a + (s - r)d]^2$$

and the covariance that enters as numerator for the expected progress is Cov (X, W) = $(1/2)pq[a + (q - p)d][a + (s - r)d]$ (Empig et al. 1972). This is another instance, therefore, where Cov(X, W) is not directly estimable from the estimated variance among testcrosses σ_f^2.

When the parent population is used as tester, p = r and q = s, variance among families is $\sigma_f^2 = (1/4)\sigma_A^2$, and the numerator of expected progress is Cov (X, W) = $2(1/4)\sigma_A^2$, which is exactly the same as that previously shown when selection is based on half-sib families and recombination from use of selfed seed. When only the best families are recombined the expected progress is

$$\Delta_G = \frac{k(1/2)\sigma_A^2}{\sigma_{\overline{HS}}}$$

Full-sib Family Selection. Under this scheme selection units X are full-sib families evaluated in replicated trials and remnant full-sib seed is used to recombine the best families. Each cycle requires two years and new full-sib families are obtained from the recombination block. If we consider selection for one sex,

$$Cov(X, W) = Cov[\overline{Y}_{i..} - \overline{Y}_{...}), W] = Cov(W, \overline{Y}_{i..}) - Cov(W, \overline{Y}_{...})$$

but Cov(W, $\overline{Y}_{i..}$) = $[1/(nr)]\Sigma_j\Sigma_k Cov(W, Y_{ijk})$ = $(1/4)\sigma_A^2$ because Cov(W, Y_{ijk}) = Cov(uncle-nephew) = $(1/4)\sigma_A^2$ and Cov(W, $\overline{Y}_{...}$) = 0. The value of Cov(X, W) is not directly estimable from the variance among full-sib families because σ_f^2 = $(1/2)\sigma_A^2 + (1/4)\sigma_D^2$, assuming no epistasis. Expected progress from selection would be

$$\Delta_G = \frac{k(1/4)\sigma_A^2}{\sigma_{\overline{FS}}}$$

where $\sigma_{\overline{FS}}$ is the square root of the phenotypic variance among full-sib family means.

Usually selection is for both sexes (c = 2 × 1/4 = 1/2) because only the best families are intercrossed for recombination; for this case expected progress is

$$\Delta_G = \frac{k(1/2)\sigma_A^2}{\sigma_{\overline{FS}}}$$

Use of prolific plants is also an alternative in full-sib family selection. Thus for each prolific plant one ear is used to produce full-sib families that are evaluated in replicated trials. A second ear for each parental plant is selfed and used for recombination of the best families. If selection is for both sexes, expected progress is the same as when remnant seeds from full-sib families are used for recombination. Hence there is no advantage in terms of expected progress if we recombine S_1 instead of full-sib families. Selection for prolificacy can result in an additional gain for yield that is, however, unpredictable.

S_1 *Family Selection.* Under this scheme the selection units are S_1 family means compared with the grand mean of all S_1 families. Remnant seeds from the selfed ears are used for recombination. Since inbreeding is involved, Cov (X, W) cannot be directly expressed as a linear function of variance components unless restrictions are imposed on either the gene frequency or the genetic model. For a one-locus situation we have

$$\text{Cov}(X, W) = 2pq[a + (q - p)d][a + (1/2)(q - p)d] = \sigma_{A_1}^2$$

if selection is for both sexes.

The above expression can be represented as $\sigma_{A_1}^2 = \sigma_A^2 + \beta_1$, where β_1 is a deviation from the additive genetic variance due mainly to dominance effects, $\beta_1 = 2pq(p - 1/2)d[a + (q - p)d]$ (Empig et al. 1972). Deviation $\beta_1 = 0$ either for $p = q = 0.5$ or for a completely additive genetic model ($d = 0$ for all loci). Thus the expected progress for S_1 family selection is

$$\Delta_G = k\sigma_{A_1}^2/\sigma_{\overline{S}_1} = k(\sigma_A^2 + \beta_1)/\sigma_{\overline{S}_1}$$

where $\sigma_{\overline{S}_1}$ is the square root of the phenotypic variance among S_1 family means.

The variance $\sigma_{A_1}^2 = \sigma_A^2 + \beta_1$ is not estimable from S_1 family trials because the variance among S_1 families estimates $\hat{\sigma}_{A'}^2 + (1/4)\hat{\sigma}_D^2 = \sigma_A^2 + \beta' + (1/4)\sigma_D^2$ (see Chap. 2). Hence expected progress can only be expressed in terms of additive genetic variance under the restriction of no dominance or gene frequency of 0.5.

S_2 *Family Selection.* By this selection scheme, S_2 families are developed by selfing S_1 plants. Selection units are S_2 family means compared with the grand mean of all S_2 families. Recombination is with remnant seed from the selfed ears. As in the preceding case of S_1 family selection, Cov(X, W) cannot be expressed as a linear function of variance components without a restriction either on gene frequency or on the genetic model. For one locus, we have

$$\text{Cov}(X, W) = 3pqa^2 + (7/2)pq(q - p)ad + (1/2)pq(q - p)^2d^2$$

after multiplying by two to accommodate selection for both male and female gametes. If we define $\sigma_{A_2}^2 = (2/3)\text{Cov}(X, W)$, then $\text{Cov}(X, W) = (3/2)\sigma_{A_2}^2 = (3/2) \cdot (\sigma_A^2 + \beta_2)$, where β_2 is a deviation from additive genetic variance due mainly to dominance effects; i.e., $\beta_2 = (10/3)pq(p - 1/2)d[a + (q - p)d]$ and $\beta_2 = 0$ either for $p = q = 0.5$ or for a completely genetic model. The expected progress is thus calculated as

$$\Delta_G = \frac{k(3/2)\sigma^2_{A_2}}{\sigma_{\overline{S}_2}} = \frac{k(3/2)(\sigma^2_A + \beta_2)}{\sigma_{\overline{S}_2}}$$

where $\sigma_{\overline{S}_2}$ is the square root of the phenotypic variance among S_2 family means. As in the case of S_1 families, $\sigma^2_{A_2} = \sigma^2_A + \beta_2$ is not estimable because the variance among S_2 family means is an estimate of $(3/2)(\sigma^2_A + \beta'') + (3/16)\sigma^2_D$, which equals $(3/2)\sigma^2_A$ only in a completely additive genetic model or gene frequency of 0.5.

Combined Selection. Combined selection makes use of two or more selection methods in the same program and can be classified into one of the following categories: (1) those combining two or more methods alternatively and (2) those using two or more methods simultaneously. For example, the modified ear-to-row method of selection is included in the first category because the procedure alternates selection among families (progeny selection) and within families (mass selection). Other combinations of selection methods are possible, e.g., combined S_1, mass, and reciprocal recurrent selection as proposed by Lonnquist (1967b, 1967c). Expected progress from a combination of selection methods is obtained by merely summing the expected progress for each individual method. In a broad sense most breeding methods would be considered as combined selection. For example, in all methods where prolific plants are required, phenotypic selection for prolificacy is done in addition to any further selection. In half-sib family selection, some mass selection is usually done when choosing parental plants. In addition to S_1 and S_2 generations, higher levels of inbreeding can be used in recurrent selection. When selected (elite) inbred lines S_n from one population are intermated to form a synthetic population, the process is similar to one cycle of recurrent selection with N generations per cycle (N > n). The effect of no planned extra selection is usually unpredictable but should be taken into account in evaluation of selection methods.

Simultaneous combinations of methods have not been studied comprehensively. Lonnquist and Castro (1967) suggested the use of combined selection through information obtained from simultaneous evaluation of half-sib and S_1 families, and results from this method have been reported (Goulas and Lonnquist 1976). Briefly, the method includes obtaining half-sib and S_1 families from the same prolific plants. They are evaluated simultaneously, providing information for selection of superior individuals. If the criterion for selection (selection unit) is the mean of both types of families, and estimates of components of variance are available, expected progress can be calculated.

Family evaluation can be included in a split-plot design. If, for example, randomized complete blocks are used, each parental plant is evaluated in a whole plot (containing one half-sib and one S_1 family) and each family type is evaluated in a split plot. However, when half-sib and S_1 families are evaluated this way, competition between families can be a serious source of experimental error because of differences of vigor (due to inbreeding of S_1 families). Border rows can be used to reduce this source of error. An alternative procedure to avoid competition effects would be to keep the half-sib

Table 6.5. Structure of the analysis of variance for half-sib (HS)
           and S_1 family evaluation in an experiment with both
           factors in strips for one environment.

Source	df	MS	E(MS)[†]
Replications	r-1		
Parents (P)	n-1	M_1	$\sigma_c^2 + 2\sigma_a^2 + 2r\sigma_p^2$
Error a (P × R)	(n-1)(r-1)	M_2	$\sigma_c^2 + 2\sigma_a^2$
Families (F)	1	M_3	$\sigma_c^2 + n\sigma_b^2 + r\sigma_{pf}^2 + nrK_f^2$
Error b (F × R)	r-1	M_4	$\sigma_c^2 + n\sigma_b^2$
Interaction (P × F)	n-1	M_5	$\sigma_c^2 + r\sigma_{pf}^2$
Error c (P × F × R)	(n-1)(r-1)	M_6	σ_c^2

[†]r is number of replications, n is number of pairs of families, and
$K_f^2 = \sum_k f_k^2$.

and the S_1 families in separate groups as a two-factor experiment with both
factors in strips (Kempthorne 1952). The analysis of variance is shown in Table 6.5.

From Table 6.5 we obtain $\sigma_p^2 = (M_1 - M_2)/(2r)$, which expresses the genetic
variance among pairs of families. The phenotypic variance among pairs of fam-
ilies is $\sigma_F^2 = \sigma_p^2 + \sigma_a^2/r + \sigma_c^2/(2r)$, which can be estimated by $\hat{\sigma}_{\overline{F}}^2 = M_1/(2r)$.
Use of border rows also is recommended, but it could increase the size of the
experiment.

Although families are evaluated in an experimental design similar to a
split plot, separate analysis for each type of family can be obtained. If
epistasis is negligible, an unbiased estimate of σ_A^2 can be obtained from half-
sib family analysis ($\hat{\sigma}_A^2 = 4\hat{\sigma}_{HS}^2$). If a completely additive model is assumed,
σ_A^2 also can be estimated from S_1 family analysis ($\hat{\sigma}_A^2 = \hat{\sigma}_{S_1}^2$). If gene frequency is 0.5 for all segregating loci, $\hat{\sigma}_{S_1}^2$ estimates $\sigma_A^2 + (1/4)\sigma_D^2$ and an estimate
of dominance variance also can be obtained.

The covariance between half-sib and S_1 families also can estimate σ_A^2
through the relation $Cov(HS, S_1) = (1/2)\sigma_A^2$ if epistasis is negligible and a
completely additive model or gene frequency of 0.5 is assumed. If no evidence
exists to support restrictions about either the genetic model or the gene frequency, an unbiased σ_A^2 estimate (from half-sib families) is preferable.

The expected progress from selection depends on which type of family is
used as the recombination unit to obtain the next generation. If the best
half-sib families are recombined, $Cov(X, W) = (3/8)(\sigma_A^2 + \beta_3)$, where $\beta_3 = (4/3) \cdot$
$pq(p - 1/2)d[a + (q - p)d] = 0$ either for $p = q = 0.5$ or for a completely additive model. Thus

$$\Delta_G = \frac{k(3/8)(\sigma_A^2 + \beta_3)}{\sigma_{\bar{P}}}$$

If remnant seeds from S_1 families are used for recombination, $Cov(X, W)$ = $(3/4)(\sigma_A^2 + \beta_3)$ and the expected progress is given by

$$\Delta_G = \frac{k(3/4)(\sigma_A^2 + \beta_3)}{\sigma_{\bar{F}}}$$

where $\sigma_{\bar{F}}$ is the square root of the phenotypic variance among family means.

The numerator $[Cov(X, W)]$ corresponds to the average value of the two methods previously discussed: half-sib family selection using S_1 seeds for recombination and S_1 family selection considering either a gene frequency of 0.5 or an additive genetic model. Expected progress from selection within families in the recombination block is the same as those given for half-sib and S_1 family selections (Table 6.12).

6.7 INTERPOPULATION SELECTION METHODS

Selection schemes designed to improve the cross between two populations are known as reciprocal recurrent selection (RRS) and all are based on the original procedure proposed by Comstock et al. (1949). All procedures have a common feature, i.e., improvement of populations by changing gene frequencies in a directed and complementary way so that a wide range of different types of gene action and interactions can be retained in the crossed population. The RRS procedure was originally designed to make maximum use of general and specific combining abilities. Its most important feature is that selection is toward the improvement of populations themselves as well as the increase of heterosis in the crossed population. When both populations are improved in this way, their cross in the first generation can be used directly for grain production. However, maximization of heterotic effects can be attained only when improved populations are used as sources of inbred lines for hybrid development. Improvement of parent populations as a source of inbred lines also can be effected through some intrapopulation selection method; but improvement would be directed only to additive genetic effects and heterosis (due to non-additive effects) is changed only by chance. RRS schemes and their general features are summarized in Table 6.6.

In all instances crossed families are evaluated in replicated trials, and components of variance are obtained from the analysis of variance. If, for example, n_1 and n_2 families are obtained from populations A and B, respectively, and are evaluated in a randomized complete block design, estimates can be obtained from an analysis as shown in Table 6.7.

6.7.1 Half-sib RRS

The selection procedure for half-sib RRS (Comstock et al. 1949) is based on families obtained from the cross between two populations, say A_0 and B_0. Crossed families are obtained in the same way as in design I described in Chap. 4, i.e., plants in population A_0 (used as males) are each crossed with a number of plants in population B_0 (used as females). Reciprocal crosses are made in the same way with different plants. At the same time the plants used as males are selfed and their S_1 seeds are used as recombination units to obtain the improved populations (A_1 and B_1). The procedure is represented as follows:

Table 6.6.   Reciprocal recurrent selection (RRS) schemes and their general features.

RRS schemes[†]	Selection unit[‡]	Recombination unit	Improved population (A_1 and B_1)
1. Half-sib RRS	HS and FS	S_1 families	progenies from $S_1 \times S_1$
2. Full-sib RRS	FS	S_1 families	progenies from $S_1 \times S_1$
3. Modified half-sib RRS-1	HS	HS families	recombined HS
4. Modified half-sib RRS-2	HS	HS families	recombined HS

[†]Scheme 1: Comstock et al. (1949); scheme 2: Hallauer and Eberhart (1970); schemes 3,4: Paterniani (1967a, 1973) and Paterniani and Vencovsky (1977, 1978).

[‡]HS: half-sibs; FS: full-sibs. In scheme 3, HS is a pooled half-sib family from a cross between an HS family from one population and the opposite population.

Table 6.7.   Structure of the analysis of variance for crossed families evaluated in a randomized complete block design in one environment.

Source	Population A as male		Population B as male	
	MS	E(MS)	MS	E(MS)
Blocks	M_A	$\sigma^2_{w_1} + n\sigma^2_1 + nf\sigma^2_{b_1}$	M_B	$\sigma^2_{w_2} + n\sigma^2_2 + nf\sigma^2_{b_2}$
Families	M_{1A}	$\sigma^2_{w_1} + n\sigma^2_1 + nr\sigma^2_{f_{12}}$	M_{1B}	$\sigma^2_{w_2} + n\sigma^2_2 + nr\sigma^2_{f_{21}}$
Error	M_{2A}	$\sigma^2_{w_1} + n\sigma^2_1$	M_{2B}	$\sigma^2_{w_2} + n\sigma^2_2$
Within	M_{3A}	$\sigma^2_{w_1}$	M_{3B}	$\sigma^2_{w_2}$

[†]Subscripts 1 and 2 denote populations A and B, respectively.

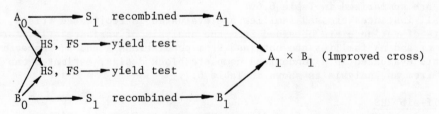

From Table 6.7, the following estimates can be determined: $\hat{\sigma}^2_{f_{12}}$ = $(M_{1A} - M_{2A})/(nr)$ and $\hat{\sigma}^2_{f_{21}} = (M_{2B} - M_{2B})/(nr)$, which also estimate $(1/4)\hat{\sigma}^2_{A_{12}}$ and $(1/4)\hat{\sigma}^2_{A_{21}}$ (as shown in Chap. 2), respectively. Expected progress is obtained as follows:

$$\Delta_G = \frac{k_1 \sigma_{f_{12}}^2}{\sigma_{\overline{F}_{12}}} + \frac{k_2 \sigma_{f_{21}}^2}{\sigma_{\overline{F}_{21}}} = \frac{k_1 (1/4) \sigma_{A_{12}}^2}{\sigma_{\overline{F}_{12}}} + \frac{k_2 (1/4) \sigma_{A_{21}}^2}{\sigma_{\overline{F}_{21}}}$$

where $\sigma_{\overline{F}_{12}}^2$ and $\sigma_{\overline{F}_{21}}^2$, the square roots of the phenotypic variances among family means, can be estimated by $M_{1A}/(nr)$ and $M_{1B}/(nr)$, respectively. When more than one environment is considered the genotype-environment interaction is included.

If gene frequency is the same in both populations, $\hat{\sigma}_{A_{12}}^2 = \hat{\sigma}_{A_{21}}^2 = \sigma_A^2$ and the above expression will be the same as that given for intrapopulation improvement, where selection is based on half-sib families and S_1 seeds from the selected parents are used for recombination.

6.7.2 Full-sib RRS

The techniques described for half-sib RRS are used for full-sib RRS except that full-sib progenies are evaluated instead of half-sib progenies, as described by Comstock et al. (1949). One main advantage is that twice as many plants can be sampled from the source populations to have the same number of progenies to evaluate as for half-sib RRS. Full-sib RRS also has the main objective of improvement of the crossed population of the two base populations. Full-sib RRS is very effective if the two source populations produce two or more female inflorescences; i.e., one ear is used to produce the S_1 seed and the second ear is used to produce the full-sib seed (Hallauer and Eberhart 1970). The procedures of full-sib RRS can be illustrated as follows:

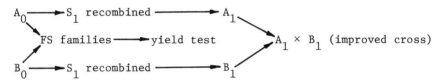

In Table 6.7, both $\hat{\sigma}_{f_{12}}^2$ and $\hat{\sigma}_{f_{21}}^2$ estimate the same variance, $(1/4) \cdot (\sigma_{A_{12}}^2 + \sigma_{A_{21}}^2) + (1/4) \sigma_{D_{12}}^2$ (according to notation used in Chap. 2). Expected progress would be

$$\Delta_G = \frac{k(1/4)(\sigma_{A_{12}}^2 + \sigma_{A_{21}}^2)}{\sigma_{\overline{F}}}$$

6.7.3 Modified RRS-1 (HS-RRS1)

A modification of the original RRS was proposed by Paterniani (1967b). Briefly, the procedure is as follows. A number of open-pollinated ears (half-sib families) are planted ear-to-row as females in a detasseling block where the male rows are plants from population B. In another isolated block, half-sib families from population B are used as females and male rows are from population A. From each block open-pollinated ears from half-sib families are harvested in bulk and evaluated in replicated trials. Remnant seed from half-sib families from populations A and B is used for recombination to form A_1 and B_1, as follows (see Paterniani and Vencovsky 1977):

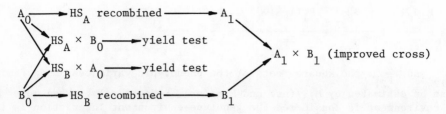

The expected progress from selection is

$$
\Delta_G = \frac{k_1(1/16)\sigma^2_{A_{12}}}{\sigma_{\overline{F}_{12}}} + \frac{k_2(1/16)\sigma^2_{A_{21}}}{\sigma_{\overline{F}_{21}}}
$$

The modified method requires three years per cycle and, because of some other advantages, it may be an efficient method to improve the population cross.

6.7.4 Modified RRS-2 (HS-RRS2)

Another modification of RRS proposed by Paterniani (1967b) makes use of prolific plants, as in the original procedure. Population A is planted in an isolated block I as females (detasseled) that are pollinated by population B (male rows). In another detasseling block II population B is used as female and population A as male. At flowering time a sample of pollen is collected from population A (male rows in block II) and used to pollinate the second ears of population A plants (female rows in block I). The reverse is done in block II; i.e., pollen collected from male rows in block I is used to polli- nate female rows in block II. In each field the first ears in the female rows are open pollinated to form half-sib families (A plants × B population in block I and B plants × A population in block II). The crossed half-sib fami- lies are evaluated in replicated trials and recombination is done with hand- pollinated ears (half-sib families) from each population. The procedure is represented as follows:

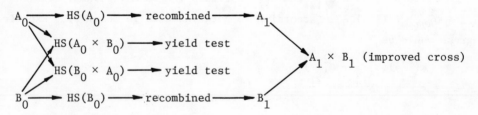

The method is relatively simple and each cycle is completed in two years. An expression that allows calculation of expected progress is given by Pater- niani and Vencovsky (in press):

$$
\Delta_G = \frac{k_1(1/8)\sigma^2_{A_{12}}}{\sigma_{\overline{F}_{12}}} + \frac{k_2(1/8)\sigma^2_{A_{21}}}{\sigma_{\overline{F}_{21}}} = \frac{k_1(1/2)\sigma^2_{f_{12}}}{\sigma_{\overline{P}_{12}}} + \frac{k_2(1/2)\sigma^2_{f_{21}}}{\sigma_{\overline{P}_{21}}}
$$

where $\sigma^2_{f_{12}}$ and $\sigma^2_{f_{21}}$ are obtained from Table 6.7.

The analyses of variance in Tables 6.3 and 6.7 for intra- and interpopu-

Table 6.8. Expected mean squares on individual plant basis for analysis of variance on plot total basis, plot mean basis, and individual plant basis.

Source	Expected mean squares[+]		
	Plot total	Plot mean	Individual plant
Families	$n^2\sigma^2 + n^2 r\sigma_f^2$	$\sigma^2 + r\sigma_f^2$	$n\sigma^2 + nr\sigma_f^2$
Error	$n^2\sigma^2$	σ^2	$n\sigma^2$

[+] n is number of plants per plot; r is number of replications; and $\sigma^2 = \sigma_w^2/n + \sigma_p^2$, where σ_w^2 is the within plot variance and σ_p^2 is the plot-to-plot environmental variance.

lation family evaluation are based on single plant data and all components of variance are estimated in the same unit. Analyses of variance are commonly performed on either plot totals or plot means and the components of variance are estimated in the corresponding units. From analyses of variance based on either plot totals or plot means, components of variance can be estimated on an individual plant basis by changing the coefficients of the expected mean squares as shown in Table 6.8.

6.8 COMPARING BREEDING METHODS

Maize breeding methods have evolved and have been modified by our increased knowledge in plant breeding and related sciences. At the present several procedures are available for the improvement of a population or a cross between two populations. Choice of the best procedure depends on the trait and population under selection, stage of the breeding program, and purpose of the breeding program. Chapter 12 gives more details on maize breeding plans and shows the peculiarities of each breeding and selection procedure.

At any stage of any breeding program, a breeder needs to know the effectiveness of a given selection procedure. Comparison of selection procedures relative to their effectiveness by use of empirical results is a difficult task because several variables are involved. Breeders have used different methods on different populations under different environments and circumstances, usually with different objectives. Also, selection criteria are not expected to be the same among breeders. Sample size, selection intensity, and relative weights when selection is for several traits are examples of breeding procedures that differ among breeders. As shown in Chap. 7, however, several programs have been conducted specifically for comparing breeding methods.

Development of quantitative genetic theory has provided the basis for comparison of relative efficiency among several breeding systems. Because of difficulties already mentioned, theoretical comparisons are useful if estimated parameters are realistic. When comparing different selection methods, two approaches may be considered: (1) that based on equal selection intensity and (2) that based on equal effective population size. The first approach is more important for comparing progress in short-term selection programs because the goal is to maximize gain within a few generations of selection. The second approach is considered if the selection program is intended to be long term; genetic variability in the population cannot be reduced drastically in a few generations if we are to expect continuous progress during the course of the

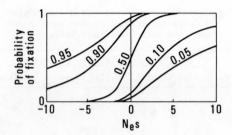

Fig. 6.3. Probability of fixation of a gene acting additively for different
           initial gene frequencies (Robertson 1960).

program. When effective size of a population can be kept at a high level,
however, differences in effective sizes resulting from different selection
procedures are irrelevant.
     Rawlings (1970) pointed out that effective size usually is not a great
problem in selection programs and that effective population sizes of 30 to 45
should be a reasonable number. Robertson (1960), on the other hand, emphasized the importance of effective size in selection programs because the total
expected progress and the half-life of a recurrent selection program are proportional to effective population size. Robertson's main conclusions were as
follows:

     1. For a single gene with selective advantage s, the chance of fixation
(expected gene frequency at the limit after an infinite number of generations)
is a function of the initial gene frequency and $N_e s$, where N_e is the effective
population size. Figure 6.3 gives the chance of fixation for genes acting additively. In artificial selection for a quantitative trait based on individual measurements, the selection differential can be expressed in k standard deviation (σ) units and the coefficient of selection s can be found by s = 2ka/
σ, where a is half the difference between the mean of the two homozygotes (on
a metric scale). Therefore, for one gene with a given frequency and effect in
a population the probability of fixation is a function of $N_e k$.
     2. The question of how long it would take to attain the selection limit
is not germane because the approach of gene frequency to the limit is asymp-

Table 6.9. Probability of fixation (Pr) of the desirable
            allele in a population where initial gene frequency
            is p and effective population size is N_e.

N_e	Initial gene frequency				
	0.1	0.2	0.3	0.4	0.5
1	0.1275	0.2480	0.3617	0.4692	0.5707
4	.1584	.2997	.4258	.5383	.6387
16	.3699	.6044	.7531	.8474	.9071
32	.5982	.8386	.9352	.9740	.9896
64	0.8385	0.9739	0.9958	0.9993	0.9999

$Pr = (1 - e^{-2N_e sp})/(1 - e^{-2N_e s})$ (Robertson 1960); selection
coefficient was given by Baker and Curnow (1969); s = 0.28
for selfing (N_e = 1) and s = 0.14 otherwise (selection of
female plants with infinite number of males, i.e., random
pollination).

Table 6.10. Gene frequency expected for 1, 5, 10, and
infinite generations of selection under
different effective population sizes for
one dominant gene, with initial gene
frequency p = 0.2 (Baker and Curnow
1969).

N_e	Generations				
	0	1	5	10	∞
1	0.2	0.219	0.236	0.237	0.258
4	.2	.212	.243	.266	0.472
16	.2	.212	.256	.304	0.948
32	.2	.212	.258	.315	0.998
64	.2	.212	.260	.322	1.000
256	0.2	0.212	0.261	0.327	1.000

totic. However, it may be useful to know how long it would take for the mean
gene frequency to get halfway to the limit or what would be the half-life of a
selection process. The half-life for any selection process for additive genes
will not be greater than 1.4N generations but may for rare recessives be $2N_e$.
The half-life of a selection process, therefore, depends directly on the effective population size.

Few results have been reported on the probability of fixation for different selection procedures. Baker and Curnow (1969) examined the consequences
of different effective population sizes on progress from selection within a
population with a specific genetic model. The probability of fixation of a
desirable allele for different initial gene frequencies is shown in Table 6.9,
which shows that ultimate probability of fixation increases for greater effective population sizes. Table 6.10 shows the increase in gene frequency over
one to ten generations for an allele with initial gene frequency p = 0.2, with
dominance.

Baker and Curnow (1969) concluded that there would be a small difference
in the frequency of the desirable homozygote when comparing effective sizes of
16 and 256 after some generations of selection, suggesting that a reasonably
rapid progress from selection can be expected with small effective population
sizes. In addition, a substantial added progress would be obtained if selection could be practiced within each of a number of replicate lines developed
out of the same original population, followed by selection of the best replicate lines.

Vencovsky and Godoi (1976) used the expected change in gene frequency in
one cycle of selection and the ultimate probability of fixation to investigate
the relative power of three selection schemes combined with varying selection
intensities: (1) selection among half-sib families: 5/100 and 15/100; (2)
selection among full-sib families: 10/100 and 30/100; and (3) half mass selection: 25/5000 and 80/16,000. All the procedures lead to equivalence of
effective population sizes, considering the first on the second selection intensity, for each procedure. Using specific computer programs for a large
number of genetic and environmental parametric combinations, the following
conclusions were obtained: Concerning the ultimate probability of fixation,
half mass selection was the most powerful method even for heritability as low
as 0.05. Full-sib family selection generally ranked second. However, the se-

lection coefficient s was generally higher for the full-sib scheme, indicating also greater expected immediate response per cycle. When the expected gain per generation was measured, half mass selection was the most efficient scheme and half-sib family selection ranked second for the selection intensities used. A lower selection pressure (500/5000) for mass selection was inefficient for immediate response, mainly under high genotype-environment interactions. Results suggest that selection programs should exploit mass selection primarily through higher selection pressure, which would provide a better balance between immediate response and long-range variability.

Expected progress from selection has been the most widely used way to compare different selection methods, and the key formula is

$$\Delta_G = (k/t)[c_i(\sigma_A^2 + \beta_i)/\sigma_{\bar{y}}]$$

where k is a function of selection intensity, c_i is a coefficient that depends on the selection method and parental control, β_i is a deviation from additive genetic variance and is different from zero only for inbred families, σ_A^2 is the additive genetic variance, $\sigma_{\bar{y}}$ is the square root of the phenotypic vari-

Table 6.11. Expected progress from several intrapopulation selection methods expressed by components of the general formula for one environment.

Method of selection	c_1[†]	c_2[†]	σ_f^2	σ_p^2	σ_w^2	β	t
			Components of $\sigma_{\bar{y}}^2$				
Mass selection	1/2	1	1	1	1	0	1
Half-sib family							
1. Modified ear-to-row							
among families	1/8	1/4	1	1/r	1/kr	0	1 or 2[#]
within families	3/8	--	1	1/r	1/kr	0	--
2. Recombining S_1	--	1/2	1	1/r	1/kr	0	3
3. Test cross‡	--	1/2	1	1/r	1/kr	--	3
Full-sib family							
1. Recombining FS	--	1/2	1	1/r	1/kr	0	2
2. Recombining S_1	--	1/2	1	1/r	1/kr	0	3
Inbred family							
1. S_1 family	--	1	1	1/r	1/kr	θ[§]	3
2. S_2 family	--	3/2	1	1/r	1/kr	$(5/3)\theta$[§]	4
Combined selection (HS+S_1)							
1. Recombining HS	--	3/8	1	1/r	1/2r	$(2/3)\theta$[§¶]	2
2. Recombining S_1	--	3/4	1	1/r	1/2r	$(2/3)\theta$[§¶]	3

[†]c_1 and c_2 refer to selection for one and two sexes, respectively.

‡c_2 = 1/2 only for a completely additive model, when β = 0.

¶Substitute σ_a^2 for σ^2 and σ_c^2 for σ_w^2 (see Table 6.5).

[#]t = 1 or 2, depending if recombination is in the same year or the next year.

§θ = 2pq(p - 1/2)d[a + (1 - 2p)d] at one locus level.

ance of the selection unit, and t is the number of years required per cycle so that the expected gain is expressed as gain per year. Using the general formula, the first comparisons among methods for intrapopulation improvement are shown in Tables 6.11 and 6.12.

Mass selection is the oldest and simplest of the schemes listed in Table 6.11. Its simplicity and the possibility of one cycle per year are the greatest advantages over other methods. The coefficient of σ_A^2 in the general formula is 1/2 (selection for one sex) or 1 (selection for both sexes), showing that the method makes use of a high proportion of additive genetic variance. Selection can be highly effective when the magnitude of genetic variance is large relative to nongenetic variances, as generally occurs with highly heritable traits. Its great disadvantage is that selection is based on individual phenotypes, resulting in a phenotypic variance among selection units greater than any other selection method. Field stratification, suggested by Gardner (1961), is intended to reduce the phenotypic variability and thus increase the expected progress and allow mass selection to be more effective for traits having lower heritabilities. Also, the possibility of selection for both sexes may increase the expected gain, as previously shown. One of the most important features of mass selection is that effective population size can be kept at a high level even if high selection pressures are used. This assures the breeder of maintaining genetic variability for several cycles and allows continuous gain from selection in long-term selection programs.

The advantage of any family selection procedure over mass selection is that genotypes are evaluated by means of a progeny test; i.e., families are evaluated in replicated trials. When experiments are replicated in a series

Table 6.12. Coefficient (c) for additive genetic variance $(\sigma_A^2)^\dagger$ relative to expected progress for several intrapopulation selection methods.

Selection method	Selection unit	Recombination unit	One sex		Both sexes[‡]	
			Among	Within	Among	Within
1. Mass	I	I	1/2	---	1	---
2. Mass	I	S_1	---	---	1	---
3. Modified ear-to-row	HS	HS	1/8	3/8	1/4	3/4
4. Modified ear-to-row	HS	S_1	1/4	1/4	1/2	1/2
5. Half-sib (testcross)[§]	HT	S_1	1/4	1/4	1/2	1/2
6. Full-sib	FS	FS	1/4	1/4	1/2	1/2
7. Full-sib	FS	S_1	1/4	1/4	1/2	1/2
8. Inbred (S_1)	S_1	S_1	1/2	1/4	1	1/2
9. Inbred (S_2)	S_2	S_2	3/4	1/8	3/2	1/4
10. Combined $(HS-S_1)$	$HS-S_1$	HS	3/16	3/8	3/8	3/4
11. Combined $(HS-S_1)$	$HS-S_1$	S_1	3/8	1/4	3/4	1/2

[†]Definition for additive genetic variance changes slightly with inbreeding and equals σ_A^2 only for gene frequency one-half (p = 0.5) or no dominance (d = 0).

[‡]Assuming same selection intensity for both sexes.

[§]Testcross using a nonrelated population as tester.

of environments genotype-environment interaction is taken into account, which is not the case for mass selection. In all instances of family selection the phenotypic variance is reduced, thus increasing the expected gain. The coefficient of σ_A^2, however, may be smaller than that for mass selection and the reduction in phenotypic variance must compensate for this difference. The advantage of family selection over mass selection increases for characters with low heritability, such as yield in maize.

For half-sib family selection several procedures can be used. For selection among and within half-sib families and one year per cycle, the coefficient of σ_A^2 is 1/8 for among-family and 3/8 for within-family selection. The among-family coefficient can be increased to c = 1/4 if recombination is done in the next year using only remnant seed from the selected families, but the gain per cycle must be divided by two to be comparable with the preceding case (c = 1/8) on a per year basis. Among-family selection assures a higher effective population size because all families rather than only selected ones are planted for recombination. When several populations are under selection in the same program, the second procedure permits the breeder to stagger yield trials and recombination phases among populations. Within-family selection is nearly as efficient as among-family selection in some instances (Webel and Lonnquist 1967), but the relative gain in each phase depends on the ratio σ_w^2/σ^2 and on selection intensity. Table 6.13 gives the expected gain for selection among half-sib families in relation to the total expected gain (among and within families) for different combinations of selection intensities for three populations. It shows that, as expected, relative efficiency of within-family selection increases for higher values of the heritability coefficient.

Another alternative for selection among half-sib families is to recombine selfed seed obtained from a second ear in the parental plants. This procedure doubles the coefficient of σ_A^2 (c = 1/2), which increases the gain per cycle but also increases the number of seasons per cycle. If recombination is done in off-season nurseries, an additional year is not necessary. In addition, this procedure forces selection for prolificacy, which may result in an addi-

Table 6.13. Expected gain in yield among half-sib families in percent of total gain (among and within populations) for different combinations of selection intensities for three populations (Ramalho 1977).

Selection intensity (%)		Cateto		ESALQ-HV 1		Paulista Dent	
Among	Within	1-HS[†]	2-HS[†]	1-HS	2-HS	1-HS	2-HS
100	2	0.0	0.0	0.0	0.0	0.0	0.0
25	8	49.8	66.5	46.3	63.3	40.8	57.9
20	10	53.6	69.8	50.1	66.7	44.5	61.6
10	20	64.3	78.3	61.0	75.8	55.5	71.4
5	40	75.3	85.9	72.6	84.1	67.8	80.1
2	100	100.0	100.0	100.0	100.0	100.0	100.0
Heritability (%)		5.8		17.5		26.2	

[†]1-HS and 2-HS refer to selection for one and for both sexes among families, respectively. In both cases, selection within families is for one sex.

tional gain per cycle. Selection based on topcrosses by use of an unrelated population as tester is similar in structure to the preceding case. The only difference is that the variance among testcrosses (half-sib families) depends also on the genetic structure of the tester population. The expected gain increases when the average gene frequencies of the favorable alleles are low in the tester. In other words, it is desirable that gene frequencies r in the tester are low enough so that the quantity (p − r) would on the average be positive and of significant magnitude (Comstock 1964). In the last two examples of half-sib family selection, effective population size is expected to be smaller than in any of the preceding cases because only selfed seed of the selected parents is represented in the recombination block.

The coefficient of σ_A^2 (c = 1/2) is greater for full-sib family selection than for half-sib selection when remnant seeds of half-sib families are used for recombination. For fixed environments the phenotypic variance among full-sibs will be greater than for half-sibs because the genetic portion of the phenotypic variance contains

$$\left(\frac{nr + 1}{2nr}\right)\sigma_A^2 + \left(\frac{nr + 3}{4nr}\right)\sigma_D^2 \quad \text{for full sibs}$$

$$\left(\frac{nr + 3}{4nr}\right)\sigma_A^2 + \left(\frac{1}{nr}\right)\sigma_D^2 \quad \text{for half-sibs, with n plants and r replications.}$$

Effective population size is expected to be smaller than for half-sibs because only the parents involved in crosses are represented in the recombination block. The full-sib family procedure also requires more work because controlled pollination is involved, whereas mass selection and half-sib family selection can be accomplished by using only open-pollinated ears in isolated plantings except for the cases when S_1 seeds are used for recombination.

Ramalho (1977) compared half-sib with full-sib family selection under realistic circumstances, using average estimates of genetic and environmental parameters. His comparisons were based on equality of selection intensity in the first instance and on equality of effective size in the second instance. Taking different combinations of selection intensities for among- and within-family selection, Ramalho showed that full-sib family selection tends to be more effective than half-sib selection (selection among and within half-sib families using remnant seed for recombination, i.e., two generations per cycle) for traits of lower heritability, which generally show coefficients of variation of greater magnitude. All comparisons were based on a total selection intensity of 2%, combined for among- and within-family selection, as shown in Table 6.14. The relative effectiveness of half-sib and full-sib fam-

Table 6.14. Effective population size (N_e) for equal selection intensity for half-sib (HS) and full-sib (FS) family selection in relation to varying selection intensities for among and within half-sibs (adapted from Ramalho 1977).

| | A. Selection intensity (i), %, for equal N_e | | | | B. Effective size for equality of i | |
| | Half-sib family | | Full-sib family | | $N_{e(HS)} = \frac{16S}{4 - 1/f}$ | $N_{e(FS)} = \frac{4S}{2 - 1/f}$ |
	Among	Within	Among	Within		
(a)	5	40	10	20	20.2	10.3
(b)	10	20	19	10	41.0	21.0
(c)	20	10	35	6	84.2	44.4
(d)	25	8	42	4	106.7	57.1

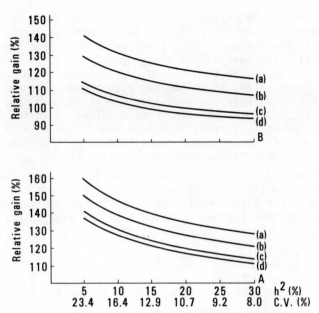

Fig. 6.4. Relative effectiveness of full-sib family selection (among and
within) in percent half-sib family selection for varying values of
heritability under the conditions: within-plot environmental vari-
ance/plot-to-plot environmental variance, σ^2_{we}/σ^2_e = 7.0, and dominance variance/additive variance, σ^2_D/σ^2_A = 0.375. Upper (B): equal
effective size; and lower (A): equal selection intensity--as specified in Table 6.14. (CV is the experimental error coefficient of
variation.) (From Ramalho 1977)

ily selection for conditions specified in Table 6.14 are shown in Fig. 6.4.

An alternative for the recombination phase in the full-sib procedure is
to use selfed seed from selected parents. The coefficient for σ^2_A in the expected gain, however, remains unchanged (c = 1/2). Use of selfed seeds for
recombination requires prolificacy in parental plants and selection for this
trait may result in additional gain. An additional year is also required unless recombination can be done in an off-season nursery.

Family selection using inbred progenies has the primary advantage of increasing genetic variability among families; i.e., c = 1 for S_1 and c = 3/2
for S_2 families. Increase in number of years per cycle has caused these
breeding systems to have limited use. If recombination can be done in an off-
season nursery, number of years per cycle can be reduced. Effective population size is expected to be the smallest among the schemes so far discussed
for a given number of selected progenies. Both S_1 and S_2 family selections
are especially recommended for traits of very low heritability, so the increase in variance among families is of primary importance to allow some progress from selection. Use of inbred families also permits selection against
undesirable recessive genes, thus providing a method for obtaining populations
more suitable for the extraction of vigorous inbred lines.

Combined selection by use of information from half-sib and S_1 progenies
simultaneously makes use of a portion of additive genetic variance that is intermediate to those used by either type of progeny individually. When recom-

bination is done using remnant seed from half-sib families the coefficient for σ_A^2 is 3/8, which is doubled (3/4) when S_1 seeds are used for recombination. In the first case a higher effective population size is assured and only two years are required per cycle. In the second case an additional year or season is required. In all cases, however, the advantage of using information from both types of progenies is that different information relative to gene action is given by each type of progeny (Goulas and Lonnquist 1976). Genes contributing to heterotic behavior are more likely to be selected in half-sib evaluation than in S_1 family evaluation, where genes with favorable additive effects receive greater emphasis. Consequently, combined selection using information from both half-sib and S_1 progenies provides for an increase in frequency of desirable alleles and allelic combinations more effectively than either half-sib or S_1 separately.

A theoretical comparison among some selection procedures also was presented (Comstock 1964), based on their effectiveness in change of gene frequency in a population. The expressions for change in gene frequency were then transformed to relative magnitudes for a comparison of relative efficiency of selection systems. Such comparisons are shown in Table 6.15.

Methods for interpopulation improvement are presented in Sec. 6.7 and the formulas to predict gain from selection are shown in Table 6.16. The original RRS procedure was proposed by Comstock et al. (1949). Reported results have shown that the method is effective (Collier 1959; Penny et al. 1963; Moll and Stuber 1971). Penny and Eberhart (1971) found very little improvement in the cross between two populations after four cycles of HS-RRS; but in a subsequent evaluation including five cycles (Eberhart et al. 1973), estimates of improvement were considerably greater than in the previous study. Despite the potential of the method, it has some inherent difficulties that may preclude its wide utilization. Some of these difficulties were outlined by Paterniani and

Table 6.15.   Expected change in gene frequency (Δ_p) per generation and relative values of selection systems for non-overdominant loci (adapted from Comstock 1964).

Method	Δ_p[†]	Relative value	
		No dominance	Complete dominance
Mass	$\dfrac{k}{2\sigma_p}A$	0.8 to 1.0	0.8 to 1.0
Full-sib family	$\dfrac{k}{4\sigma_p}A$	1.0	1.0
Test progeny (1)[‡]	$\dfrac{k}{6\sigma_p}A$	0.67	0.67
Test progeny (2)[§]	$\dfrac{k}{6\sigma_p}[A + 2(p-r)d]$	0.67	0.67 to 1.0
Selfed progeny	$\dfrac{k}{3\sigma_p}[A + (P-1/2)d]$	1.33	1.33 to 2.0

[‡]Parental population used as tester; [§]unrelated population used as tester; and [†]$A = [a + (q - p)d]$.

Table 6.16. Coefficient (c) for additive genetic variance (σ_A^2) relative to
expected progress for interpopulation selection methods.

Selection method[†]	Selection unit	Recombination unit[‡]	One sex		Both sexes	
			$\sigma_{A_{12}}^2$	$\sigma_{A_{21}}^2$	$\sigma_{A_{12}}^2$	$\sigma_{A_{21}}^2$
1. HS-RRS	HS-FS	S_1	--	--	1/4	1/4
2. FS-RRS	FS	S_1	--	--	1/4	1/4
3. HS-RRS 1	HS	HS	1/16	1/16	--	--
4. HS-RRS 2	HS	HS	1/8	1/8	--	--

[†]Methods 1, 3, and 4 have two components in the formula for expected
progress. Method 2 has only one component and the numerator is

$$(1/4)\sigma_{A_{12}}^2 + (1/4)\sigma_{A_{21}}^2 = 1/4(\sigma_{A_{12}}^2 + \sigma_{A_{21}}^2) = (1/2)\sigma_{A_{(12)}}^2$$

See Chap. 2.

[‡]See Table 6.6.

Vencovsky (1977) as follows: (1) Selfing and outcrossing at the same time involve considerable labor, which reduces number of genotypes tested. (2) A
sample of four or five female plants to be crossed with each male plant may
not be adequate, thus reducing the accuracy of male evaluations. For this
reason it should be more convenient to self the plants in one year and make
the crosses in the following year using a detasseling block, but this procedure increases the length of each cycle by one year. (3) A more effective recombination among selected S_1 families may require an additional year and thus
the interval between cycles may become five years. (4) If each cycle interval
becomes greater the gain is reduced when expressed on a per year basis. If
cycle length is five years, tests are carried out in only one year, and selection is strongly affected by the season of the particular year. When cycle
interval is three years, the problems are mainly characterized by a less effective S_1 family recombination and by an inadequate number of females per
male.

The methodology described by Penny and Eberhart (1971) includes ten female plants for each male, which actually increases accuracy of genotype evaluation through half-sib families. A less effective S_1 recombination is also a
problem in FS-RRS, where each plant genotype is evaluated in a cross with another genotype. However, individual genotypes for each cross can be multi-
plied by selfing in each population to maintain the gametic array from each
genotype. One advantage of this procedure is that high performing families
can be reproduced and inbred lines for hybrid development extracted from them
through the procedure suggested by Hallauer (1967) and Lonnquist and Williams
(1967). The combination of population improvement development makes FS-RRS an
integrated method of population improvement and inbred line development. Options for a wider utilization of this method were described by Hallauer and
Eberhart (1970).

A comparative study between HS-RRS and FS-RRS was reported by Jones et
al. (1971), using theoretical comparisons of expected change in gene frequency
and computer simulation. The relative efficiency of FS-RRS over HS-RRS is
given by $\Delta_{GAF}/\Delta_{GAH} = k_F\sigma_{\overline{PF}}/(k_H\sigma_{\overline{PH}})$, where Δ_{GAF} and Δ_{GAH} are the expected gene

frequency changes in population A for FS-RRS and HS-RRS, respectively. Using realistic parameter estimates, it was found that $\sigma_{\overline{PF}}/\sigma_{\overline{PH}} = 1.18$, so the selection differential for FS-RRS would need to be 1.2 times greater than for HS-RRS to give a similar response. Simulation results suggested that FS-RRS has an advantage over HS-RRS at lower selection intensities and when the environmental variance is large relative to the total genetic variance. As selection intensity increases this advantage decreases.

Vencovsky (1977) made a comparison between FS-RRS and HS-RRS-2 relative to effective population size and expected progress. Effective population sizes for the FS-RRS and HS-RRS-2, respectively, are $N_e \simeq 2p_{FS}T/(2 - p_{FS})$ and $N_e \simeq (8p_{HS}T)/(4 + 3/\overline{M} - 1/\overline{F})$, where p_{FS} and p_{HS} are the proportion of selected families in each method, T is the total number of families tested, and $\overline{M}$ and $\overline{F}$ are the average number of males and females per male plant in half-sib families in the crossing block. A comparison between the two formulas leads to the conclusion that to equalize the effective population size, p_{FS} must equal $2p_{HS}/(p_{HS} + 0.25D)$, where D is the denominator of the population size of HS-RRS-2. In terms of expected progress, relative efficiency HS-RRS-2/FS-RRS $= 0.75k_{HS}\sigma_{\overline{FS}}/(k_{FS}\sigma_{\overline{HS}})$. Using parameters approaching a realistic situation, values for $\sigma_{\overline{FS}}/\sigma_{\overline{HS}}$ were found to be in the range of 1.23 to 1.12, which is a good approximation to that by Jones et al. (1971). Taking $\sigma_{\overline{FS}}/\sigma_{\overline{HS}} = 1.18$ and $p_{\overline{HS}} = 0.17$ (proportion of selected families in HS-RRS-2), the relative efficiency HS-RRS-2/FS-RRS $= (0.75)(1.489)/[(1.159)(1.18)] = 1.137$ for an approximation to equality of effective population size. This means that under the given conditions the expected progress for HS-RRS-2 would be 13.7% greater than that expected for FS-RRS. Such comparisons take into account differences in cycle duration, which are three years for FS-RRS and two years for HS-RSS-2. Another advantage of HS-RRS-2 over the two other methods is that larger samples are used in the crossing phase. When the comparison is based on equal selection intensities (but different effective population sizes), $\Delta_{GHS}/\Delta_{GFS} = (0.75)(1.18) = 0.885$; i.e., HS-RRS-2 would be 11.5% less effective than FS-RRS in changing the mean of the crossed population.

Paterniani and Vencovsky (1977) showed that reciprocal recurrent selection based on testcrosses of half-sib families (HS-RRS-1) makes use of a smaller portion of the additive genetic variance of the population cross than does the reciprocal recurrent selection based on prolific plants (HS-RRS-2). Consequently progress from selection is expected to be smaller for HS-RRS-1 but effective population size can be maintained at a higher level. It was shown for HS-RRS-1 that effective population size for each parental population is given by $N_e = 16S/(4 + 3/M - 1/F)$, where S is the number of half-sib families selected, M is the number of remnant seeds taken per selected family for male rows, and F is the number of detasseled plants per family from which ears are taken for the next cycle.

6.9 INCREASING GAIN FROM SELECTION

From the general prediction formula, there are several ways to increase the progress from selection:

1. *Increasing selection pressure.* Selection differential (k) is a direct function of the proportion of selected units (i.e., selection intensity) so that k increases as the selected group decreases in size, leading to a

greater expected progress. However, selection intensity must be reasonably
chosen because genetic variability can be reduced drastically with a very high
selection pressure. If the program is long term, attention must be directed
to the preservation of genetic variability, assuming a low but long-term gain
from selection. A relative comparison of expected progress for short, intermediate, and long-term selection programs can be seen in Fig. 1.3. Choosing
an appropriate selection intensity depends also on population size. Low selection intensity in small populations may lead to drastic changes in population structure due to small effective population size resulting in some inbreeding effect.

   2. *Adjusting the coefficient of σ_A^2*. The value of c depends on the selection method used. For example, c = 1/2 for mass selection and c = 1/8 for
half-sib family selection if only female gametes are selected. However, if
both sexes are selected at the same selection intensity, then c assumes values
of 1 and 1/4, respectively. Thus one can conclude that for any given method,
parental control is an important way to increase progress from selection.

   3. *Increasing genetic variability*. Genetic variability is determined at
the time the population is either formed or chosen to undergo selection. Development of composite varieties from genetically divergent populations is
very important as the first step in a breeding program (Eberhart et al. 1967).
Once a population has been developed, the influence of σ_A^2 on expected gain can
be controlled only by selection (parental control) or by eventually limiting
the range of environments.

   4. *Controlling environmental effects*. Gain from selection increases directly by decreasing phenotypic variance among selection units. Considering a
fixed amount of genetic variability, phenotypic variance can be decreased in
several ways, most of them related to improvement of experimental techniques.
For individual plant selection, phenotypic variance may be decreased by control of all factors that cause experimental error: (a) Choose uniform soils
(in some soils, heterogeneity is an inherent property); topography; geographical conditions; or any factor that may limit normal growth, such as acidity,
salinity, organic matter. In some special cases limited environmental conditions are representative of the environment for which the population is being
selected. Also, cyclic crop rotation using uniform crops is a useful procedure to improve soil homogeneity. (b) Uniform soil preparation, fertilization, pesticide treatment (insecticides, herbicides, etc.), and other cultural
practices should be used to minimize the differences because of cultural practices. (c) Care in data collection, in analysis, and in handling experimental
material should be used. (d) Field stratification in small blocks and selection within blocks so that environmental variation among blocks is separable
from phenotypic variance (among individuals within blocks) was suggested by
Gardner (1961).

   For family selection, an increase in number of replications and in number
of plants per plot leads to a decrease in phenotypic variance and consequently
an increase in genetic gains. Number of environments (locations or years, or
both) also influences the magnitude of phenotypic variance. Such effects can
be visualized in the following formula for phenotypic variance among family
means:

$$\sigma_{\overline{F}}^2 = \sigma_f^2 + \sigma_{fe}^2/e + \sigma^2/(re)$$

where f, e, and r refer to number of families, environments, and replications,
respectively. An increase in number of environments decreases $\sigma_{\overline{F}}^2$ because of

reduction in σ_{fe}^2 and σ^2.   Thus if selection is conducted for a population of environments, number of environments is the most powerful factor in reducing phenotypic variance.   Expected gain will be affected by genotype-environment interactions within the defined population of environments.   If only one environment is considered and selection is directed to a population of environments, expected response tends to be overestimated.   Usually the extensiveness of selection experiments is limited by space, amount of work, and available resources and facilities; an increase in number of environments usually must be at the expense of number of replications per environment.   If genotype-environment interaction is of low magnitude (as known by previous studies or some other evidence), increase in e at the expense of r makes little difference in reduction of phenotypic variance.   If number of replications is reduced to one, $\sigma^2 + \sigma_{fe}^2$ are confounded, leading to an unknown estimate of experimental error and reducing the precision of family mean estimates, particularly if genotype-environment interaction is large.

The effect of number of plants per plot in decreasing phenotypic variance has an asymptotic pattern and little change is obtained after a certain limit. For example, Eberhart (1970) found that very little increase in gain would be obtained with more than 15 or 20 plants per plot in two replications and four locations.   But number of plants per plot cannot be decreased a great deal because sampling of genetic material is also of concern and may be an important source of experimental error.   Increase in number of replications and locations also has a limit beyond which the gain in precision is very low (Eberhart 1970).

Use of adequate experimental designs is another way to reduce phenotypic variability in family selection by removing some of the environmental variation.   For example, lattice designs may be used rather than randomized blocks when a large number of families is evaluated, as is usually done; lattice efficiency demonstrates that environmental variation exists within replications and can be removed to lower the experimental error.

6.10   CORRELATION BETWEEN CHARACTERS AND CORRELATED RESPONSE

Correlation, measured by a correlation coefficient, is important in plant breeding because it measures the degree of association, genetic or nongenetic, between two or more characters.   If genetic association exists, selection for one trait will cause changes in other traits--this is called correlated response.   The cause of correlation can be genetic and/or environmental.   Genetic causes may be attributed to pleiotropism and/or linkage disequilibrium. Pleiotropism occurs when one gene affects simultaneously several physiological pathways, resulting in influence over several observed characters.   Linkage disequilibrium refers to genes which show a tendency to being transmitted together within a population.   If two nonallelic genes, say A and B, with frequencies p_A and p_B in the population, are included in the gametes, the probability of their being transmitted together is $p_{AB} = p_A p_B$ with linkage equilibrium.   If the genes have linkage disequilibrium, the probability that they are included in the same gamete is more or less than $p_A p_B$, and linkage disequilibrium is given by $\Delta_{AB} = p_{AB} - p_A p_B$.   Linkage disequilibrium tends to be dissipated over generations in a random mating population and the rate of dissipation depends on how close the genes are located on the chromosomes or, in other words, on the recombination rate between the two genes. Given a recombination rate r_1 for frequency of recombinants and r_0 for nonrecombinants, where $r_0 = 1 - r_1$, the probability of genes coming together in any generation t of random mating is $p_{AB_t} = r_0 p_{AB_{t-1}} + (1 = r_0) p_A p_B$ and the amount of linkage disequilibrium is $\Delta_{AB_t} = p_{AB_t} - p_A p_B = r_0 \Delta_{AB_{t-1}}$ (Cockerham 1956). Thus the initial amount of linkage disequilibrium is dissipated at the rate of the recombination fraction of each generation of random mating. The approach to linkage equilibrium is gradual and asymptotic; it is faster when the recombination fraction approaches its maximum value ($r_1 = 1/2$) and it is slower for closely linked genes. When genes are not closely linked, linkage disequilibrium is not an important cause of correlation between characters in random mated populations. In such cases the existence of genetic correlations is mostly attributed to pleiotropic effects.

Environmental correlations also exist because measurements of several traits are taken from the same individual or from the same family. For example, a positive environmental correlation is expected to occur between plant height and ear height in the same plants because a microenvironment that favors plant height also increases ear height, and vice versa. When two traits are evaluated by the average (or total) of a family, the environmental deviation in a given plot containing the given family affects all individuals of that plot and causes an environmental correlation of characters among plots.

The linear correlation coefficient r between variables, say trait 1 and trait 2, is calculated by the ratio, $r = \hat{Cov}(1, 2)/(\hat{\sigma}_1 \hat{\sigma}_2)$. If the two variables are two characters in a maize plant, Cov(1, 2) is the covariance between the two characters and σ_1 and σ_2 are their standard deviations. The correlation coefficient is a measure of the degree of association between two characters or the degree to which two characters vary together. It is an estimate of ρ, a parameter of the bivariate normal distribution. The quantity $r\sqrt{n-2}/\sqrt{1-r^2}$ is distributed as t with $(n - 2)$ df and is used to test the null hypothesis that $\rho = 0$ when r is calculated from paired values.

In plant breeding, genetic and phenotypic correlations are important. Genetic correlations are related only to genetic causes. The expression *genetic correlation* is more appropriate when all genetic effects are involved (broad sense) and has a wider use in homozygous self-pollinated and in apomictic species. On the other hand, *additive genetic correlation* (or simply *additive correlation*), involving only additive effects, is more appropriate to cross-pollinated species, such as maize, mainly because the information from the correlation is used in connection with recurrent selection. Phenotypic correlation involves both genetic and environmental effects.

Estimation of genetic and phenotypic correlations is based on components of variances and covariances that are estimated from analyses of variance and covariance, respectively. Covariance components are obtained in the same manner as variance components, because for any experimental design the coefficients of expected mean products are the same as those for expected mean squares (Mode and Robinson 1959). Genetic components of covariance are related to the covariance of relatives in the same way as genetic components of variance and thus can be estimated by the same methods, which are described in Chap. 4. For example, in an analysis of half-sib families the covariance between half-sibs involving two characters is an estimate of one-fourth the additive genetic covariance between them [(1/4)Cov A_{12}]. An example of estimation of additive genetic variances and covariances is given in Table 6.17, where data and analysis are on an individual plant basis.

From Table 6.17 the following components of genetic variance and covariance can be estimated: $\hat{\sigma}_{A_1}^2 = 4\hat{\sigma}_{f_1}^2 = 4(103.74 - 24.57)/30 = 10.556$; $\hat{\sigma}_{A_2}^2 = 4\hat{\sigma}_{f_2}^2 = 4(15.36 - 5.57)/30 = 1.305$; and Cov $A_{12} = 4(16.06 - 4.91)/30 = 1.487$. Additive genetic correlation is then calculated by $r_A = 1.487/[(1.142)(3.249)]$

Table 6.17.   Analysis of variance and covariance for two tassel traits, number of branches (subscript 1), and weight (subscript 2) for half-sib families in a maize population (Geraldi et al. 1975).

Source	df	MS_1	MP_{12}	MS_2	$E(MS)^{+}$
Reps/Expt.	4	123.34	10.73	7.35	
Families/Expt.	38	103.74	16.06	15.36	$\sigma_w^2 + n\sigma_p^2 + nr\sigma_f^2$
Pooled error	76	24.57	4.91	5.57	$\sigma_w^2 + n\sigma_p^2$
Within[+]	1080	21.36	5.17	4.22	σ_w^2

[+] Or E(MP), by changing variance to covariance.

[+]Estimated from 40 families and n = 10 plants/family.

= 0.401.   Phenotypic variances $(\hat{\sigma}_{P_1}^2$ and $\hat{\sigma}_{P_2}^2)$ are estimated by $\hat{\sigma}_f^2 + \hat{\sigma}_p^2 + \hat{\sigma}_w^2$ and the phenotypic covariance by Cov P_{12} = Cov f + Cov p + Cov w; hence the following values are found: $\hat{\sigma}_{y1}^2$ = 24.320, $\hat{\sigma}_{y2}^2$ = 4.681, and $\hat{Cov}\ y_{12}$ = 5.517.   Using these estimates the phenotypic correlation coefficient is found to be r_P = 0.516.

When the experiment includes only two replications, components of genetic variance and covariance can easily be found by using a *cross covariance* technique as suggested by Falconer (1960).   Denoting by I_1 and I_2 the measurements (half-sib family values in the example given) of traits 1 and 2 in one replication and by II_1 and II_2 the measurements of the same traits in another replication, the following relations for half-sib families hold:

$$\hat{Cov}(I_1,\ II_1) = \hat{\sigma}_{f_1}^2 = (1/4)\hat{\sigma}_{A_1}^2 \qquad \hat{Cov}(I_2,\ II_2) = \hat{\sigma}_{f_2}^2 = (1/4)\hat{\sigma}_{A_2}^2$$

$$(1/2)[\hat{Cov}(I_1,\ II_2) + \hat{Cov}(I_2,\ II_1)] = \hat{Cov}\ f_{12} = (1/4)\hat{Cov}\ A_{12}$$

Additive genetic correlation is important in selection programs because it gives information about the degree of association between two traits by way of additive or breeding values of individuals, which are the effects that can be changed by selection.   In other words, selection for one trait will cause a change in its mean through additive effects of genes of selected individuals. If another trait is correlated additively to the first, selection will cause an indirect change in the mean of the second trait.   Hence the basic question is: If selection is for one trait, to what extent will it change the second trait?   It depends on the degree of association between the two traits.   This indirect change is known as correlated response and can be predicted in the same way as direct response for one trait.   The prediction procedure also involves a regression problem.   Denoting by X_1 a measure of trait 1 in the selected individual or family (selection unit) and by W_1 a measure of the same trait in an individual from the improved population genetically related to X_1, then for a unit deviation in X a deviation of $b_{W_1 X_1}$ (regression coefficient of W_1 on X_1) is expected in W_1; this regression has been shown to be $c\sigma_{A_1}^2/\sigma_{P_1}^2$. Denoting by W_2 a measure of trait 2 in the same individual in the improved population, then for each unit deviation in breeding value of the selected trait there is an expected deviation of $b_{W_2W_1}$ in breeding value of the second trait. The total change in the second trait or the correlated response is

$$CR_2 = \Delta_{G_1} b_{W_2W_1} = [s(c\sigma^2_{A_1})/\sigma^2_{P_1}](\text{Cov } A_{12}/\sigma^2_{A_1}) = sc(\text{Cov } A_{12})/\sigma^2_{P_1}$$

If truncation selection is used, this formula changed to $CR_2 = kc \text{ Cov } A_{12}/\sigma^2_{P_1}$, where Cov A_{12} is the additive covariance between the two traits, $\sigma^2_{P_1}$ is the phenotypic variance of the selection unit, and k is a function of the selection intensity.

For example, considering mass selection for one sex, $CR_2 = k(1/2)\text{Cov } A_{12}/\sigma_{P_1}$ can be written as

$$CR_2 = k(1/2)[\text{Cov } A_{12}/(\sigma_{A_1}\sigma_{A_2})](\sigma_{A_1}\sigma_{A_2}/\sigma_{P_1}) = (1/2)kr_{A_{12}}h_1\sigma_{A_2}$$

where h_1 is the square root of heritability for trait 1.

In the sample given in Table 6.17, the correlated response for tassel weight under mass selection for number of branches at a selection intensity of 10% ($k = 1.755$) is estimated by $CR_2 = 1.755(1/2)(1.487/24.320) = 0.265$ g, which represents a change of 3.6% of the original mean ($m_1 = 7.37$ g). For among and within half-sib family selection (one cycle per year) the expected correlated response would be $CR_2 = 1.400(1/8)(1.487/1.860) + 1.755(3/8)(1.487/4.622) = 0.140 + 0.212 = 0.352$ g, for a selection intensity of 20% among families and 10% within families. This change represents a correlated response in tassel weight of about 4.8% of the original mean.

This example also illustrates another important point relative to correlated traits--the possibility of indirect selection when the primary trait is difficult to measure or evaluate. It has been suggested that selection for smaller tassels would increase plant efficiency in maize (Hunter et al. 1969, 1973; Buren et al. 1974; Mock and Schuetz 1974), but direct selection for tassel weight would be difficult. The number of branches per tassel can be evaluated more easily and with a relatively high degree of accuracy. The merit of indirect selection relative to direct selection on trait 2 is measured by the ratio of expected correlated response over direct response, which is $r_A k_1 h_1/(k_2 h_2)$, where r_A is the additive genetic correlation, k_1 and k_2 are the selection differentials in standard units, and h_1 and h_2 are the square roots of heritabilities for traits 1 and 2, respectively (Falconer 1960). Indirect selection has an advantage over direct selection when $r_A h_1 > h_2$, i.e., the secondary trait has a higher heritability than the desired character and the additive correlation between them is high. In our example (Table 6.17) indirect selection for tassel weight would not show greater progress than direct selection because $r_A h_1 < h_2$. However, indirect selection should be successful because tassel weight is more difficult to evaluate, additive genetic correlation with number of branches is moderate ($r_A = 0.40$ in our case), and the heritability coefficient for number of branches ($h^2 = 0.43$) is substantially higher than for tassel weight ($h^2 = 0.28$). Note that heritability coefficients, as well as additive genetic correlation, depend on the population under selection and on environmental conditions, suggesting that the advantage of indirect selection must be investigated for each particular situation. Information from previous studies relative to heritability, types of gene action, and environmental conditions may serve as a general guide for the breeder's decisions.

6.11 SELECTION FOR MORE THAN ONE TRAIT

In practical maize breeding procedures, selection for more than one trait is common. The trait of primary importance usually is grain yield, but several other agronomic traits such as plant and ear height, lodging resistance, and disease and insect resistance must be included if the material is to be useful to the commercial grower. Alternatives for the selection of several traits are (1) tandem selection, (2) independent culling levels, and (3) selection index.

Tandem selection emphasizes selection for only one trait for a number of generations. One problem of tandem selection is determining the number of generations of selection to be devoted to each trait. If a high selection intensity is used, genetic variability decreases rapidly and gain from selection is expected to be reduced after a few generations. Thus the desired limit for each trait, the importance of the trait, and its heritability must be determined before tandem selection is initiated. For example, selection for grain yield would require more generations than selection for plant or ear height. Information about correlated response also is important because a long-term change in one trait may cause an indirect but undesirable change in another important trait. For example, if grain yield and ear height are positively correlated, a long-term selection for grain yield would increase ear height substantially--this has been recognized to be undesirable under the current concepts of maize ideotypes. An alternative would be to alternate selection-- e.g., M_1 cycles for trait X followed by N_1 cycles for trait Y and again M_2 cycles for trait X and so on (Turner and Young 1969). Expected changes, direct or indirect, under tandem selection can be visualized from previous sections of this chapter. Tandem selection seems to be useful when the relative importance of each trait changes throughout the years (not the case for most maize traits). If genetic correlations do not exist between say yield and disease and insect resistance, tandem selection can be used effectively to increase the level of disease and insect resistance before selection for yield is initiated.

Independent culling is selection at a given intensity for several traits in the same generation but in sequence for each trait. Suppose that data from 500 half-sib families are to be used as the basis for family selection. Usually these data are family means from replicated trials. The breeder can first select the 200 best families (40%) based on yield. From this sample of 200, a selection intensity of 50% (100 families) can be used for ear height and followed with a 50% selection intensity used for lodging resistance. The total selection intensity would be $0.40 \times 0.50 \times 0.50 = 0.10$ or 10%, and only the 50 best families would be used for recombination. This procedure has not been widely used in maize breeding on a formal basis but is often used in applied breeding programs by compromising on different traits.

Use of a selection index in its broad sense seems to be common. In most applied selection procedures, breeders use an intuitive selection index. By this procedure selection for several traits is made simultaneously and breeders' decisions are based on the relative weights they give to each trait. Visual acuity and experience will improve their decisions. The inherent subjectivity of the selection process enables breeders to put into practice their ability to recognize the desirable genotypes. In this sense plant breeding has been considered an art rather than a scientific method. Genotype evaluation is based on individual plant observation by phenotypic selection and selection efficiency depends on how efficiently breeders apply their empirical weights for several traits. Selection based on family evaluation in replicated trials also requires accuracy in the evaluation of several characters to

allow selection to be efficient. Some quantitative characters, such as yield
and plant or ear height, can be measured directly but a metrical evaluation is
not possible for some traits, such as lodging resistance and disease resistance. For these traits a scale (e.g., from 0 to 5) is commonly used and the
accuracy of the evaluation also depends on the breeder's experience and the
amount of compromise made for different traits. If the progeny is exceptionally high yielding, the breeder may select it even though the level of some
other trait may not be as high as desired. The family means from replicated
trials are further examined. It is also important to observe family performance in each individual replication for all traits: With four replications a
given family can have an outstanding performance in three of them but a very
poor performance in the fourth, resulting in a relatively low average. This
may indicate some abnormality for that family in the fourth replication and
previous observation in the experimental field may serve as auxiliary information. Also, consistency of family performance over replications is an important criterion for selection. Another detail has to do with stand variation.
If correction is made to compensate for stand variations and the stand in several replications is consistently low, a given family's mean may be overestimated. In addition, consistently low stand may be an indication of some genetic abnormality. Use of computer programs to list family means and data
from individual replications for all traits may facilitate breeders' work, but
decisions about selection still are based on their ability to give to the several traits the appropriate weights they have visualized.

The optimum selection index, first used by Smith (1936) in plants and
later by Hazel and Lush (1942) in animals, has not been used extensively in
maize breeding. Several authors have reported that use of a selection index
would improve selection efficiency relative to selection based on only one
trait (Laible and Dirks 1968; Wolff 1972; Martin and Salvioli 1973; Kauffman
and Dudley 1979; St. Martin 1980). The superiority of the index was reported
by Young (1961) when contrasted with tandem selection and independent culling
level. It was concluded that the superiority of the index increases with increasing number of traits under selection but decreases with increasing differences in relative importance. Its superiority was at a maximum when the
traits considered were equally important. Gain from selection for any given
trait is expected to decrease as additional traits are included in the index,
so the choice for traits to be included must be done objectively.

Use of a selection index is most suitable for animals and crop species
where the relative value of each character is readily determined through its
economic value. In maize, yield is the trait of primary importance, but other
traits (e.g., stalk lodging) also have a direct effect on yield when mechanical harvesting is used. Other traits are not measured directly as is yield,
but they affect final yield so that assignment of relative weights is a subjective task. Also, precision of variance and covariance estimates are generally low, thus limiting greater use of a selection index. In the near future
highly improved populations will require better techniques for further improvement and a selection index may be useful when high precision parameter
estimates that can enhance expected progress are available. The computational
procedures for the use of selection indexes are given by many authors such as
Smith (1936), Hazel and Lush (1942), Hazel (1943), Robinson et al. (1951),
Kempthorne (1957), Brim et al. (1959). For this reason only a brief presentation of the subject is given.

The phenotypic value of an individual can be represented by $P = G + E$ for
each trait. When considering several traits, it is desirable to choose individuals with the best combination of these traits. The basis for such a selection is the selection index, which takes into account a combination of

traits according to their relative weights. Thus each individual has an index value (score) and selection is based on this value. It is possible that the highest yielding individuals will not get the highest scores in the use of a selection index. Because grain yield in maize is the most important trait that has a direct economic value, relative weights for other traits should be given subjectively. Note that even if a weight of zero is given for all traits except yield (weight 1), they will contribute to the optimum index if they are correlated with the primary trait. Such a situation is illustrated by Kempthorne (1957).

The genotypic value of an individual considering several traits is H_j = $a_1 G_{1j} + a_2 G_{2j} + \ldots + a_i G_{ij}$, where a_i ($i = 1, 2,\ldots, n$) is the relative weight and G_{ij} is the genotypic value of the jth individual for the ith trait. The objective is to make a selection so that the above combination of characters (H_j) is the most desirable. Because the G_{ij} are not known but are evaluated through P_{ij} (phenotypic values), H_j must be evaluated by an index I_j based on phenotypic values such that the correlation between H_j and I_j (denoted by ρ_{HI}) is as great as possible. The maximization of ρ_{HI} leads to the following set of equations: GA = PB, where G is a matrix of genetic variances and covariances, A is a vector of a values (relative weight), P is a matrix of phenotypic variances and covariances, and B is a vector of unknown b values. The solution of the set of equations leads to the estimation of b values that give the highest correlation with H_j when applied to the index I_j. The b values also maximize the expected progress by selection based on indexes (Turner and Young 1969). For a more detailed understanding, the reader can refer to the authors mentioned.

Some modifications have been suggested for use of a selection index. Kempthorne and Nordskog (1959) presented a theory for use of a restricted selection index; i.e., an index constructed such that for N = n_1 + n_2 traits involved, n_1 desirable traits are to be changed as efficiently as possible and the n_2 remaining desirable traits are to be unchanged. Tallis (1962) generalized the theory of a restricted selection index for the case where it is desirable to improve n_1 traits without limit and n_2 traits only to a predetermined limit. Williams (1962) suggested the use of a base index that differs from the optimum index because the traits are weighted directly by their economic values. Pesek and Baker (1969) used the concept of "desired genetic gains" to overcome the problem of assigning relative economic weights in crop species. Details for the application of the modified method were presented by Pesek and Baker (1970). A comparison among three different selection indexes was reported by Suwantaradon et al. (1975) in maize: Conventional indexes did not give satisfactory improvement for all traits. Base indexes were 95% and 97% as efficient as conventional selection indexes when relative economic weights were used. Modified selection indexes (based on desired gains) were shown to be 46% and 61% as efficient as conventional indexes for the specified desired gains and relative economic weights considered.

REFERENCES

Baker, L. H., and R. N. Curnow. 1969. Choice of population size and use of variation between replicate populations in plant breeding selection programs. *Crop Sci.* 9:555-60.

Brim, C. A.; H. W. Johnson; and C. C. Cockerham. 1959. Multiple selection criteria in soybeans. *Agron. J.* 51:42-46.

Buren, L. L.; J. J. Mock; and I. C. Anderson. 1974. Morphological and physiological traits in maize associated with tolerance to high plant density. *Crop Sci.* 14:426-29.

Cockerham, C. C. 1956. Effects of linkage on the covariances between rela-
tives. *Genetics* 41:138-41.
————. 1961. Implications of genetic variances in a hybrid breeding pro-
gram. *Crop Sci.* 1:47-52.
Collier, J. W. 1959. Three cycles of reciprocal recurrent selection. *Proc.
Annu. Hybrid Corn Ind. Res. Conf.* 14:12-23.
Compton, W. A., and R. E. Comstock. 1976. More on modified ear-to-row selec-
tion in corn. *Crop Sci.* 16:122.
Comstock, R. E. 1964. Selection procedures in corn improvement. *Proc. Annu.
Hybrid Corn Ind. Res. Conf.* 19:87-94.
Comstock, R. E.; H. F. Robinson; and P. H. Harvey. 1949. A breeding proce-
dure designed to make maximum use of both general and specific combining
ability. *Agron. J.* 41:360-67.
Cress, C. E. 1966. Heterosis of the hybrid related to gene frequency differ-
ences between two populations. *Genetics* 53:269-74.
Dobzhansky, T. 1951. *Genetics and the Origin of Species*. Columbia Univ.
Press, New York.
Eberhart, S. A. 1970. Factors affecting efficiencies of breeding methods.
African Soils 15:669-80.
Eberhart, S. A.; S. Debela; and A. R. Hallauer. 1973. Reciprocal recurrent
selection in the BSSS and BSCB1 maize populations and half-sib selection
in BSSS. *Crop Sci.* 13:451-56.
Eberhart, S. A.; M. N. Harrison; and F. Ogada. 1967. A comprehensive breed-
ing system. *Züchter* 37:169-74.
Empig, L. T.; C. O. Gardner; and W. A. Compton. 1972. Theoretical gains for
different population improvement procedures. Nebraska Agric. Exp. Stn.
Bull. MP26 (revised).
Falconer, D. S. 1960. *Introduction to Quantitative Genetics*. Ronald Press,
New York.
Fisher, R. A., and F. Yates. 1948. *Statistical Tables for Biological, Agri-
cultural, and Medical Research*, 3d ed. Oliver & Boyd, Edinburgh.
Gardner, C. O. 1961. An evaluation of effects of mass selection and seed ir-
radiation with thermal neutrons on yields of corn. *Crop Sci.* 1:241-45.
Gardner, C. O., and S. A. Eberhart. 1966. Analysis and interpretation of the
variety cross diallel and related populations. *Biometrics* 22:439-52.
Geraldi, I. O.; J. B. Miranda, Fo.; and E. Paterniani. 1975. Estimativas de
parâmetros genéticos e fenotipicos em caracteres do pendão de milho (*Zea
mays* L.). Rel. Cient. Inst. Genét. (ESALQ-USP) 9:87-91.
Goulas, C. K., and J. H. Lonnquist. 1976. Combined half-sib and S₁ family
selection in a maize composite population. *Crop Sci.* 16:461-64.
Hallauer, A. R. 1967. Development of single-cross hybrids from two-eared
maize populations. *Crop Sci.* 7:192-95.
————. 1974. Heritability of prolificacy in maize. *J. Hered.* 65:163-68.
Hallauer, A. R., and S. A. Eberhart. 1970. Reciprocal full-sib selection.
Crop Sci. 10:315-16.
Hallauer, A. R., and J. H. Sears. 1969. Mass selection for yield in two va-
rieties of maize. *Crop Sci.* 9:47-50.
Hazel, L. N. 1943. The genetic basis for constructing selection indexes.
Genetics 28:476-90.
Hazel, L. N., and J. L. Lush. 1942. The efficiency of three methods of se-
lection. *J. Hered.* 33:393-99.
Hunter, R. B.; C. G. Mortimore; and L. W. Kannenberg. 1973. Inbred maize
performance following tassel and leaf removal. *Agron. J.* 65:471-72.
Hunter, R. R.; T. B. Daynard; L. J. Hume; J. W. Tanner; J. L. Curtis; and L.
W. Kannenberg. 1969. Effect of tassel removal on grain yield of corn
(*Zea mays* L.). *Crop Sci.* 9:405-6.

Jones, L. P.; W. A. Compton; and C. O. Gardner. 1971. Comparisons of full- and half-sib reciprocal recurrent selection. *Theor. Appl. Genet.* 41:36-39.

Kauffman, K. D., and J. W. Dudley. 1979. Selection indices for corn grain yield, percent protein, and kernel depth. *Crop Sci.* 19:583-88.

Kempthorne, O. 1952. *Design and Analysis of Experiments.* Wiley, New York.

———. 1957. *An Introduction to Genetic Statistics.* Wiley, New York.

Kempthorne, O., and A. W. Nordskog. 1959. Restricted selection indices. *Biometrics* 15:10-19.

Laible, C. A., and V. A. Dirks. 1968. Genetic variances and selection value of ear number in corn (*Zea mays* L.). *Crop Sci.* 8:540-43.

Lerner, I. M. 1958. *The Genetic Basis of Selection.* Wiley, New York.

Lonnquist, J. H. 1949. The development and performance of synthetic varieties of corn. *Agron. J.* 41:153-56.

———. 1964. Modification of the ear-to-row procedure for the improvement of maize populations. *Crop Sci.* 4:227-28.

———. 1967a. Mass selection for prolificacy in maize. *Züchter* 37:185-87.

———. 1967b. Intrapopulation improvement: Combination S_1 and HS selection. Maize 5, CIMMYT.

———. 1967c. Interpopulation improvement: Combined S_1, mass, and reciprocal recurrent selection. Maize 6, CIMMYT.

Lonnquist, J. H., and M. Castro G. 1967. Relation of intra-population genetic effects to performance of S_1 lines of maize. *Crop Sci.* 7:361-64.

Lonnquist, J. H., and N. E. Williams. 1967. Development of maize hybrids through selection among full-sib families. *Crop Sci.* 7:369-70.

Martin, G. O., and R. A. Salvioli. 1973. A study of the association between yield components and a selection index in maize (*Zea mays* L.). *Plant Breed. Abstr.* 43:216.

Mock, J. J., and S. Schuetz. 1974. Inheritance of tassel branch number in maize. *Crop Sci.* 14:885-88.

Mode, C. J., and H. F. Robinson. 1959. Pleiotropism and the genetic variance and covariance. *Biometrics* 15:518-37.

Moll, R. H., and C. W. Stuber. 1971. Comparisons of response to alternative selection procedures initiated with two populations of maize (*Zea mays* L.). *Crop Sci.* 11:706-11.

———. 1974. Quantitative genetics: Empirical results relevant to plant breeding. *Adv. Agron.* 26:277-313.

Moll, R. H.; J. H. Lonnquist; J. V. Fortuno; and E. C. Johnson. 1965. The relationship of heterosis and genetic divergence in maize. *Genetics* 52:139-44.

Paterniani, E. 1967a. Interpopulation improvement: Reciprocal recurrent selection variations. Maize 8, CIMMYT.

———. 1967b. Selection among and within families in a Brazilian population of maize (*Zea mays* L.). *Crop Sci.* 7:212-16.

———. 1973. Recent studies on heterosis. In *Agricultural Genetics*, R. Moav, ed., pp. 1-22. Natl. Counc. Res. Dev., Jerusalem.

Paterniani, E., and R. Vencovsky. 1977. Reciprocal recurrent selection in maize (*Zea mays* L.) based on testcrosses of half-sib families. *Maydica* 22:141-52.

———. 1978. Reciprocal recurrent selection based on half-sib progenies and prolific plants in maize (*Zea mays* L.). *Maydica* 23:209-19.

Paterniani, E.; A. Ando; J. B. Miranda, Fo.; and R. Vencovsky. 1973. Efeitos de raios gama no comportamento e na variância de progênies de meios irmãos em milho. Rel. Cient. Inst. Genét. (ESALQ-USP) 7:161-67.

Penny, L. H., and S. A. Eberhart. 1971. Twenty years of reciprocal recurrent selection with two synthetic varieties of maize (*Zea mays* L.). *Crop Sci.*

11:900-903.

Penny, L. H.; W. A. Russell; G. F. Sprague; and A. R. Hallauer. 1963. Recurrent selection. In *Statistical Genetics and Plant Breeding*, W. D. Hanson and H. F. Robinson, eds., pp. 352-67. NAS-NRC Publ. 982.

Pesek, J., and R. J. Baker. 1969. Desired improvement in relation to selection indices. *Canadian J. Plant Sci.* 49:803-4.

————. 1970. An application of index selection to improvement of self pollinated species. *Canadian J. Plant Sci.* 50:267-76.

Ramalho, M. A. P. 1977. Eficiência relativa de alguns processos de selecão intrapopulacional no milho baseados em familias não endógamas. Tese de Doutoramento, ESALQ-USP, Piracicaba, Brazil.

Rawlings, J. O. 1970. Present status of research on long and short term recurrent selection in finite populations--choice of population size. Proc. Second Meet. Work. Group Quant. Genet., sect. 22. IUFRO, Raleigh, N.C. Pp. 1-15.

Robertson, A. 1960. A theory of limits in artificial selection. *Proc. R. Soc.* 153:234-49.

Robinson, H. F.; R. E. Comstock; and P. H. Harvey. 1951. Genotypic and phenotypic correlations in corn and their implications in selection. *Agron. J.* 43:282-87.

————. 1955. Genetic variances in open pollinated varieties of corn. *Genetics* 40:45-60.

St. Martin, S. K. 1980. Selection indices for the improvement of opaque-2 maize. Ph.D. dissertation, Iowa State Univ., Ames.

Smith, H. F. 1936. A discriminant function for plant selection. *Ann. Eugen. London* 7:240-50.

Suwantaradon, K.; S. A. Eberhart; J. J. Mock; J. C. Owens; and W. D. Guthrie. 1975. Index selection for several agronomic traits in the BSSS2 maize population. *Crop Sci.* 15:827-33.

Tallis, G. M. 1962. A selection index for optimum genotype. *Biometrics* 18:120-22.

Turner, H. N., and S. S. Y. Young. 1969. *Quantitative Genetics in Sheep Breeding*. Cornell Univ. Press, Ithaca, New York.

Vencovsky, R. 1969. Genética quantitativa. In *Melhoramento e Genética*, W. E. Kerr, ed., pp. 17-38. Univ. São Paulo, São Paulo, Brazil.

————. 1977. Effective size of monoecious populations submitted to artificial selection. Institute de Genetica, ESALQ-USP, Piracicaba, Brazil.

Vencovsky, R., and C. R. M. Godoi. 1976. Immediate response and probability of fixation of favorable alleles in some selection schemes. *Proc. Int. Biom. Conf.* Boston, Mass. Pp. 292-97.

Webel, O. D., and J. H. Lonnquist. 1967. An evaluation of modified ear-to-row selection in a population of corn (*Zea mays* L.). *Crop Sci.* 7:651-55.

Williams, J. S. 1962. The evaluation of a selection index. *Biometrics* 18:375-93.

Wolff, F. 1972. Mass selection in maize composites by means of selection indices. *Meded. Landbouwhogesch. Wageningen* 72:1-80.

Young, S. S. Y. 1961. A further examination of the relative efficiency of three methods of selection for genetic gains under less restricted conditions. *Genet. Res.* 2:106-21.

7

Selection:
Experimental Results

Maize selection, as planned experiments in the United States (and probably in the world), started at the end of the nineteenth century. The first methods (mass selection and ear-to-row selection) were soon abandoned because of their supposed lack of efficiency in improving complex traits, such as grain yield. Some other methods and modifications of the previous ones have been used during the twentieth century for improvement of populations, mainly after recurrent selection was recognized as a useful method to increase the frequency of favorable genes in the populations (Jenkins 1940; Hull 1945; Comstock et al. 1949; Lonnquist 1949; Sprague and Brimhall 1950; Jenkins et al. (1954). Also, improvement of populations was recognized as an imperative to increase the probability of obtaining superior hybrids.

Results of many selection experiments have been reported in the literature and the most important information is generally related to one or more of the following items: (1) effectiveness of selection method for the trait or traits under selection; (2) correlated response of unselected traits; (3) change in genetic variability as measured by additive genetic variance, heritability coefficient, and genetic coefficient of variation estimates; (4) change in combining ability with specific or nonspecific testers; (5) change in inbreeding depression, as evaluated by the performance of inbred progenies; (6) change in heterosis either for intra- or interpopulation improvement; (7) comparison of relative effectiveness of different selection methods; (8) effect of environment and genotype-environment interaction; (9) effect of random drift and inbreeding in small populations; (10) value of improved populations as sources of inbred lines for hybrid development; (11) value of environmental control and other techniques (isolation, parental control, etc.) as means for improving selection efficiency; (12) value and efficiency of the selection index; (13) inferences about predominant types of gene action; (14) goodness-of-fit of quantitative genetic models, e.g., expected progress from selection; and (15) value of different testers for population improvement.

7.1 MEASURING CHANGES FROM SELECTION

The objective of most selection experiments is to obtain improved varieties or hybrids; and the success itself is a measure of selection effectiveness, which varies with the way by which effectiveness is quantified. Some researchers use the linear regression coefficient from the performance of populations on the number of selection cycles and the gain is given on a per cycle basis; it can be expressed in the original units, in percent of the observed original mean, or in percent of the original mean predicted by the linear regression (which is a good reference point only when there is a good fit

of the response to selection and the linear regression model). The gain also
can be expressed on a per year basis by dividing the gain per cycle by the
number of years required to perform one cycle of selection; this is the most
appropriate unit for comparing selection methods when the methods require different numbers of years per cycle. In some instances the gain is expressed in
terms of the selection unit, e.g., the gain in performance of S_1 lines in the
S_1 family selection or the gain in performance in the population $\times$ testcross
in recurrent selection using a nonrelated tester. In reciprocal recurrent selection, the most important information is the gain in the crossed population,
although changes in the parental populations and in heterosis also are useful
if they are not confounded with inbreeding. Another possible measure of the
gain from selection is the change in the mean of selection units (families)
evaluated in different years and adjusted to a check (constant through cycles)
performance. In some instances (usually commercial hybrids) the performance
of improved populations or the gain itself is merely expressed in percent of
check performance. For any unit by which the gain can be expressed, the most
precise evaluation of the gain from selection is obtained when the original
population, the improved populations representing each cycle, and some commercial hybrids or varieties used as checks are represented in yield trials replicated over environments. However, for simplicity some preliminary results
during the selection program can be reported from the available data, e.g.,
the performance of family means expressed in percent of checks in each cycle.

The simple or multiple regression model was suggested by Eberhart (1964)
to estimate the rate of response for continued selection. For one population
undergoing selection, the simplest linear model is

$$Y_i = \mu_0 + \beta_1 X_i + \delta_i$$

where Y_i are observed means over cycles of selection (i = 0, 1,..., c); μ_0 estimates the original population mean; β_1 is the linear regression coefficient,
which expresses the rate of gain per cycle of selection; X_i are cycles of selection; and δ_i are deviations from the linear model. The least squares regression analysis, as given by Anderson and Bancroft (1952) or Steel and Tor-
rie (1960), is used for estimation of parameters ($\hat{\mu}_0$ and $\hat{\beta}_1$) and to partition
the variation among populations into sums of squares due to linear regression
and deviations from the model. If deviations from the model are shown to be
nonsignificant, $\hat{\beta}_1$ provides a good estimation of gain from selection (gain per
cycle) and the sum of squares due to $\hat{\beta}_1$, calculated as $R(\hat{\beta}_1) = R(\hat{\mu}_0, \hat{\beta}_1)$
$- R(\hat{\mu}_0)$, provides a test for significance of the observed changes due to se-
lection. If evaluation of effect of quadratic regression is desired, the model is

$$Y_i = \mu_0 + \beta_1 X_i + \beta_2 X_i^2 + \delta_i$$

where β_2 is the quadratic regression coefficient.

A somewhat different model can be used to fit population means resulting
from different selection procedures in the same base population. A linear regression coefficient must be assigned to each selection procedure for estimation of parameters and analysis of variance. A model for fitting two differ-
ent procedures is: $Y_{ij} = \mu_0 + \beta_{11} X_{i1} + \beta_{12} X_{i2} + \delta_{ij}$, or, in general,

$$Y_{ij} = \mu_0 + \sum_j \beta_{1j} X_{ij} + \delta_{ij}$$

where j = 1, 2,..., m refers to the method of selection. When only one base
population is used there is only one parameter μ_0 to express the original population mean. Examples of maize selection programs that fit this model are

(a) divergent recurrent selection in one population; and (b) recurrent selection involving different methods, e.g., S_1 and testcross family selection. In the analysis of variance, hypotheses that can be tested are:

Hypothesis	Interpretation of hypothesis	Sum of squares
$H_0: \beta_{11} = 0$	Ineffective for procedure 1	$R(\beta_{11}) = R(\mu_0, \beta_{11}, \beta_{12}) - R(\mu_0, \beta_{12})$
$H_0: \beta_{12} = 0$	Ineffective for procedure 2	$R(\beta_{12}) = R(\mu_0, \beta_{11}, \beta_{12}) - R(\mu_0, \beta_{11})$
$H_0: \beta_{11} = \beta_{12} = \beta$	Equally effective for procedures 1 and 2	$R(\beta_{11} = \beta_{12}) = R(\mu_0, \beta_{11}, \beta_{12}) - R(\mu_0, \beta)$
$H_0: \beta_{11} = \beta_{12} = \beta'$	Equally effective in opposite directions for procedures 1 and 2	$R(\beta_{11} = -\beta_{12}) = R(\mu_0, \beta_{11}, \beta_{12}) - R(\mu_0, \beta')$

An example is given in Table 7.1, using data from ten cycles of divergent mass selection for ear length (Cortez-Mendoza and Hallauer 1979). Deviations from regression were nonsignificant in the analysis of variance after fitting the linear regression. For testing the four hypotheses, the reduced models were used to solve the normal equations for determining sums of squares for each formulated hypothesis (Table 7.1). Response to divergent mass selection for increased and decreased ear length was significant, and rates of response also were significantly different for increased and decreased ear length.

More complex models also can be used for interpopulation improvement when information about the effects of selection on populations themselves, on the crossed population, and on heterosis in the crossed population are desired. A similar model can be applied when two populations are improved independently by some method of intrapopulation recurrent selection. For example, results reported by Paterniani and Vencovsky (1977) on a modified reciprocal recurrent selection program were analyzed on the basis of a more complex model that included the mean of the base populations, heterosis in the original cross, changes in the means of populations themselves, changes in heterosis in crosses between advanced cycles, and the effect of reciprocal crosses. The use of such a model and results are shown in Table 7.2.

Table 7.1. Hypothesis tested, sums of squares, and mean squares for evaluating ten cycles of divergent mass selection for ear length in maize (Cortez-Mendoza and Hallauer 1979).

Hypothesis	Estimates	$R(\hat{\mu}_0, \hat{\beta}_{1j})$[+]	$R(\hat{\mu}_0, \hat{\beta}_{11}, \hat{\beta}_{12}) - R(\hat{\mu}_0, \hat{\beta}_{1j})$	df	Mean square	F-value
$\beta_{11} = 0$	$\hat{\mu}_0 = 20.91$ $\hat{\beta}_{11} = -0.82$	3925.04	9.18	1	9.18	42.9
$\beta_{12} = 0$	$\hat{\mu}_0 = 16.77$ $\hat{\beta}_{12} = 0.70$	3898.27	35.95	1	35.95	168.0
$\beta_{11} = \beta_{12} = \beta$	$\hat{\mu}_0 = 19.52$ $\hat{\beta} = -0.16$	3834.06	100.16	1	100.16	468.1
$\beta_{11} = -\beta_{12} = \beta'$	$\hat{\mu}_0 = 18.66$ $\hat{\beta}' = 0.48$	3931.42	2.80	1	2.80	13.1
Pooled error				436	0.21	

[+]$\hat{\beta}_{1j}$ refers to $\hat{\beta}_{11}$, $\hat{\beta}_{12}$, $\hat{\beta}$, and $\hat{\beta}'$.

Table 7.2. Expected and observed means (yield, t/ha), based on a
multiple regression model, of populations (P: Piramex; C:
Cateto) and population crosses; and estimates of parameters after reciprocal recurrent selection based on testcross of half-sib families (adapted from Paterniani and
Vencovsky 1977).

Population	Model for expected mean[†]	Observed	Expected[‡]
P_0	m_p	5.641	5.727
C_0	m_c	4.109	4.023
P_1	$m_p + g_p$	6.012	5.927
C_1	$m_c + g_c$	4.236	4.323
$P_0 \times C_0$	$(1/2)(m_p + m_c) + h + r$	5.834	5.822
$C_0 \times P_0$	$(1/2)(m_p + m_c) + h - r$	5.799	5.724
$P_1 \times C_0$	$(1/2)(m_p + g_p + m_c) + h + 1/2 g_h + r$	5.754	6.015
$C_1 \times P_0$	$(1/2)(m_p + m_c + g_c) + h + 1/2 g_h - r$	4.530	5.967
$P_1 \times C_1$	$(1/2)(m_p + g_p + m_c + g_c) + h + g_h + r$	6.505	6.258
$C_1 \times P_1$	$(1/2)(m_p + g_p + m_c + g_c) + h + g_h - r$	5.998	6.160

Estimates: $\hat{m}_p = 5.727$; $\hat{m}_c = 4.023$; $\hat{h} = 0.898$; $\hat{r} = 0.049$; $\hat{g}_p = 0.200$; $\hat{g}_c = 0.300$;
$\hat{g}_h = 0.185$

[†] g_p, g_c: expected gain in performance of populations per se, due to
selection; h: expected heterosis in original cross; g_h: expected
change in heterosis; and r: expected deviation in reciprocal
crosses.

[‡] Deviations from model were nonsignificant; $R^2 = 0.69$.

Least squares multiple regression procedures have not been used extensively for evaluating selection response because either the potential and usefulness of the procedures were ignored or use of desk calculators for analyzing complex models was not feasible. However, as pointed out by Eberhart
(1964), availability of general multiple regression computer programs permit
the use of very large X'X matrices and estimation of parameters and analysis
of variance can be performed easily.

In most selection experiments in which prediction of gain was possible,
predicted values tended to be greater than observed gains. Several factors
may account for the discrepancies between predicted and observed values.
Probably the most important factors are sampling error of predictors and genotype-environment interaction effects. Prediction of gain from selection is
based on the heritability coefficient, which is a ratio of variances. It also
is well known that variance estimates are generally associated with large errors, which are reflected in the error of predicted gains. One way to reduce
sampling error would be to average the estimates of several cycles. The procedure introduces a bias because the genetic situation is expected to change
from cycle to cycle, but if the bias is small relative to the reduction in
sampling error the average may be closer to the true value than are individual
predictions (Moll 1959). Genotype-environment interaction causes an upward

bias in the predicted gain, as already shown in Chap. 6. Other possible causes of discrepancies may be random drift and inbreeding in small populations,
linkage disequilibrium, and epistasis. In spite of the fact that predicted
gains are generally greater than observed gains, a reasonably good agreement
through cycles of selection has been observed in some instances.

Comparison of relative effectiveness of different selection methods pre-
sents some difficulties (see Chap. 6). In some instances comparisons between
methods were possible but some results are contradictory. Effectiveness of
different selection methods can, however, be compared with results that are
summarized from the literature in the following pages.

7.2 IMPROVEMENT FROM INTRAPOPULATION SELECTION
7.2.1 Mass (or individual) Selection

Mass selection has been shown to be effective in modifying highly heritable traits in maize. Smith (1909) first reported results from divergent selection for ear placement in the Leaming open-pollinated variety of maize.
Selection for high and low ear placement was in adjacent blocks planted ear-
to-row where selection for ear height was only within the best yielding rows.
Results for two traits are summarized in Table 7.3.

Because the data were obtained in different years, the most adequate
measure of effect of selection on ear height is provided by the difference between high and low selection blocks. The difference increased at a rate of
15.7 cm per generation, as measured by the linear regression coefficient. Under the hypothesis that selection for high and low ear height was equally effective, the change would be about 7.9 cm per generation in each direction.
The effect of selection on yield cannot be measured because high yield was
taken into account in selection for low and high ear height. Nevertheless,
3 years of data indicated no great differences between the two strains for
divergent selection for ear height.

Selection for ear length was concluded to be ineffective by Williams and
Welton (1915); they stated, "It appears that within a variety, at least, the
length of an ear of corn is largely a matter of environment and cannot be expected to influence materially succeeding generations." This conclusion was
qualified as unjustified by Sprague (1966), because ear length is a highly
heritable trait (see Table 5.1).

Kiesselbach (1922) reported results of several selection experiments.
Selection for high, medium, and low ear height was conducted in contiguous
blocks for 5 years and resulted in average ear heights of 137, 119, and 112

Table 7.3. Direct response to divergent selection for ear height placement and
indirect response for yield of the maize variety Leaming (Smith 1909).

Year	Ear height (cm)			Yield (q/ha)	
	High	Low	Difference	High	Low
1903	143.2	108.7	34.5	---	---
1904	127.8	97.3	30.5	---	---
1905	160.8	105.7	55.1	---	---
1906	143.8	64.8	79.0	45.4	45.6
1907	183.9	84.3	99.6	40.5	42.9
1908	145.5	58.7	86.8	38.2	37.3

cm, respectively, with very small differences in yield (32.4, 33.3, and 33.6 q/ha). Similarly, selection for low and high lodging resulted in 22% and 29% (on the average) of lodged plants, respectively. Selection for three levels of erectness of ears (erect, medium erect, and drooping ears) resulted in 42%, 29%, and 21% erect ears as averaged over the period. During 6 years of selection for ear type relative to height, diameter, and texture (smooth or rough) there was not a clear trend of selection when comparisons were made for yield. Two years of selection for deep-grained, rough ears and shallow-grained, smooth ears were compared for yield with the original Hogue's Yellow Dent open-pollinated variety; on the average the grain yields were, respectively, 40.2, 43.7, and 40.5 q/ha. Additional data for 6 years of selection for long, smooth ears of Reid Yellow Dent open-pollinated variety contrasted for yield with standard, medium rough ears of the same variety showed that the smooth type surpassed the rough type by 2.5 q/ha. In all experiments for ear type for 12 different years the long, smooth ears surpassed all others in yield. However, the author concluded from limited evidence that ear type considerations are rather neutral in their relation to yield.

Smith and Brunson (1925) concluded that mass selection was as effective as ear-to-row selection in improving yield over ten cycles of selection in the Reid Yellow Dent variety. The average performances over ten years were 106.8% and 109.3% of the original variety for the strains maintained by mass selection and by ear-to-row selection for high yield, respectively. In either case the gain was very small.

Lonnquist (1952) and Lonnquist and McGill (1956) reported the effect of visual selection for desirable plants in advancing four synthetic varieties of maize. They concluded that a slight improvement in yield was observed for all synthetics and that gain in yield was accompanied by slightly later maturity. It is quite possible that in advancing synthetic varieties or variety composites many breeders around the world have observed some gain in yield and other traits by simple or visual mass selection; e.g., adaptive mass selection in semiexotic populations resulted in an increase in yield of about 5% per year and also an increase in prolificacy (Mathema 1971). Also, effective selection for early silking was reported by Hallauer and Sears (1972) in a program for integrating exotic germplasm in the Corn Belt maize program. Paterniani (1974b) observed very little increase in yield (2.3%) in advancing a cross between two populations by two generations of mass selection. Thus mass selection seems to be very useful in adapting exotic or semiexotic populations or in advancing generations of synthetic varieties or composites. Mild selection for important quantitative traits like lodging resistance, disease resistance, prolificacy, ear height, and other plant and ear characters, as well as for some qualitative traits, will surely result in some improvement in the population and possibly some increase in yield. Therefore mass selection is the most suitable method in these cases to effect some improvement without drastic loss of genetic variability because of the large effective population size that can be maintained. Populations obtained by this method are usable as base populations in selection experiments (including mass selection) that use higher selection pressures.

The procedure used by Sprague and Brimhall (1950) and Sprague et al. (1952) for selection of oil content in maize kernels is a type of mass selection with recombination of remnant selfed seed. This procedure follows: A given number of plants was selfed and the seed was analyzed for oil content. A sample of ears with the highest oil percentage was grown ear-to-row and all possible intercrosses were made by hand, providing seed for the next cycle of selection. This was called the recurrent series. For comparison, a sample of the selfed ears was used to continue selfing and selection (selfing series).

Results reported by Sprague et al. (1952) in the population Iowa Stiff Stalk
Synthetic show that the recurrent series was more effective in increasing oil
content in the kernel than the selfing series. In the recurrent series the
change was from 4.97% to 7.00% or 0.41% per year, while in the selfing series
the mean oil percentage increased from 4.97 to 5.62 during five generations,
an average of 0.13% oil per year. The procedure used by Jenkins et al. (1954)
did not include selfing selected plants. However, pollination was controlled
because a sample of pollen of selected plants was collected and mixed in apparently equal proportions and used to pollinate the same selected plants, resulting in selection for both sexes. Results after three cycles of selection
demonstrated that this recurrent selection method was effective to increase
the frequency of genes for resistance to northern leaf blight, *Helminthospori-*
um turcicum Pass.

Mass selection for yield in maize was more effective after the modification introduced by Gardner (1961). After four cycles of mass selection in the
Hays Golden open-pollinated variety (control and irradiated), using the grid
system to minimize the effect of environment, an average gain per year of
3.93% over the parental control variety was obtained--resulting in total gain
of about 15%. Lonnquist (1961) reported that an additional cycle in the same
variety showed evidence of continued improvement, increasing total gain to
about 19%. Subsequent reports of the same experiment (mass selection in Hays
Golden variety) indicated additional gain after continued selection. Thus,
Lonnquist et al. (1966) reported a 12.7% improvement for yield after six cycles or a rate of 2.1% per cycle. Gardner (1968) observed a gain of 2.7% per
cycle after 10 generations. In later reports Gardner (1969a, 1973) showed a
gain of 3.0% per cycle after 15 generations, as predicted by the linear regression over cycles. This was in excellent agreement with the expected 3.08%
response.

Gardner (1976) reported results from 19 generations in the control and
irradiated Hays Golden variety. Both selected populations tended to plateau
at about generation 13 (Fig. 7.1). The trend of response did not depart significantly from a linear regression. At generation 17 yields started to decrease. Gardner (1977) presented some possible causes of the sudden decrease
in generation 17, summarized as follows. The original population may have
been contaminated or unintentionally selected during its periodical (one in
each four years) multiplication, but there was no evidence that the population
had changed. Another hypothesis is that the shift from isolated nurseries

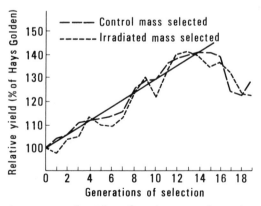

Fig. 7.1.   Response to mass selection for increased grain yield in Hays Golden
            (Gardner 1976).

with uniform soil easy to water uniformly to extremely variable soils and other environmental factors may have precluded an accurate identification of superior genotypes. Severe heat and drought in the last cycles of selection ac-
centuated that problem. Also, previous selection could lead to favorably
linked epistatic combinations of genes for which selection was no longer effective in extreme environmental variations, leading to reduced yields. Still
another hypothesis is related to genotype-environment interaction in the later
years of testing. It was observed during the experiment that Hays Golden has
the capacity to perform well under less favorable conditions, but plasticity
of response is not found in the selected material. Thus the selected generations would interact more with environment than the original Hays Golden variety, leading to a relative decrease in yield of the selected material.

Some other experiments have shown encouraging results from mass selection
for yield. Johnson (1963) obtained an increase of 33% after three cycles in a
tropical variety, an average of 11% per cycle. Also, Eberhart et al. (1967)
increased yield of Kitale Composite A from 51.2 to 55.0 q/ha, a gain of 7.4%
in one cycle of selection. Vencovsky et al. (1970) reported a progress of
3.8% and 1.7% per cycle in two Brazilian populations, Paulista Dent (five cycles) and Cateto M. Gerais (three cycles), respectively. Arboleda-Rivera and
Compton (1974) developed improved populations through mass selection for three
different seasonal conditions. Selection in the rainy season increased grain
yield and prolificacy at rates of 10.5% and 8.8% per cycle, respectively, when
tested in rainy seasons (direct response). However, when tested in dry seasons (indirect response) the increases were only 0.8% and 1.0% per cycle, respectively. Selection in dry seasons resulted in a direct response for yield
of 2.5% per cycle, which increased to 7.6% per cycle when tested in rainy seasons; similarly, the gain in ears per plant was 4.4% and 11.4%, respectively.
In the population tested under both seasons, rainy and dry, gains for yield
were 5.3% and 1.1% per cycle, respectively, and for prolificacy the respective
gains were 7.0% and 3.3% per cycle.

A similar experiment was conducted by Rodriguez et al. (1976). Mass selection in population MB-21 was performed in three ways: selection in the
less favorable season (MB-21-A); selection in the more favorable season (MB-21
-B); and alternate selection in both seasons (MB-21-AB). The average gains of
yield per cycle were 3.3% (two cycles), -1.0% (two cycles), and 4.9% (four cycles), respectively. The greater increase in prolificacy (1.83% per cycle)
also was observed in the population selected in both seasons (MB-21-AB),
whereas decreases in prolificacy of 0.35% and 3.17% per cycle were obtained
from selections in MB-21-A and MB-21-B, respectively. It was inferred that
less favorable environmental conditions allowed greater responses to selection.

Genter (1976b) reported that after ten cycles of mass selection in a composite of Mexican races, yield increased 171%, with a gain per cycle (regression coefficient) of 19.1%. Increase in yield was accompanied by a decrease
in plant height, ear height, days to silk, grain moisture, smutted plants, and
days from pollen shed to silk emergence; no change was detected for root lodging but stalk lodging increased. Josephson and Kincer (1976) reported results
of 14 generations of mass selection for yield in the variety Jellicorse; no
increase was observed in the first 4 generations, maximum increase per generation was 13.1%, and no increase was shown beyond the tenth generation. Similar increases in yield were obtained by Osuna-Ayalla (1976) in two populations
after six cycles of selection. Estimated gains per cycle (linear regression
coefficient) were 2.82% and 3.45% for Dent Composite and Flint Composite, respectively.

Other short-term selection experiments have demonstrated the effectiveness of mass selection. Miranda et al. (1971) reported results of selection for high yield (two cycles) and for low yield (one cycle) in selection either for one sex or for both sexes in the population Dent Composite. Two cycles of selection for high yield were not effective (1.7% increase) when based only on the female parent (one sex) but were effective (7% increase) when based on both sexes. One cycle of selection for low yield, however, was not effective when selecting for both sexes but was highly effective (15.7% decrease) when selecting for only one sex. In spite of discrepancies in results, a combined least squares analysis showed that selection for only male gametes (effect of detasseling the poorest plants) was effective as well as was selection for only female gametes (no detasseling). The limited number of cycles precludes any general conclusion. Hakim et al. (1969) obtained an average progress of 4% in only one cycle of selection in the Philippines; when the selected material was evaluated in the same season the gain was 9% over the original. Also, one cycle of mass selection resulted in significant increases by two methods of mass selection reported by Moro et al. (1974). Ordinary stratified mass selection resulted in progress of 11.6%. In the second procedure a constant genotype represented by a double-cross hybrid was planted in alternate hills of the white opaque-2 variety (the population under selection). At harvest the yield of each opaque-2 plant was compared to the yield of the adjacent double-cross hybrid plant, and a progress of 5.6% resulted by this selection procedure. Theoretical considerations suggest that a greater efficiency of the second method relative to stratified mass selection is attained whenever variation among plants of the constant genotype is smaller than environmental variation among strata. A full advantage would be attained by using a single cross as the constant genotype. Simple mass selection in an opaque-2 population conducted by Palma and Burbano (1976) resulted in an increase of 3.3% and 8.2% in the first and second cycles, respectively.

A less pronounced response to mass selection for grain yield has been reported by some authors. Hallauer and Wright (1967) obtained an increase of 4.5% after three cycles for the open-pollinated variety Iowa Ideal; an increasing trend in grain moisture, root lodging, and ear droppage also was observed. In a subsequent report (Hallauer and Sears 1969), small progress was obtained after selection in Krug (six cycles) and Iowa Ideal (five cycles) with a nonsignificant increasing rate of about 1.5% per cycle. Darrah et al. (1972, 1978) obtained rates of progress of 0.9 and 0.38 q/ha (0.8%) at 2% selection intensity after two and six years of mass selection, respectively, in the population Kitale Composite A. At a 10% selection intensity the rates of progress were 0.8 and 0.53 q/ha (1.1%) for two and six years of selection, respectively. In a population of Nigerian Composite B, progress of 16% was obtained by Obilana (1974) after four cycles of selection for yield. A summary of results obtained by three or more cycles of mass selection for grain yield is presented in Table 7.4.

Mass selection for traits other than yield has been shown to produce effective change in selected traits and correlated responses in other traits. Lonnquist (1967) showed that selection for prolificacy in the Hays Golden variety gave an increase in the number of ears per plant and in grain yield; a gain of 6.28% per cycle in yield was observed after five cycles of selection at 5% selection intensity, so total gain is comparable with that reported by Gardner (1969b) after ten cycles of mass selection at 10% selection intensity in the same variety. Torregroza and Harpstead (1967) also reported effectiveness of divergent mass selection for prolificacy; selection for increasing prolificacy resulted in 14% greater yield and 28% more ears per plant, while selection for single ear plants resulted in 5% decrease in yield and 7% decrease in ears per plant. Results from a similar experiment reported by Tor-

Table 7.4. Progress from mass selection for yield (three or more cycles) expressed in percent of the original mean for several populations under varying selection intensities.

Population	Selection intensity (%)	Number of cycles	Average gain/cycle (%)	Reference
Hays Golden	10.0	15	3.0	Gardner (1976)
Tropical	4.7	3	10.3	Johnson (1963)
Paulista Dent	20.0	5	3.8	Vencovsky et al. (1970)
Cateto M. Gerais	20.0	3	1.7	"
Jellicorse	---	14	0.9	Josephson and Kincer (1973)
Dent Composite	10.0	6	2.8	Osuna-Ayalla (1976)
Flint Composite	10.0	6	3.4	"
Krug	7.5	6	1.6	Hallauer and Sears (1969)
Iowa Ideal	7.5	5	1.4	"
Kitale Composite A	2.0	6	0.8	Darrah et al. (1978)
Kitale Composite A	10.0	6	1.1	"
Nigerian Composite B	10.0	4	1.9	Obilana (1974)
Composite of Mexican races	6.0	10	19.1	Genter (1976b)
Mezcla Varietales Amarillos[†]	5.0	3	2.5 (7.6)	Arboleda-Rivera and Compton (1974)
Mezcla Varietales Amarillos[†]	5.0	3	0.8 (10.5)	"
Mezcla Varietales Amarillos[†]	5.0	3	1.1 (5.3)	"
MB-21[‡]	5.0	2	3.3	Rodriguez et al. (1976)
MB-21[‡]	5.0	2	-1.0	"
MB-21[‡]	5.0	4	4.9	"

[†]Selected for dry, rainy, and both seasons, respectively; the respective gains refer to test in dry season or in wet season (in parentheses).

[‡]Selection for less favorable, more favorable, and both seasons, respectively; gains refer to average test over both seasons in one year.

regroza (1973) after 11 generations of selection showed 48% increase in prolificacy and 35% increase in yield in selection for high prolificacy; on the
other hand, decreases of 16% in prolificacy and 7% in yield resulted from selection for single ear plants.   Torregroza et al. (1976) evaluated advanced
intervarietal crosses MB-51 and MB-56 after nine and four cycles of mass selection for prolificacy, respectively.   The average response was an increase
from 1.29 (original) to 1.62 (ninth cycle) ears per plant in MB-56.   Gains per
cycle (regression coefficients) were 3.38 ± 0.30% and 2.0 ± 0.77% for MB-51
and MB-56, respectively; the corresponding gains per cycle for yield were 5.45
± 1.17% and 3.00 ± 1.73%, respectively.   Kincer and Josephson (1976) reported
a 13.2% increase in ears per plant and 33.1% increase in yield after 5 generations of selection for prolificacy in a program started after 9 generations of
selection for yield in the variety Jellicorse.

Mass selection for ear height has shown a good rate of progress in some
selection programs.   Vera and Crane (1970) reported a rate of gain of about
4.5% per cycle after two cycles of selection for low ear height in two populations.   Subsequently, Acosta and Crane (1972) observed a reduction of about
24% in ear height in both selected populations after four cycles at a selection intensity of 20%.   Josephson and Kincer (1973) reported a reduction in
ear height of 18.3 cm in a late synthetic and 25.9 cm in an early synthetic
after 10 generations of selection for lower ear height.   Subsequently, Josephson et al. (1976) reported that the reduction in ear height was 3.18 cm per
cycle in both synthetics after 12 generations of selection for lower ear
height.   Changes in other traits also were observed in these reports.   In each
case reduction in ear height was followed by reduction in plant height and
yield.   A positive association between yield and ear height or plant height
also was observed when selection was for yield by Gardner (1969a), Hallauer
and Sears (1969), and Darrah et al. (1972).

Preliminary results of mass selection for divergent ear length was reported by Hallauer (1968); selection was effective in separating the original
population (Iowa Long Ear Synthetic) into short ear and long ear subpopulations.   After ten generations, Cortez-Mendoza and Hallauer (1979) observed an
increase of 0.32 cm (1.6%) and a decrease of 0.64 cm (3.2%) per year in the
long ear and short ear subpopulations, respectively; the asymmetry of response
was probably due to the higher frequencies of genes for long ear length in the
original population, which was synthesized from long ear inbred lines.

Time of flowering was shown to be a trait that can be changed effectively
by mass selection.   Troyer and Brown (1972) selected three late semiexotic
synthetics (adapted approximately to 39° latitude) for early flowering.   In a
further report (Troyer and Brown 1976) selection also was effective in seven
southern Iowa adapted (41.5° latitude) synthetics.   In a third experiment
(Troyer 1975), selection in three synthetics adapted to 44° north latitude
from four cycles of selection resulted in significant changes in several
traits.   A summary of results of these three experiments is presented in Table
7.5.

Troyer (1976) also selected the earliest 2% of plants to flower in 18 F_2
populations.   In one group of 8 populations (whose parents had similar flowering dates), the effect of selection was evaluated after sibmating two cycles
and selfing the third cycle and then tested in crosses with Iowa inbred B14.
Selection effect per cycle averaged 4% selection index increase, 340 kg/ha
yield increase, 0.6% grain moisture decrease, and 0.6 days less to flower.
The second group (10 F_2 populations from five elite lines), selected in the
same manner, was evaluated in crosses with all possible (15) double crosses
among the lines; selection effect per cycle averaged 250 kg/ha yield increase,
1% grain moisture decrease, 3.7% stalk breakage increase, 7.0 cm plant height

Table 7.5. Results of selection for early flowering expressed as the gain
on average basis (a) or by the linear regression coefficient
(b) in percent of the original population for several traits
(adapted from Troyer and Brown 1972, 1976; Troyer 1975).

Latitude of adaptation		Days to silk	Kernel moisture	Plant height	Ear height	Silk delay	Yield	Broken stalks
(I) 39°[+]	(a)	-2.3	-3.2	-2.1	-5.2	- 7.6	1.9	47.2
	(b)	-2.4	-3.3	-1.9	-3.2	- 7.1	2.2	47.0
(II) 41.5°	(a)	-2.3	-4.4	-1.9	-3.0	-12.4	1.8	3.5
	(b)	-2.2	-4.2	-1.8	-2.8	-11.4	2.0	31.4
(III) 44°	(a)	-1.1	-4.0	-1.4	-2.6	- 7.3	-3.2	11.4
	(b)	-1.1	-3.7	-1.5	-2.8	- 6.8	-3.3	10.0

+
(I) Three late semiexotic synthetics; (II) seven southern Iowa adapted
synthetics; and (III) three synthetics adapted to 44° latitude.

decrease, 1.2 days less to flower, and 0.1 day less silk delay. Obilana
(1974) detected no improvement in selecting for earliness when evaluation was
over three environments. However, when evaluation was in only one location,
days to tassel were reduced from 52 to 46, a gain of 11.5% in two cycles. The
third cycle gave an additional gain of 5.8% in the same location. No significant increase was observed in the number of ears per plant, plant and ear
height, and grain yield.

Selection for several other traits has been reported in the literature.
Ariyanayagan et al. (1974) observed a change in leaf angle of about 3.8° and
10.2° per generation of selection (measured by two methods of leaf angle determination) in a bidirectional phenotypic selection for this trait; types selected toward erect leaves were shorter in plant height, later in maturity,
greater in lodging resistance, and greater in light transmissibility. Grain
yield variations attributable to leaf angle differences were small and nonsignificant. Zuber et al. (1971) reported that the percentage of ears with earworm damage was reduced 2.8% per generation after ten generations of selection
for earworm (*Heliothis zea* Boddie) resistance in two synthetics. Dudley and
Alexander (1969) observed changes in yield and plant and ear heights after selection for good seed set and agronomic type in four autotetraploid synthetics. Padgett et al. (1968) obtained about 2% gain per cycle in weight per 100
kernels in each direction after nine cycles of divergent mass selection for
seed size. Hanson (1971) selected for high and low stalk volume by use of
phenotypic selection within full-sib families of the open-pollinated variety
Jarvis. After six cycles the ratio between performance of high vs. low stalk
volume was 1.42 for fresh weight and 1.32 for dry weight. It was concluded
that selection modified stalk and leaf production primarily and root production secondarily. Net photosynthetic rates were not associated with differential productivity, supporting the hypothesis that net photosynthetic rate
did not play a major role in determining differential productivity within the
adapted population.

Subsequently Hanson (1973) reported the effect of selection for stalk
volume on the efficiency and number of chloroplasts. Selection for high stalk
volume resulted in less chlorophyll per gram of fresh weight but more total
chlorophyll than selection for low volume. An identical pattern was found for
the measure of DNA. Apparently selection resulted in similar chloroplast num-

ber per cell but selection for high stalk volume produced more chloroplasts per leaf. No apparent differences in physical characteristics of chloroplasts were found between selections.

7.2.2 Half-sib Family Selection

Use of half-sib families as progeny tests for maize improvement was introduced in 1896 by Hopkins in a program for quality improvement of the maize kernel. The open-pollinated variety Burr's White was used as the base population and this program has continued until the present. The program started with 163 open-pollinated ears analyzed for oil and protein content from which four strains were selected: high protein, high oil, low protein, and low oil (Hopkins 1899). Each strain was grown in a separate isolated field and selected according to ear-to-row procedure during 9 generations. Approximately 20% of the ears analyzed were selected. From generations 10 to 25, alternate rows in each breeding plot were detasseled and 20 ears for analysis were taken from the 6 highest yielding rows. Four ears per row were saved. In 1921 the system was again altered because yielding ability was disregarded and 2 seed ears were chosen from each of 12 detasseled rows. After generation 28 a system of intrastrain reciprocal crossing between substrains was introduced and natural pollination was replaced by controlled hand pollination (Leng 1962). This breeding procedure has been continued to the present. Selection for oil and protein content was effective during the course of the experiment. The first report of selection results was presented by Hopkins after two cycles of selection. Subsequently, periodical reports showed continued progress from selection in the four strains (Smith 1909; Winter 1929; Woodworth et al. 1952; Leng 1962). A report by Dudley et al. (1974) showed that after 70 generations of selection the mean oil content of the high oil and low oil strains were 16.6% and 0.4%, which are 354.8% and 8.5%, respectively, of the oil content in the original variety. Mean protein content in the high protein and low protein strains was 26.6% and 4.4% representing, respectively, 244% and 40.4% of the original variety protein content.

After generation 37 a system of random mating or nonselection was compared with the regular breeding system for eight generations (1934 to 1941). During this period neither continued selection nor the relaxation of selection in the high protein strain produced any significant changes. In the low protein strain, selection produced significant progress for low protein content but no change was observed by random mating. The same pattern was observed in the high oil strain, i.e., significant change by selection but no change by random mating during the first four generations. After four generations a dramatic increase in oil content occurred in both strains. The selected population returned to the previous level in the following year, but the nonselected population remained relatively high in the subsequent years (Leng 1962).

After generation 48, reverse selection was initiated in all strains in parallel with the normal selections. At generation 70, a 5.60% change was observed in the reverse high oil strain (8.85%) but only 1.62% was observed in the reverse low oil strain (2.38%) as compared with the oil content before reverse selection of 13.45% and 0.76%, respectively. The reverse high protein strain had 8.5% and the reverse low protein strain had 9.6% protein after 22 cycles of selection, as compared to 19.2% and 5.1% protein, respectively, before reverse selection began. After seven generations of reverse selection the reverse high oil strain was again subdivided, including selection toward high oil (switchback selection). At generation 70 the mean oil content had increased again to 14.02%. Dudley and Lambert (1969) reported good agreement between the average expected and the actual gain in the four strains. Results

Table 7.6. Predicted and observed response per generation from selection
            for oil and protein content over 65 generations (Dudley and
            Lambert 1969).

Direction	Oil content, %		Protein content, %	
	Predicted	Observed	Predicted	Observed
High	0.13	0.16	0.26	0.22
Low	-0.02	-0.07	-0.16	-0.09

are summarized in Table 7.6.

Dudley (1976) reported that after 76 generations, gains from selection
for high oil and low oil strains were 279% and 92% of the original population
mean. Similarly, in the high protein and low protein strains, gains of 133%
and 78%, respectively, were achieved. Dudley also analyzed several other genetic aspects of the Illinois oil and protein selection experiments; he concluded that enough genetic variation was present to expect future response to
selection.

Some results from use of half-sib family selection (ear-to-row) were reported in the early part of this century. The first experiments were encouraging, but further disappointing results caused the method to be considered
ineffective in improving grain yield in maize. Smith (1909) used the ear-to-
row system to select for high and low ear heights. The first criterion was
selection for yield among rows and the second the height of ear for selection
(phenotypic) within rows. Alternate rows were detasseled in a block of 24
rows and four seed ears were taken from each of the best 6 detasseled rows.
The effect of selection on yield seemed to be nonsignificant as indicated by
data from three years of testing (see Table 7.3). Some recent results have
shown that selection for low ear height usually is followed by a decrease in
yield. In Smith's experiment selection for low ear height was especially effective. Hence it is plausible to assume that primary selection for yield was
compensated for by any downward change caused by selection for low ear height,
although no precise comparison is possible from the data. Smith also selected
for erect and drooping ears, using the same procedure as for ear height selection. He concluded that after five years the trait had changed to a considerable extent, while no apparent change was observed in yield.

Williams (1907) reported the results of several ear-to-row tests for
grain yield. He also pointed out the advantage of using remnant seed from selected rows instead of selecting directly in the ear-to-row test. By use of
remnant seeds it was shown that the Ohio Standard Leaming open-pollinated variety gave a yield of 4.5 q/ha in excess of the original Leaming. On the other hand, selection within the highest yielding rows of the Clarage open-pollinated variety gave an excess of 1.8 q/ha over the original stock. In both instances comparisons were made in adjacent blocks without replications; consequently their precision was very poor. Hartley (1909) reported that ears selected from high-yield breeding rows yielded 11.3 q/ha, or 16%, in excess over
yields from a general field of maize that was planted in alternate rows to the
selected ears from the previous year.

Some comparisons of grain yield were also presented by Noll (1916). It
was observed that seeds from high yielding ear-rows produced less than seeds
from the general field in a selection experiment with College White Cap open-
pollinated variety. In a similar experiment with the 90-Day Clarage variety,
both sources produced practically equivalent yields. In later experiments involving 90-Day Clarage, seeds from different rows were planted separately. In
a test in 1913, the greatest gain was 2.9 q/ha, while two out of six plots

yielded less than field-grown seed. In 1915 all plots from high yielding rows
outyielded field-grown seed, and in one instance the excess was 8.0 q/ha. In
1914 remnant seed of the best ears was planted and crossed and five ears from
each cross were planted ear-to-row in the variety field. In both instances
the block of field-grown seed used as the check was superior in yield, showing
no progress because of selection.

Kiesselbach (1922) reported some results of the ear-to-row system at the
Nebraska Experiment Station. The following procedures were used to increase
high yielding strains: (1) continuous ear-to-row selection within the most
productive ear-to-row strains, (2) increasing highest yielding original ears
by using remnant seeds in isolated blocks to avoid contamination, (3) mixing
several productive ear-to-row strains and increasing thereafter in a single
isolation block, and (4) natural crossing of high yielding ear-to-row strains.
Ear-to-row selection in Hogue's Yellow Dent showed no great differences in a
comparative test over seven years (1911 to 1917). Average yields in quintals
per hectare were 33.4 for continuous selection, 29.9 for a single high yielding strain increased by using remnant seeds, 34.5 for a composite of four
strains, 34.2 for the intercrossing of four strains, and 33.6 for the original
Hogue's Yellow Dent. Selection in the Nebraska White Prize variety, using the
same procedures, averaged 39.7, 38.1, and 40.7 q/ha using procedures (1), (3),
and (4), respectively; the original population yielded 40.0 q/ha.

Smith and Brunson (1925) used the ear-to-row procedures in a selection
program for high and low yield in a Reid Yellow Dent variety. The same population also was maintained by mass selection in an isolated block. After ten
years the high yielding strain selected by ear-to-row averaged 39.1 q/ha and
the low yielding strain 32.5 q/ha; these values represent 109.3% and 90.9% of
the original variety. Mass selection resulted in a 6.8% increase over the
original variety in the same period. It was concluded that ear-to-row selection for low yield was more effective than selection for high yield. However,
precision of the comparisons was very poor, which was common in early maize
experiments.

In general, lack of adequate field plot techniques for selection and for
comparison of results from selection, associated with other factors such as
inbreeding (due to small populations) and lack of isolation, caused the ear-
to-row method to be regarded as powerless for improvement of maize yield. Only after modifications introduced by Lonnquist (1964) was this method again
regarded as very promising for population improvement.

The modified ear-to-row method is based on among and within half-sib family selection, and the first results were published by Paterniani (1967). After three cycles of selection within the population Paulista Dent, the improved material yielded 42% in excess over the original population with a regression coefficient of 13.6% over selection cycles. Webel and Lonnquist
(1967) reported a gain of 9.4% per cycle relative to the parental variety Hays
Golden after four cycles of selection. The first cycle was obtained from selection among and within full-sib families, since the tested progenies resulted from crosses between individual plants within the population; half-sib selection started with open-pollinated ears obtained from the recombination
block of full-sib families. In the second cycle, data from the crossing (recombination) block were used as additional information for among-family selection. In subsequent cycles the crossing block was not the source for family
performance data, thus providing better opportunities for selection within
families because individual plants could be evaluated and selected for grain
yield.

Other reports also have shown that half-sib family selection (modified
ear-to-row procedure) is an effective method for the improvement of maize pop-

ulations. Eberhart et al. (1967) reported results of two cycles of ear-to-row (selection only among families) in three populations; selection was effective for Kitale II (2.8% per cycle) and for Ec573 (11.4% per cycle) but not for Kitale Composite A-synthetic 1. Paterniani (1969) reported a gain per cycle of about 3.8% (linear regression coefficient) in the population Piramex. Darrah et al. (1972) reported gains varying from 0.9 to 3.2 q/ha per year in four populations (KII, Ec573, H611, and KCA). Darrah et al. (1978) observed different trends in response to selection in the populations KII, Ec573, and H611: 0.83, 2.59, and -0.43 q/ha per year, respectively, after six years of selection. The population Kitale Composite A (KCA) gave a response of 1.9 q/ha per year (5.2%) in ten years as measured by the average response in four selection experiments (Darrah 1975). Hakim et al. (1969) obtained 6% improvement in one cycle of selection in the Philippines.

Troyer et al. (1965) reported effectiveness of ear-to-row selection for adaptation of an introduced population; over six years he observed an increase in yield, a decrease in grain moisture, and a decrease in stalk and root lodging. Paterniani (1974a) obtained an increase in yield of about 5% after two cycles of mass selection and one cycle of among and within half-sib family selection in the population ESALQ-HV1. Relatively small progress was obtained by Lima et al. (1974) after two cycles of selection in two populations; progress of about 3% and 2% was obtained for the Flint Composite and Dent Composite, respectively, where selection also was based on other traits (ear height and stalk lodging). Paterniani (1974a) obtained a substantial gain (about 35%) after one cycle of selection in Piranão, an open-pollinated brachytic population. Sevilla (1975) evaluated eight cycles of selection in the variety PMC-561, obtaining an average gain of 9.5% per cycle. Segovia (1976) reported a 3.2% gain per cycle after three cycles of selection in the variety Centralmex, but no additional gain was detected in the following three cycles.

Selection in the Hays Golden population, first reported by Webel and Lonnquist (1967), was continued for 10 generations and results were reported by Compton and Bahadur (1977). An observed response (linear regression estimate) of 5.26% per generation was in good agreement with the expected 4.87% response. Gardner (1976) reported results up to 12 generations and a curvilinear response over all cycles was observed, as shown in Fig. 7.2. The line-

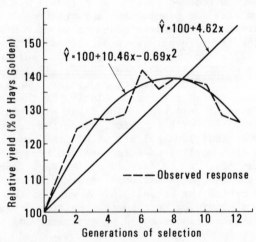

Fig. 7.2. Response to half-sib family (modified ear-to-row) selection in Hays Golden (Gardner 1976).

ar regression estimate of the gain per cycle was 4.62%, which is in good
agreement with the predicted response of 4.53%; but the trend of response was
clearly not linear with a better fit to a quadratic response curve.   The relative decrease in yield in the last generations was not expected.   Decrease in
additive genetic variance is not a sufficient explanation because it could
cause the population to level but not to decrease in yield.   A change in selection criterion in generation 8 by using a selection index based on lodged
plants and dropped ears might account for lack of continued response.   Additional data showed an average gain of 1.8% per cycle in the population IAC-1
(Miranda et al. 1977).   A summary of results from some half-sib selection experiments for yield is presented in Table 7.7.

A type of half-sib family selection was reported by Moll (1959).   According to this procedure, several plants in a population were used as males and
each was crossed with four plants as females; the best half-sib families that
identified the best male plants were recombined.   One, two, and three cycles
of selection in the populations Indian Chief, Jarvis, and (C121 × NC7)F_2 resulted in 11%, 7%, and 17% yield increase, respectively, over the original
populations.

Selection based on topcrossing with a tester is also a type of half-sib
family selection.   This procedure is commonly known as recurrent selection
(RS) for combining ability.   When the tester has a broad genetic base (open-
pollinated, synthetic, or composite varieties) recurrent selection is for general combining ability, including also the case where the parental population
is used as tester.   Lonnquist (1949) selected S_1 lines based on their performance as topcrosses with the parental variety Krug.   The topcrosses represent
half-sib families, which are evaluated in replicated trials, and remnant S_1
seeds of the male plants from the best topcrosses are planted for recombination.   Although the half-sib families are obtained from the cross, S_1 progenies × populations, they are expected to have the same array of genotypes as
S_0 plants × population.   There is an advantage in the first procedure because
evaluation of S_1 progeny itself is possible, providing the basis for a more

Table 7.7.   Effect of half-sib family (modified ear-to-row) selection (three or
more cycles) on yield in several populations.

Population	Selection intensity, %	Number of cycles	Gain per cycle, %	Reference
Paulista Dent	15.0	3	13.6	Paterniani (1967)
Piramex	23.7	4	3.8	Paterniani (1969)
Hays Golden[+]	20.0	4	9.4	Webel and Lonnquist (1967)
Hays Golden[+]	20.0	10	5.3	Compton and Bahadur (1977)
Hays Golden[+]	20.0	12	4.6	Gardner (1976)
Kitale Composite A	---	6	2.2	Darrah (1975)
Centralmex	22.5	3	3.2	Segovia (1976)
IAC-1	15.5	7	1.8	Miranda et al. (1977)

[+]Refers to subsequent reports of the same selection program.

accurate evaluation than the phenotypic appearance of individual S_0 plants. Results obtained by Lonnquist (1949) indicated a significant change in yield after one cycle of selection. Synthetic 2 of the first cycle of selection for high and low yield produced 142% and 85%, respectively, of the parental variety in the first year test and 118% and 88% in the second year. In the following year, after visual mass selection, the low yield Synthetic-3 equaled the parental variety, while the high yield Synthetic-3 was 27% greater in yield.

Lonnquist (1951) reported that the high and low yielding synthetics obtained by recurrent selection, using the parental population as tester, also were differentiated in relation to combining ability with the single cross WF9 × M14. Lonnquist and McGill (1956) selected S_1 lines based on topcrosses with the parental variety in three populations (Krug, Reid, and Dawes open-pollinated varieties) and obtained increases in yield of 22%, 9%, and 9%, respectively. In two cycles of selection the relative performance of the four populations in percent of the double cross US13 were 87%, 86%, 85%, and 72% in the first cycle and 98%, 95%, 102%, and 88% in the second cycle for the populations Krug, Reid, and Synthetics A and B, respectively. Crosses within each cycle in all possible combinations averaged 68.0 q/ha with 18.8% moisture in the first cycle and 71.9 q/ha with 19.2% moisture in the second cycle. The commercial hybrid US13 yielded 68.0 q/ha with 19.0% moisture. Lonnquist and Gardner (1961) reported that one cycle of selection resulted in an increase from 57.3 to 60.4 q/ha and from 54.3 to 56.7 q/ha in the populations Krug and Nubold Reid, respectively. There was also an increase in the cross between the two populations from 59.1 to 67.1 q/ha. Heterosis relative to midparent also increased from 6.0% to 14.6%, showing that the improvement of the populations was based mainly on additive effects.

Eberhart et al. (1967) used the procedure described by Lonnquist (1949) with the variety Kitale. An evaluation of this experiment was presented by Eberhart and Harrison (1973); two cycles of selection resulted in 5.2 q/ha (15%) increase in yield at a level of 34.6 q/ha. In better environments the original Kitale yielded 65 q/ha, and the improved material yielded 9.4 q/ha (14%) more than the original variety. In poor environments where the original variety yielded only 15 q/ha, the improved material was expected to yield 2.4 q/ha (16%) higher.

Lonnquist (1952) reported results of two cycles of recurrent selection in two populations using nonrelated synthetics as testers. In the first and second cycles the Krug variety yielded 82.4% and 88.4% of the double cross US13, respectively, followed by a slight increase in stalk lodging and grain moisture. Similarly, Synthetic A yielded 81.4% and 99.5% of US13 in the first and second cycles, respectively, followed by a slight increase in grain moisture but no apparent change in stalk lodging. Lonnquist (1961) reported that three cycles of selection in five populations (Krug, Reid, Synthetics A and B, and SSS) resulted in an average yield increase of 7.2 q/ha, whereas the intercross within cycles also increased yield 7.2 q/ha. In another comparison involving two cycles in three populations (Krug and Synthetics A and B), an average increase of 3.9 q/ha was obtained. After four cycles an average increase of 3.4 q/ha per cycle was observed for the three populations; the intercrosses showed an average increase of 3.5 q/ha (Lonnquist 1963). Thompson and Harvey (1960) reported that the mean yield of the topcrosses increased from 77.8% to 98.8% relative to check means after five cycles of recurrent selection in Synthetic A.

Recurrent selection for specific combining ability is selection based on testcrosses with a narrow base tester. Sprague and Russell (1957) selected two populations (Lancaster and Kolkmeier) to study selection for combining

ability with a specific tester (inbred line, Hy). The first cycle showed a
gain of about 6% in the population x tester performance for each population
and in the cross between the two populations. After two cycles the gains per
cycle (linear regression coefficient) were 4.3%, 12.8%, and 15.1% for the
crosses, Lancaster x Hy, Kolkmeier x Hy, and Lancaster x Kolkmeier, respectively. Sprague et al. (1959) reported that selection for specific combining
ability with the inbred line Hy gave an increase in testcross yields of about
4.2% and 14.7% per cycle, respectively, relative to the original population
testcross. Yields of the populations themselves decreased slightly, but the
cross between the two populations increased at a rate of 7.9% per cycle.
Lonnquist (1961) reported results from selection in Krug using the single
cross WF9 x M14 as tester. After three cycles the data indicated a gain per
cycle (regression coefficient) of about 3.6% with a slight decrease in grain
moisture and a slight increase in stalk lodging.

Penny (1959) and Penny et al. (1962) obtained significant progress in selecting two populations for combining ability with inbred tester B14. After
two cycles of selection for high and low yield, changes in yield of the populations themselves were 7.1% and 4.3% per cycle after selection for high yield
and -17.4% and -7.1% after selection for low yield for Alph and (WF9 x B7),
respectively. The cross Alph x B14 changed about 7.2% per cycle for high
yield, but selection for low yield also increased slightly (1% per cycle).
The cross, (WF9 x B7) x B14, showed a decrease when selected for low yield
(-3.5% per cycle) but a very small increase for high yield (1.1% per cycle).
The cross between the two populations indicated an increase (3.8% per cycle)
in high x high and a decrease of 7.1% per cycle in low x low. When all data
were considered the predominant type of selection appeared to have been for
genes exhibiting partial to complete dominance and largely additive effects.

Additional results of selection in Alph and (WF9 x B7) for combining
ability with B14 were presented by Russell et al. (1973). After five cycles
of selection, progress from selection was by changes in the yield of the testcrosses [Alph x B14 and (WF9 x B7) x B14], the populations themselves, and in
crosses with related and unrelated testers. The gains per cycle as measured
by linear regression coefficients in percent of the original population means
are shown in Table 7.8. The table shows that several traits had an increasing
trend in most instances. The population cross had an increase of 4.09 q/ha
per cycle. Because the gain from selection was significant for the specific
tester B14 as well as for unrelated testers, it was concluded that selection
was mainly for general combining ability (mostly additive) effects.

Less pronounced selection effects were observed by Walejko and Russell
(1977) in evaluating five cycles of selection in the populations Kolkmeier and
Lancaster for combining ability with the inbred line Hy (Table 7.9). Yield was
the primary trait in selection, and many of the C0 vs. C5 comparisons for yield
showed significant differences. Despite the small increase or decrease in
observed yield in the populations themselves, it was concluded that the recurrent selection program was successful in increasing frequencies of genes affecting yield because the testcrosses with the inbred line Hy and the population cross showed significant changes in yield. They concluded that recurrent
selection from use of an inbred line as tester seems to be an efficient method
for improving breeding populations and that the tester can be replaced by another line without causing loss of improvement achieved.

Horner et al. (1976) reached the same conclusion in a population selecting for combining ability with the single cross F44 x F6. After seven cycles
of selection the testcross performance showed 18% more grain yield, 9% lower
ear height, and 35% less lodging. Performance was similar when the same selected populations were crossed with an unrelated synthetic, showing that the

Table 7.8. Linear regression coefficients (gain per cycle for five cycles) in percent of the original mean of populations themselves and in crosses with related and unrelated testers (adapted from Russell et al. 1973).

Population		Yield (q/ha)[†]	Ears per[‡] 100 plants	Moisture (%)[‡]	Plant height (cm)[‡]	Ear height (cm)[‡]
Alph	per se	5.4	3.4	2.4	1.8	1.8
	Alph CO	6.0	2.9	1.4	2.2	2.2
	B14	4.4	2.0	0.9	0.7	0.8
	BSBB	6.7	3.2	1.3	1.7	2.3
(WF9 × B7)	per se	3.8	4.1	1.3	1.2	1.1
	Alph CO	3.6	0.9	-0.6	0.7	0.8
	B14	1.8	-0.8	-1.0	0.6	1.3
	BSBB	2.4	0.8	-0.2	1.0	1.0

[†]In percent of the observed original mean.

[‡]In percent of the estimated original mean.

Table 7.9. Estimated mean (a) and gain per cycle (b: linear regression coeffocient)[+] for two populations, their cross, and
in crosses with the specific tester Hy (Walejko and
Russell 1977).

Population	Yield(q/ha) a	b	Moisture(%) a	b	Lodging(%) Root a	b	Stalk a	b
Kolkmeier	44.0	1.6	28.6	-0.3	24.3	-10.4	11.3	10.7
Lancaster	47.7	-2.0	22.6	-0.5	18.2	-3.3	26.3	-2.0
Kolk. × Lanc.	59.9	4.1	25.2	-0.8	19.7	-4.4	16.9	3.3
Kolk. × Hy	63.2	4.4	26.7	-1.0	19.2	-2.9	9.1	9.4
Lanc. × Hy	67.2	3.3	24.5	-0.9	11.5	-2.4	15.1	-4.4

[+]In percent of the estimated mean (a).

narrow base tester was effective for improving general as well as specific
combining ability. Eberhart et al. (1973) reported on seven cycles of selection in Iowa Stiff Stalk Synthetic using the double cross Ia13 as tester. Increase in yield in the testcross was linear at a rate of 1.65 q/ha (2.6% per
cycle) but the population itself increased at a lower rate (1.4% per cycle).
Selection for earworm resistance was shown to be ineffective by Widstrom et
al. (1970), as indicated by the linear increase in earworm rating (b = 0.07
± 0.05) after four cycles of selection in a synthetic using a single cross as
tester. A summary of results from recurrent selection for general and specific combining ability for yield is shown in Table 7.10.

7.2.3 Full-sib Family Selection

Full-sib families are obtained from plant-to-plant crosses, thus providing better parentage control for selection of superior families. Plant-to-
plant crosses were used by A. E. Blount about 1868, as reported in the 1936
USDA Yearbook. Blount reported on 100 to 300 desirable plants that were pollinated with pollen from other desirable plants selected as males. Apparently
there was no further test of the paired crosses, so the mixture of all crosses
was similar to phenotypic (mass) selection with control of the pollen parent.
The Blount White Prolific variety resulted from this selection and was widely
distributed.

Harland (1946) suggested a method by which paired crosses were tested in
replicated trials followed by selection of the best ones and recombination in
detasseling blocks. Using this procedure with a local variety never selected
before, development of an improved variety yielding more than 62.8 q/ha was
reported. Lonnquist (1961) reported that selection of S_0 x S_0 crosses in Krug
III Synthetic resulted in an increase of 3.5% for high yield and a decrease of
6.1% for low yield (relative to the parental population) in one year of testing. In the following year's test the performances of the first cycle for
high and low yield were 114% and 101% of the parental population, respectively. The average of two years of testing indicated an increase of 8.9% for
high-yield selection and a decrease of 2.6% for low-yield selection. At the
same time a system of paired crosses was proposed; each S_0 plant was tested
with two other genotypes instead of one. The system was called chain crosses
and is in sequence as follows: 1 x 2, 2 x 3,..., (n - 1) x n, n x 1. It also
was suggested to select plants on the basis of their average effect in the two

Table 7.10. Effect of selection (gain per cycle) for combining ability on yield of populations per se and in testcrosses.

Population	Tester[+]	No. of cycles	Gain per cycle (%)				Reference
			Popul. per se	‡	Test-cross	‡	
Krug (high)	B	1	17.7	---	---	---	Lonnquist (1949)
Krug (low)	B	1	-12.0				
Krug	B	2	6.8	(1)	---	---	Lonnquist (1952)
Syn. A	B	2	18.1	(1)	---	---	Lonnquist (1952)
Krug	B	1	22.0	---	---	---	Lonnquist and McGill (1956)
Reid	B	1	9.0	---	---	---	" "
Dawes	B	1	9.0	---	---	---	" "
Krug	B	2	11.0	(1)	4.8	(1,2)	" "
Reid	B	2	9.0	(1)	1.5	(1,2)	" "
Syn. A	B	2	17.0	(1)	5.5	(1,2)	" "
Syn. B	B	2	16.0	(1)	11.3	(1,2)	" "
Lancaster	N	2	---		4.3	---	Sprague and Russell (1957)
Kolkmeier	N	2	---		12.8	---	" "
Lancaster	N	2	-1.0	---	4.3	---	Sprague et al. (1959)
Kolkmeier	N	2	-0.6	---	14.7	---	"
Alph	N	2	2.8	---	6.3	---	Penny (1959)
(Wf9 × B7)F2	N	2	10.3	---	1.3	---	"
SS Syn.	N	4	---	---	1.3		
Krug	B	3	3.9	(3)	4.9	(2,3)	Lonnquist (1961)
Reid	B	3	8.6	(3)	4.9	(2,3)	"
Syn. A	B	3	-0.5	(3)	4.1	(2,3)	"
Syn. B	B	3	9.0	(3)	8.0	(2,3)	"
SS Syn.	B	3	13.9	(3)	6.2	(2,3)	"
Krug	N	3	---	---	3.6	---	"
Krug	B	1	5.4	---	8.9	(4)	Lonnquist and Gardner (1961)
Nubold	B	1	4.5	---	4.0	(4)	" "
Alph (high)	N	2	7.2	---	7.2	---	Penny et al. (1962)
Alph (low)	N	2	-17.7	---	1.0	---	"
(Wf9 × B7) (high)	N	2	4.4	---	1.1	---	"
(Wf9 × B7) (low)	N	2	-7.3	---	-3.5	---	"
3 populations	B	4	6.4	(5)	5.2	(2)	Lonnquist (1963)
Kitale	B	2	7.5	---	---	---	Eberhart and Harrison (1973)
Alph	N	5	5.9	---	4.6	---	Russell et al. (1973)
Wf9 × B7	N	5	4.4	---	1.8	---	"
Lancaster	N	5	-2.0	---	3.3	---	Walejko and Russell (1977)
Kolkmeier	N	5	1.6	---	4.4	---	"
FSB (HT)	N	5	---	---	3.5	(6)	Horner et al. (1976)

[+]B: broad base, N: narrow base. ‡(1): From first to second cycle in percent of US13; (2): average performance in all possible cross combinations within cycles; (3): from first to third cycle in percent of estimated original mean; (4): average performance in a 12 × 12 diallel cross; (5): average gain (linear regression) in three populations (Krug and Synthetics A and B) in percent of estimated original mean; (6): seventh cycle yield 3.3% in excess over C_5.

crosses and then to select the best cross within each pair selected in the first phase. Selection for high and low yield in Krug III gave 10.6% increase and 4.9% decrease, respectively, after one cycle of selection evaluated in two years. Selection in the chain series therefore seemed to be more effective than the paired cross to improve yield.

Robinson and Comstock (1955) reported that one cycle of selection in four populations (CI21 x NC27, NC34 x NC45, Jarvis, and Weekly) resulted in an average yield increase of 9.8% over check yields. An additional cycle in the first two populations did not show any further increase. Moll and Robinson (1966, 1967) reported that three cycles of full-sib selection resulted in yield increases of 3.6% and 2.1% in the populations Jarvis and Indian Chief in two years of testing. The cross between the two populations increased 3.8% in yield, as indicated by one year of testing. Moll and Stuber (1971) obtained

Table 7.11. Average yield, plant height, and ear height of subpopulations
resulting from five selection criterion (adapted from Moll
et al. 1975).

| Trait | Selection criterion[†] | | | | | Jarvis |
	A	B	C	D	E	
Yield (g/plant)	249.5	234.5	247.2	249.5	250.4	239.5
Plant height (cm)	284.5	264.1	271.8	285.7	293.6	282.5
Ear height (cm)	126.5	110.2	115.3	128.5	136.1	124.5

[†]A, B: single trait selection for yield and ear height, respectively; C, D:
restricted selection index with desired change of -5.1 and +5.1 cm in ear
height, respectively; E: selection index for maximum expected change in
yield.

significant increases in yield after six cycles of full-sib family selection.
The gain per cycle in grams per plant was 9.47, 6.81, and 9.76 for the populations Jarvis, Indian Chief, and Jarvis x Indian Chief, respectively. At a
planting density of 50,000 plants/ha, those increases correspond to 4.74,
3.40, and 4.88 q/ha. The variety cross yield increased at a rate of 7.21 g
per plant (3.60 q/ha). Moll et al. (1975) evaluated five criteria for one cycle of divergent selection through full-sib families from remnant seed for recombination in the variety Jarvis. The average effects of selection for each
criterion are shown in Table 7.11. Data show that selection resulted in significant differences from the original variety for both yield and ear height
for all selection criteria employed.

Four cycles of selection among full-sib families were completed by Jinahyon and Moore (1973). They observed that yield increased at a rate (regression coefficient) of about 7.9% per cycle; there was a decreasing trend in
plant and ear heights and plant lodging decreased from 24% to 8% in four cycles. In an F_2 population, four cycles of full-sib family selection resulted
in an increase in yield of 34% or a gain per cycle (regression coefficient) of
8.6% over the parental population, as reported by Genter (1976a). Equally effective selection was reported by Compton (1977) in the variety Krug. After
four cycles of full-sib family selection by use of a selection index called
"harvest-able yield" (index = yield x undropped ears x upright plants), the
response was in the right direction for all traits in question: Yield increased from 64.1 to 72.9 q/ha (b = 2.9 ± 1.2 q/ha per cycle) and the index
value increased from 49.2 to 59.7 q/ha (b = 2.8 ± 0.6 q/ha per cycle); percentage of undropped ears and upright plants increased and moisture decreased
slightly. A summary of results from full-sib family selection is shown in Table 7.12.

7.2.4 Inbred Family Selection

Selfing was first used in recurrent selection systems to maintain tested
genotypes. It has been shown that variability among families increases with
inbreeding, so use of inbred families in a recurrent selection system is mainly for characters of low heritability. After phenotypic recurrent (mass) selection was shown to be effective to increase resistance to *Helminthosporium
turcicum* (Jenkins et al. 1954), S_1 family selection was used by Bojanowski
(1967) in an attempt to increase resistance to smut (*Ustilago maydis*). Selection in a selfing series also was conducted to provide a basis for comparison.
However, results were disappointing relative to effectiveness both of continuous inbreeding and selection and of recurrent selection by means of S_1 fami-

Table 7.12. Effect of full-sib family selection on yield of several populations.

Population	Number of cycles	Gain per cycle (%)	Reference
Krug (high)[†]	1	8.9	Lonnquist (1961)
Krug (low)[†]	1	-2.6	"
Krug (high)[‡]	1	10.6	"
Krug (low)[‡]	1	-4.9	"
Jarvis	6	3.5	Moll and Stuber (1971)
Indian Chief	6	2.8	" "
(Jarvis × Indian Chief)[§]	6	2.5	" "
(Jarvis x Indian Chief)-syn.	6	2.8	" "
(CI21 × NC7)	10	4.0	" "
Cupurico × Flint Composite	4	7.1	Jinahyon and Moore (1973)
(Va17 × Va29)F_2	4	9.3	Genter (1976a)
Krug[¶]	4	4.7* (5.8)	Compton (1977)

[†]Divergent selection of full-sib families obtained by paired crosses; [‡]divergent selection of full-sib families obtained by chain crosses; [§]cross between populations over cycles of intra-population improvement; [*]single trait (yield) selection; [¶]gain per cycle on yield and on "index" (in parentheses), respectively.

lies. Further reports have demonstrated the effectiveness of inbred family selection for improvement of quantitative traits.

Penny et al. (1967) selected five populations for European corn borer (*Ostrinia nubilalis* Hübner) resistance during three cycles. It was observed that two cycles of selection were sufficient to shift frequencies of resistance genes to a high level in all varieties and that three cycles produced essentially borer-resistant varieties.

Jinahyon and Russell (1969a, 1969b) evaluated three cycles of recurrent selection to improve stalk rot (*Diplodia zea* [Schw.] Lev.) resistance in the open-pollinated variety Lancaster. Progress of improvement was determined in the cycle populations themselves--population testcrosses with WF9 x Hy and Os420 x 187-2 and diallel crosses of the CO, C1, C2, and C3 populations. All methods of evaluation based on artificial inoculation with *Diplodia zea* showed significant improvement for stalk rot resistance. Correlated changes with selection for stalk rot resistance were greater plant vigor, later maturity, better disease resistance, greater stalk strength, and greater yields in hybrid crosses.

Jinahyon and Moore (1973) observed an increase in yield of 8.3% per cycle after two cycles of S_1 family selection in Thai Composite; stalk lodging in CO decreased from 53% to 17% in the second cycle. They also observed slight decreases in plant and ear height and no changes in the number of days to silk.

Scott and Rosenkranz (1974) used three variations of inbred family selection to increase resistance to corn stunt: (1) selecting the best plant in the best 10 of 463 S_1 progeny rows, (2) selecting the best 23 progenies derived from the original 463 S_1 progenies, and (3) selecting the best 10 of 100 S_1 progenies that were evaluated in a replicated test. Each method of selection was effective, but the most effective was that based on replicated tests (method 3); after one cycle of selection 21% of the plants in the improved population were infected with corn stunt compared with 52% diseased plants in the original population.

Mock and Bakri (1976) obtained some progress in selecting for cold tolerance. Percent of emergence and dry weight increased 8.4% and 0.6 kg per cycle, respectively, in the population BSSS13(SCT). On the other hand, those traits did not respond consistently to selection of S_1 families in the population BSSS2(SCT). Emergence index was not changed by selection in either population. Several other programs of inbred family selection involved comparisons with other selection schemes, and these are mentioned later.

7.2.5 Comparisons among Different Intrapopulation Selection Methods

The relative importance of selection for specific and for general combining abilities was studied by Lonnquist and Rumbaugh (1958). From the variety Krug (KI), a sample of 152 S_0 plants was selected and topcrossed with the single cross WF9 x M14; S_1 seed from the best 31 testcrosses was intermated to form the synthetic KII. Because KII showed no difference from KI, a new sample of 91 S_0 plants was selfed and topcrossed with the parental KI as tester. At the same time 30 S_1 lines from the first sample were topcrossed with KI to make 121 topcrosses available for yield testing. Sixteen of 121 topcrosses were selected and the KII(s) synthetic was obtained after recombination of remnant S_1 seeds. The two second cycle synthetics, KII (based on narrow base tester) and KII(s) (based on broad base tester), were subsequently compared in performance trials; they yielded 95.0% and 98.5% of the double cross US13, which yielded 62.6 q/ha. Thus selection for general combining ability apparently was more effective in selection of lines having greater additive genetic effects and produced a greater population yield than selection based on the narrower genetic base tester.

A different conclusion was presented by Horner et al. (1963) using a narrow base tester F6 and the parental broad base population Florida 767 as testers. Performance of testcrosses in percent of the double-cross check hybrid Dixie 18 was greater for the narrow base tester in the first three cycles but about 6% greater for the broad base tester in the fourth cycle; the linear rate of increase was 1.7% and 3.4% of Dixie 18 per cycle for the narrow base and broad base testers, respectively. Composites formed by intercrossing lines in the first three cycles were crossed with 11 different testers; in the F6 series grain production increased significantly (from 96.3% to 102.8% relative to the grand mean), but no significant change (98.6% to 100.8%) was observed in the broad base tester series when crossed with the same testers. The combining ability of the two composites with the inbred line F6 showed an average gain in yield of 6.5% per cycle in the F6 series but only 1.5% in the broad base tester series. The conclusion was that recurrent selection for combining ability with an inbred line tester was more effective than with a broad base tester in improving grain yield in maize.

Penny (1968) also compared narrow base (double cross) vs. broad base (synthetic) testers in a selection experiment with Iowa Stiff Stalk Synthetic. After three cycles of selection involving the broad base tester and six cycles involving the narrow base tester, no difference was observed in gains per cycle, which were 1.4% and 1.8% per cycle, respectively. A composite formed from elite inbred lines showed an increase in yield of 5% over the original variety. However, time and effort required to develop inbred lines causes this system not to be comparable with the former ones.

Lonnquist (1968) compared recurrent selection based on an unrelated tester (BIII synthetic) and on the parental population (KIII synthetic) as tester. He also included selection based on S_1 lines themselves. An increase of 15%

relative to KIII was observed in the derived population when the parental population (KIII) was used as tester. The population derived from selection of S_1 lines themselves showed an increase of 4% in grain yield, whereas no apparent gain resulted from selection based on testcrosses with an unrelated tester. In a previous experiment (Lonnquist and Lindsey 1964) a comparison was made between recurrent selection based on S_1 lines themselves and on testcrosses with an unrelated population (BIII synthetic) as the tester. Results indicated that both evaluation procedures were effective with a slight advantage for the use of an unrelated tester.

Horner et al. (1969) compared efficiencies of a narrow base tester (inbred line, F6) and a broad base tester (parental population) and also included selection based on S_2 families themselves for higher grain yields. After three cycles of selection, evaluation was based on (1) random mated populations, (2) selfed populations (bulk of S_1 lines), and (3) crosses with 11 unrelated testers. Differences among methods were detected only in the second and third cycles. The highest yielding random mated population was obtained using the parental population as tester, whereas the highest yielding selfed population was produced by selection of S_2 progenies themselves. The average combining ability with unrelated testers increased significantly (5.2%) for the three methods, but there were no significant differences among them. The inbred tester method was effective in increasing yield of selfed populations, but it was inferior to the other two methods for population improvement in overall evaluation. In addition, it was concluded that both the S_2 progeny method and the parental tester method were effective but in different ways. Selection based on inbred families places relatively more emphasis on contributions of homozygous loci; the parental tester method emphasizes the contribution of heterozygous loci.

Using the methods described above, Horner et al. (1973) subsequently compared results of five cycles of selection. General combining ability, evaluated through testcrosses with two broad base testers, showed a significant linear increase for all methods. Selection based on the inbred tester was significantly more effective than the two other methods, showing a gain in grain yield of 4.4% per cycle. The parental tester and S_2 progeny methods showed gains of 2.4% and 2.0%, respectively. Performance of the random mated populations after adjustment for inbreeding depression also showed a linear increase, but differences among the three methods of selection were not significant. Such results and the result reported by Lonnquist (1968) suggest that selection based on inbred families (S_1 or S_2 lines) has been less effective than expected. The S_2 progeny method, particularly, is theoretically a more effective method for changing frequencies of genes having additive effects than are the testcross methods (Horner et al. 1969).

Comparisons between S_1 line and testcross performance as a basis for selection have been presented in several other reports. Koble and Rinke (1963) found significant correlations between S_1 lines and topcrosses with a related and an unrelated tester for yield and several other traits. Genter and Alexander (1966) compared S_1 lines themselves and testcross performance as a basis for selection in the population CBS. Inbred lines obtained from the population selected on the basis of S_1 progeny performance showed an increase of 31.4% in yield after two cycles; yield of S_1 lines from the population selected on the basis of testcrosses (two unrelated single-cross testers) increased at a lower rate, i.e., 17.9% in two cycles.

Duclos and Crane (1968) evaluated one cycle of selection through S_1 line performance in topcross with three synthetics; S_1 lines from the original population and from the derived populations based on S_1 lines and on topcross (double cross as tester) performance yielded 31.9%, 40.7%, and 36.9% of the

Table 7.13. Effect of selection (gain per cycle) on yield of populations themselves and in testcrosses in comparative studies of recurrent selection.

Population	Type of family[†]	No. of cycles	Gain per cycle Popul. per se	[‡]	Testcross	[‡]	Reference
Florida 767	BT	4	---	---	3.4	(1)	Horner et al. (1963)
	NT	4	---	---	1.7	(1)	"
Krug III	BT	1	13.8	---	---	---	Lonnquist and Lindsey (1964)
	S_1	1	10.9	---	---	---	" "
CBS	BT	2	9.0	(2)	---	---	Genter and Alexander (1966)
	S_1	2	15.7	(2)	---	---	" "
BSSS	BT	3	1.4	---	7.4	---	Penny (1968)
	NT	6	1.8	---	2.6	---	"
Krug III	BT(P)	1	15.0	---	---	---	Lonnquist (1968)
	BT(U)	1	4.0	---	---	---	"
	S_1	1	-1.0	---	---	---	"
Purdue Ex-syn. 1	BT	1	25.4	(2)	12.4	(3)	Duclos and Crane (1968)
	S_1	1	38.7	(2)	8.1	(3)	" "
Krug Yellow Dent	FS	1	15.0	---	---	---	Silva and Lonnquist (1968)
	S_1	1	11.0	---	---	---	" "
Composite	BT(P)	1	8.1	---	---	---	Horner et al. (1969)
	NT	1	-3.2	---	---	---	"
	S_2	1	-2.6	---	---	---	"
Syn. A	BT	2	2.7	---	18.0	(4)	Carangal et al. (1971)
	S_1	2	4.6	---	15.7	(4)	"
BSK	BT	4	3.9	---	1.7	(5)	Burton et al. (1971)
	S_1	4	1.9	---	2.5	(5)	"
Florida 767	BT	4	---	---	16.7	(7)	Horner et al. (1973)
	NT	4	---	---	19.6	(7)	"
	S_2	4	---	---	47.7	(6,7)	"
VLE	BT	2	-1.3	---	0.8	---	Genter (1973)
	S_1	2	1.3	---	0.6	---	"
VCBS	BT	2	1.4	---	2.1	---	"
	S_1	2	6.7	---	3.6	---	"

[†]BT: broad base tester (P: parental; U: unrelated tester); NB: narrow base tester; S_1 and S_2: selfed families; FS: full-sib families. [‡](1) In percent of Dixie-18; (2) evaluated through S_1 line performance; (3) topcross performance; (4) gain in testcross performance with Min 707 from first to second cycle in percent of first cycle; (5) performance in testcrosses; corresponding gains in selfed populations are 3.9% and 9.6% per cycle, respectively; (6) S_2 line performance per se; (7) in crosses with two unrelated testers the increases were 2.4%, 4.4%, and 2.0% per cycle, respectively.

checks. Topcrosses with unrelated testers yielded 85.7%, 85.4%, and 88.7% of the checks, respectively. Results of one cycle of selection indicated that selection based on S_1 line performance produced the best yielding S_1 lines and selection based on topcross performance with a double-cross tester produced the best topcross yields with unrelated testers. Populations obtained by intermating top lines in each method showed that both methods resulted in highly significant yield improvement. The second cycle of selection, however, was not effective.

Carangal et al. (1971) evaluated two cycles of selection based on two types of families. In the first cycle there was a superiority of 42.6% of testcrosses over S_1 lines for yield, which has no meaning because differences of inbreeding were involved. In the second cycle, yields of S_1 lines from selection based on S_1 performance were not different from those derived from selection based on testcross performance. However, yield of testcrosses in the population selected for combining ability (based on testcrosses) was slightly greater (2%) than of testcrosses from selection based on S_1 performance. Advanced populations obtained by S_1 progeny and testcross performance exceeded the parental variety by 4.6% and 2.7%, respectively.

Burton et al. (1971) evaluated four cycles of selection in Krug Hi I Synthetic 3 based on S_1 progeny performance and on testcrosses with an unrelated double cross as tester. A bulk of S_1 lines from the populations derived by the two methods (S_1 and testcross) had better yield increases for the selfing series (38.7%) than for the testcross series (12.0%). Evaluation of selection effects in testcrosses with four single crosses showed that in the selfing series the increase in testcross performance was 10.6% after four cycles of selection, whereas in testcross selection only 5.7% yield increase was observed.

Genter (1973) reported results of two cycles of selection for yield in two populations; no progress was observed in either testcross (related or unrelated tester) or S_1 selection in the VLE population. Selection based on S_1 performance was effective in improving VCBS population (14.3% over two cycles), but selection based on testcrosses showed a nonsignificant increase of 2.7% over the parental variety.

A comparison between full-sib family selection and S_1 family selection (Silva and Lonnquist 1968) showed that both selected populations were significantly higher yielding than the original variety Krug. The observed gains were 11.2% and 15.0% for S_1 selection and full-sib selection, respectively, at selection intensities of 4% and 33%. A summary of results of several selection programs involving comparisons among the testcross and selfing methods are presented in Table 7.13.

7.2.6 Combination Selection

Thompson (1972) used three different methods in a divergent selection program for lodging resistance. To increase lodging resistance three cycles of testcross evaluation (cycles 1, 4, and 5), one cycle of S_1 family selection (cycle 2), and three cycles of full-sib family selection (cycles 3, 6, and 7) were used. Selection to increase lodging was based on one cycle (first) of testcross evaluation followed by five cycles of full-sib family selection. Two populations (Synthetics 8 and 9) were used and selection progress was evaluated by crosses between comparable cycles of the two synthetics, e.g., second cycle (Synthetic 8) x second cycle (Synthetic 9). Compared with the cross of the original unselected synthetics as 100%, selection to increase and decrease lodging showed 28% and 228% as many erect plants, respectively, in the last cycle. Results of the first cycles of selection were previously reported by Thompson (1963).

Goulas and Lonnquist (1976) used combination selection in the sense that selection was based simultaneously on performance of two types of families, half-sibs and S_1 progenies. By selecting on the mean performance of half-sib families and S_1 progenies obtained from the same plants a significant improvement was obtained. After two cycles yield increased 24% over the parental variety. There was also a decrease in grain moisture (6% in two cycles) and an increase of 7% in ear height.

7.3 IMPROVEMENT FROM INTERPOPULATION SELECTION

Several selection programs for interpopulation improvement were initiated after Comstock et al. (1949) proposed reciprocal recurrent selection (RRS). At Iowa State University an RRS program was initiated in 1949 in the populations Iowa Stiff Stalk Synthetic (BSSS) and Corn Borer Synthetic No. 1 (BSCB1). After two cycles of selection Penny (1959) reported a gain of 5.1% per cycle in the cross between the two populations. After four cycles of selection Penny (1968) reported that increase in yield in the crossed population was from 60.8 to 65.2 q/ha, a gain of 1.8% per cycle in a two-year evaluation test. A least squares analysis of several experiments presented by Penny and Eberhart (1971) indicated very little progress (1.18 q/ha) in the population cross. The parental population BSSS showed a small increase (1.38 q/ha per

cycle) and BSCB1 a small decrease (-0.64 q/ha per cycle) in yield after three
cycles of selection. Similar changes were observed by Hallauer (1970) in
evaluating the same populations and population crosses.

A more comprehensive evaluation after five cycles of selection was presented by Eberhart et al. (1973). An increase in yield of 2.73 q/ha (4.5%)
per cycle was detected in the population cross. The population BSSS selected
for combining ability with double cross Ia13 as tester also showed an increase
in yield (2.31 q/ha or 3.8% per cycle) when crossed with corresponding cycles
of BSCB1 from the RRS program. Changes in yields of the populations themselves were very small, i.e., 0.9% and 0.4% per cycle for BSCB1 and BSSS.
There was also an increase in ears per 100 plants in the populations and population cross. Stalk lodging was reduced in nearly all improved strains and in
their crosses. A slight decrease in plant height was observed in the populations and a slight increase in the population cross. Increase in yield
through RRS in the population cross without improvement in the parental populations or in topcross to an unrelated tester was consistent with expectations
from changing gene frequencies at loci involving overdominant (or pseudoover-
dominant) gene action.

Results of an RRS program involving the varieties Jarvis and Indian Chief
at North Carolina State University were reported by Moll (1959). After one
cycle of selection mean yield of the testcrosses changed from 8% less to 2%
less than the mean of commercial double crosses; prediction for the second cycle suggested that testcross yields would be 7% greater than double-cross
yields. Results of three cycles were subsequently reported by Moll and Robinson (1966, 1967). Increases in yield were 0.67%, 0.16%, and 2.86% per cycle
in Jarvis, Indian Chief, and Jarvis x Indian Chief, respectively, in the second cycle. In the third cycle, greater increases in yield were detected in
the populations per se, i.e., 4.3% and 1.7% per cycle in Jarvis and Indian
Chief, respectively. A smaller increase (0.8% per cycle) was detected in the
population crosses.

A full-sib family selection program in the same varieties showed an increase of 3.6%, 2.1%, and 3.8% per cycle in Jarvis, Indian Chief, and Jarvis
x Indian Chief, suggesting that in three cycles intrapopulation selection was
more effective than RRS for improvement of the variety cross; however the selection differential was greater in the former, so the effectiveness of selection was nearly equal in both methods. Heterosis did not seem to have been
changed by the intrapopulation improvement, whereas an increase in heterosis
was observed in the first two cycles in RRS followed by a decrease in the
third cycle.

Finally, Moll and Stuber (1971) evaluated six cycles of the same RRS program. Total changes in yield were 14%, 7%, and 21% for Jarvis, Indian Chief,
and Jarvis x Indian Chief, respectively, over the original populations and
population cross. Responses of both varieties to full-sib family selection
were 2.1 times greater than their responses to RRS, but the variety hybrid in
RRS showed 1.3 times greater response than in full-sib family selection. Accumulated responses in the variety composite after six cycles was 20.3% above
the mean of the unselected populations, which is approximately the heterosis
in the variety cross. Therefore six cycles of selection were necessary in the
variety composite to attain the yield level equivalent to the original variety
cross, i.e., the starting point of RRS. In other words, six cycles were necessary to recover the loss of one-half the heterosis that is expected to occur
in the composite of a variety cross. Evaluation of other traits showed that
increases in yield were generally accompanied by decreases in plant and ear
heights and increases in tillers and ears per plant. An additional evaluation
was reported by Moll et al. (1978) for six and eight cycles of selection.

Table 7.14. Progress per cycle (percent of original mean) through full-sib family (FS)
            and reciprocal recurrent selection (RRS) in two populations and their cross
            in four independent evaluations.

Population	1st (3 cycles)[†]		2nd (3 cycles)[‡]		3rd (6 cycles)[§]		4th (8 cycles)[§]	
	FS	RRS	FS	RRS	FS	RRS	FS	RRS
Jarvis	3.6	4.3	3.5	2.3	3.6	2.7	3.6	2.5
Indian Chief	2.1	1.7	2.8	1.2	2.7	0.9	3.2	1.4
Jarvis × Indian Chief	2.8	0.8	2.5	3.5	2.4	3.1	2.2	3.2

[†]Moll and Robinson (1966); [‡]Moll and Stuber (1971); [§]Moll et al. (1978).

Full-sib family selection was more effective than RRS for intrapopulation improvement, but RRS was more effective for the improvement of the variety
cross. Results from four independent evaluations of the full-sib and RRS programs for Jarvis and Indian Chief are included in Table 7.14.

An RRS program was initiated in Texas by J. S. Rogers. Results of the
first two cycles were reported by Collier (1959) and Douglas et al. (1961).
Increases in yield were at rates of 10.0% and 1.8% per cycle in the parental
populations Ferguson's Yellow Dent and Yellow Surecropper. Yield of the population cross increased at a rate of 5.8% per year. Thompson and Harvey (1960)
reported that average yield of topcrosses in an RRS program involving Synthetic A and Synthetic G increased from 82.5% to 94.0% of check yields from the
first to the third cycle. Torregroza et al. (1972) obtained 4.5% and 15.0%
gain per cycle in the populations Harinoso Mosquera and Rocamex V7, respectively, after two cycles of RRS. The population cross yielded 32% and 34% in
excess over the original cross in the first and second cycles, respectively.

Results of two cycles of RRS in Kenya were reported by Darrah et al.
(1972). The parental populations KII and Ec573 increased in yield at a rate
of 0.6 and 3.0 q/ha per year, respectively. The population cross showed an
increase of 3.3 q/ha per year. After three cycles Darrah et al. (1978) obtained a gain of 2.09 q/ha per year in the population cross and 0.97 q/ha per
year in the parental population Ec573 but a very small change (-0.02 q/ha per
year) in KII population.

Gevers (1974) reported results of three cycles of RRS in the populations
Teko Yellow and Natal Yellow Horsetooth, using two methods for sampling the
male parents before crossing. When the pollen parents were taken at random in
the population, grain yield showed an increase of 7.5%, 7.4%, and 5.8% per cycle in Teko Yellow, Natal Yellow Horsetooth, and Teko Yellow x Natal Yellow
Horsetooth, respectively. When male parents were selected for agronomic
traits, however, changes in yield were 7.1%, -0.5%, and 3.3%, respectively.
It was concluded that random sampling of parents before testcross evaluation
led to greater progress in the populations themselves and in the population
cross than when parents were selected for agronomic traits. Heterosis increased at similar rates in both methods.

Thomas and Grissom (1961) evaluated two cycles of RRS in popcorn. The
gains per cycle were 10.7% increase for popping volume, 6.9% increase in
yield, and 37.8% decrease in root lodging where evaluations were in percent of
check hybrids; RRS was effective in improving several traits in the crossed
population.

Jugenheimer and Bhatnagar (1962) reported results of two cycles of RRS
for plant lodging. In one population, percent of erect plants shifted from

Table 7.15. Effect of reciprocal recurrent selection on yield of populations themselves and on population cross.

Parental populations A	B	Method[+]	No. of cycles	Gain per cycle (%) A	B	A×B	[‡]	Reference
Yellow Surecropper	Ferguson Y. Dent	HS-RRS	2	10.0	1.8	5.8		Collier (1959); Douglas et al. (1961)
Syn. A	Syn. G	HS-RRS	3	---	---	5.8	(1)	Thompson and Harvey (1960)
Population A	Population B	HS-RRS	2	---	---	7.1	(2)	Thomas and Grissom (1961)
Jarvis	Indian Chief	HS-RRS	3	4.3	1.7	0.8	(3)	Moll and Robinson (1967)
Stiff Stalk Syn.	Corn Borer Syn.	HS-RRS	4	2.5	-1.1	1.8	(3)	Penny and Eberhart (1971)
Jarvis	Indian Chief	HS-RRS	6	2.3	1.2	3.5		Moll and Stuber (1971)
Harinoso Mosquera	Rocamex V7	HS-RRS	2	4.5	15.0	17.0		Torregroza et al. (1972)
Stiff Stalk Syn.	Corn Borer Syn.	HS-RRS	5	0.4	0.9	4.5		Eberhart et al. (1973)
Kitale II	Ecuador 573	HS-RRS	3	-0.1	5.0	7.0		Darrah et al. (1978)
Teko Yellow	N.Y. Horsetooth	HS-RRS	3	7.5	7.4	5.8	(4)	Gevers (1974)
BS-10	BS-11	FS-RRS	3	6.0	5.7	3.2		Hallauer (1977)
Dent Composite	Flint Composite	HS-RRS1	1	13.6	-7.8	3.7		Paterniani (1971)
Piramex	Cateto	HS-RRS1	1	3.5	7.5	7.5		Paterniani and Vencovsky (1977)
Dent Composite	Flint Composite	HS-RRS2	2	---	---	3.5		Paterniani and Vencovsky (1978)
Jarvis	Indian Chief	HS	6	2.7	0.9	3.1		Moll et al. (1978)
Jarvis	Indian Chief	HS	8	2.5	1.4	3.2		Moll et al. (1978)

[+](1) From first to third cycle in percent of checks; (2) popcorn population selected for popping volume, yield, and lodging resistance; (3) percent of observed means; (4) for random sampling of male parents—gains were 7.4%, -0.5%, and 3.3% per cycle, respectively, when parents were selected for agronomic traits.

[‡]See Chap. 6.

71% to 83% and 29% when selected for low and high lodging, respectively. In the other population, selection for low and high lodging resulted in changes from 62% to 81% and 32% erect plants, respectively.

Reciprocal full-sib selection (FS-RRS) was first evaluated by Hallauer (1973). After one cycle of selection involving populations Iowa Two-ear Synthetic (BS10) and Pioneer Two-ear Composite (BS11), a gain of 10.1% was realized in the cross C1 x C1 over C0 x C0. After three cycles of selection Hallauer (1975a) reported that full-sib progenies were at a higher yield level than those from the original populations. Full-sib progenies of the original populations averaged four standard deviations below the mean of check hybrids. After one cycle of selection full-sib progenies were only one standard deviation below the mean yield of six check hybrids. No improvement was observed in the second cycle, during which mechanical harvesting was used. Hallauer (1977) reported that after three cycles of reciprocal full-sib selection, yield increased 18%, 17%, and 9.7% in BS10, BS11, and BS10 x BS11, respectively.

Paterniani (1971) reported that a modification of RRS, here called HS-RRS2 (Chap. 6), resulted in 13.6% yield increase in the Dent Composite population and 7.8% decrease in the Flint Composite population after the first cycle of selection. The population cross increased 3.7% in the first cycle and 6.4% over the original cross after the second cycle (Paterniani 1974c). Paterniani and Vencovsky (1978) reported a 3.5% increase per cycle in the population cross after two cycles of the HS-RRS2 program.

Paterniani and Vencovsky (1977) obtained a 7.5% gain in the cross between populations Cateto and Piramex after one cycle of RRS based on testcrosses of half-sib families (HS-RRS-1). A least squares analysis involving the population means indicated that observed progress was significant and deviations from the model were not significant. In addition, progress in the population cross (7.5%) was partitioned into two portions: 4.3% due to changes in population means and 3.2% due to change in heterosis in the population cross. A summary of results from interpopulation improvement is presented in Table 7.15.

7.4 GENERAL EFFECTS OF SELECTION

Important information can be obtained from selection experiments besides the direct effect of selection on yield and other traits, e.g., relative changes in heterosis and combining ability effects when the population is crossed with specific or nonspecific populations or testers. Lonnquist (1951) selected for high and low yields using the parental population Krug as tester. High and low yield synthetics were topcrossed with the single cross WF9 x M14, and topcross yields were found to be 1.7 and 9.7 q/ha, respectively, below the tester mean. It was concluded that after only one cycle of recurrent selection the original Krug population was separated into two distinct groups with respect to combining ability with an unrelated narrow base tester.

Sprague and Russell (1957) used half-sib family selection in populations Lancaster and Kolkmeier for combining ability with the inbred line Hy. After two cycles, the cross Lancaster x Kolkmeier showed a rate of increase greater than either of the populations themselves. Results were in agreement with those expected on the basis of partial to complete dominance of genes controlling yield. The same conclusion was presented by Sprague et al. (1959) in a further study involving the same populations.

Lonnquist and Gardner (1961) obtained increases in yield of 5.4% and 4.5% after one cycle of recurrent selection in the varieties Krug and Nubold Reid.

The population cross yield increased 13.5% and heterosis increased from 6.0% to 14.6% in one cycle. Crosses between advanced and original populations showed heterosis of 8.3% and 8.7% of midparent values for Krug and Nubold Reid, respectively. Results indicated that improved yields of the derived populations were the results of additive gene action present in the parental varieties and that genes were in the partial to complete dominance range, thereby resulting in an increase in heterosis. Similar conclusions were reached by Penny et al. (1962) by use of half-sib family selection in Alph and WF9 x B7 for combining ability with the inbred line B14. They observed that crosses between populations in two cycles of selection produced significantly greater yields than the original cross. Selection for poorer yield was effective in the cross between populations in both cycles.

In crosses between three populations within cycles, Lonnquist (1963) obtained an average increase in yield that was of the same magnitude as the average increase in yield of the populations themselves. These results indicated that no apparent change occurred in nonadditive genetic effects and progressively greater yields in the intercrosses were directly related to improvement shown in the parent populations through selection based on additive genetic variation.

Horner et al. (1963) selected for specific combining ability (inbred line F6 as tester) and for general combining ability (parental population Fla. 767 as tester) during four cycles. Average performance of the selected populations in combination with 11 unrelated testers showed a significant increase in the SCA series but not in the GCA series. In the same manner, SCA with the inbred tester was increased significantly but GCA was not.

Moll and Robinson (1967) compared full-sib family selection and RRS involving the varieties Jarvis and Indian Chief. After three cycles, full-sib selection resulted in greater population performance than RRS and the population cross increased at a higher rate in full-sib selection. Subsequent reports by Moll and Stuber (1971) and Moll et al. (1978) showed that heterosis of the variety cross from full-sib family selection decreased from the first to the sixth (3.0%) and eighth (12.1%) cycles, but increased after six (8.5%) and eight (11.3%) cycles of RRS.

Horner et al. (1969) evaluated three cycles of recurrent selection based on yield of topcross progeny (inbred tester and population as tester) and of S_2 progenies. The improvement in combining ability with 11 unrelated testers was increased significantly but there were no differences among methods. Vencovsky et al. (1970) evaluated five cycles of mass selection in Paulista Dent (PD) and three cycles in Cateto M. Gerais (CMG) varieties. Heterosis in the cross between the two populations increased in the first cycle but decreased in the third cycle. Greatest heterosis was observed in the cross of PDIII $\times$ GMGI, but the greatest yielding cross was PDV $\times$ GMGIII with no significant heterosis. Such results showed that improvement in the population cross was mainly at the expense of improvement in the populations themselves through additive genetic effects within populations.

Burton et al. (1971) showed that selection in a BSK population through testcross with a double-cross tester and through S_1 lines per se both resulted in an increase in mean yield and in GCA. However, S_1 selection was more effective. The cross between the advanced populations exhibited heterosis, indicating that the two methods developed populations that differed in gene frequency. Johnson and Salazar (1967) reported that three cycles of mass selection increased average combining ability of a Cuban Yellow Flint variety about 20% with three inbred lines. The same rate of improvement (from 33.6 to 40.5 q/ha) was observed in the variety itself.

Horner et al. (1973) compared four cycles involving three methods of recurrent selection. GCA evaluated through testcrosses with unrelated testers
was increased significantly by all methods. Use of narrow base (inbred line)
tester was more effective, indicating that the inbred line was homozygous recessive at many important loci. The inbred tester had greater testcross variances and permitted more successful selection of dominant favorable alleles
than a broad base tester, which probably has intermediate gene frequencies at
most loci. Genter (1973) detected no significant selection effect either by
S_1 progeny or testcross evaluation in the population VLE. Testcross selection
in VCBS was effective in increasing yield in crosses but not in the population
itself, whereas S_1 selection increased both combining ability and population
yield.

Eberhart et al. (1973) reported that for RRS involving BSSS and BSCB1,
heterosis increased from 15% in C0 x C0 to 37% in C5 x C5. Similarly, heterosis increased to 34% in the cross between BSCB1 in the fifth cycle of RRS and
BSSS in the seventh cycle using the double cross Ia13 as tester. Gevers
(1974) observed that heterosis increased from 6% in the original population
cross to 10.3% and 11.0% after three cycles of RRS using two procedures for
selecting male parents, random selection and selecting for agronomic traits,
respectively.

Genter and Eberhart (1974) evaluated the performance of six original and
advanced populations in diallel crosses. Significant improvement in the average cross performance was obtained for VCBS(HT), NHG, PHCB, and PHWI populations but not for BSK and BSSS.

Results obtained by Horner et al. (1976), after selection for SCA with a
single-cross tester, showed that selection was as effective for improving GCA
as it was for SCA. They speculated about the possibility of changing testers

Table 7.16. Changes in midparent (MP) and variety cross (VC) mean yield and in
heterosis (h) after intra- or interpopulation recurrent selection.

Method[+]	No. of cycles	Gain per cycle				h (% MP) in C0	Reference
		in % of C0			h[‡] (% MP)		
		MP	VC	h			
RRS	2	5.9	5.8	5.3	-0.14	24.6	Collier (1959)
HT	2	5.9	2.9	-3.3	3.87	47.4	Penny (1959)
HT-high	2	5.9	3.8	-0.3	-2.73	48.1	Penny et al. (1962)
HT-low	2	-13.0	-7.2	4.4	11.21	48.1	" "
RRS	3	2.9	0.8	-12.0	-6.88	16.7	Moll and Robinson (1967)
FS	3	4.4	3.7	-0.4	-2.13	16.7	" "
HS	3	3.1	6.1	47.2	2.92	11.2	Paterniani (1968)
M	3	3.9	2.7	-13.0	-1.09	7.4	Vencovsky et al. (1970)
HT	5	5.2	7.3	12.6	1.98	39.4	Russell et al. (1973)
RRS	5	0.5	4.2	24.8	4.20	14.4	Eberhart et al. (1973)
HT	2	0.2	3.7	29.9	4.07	12.8	Genter (1973)
S_1	2	4.2	3.4	-2.5	-0.76	12.8	"
RRS[1]	3	7.5	5.8	-7.8	-1.26	5.9	Gevers (1974)
RRS[2]	3	3.3	3.3	2.8	0.13	5.9	"
HT	5	-0.2	3.6	15.4	5.10	25.4	Walejko and Russell (1977)
RRS-1	1	5.1	7.5	20.6	2.71	18.4	Paterniani and Vencovsky (1977)
FS-RRS	3	5.9	3.2	-24.2	-2.49	9.7	Hallauer (1977)
HS	6	4.0	-0.7	-12.9	-5.3	39.2	Darrah et al. (1978)
RRS	3	2.2	7.1	19.4	6.3	39.2	"

[+]RRS: reciprocal recurrent selection; RRS[1] and RRS[2]: males randomly sampled and
males selected, respectively; RRS-1: RRS based on testcross of half-sib families;
FS: full-sib family selection; FS-RRS: full-sib reciprocal recurrent selection;
HT: half-sib (testcross) family selection; HS: half-sib (modified ear-to-row)
family selection; S_1: S_1 family selection.

[‡]h = (VC - MP); h (% MP) = 100 h/MP.

in a recurrent selection program with little loss in accumulated improvement. Such results, as well as those reported by other authors (Sprague et al. 1959; Lonnquist 1961; Russell et al. 1973) have suggested that selection with a specific tester has been effective primarily for additive genetic effects. Results obtained by Horner et al. (1973) showed that the inbred line F6 was nearly twice as effective as a broad base tester or S_2 progeny selection for improving frequencies of genes having additive effects.

Walejko and Russell (1977) reported results of five cycles of selection for SCA with the inbred line Hy. Yield changes observed for the population crosses (Lancaster x Kolkmeier) from C0 x C0 to C5 x C5 were at rates of 2.45 ± 0.45 q/ha. Changes in yield of testcrosses with nonspecific testers also were similar to those of the Hy testcrosses. A comprehensive evaluation of selection effects suggested that selection was effective for increasing gene frequency of favorable alleles in both populations, and that gene action for yield heterosis was from genes having additive effects with partial to complete dominance.

Changes in midparent and variety cross mean yields and in heterosis are summarized in Table 7.16 for several intra- and interpopulation selection experiments. Selection for high yield increased midparent and variety cross yields in all selection schemes except that reported by Walejko and Russell (1977), where selection was effective for increasing the variety cross yield but not the midparent yield. About 50% of the studies showed a decrease in intervarietal heterosis. RRS schemes were designed to maximize selection for GCA and SCA effects and therefore increase the variety cross mean through selection for nonadditive genetic effects. The limited amount of available experimental data, however, does not show consistent trends in the effectiveness of selection for increase of heterosis in the variety cross. In some instances the apparent failure of RRS to increase heterosis may be attributable to differences in magnitude of effects of genes controlling yield. If genes of larger effects are mostly additive, selection would change primarily the frequencies of such genes and selection for allelic (dominance) or nonallelic (epistasis) interaction would be less effective in the first generations.

A more extensive study (Genter and Eberhart 1974), summarized in Table 7.17, involves 20 variety crosses among original and advanced cycles of intrapopulation selection. General results do not depart much from those shown in Table 7.16. In only 1 case midparent yield decreased after selection, while 6 out of 20 showed a decrease in the variety cross yield. Heterosis in absolute values decreased in 8 of the 20 instances after selection. Heterosis measured in percent of midparent decreased in 50% of the crosses. Despite the limited amount of information, results suggest that changes in heterosis after intrapopulation selection is largely a matter of chance.

Theory and empirical results have shown that selection directly affects the variability within populations. The extent of changes in variability depends on several factors, e.g., selection intensity, initial amount of variability (actual and potential), linkage disequilibrium, and rate of recombination. Researchers have shown that environmental factors may affect the chiasma frequency and consequently the recombination rate, so the gradual liberation of potential variability also depends on environmental conditions to some extent (Mock 1973).

One of the first reports on changes in variability as a result of selection was reported by Winter (1929) after selection for protein and oil in Burr's White in the period of 1896 to 1924 in Illinois. The measure of variability through three different methods (Weinberg's formula, standard deviation, and extramodal coefficient) showed an increasing trend for high protein and high oil strains but a decreasing trend in low protein and low oil

Table 7.17. Changes in midparent (MP) and variety cross (VC) mean yield and
in heterosis (h) after intrapopulation recurrent selection
(adapted from Genter and Eberhart 1974).

| Populations[†] | | Methods[‡] | | No. of cycles | | Total gain over cycles in % of CO | | | h[§] | h (% MP) |
A	B	A	B	A	B	MP	VC	h	(% MP)	in CO
VCBS	BSK	HT	S_1	3	4	16.2	18.4	30.6	1.5	13.0
VCBS	BSSS	HT	HT	3	7	8.9	18.0	73.9	9.7	16.3
VCBS	NHG	HT	M	3	12	17.0	8.1	-44.5	-8.9	17.0
VCBS	PHCB	HT	M	3	9	14.9	23.9	98.3	8.8	12.1
VCBS	PHWI	HT	M	3	9	8.4	17.8	117.8	9.4	9.4
VCBS	BSK	S_1	S_1	4	4	16.0	18.2	30.6	1.6	13.0
VCBS	BSSS	S_1	HT	4	7	8.7	19.3	84.8	11.4	16.3
VCBS	NHG	S_1	M	4	12	16.8	15.9	11.0	-0.9	17.0
VCBS	PHCB	S_1	M	4	9	14.7	23.9	100.0	9.0	12.1
VCBS	PHWI	S_1	M	4	9	8.2	29.3	149.5	12.2	9.4
BSK	BSSS	S_1	HT	4	7	2.8	-14.6	-89.0	-20.9	23.4
BSK	NHG	S_1	M	4	12	11.2	-0.2	-75.7	-11.8	15.1
BSK	PHCB	S_1	M	4	9	8.9	-6.5	-72.9	-17.4	23.2
BSK	PHWI	S_1	M	4	9	2.7	-5.4	-77.6	-8.7	11.2
BSSS	NHG	HT	M	7	12	4.3	-4.7	-37.4	-11.0	27.6
BSSS	PHCB	HT	M	7	9	2.0	1.3	-2.1	-0.8	19.2
BSSS	PHWI	HT	M	7	9	-3.4	3.3	45.9	-8.0	15.7
NHG	PHCB	M	M	12	9	10.1	12.7	29.8	2.7	14.8
NHG	PHWI	M	M	12	9	4.2	8.1	85.7	4.0	5.1
PHCB	PHWI	M	M	9	9	1.9	-1.1	-24.5	-3.4	13.3

[†]VCBS: Virginia Corn Belt Southern Synthetic; BSK: Krug Hi I Synthetic 3;
BSSS: Stiff Stalk Synthetic; NHG: Nebraska Hays Golden; PHCB: Pioneer Hi-
Bred Corn Belt Synthetic; PHWI: Pioneer Hi-Bred West Indian Synthetic.

[‡]HT: half-sib (testcross) family selection; S_1: S_1 family selection; M: mass
selection.

[§]h = (VC - MP); h (% MP) = 100 h/MP.

strains. Variability measured through the coefficient of variation showed a
decreasing trend in high oil and high protein strains because the mean of the
traits increased in both cases. In the same way the coefficient of variation
showed an increasing trend in low protein and low oil strains because the mean
decreased in both cases during the selection period. Leng (1962) demonstrated
the existence of variability after 48 generations as indicated by effective
reverse selection in all four strains. Dudley and Lambert (1969) estimated
the genetic variability in five strains of the Illinois selection program.
Estimates of variance among half-sib families were significant in all cases
but were greater in the strains selected for decreasing traits (low oil and
low protein). A direct comparison of different estimates was not available
because of nonhomogeneity of error variances. For this reason a rough measure
of heritability on a single-plot basis was presented for comparisons. Results
are summarized in Table 7.18.

Heritability for nonselected characters was expected to be greater than
for selected characters in each population, because changes in variability
without selection would be attributable only to inbreeding whereas selected

Table 7.18. Estimates of variance among half-sib families ($\hat{\sigma}_g^2$) and heritability on plot basis $(\hat{\sigma}_g^2/\hat{\sigma}_p^2)^+$ (adapted from Dudley and Lambert 1969).

Population	Oil		Protein	
	$\hat{\sigma}_g^2$	$\hat{\sigma}_g^2/\hat{\sigma}_p^2$ (%)	$\hat{\sigma}_g^2$	$\hat{\sigma}_g^2/\hat{\sigma}_p^2$ (%)
Illinois high oil (IHO)	0.0724	12.5	0.1082	11.4
Illinois low oil (ILO)	.0008	16.7	.0859	27.7
Illinois high protein (IHP)	.0438	39.8	.1726	20.5
Illinois low protein (ILP)	.0090	15.5	.0538	23.0
Illinois high protein (HN)	0.0210	30.0	0.3136	27.3

$^+\hat{\sigma}_p^2$ is phenotypic variance on a plot basis, i.e., $\hat{\sigma}_g^2 + \hat{\sigma}_{gy}^2 + \hat{\sigma}^2$.

characters would have reduced variability because of both inbreeding and selection. This seemed to have occurred in populations ILO, IHP, and IHP(HN). Heritability for oil content in ILO was slightly greater than in IHO. Variability in ILO, however, was due to the variation in percent of germless kernels; such variability would have limited usefulness for selection toward less oil content.

Robinson and Comstock (1955) estimated additive and dominance variances in several hybrid populations and open-pollinated varieties in different cycles of selection. As a result of selection a decrease in the additive genetic variance estimate was observed in the populations of Jarvis, NC34 x NC45, and CI21 x NC7, but a small increase occurred in the Weekley variety. Subsequently, Moll and Robinson (1966) reported estimates of additive genetic variance in several cycles of intrapopulation selection and of variance among testcrosses in RRS. Results are summarized in Table 7.19. Intrapopulation

Table 7.19. Additive genetic variance ($\hat{\sigma}_A^2$) and testcross progeny ($\hat{\sigma}_g^2$) estimates in several cycles of selection in intra- and interpopulation improvement, respectively (adapted from Moll and Robinson 1966).

Population	Cycles of selection						
	0	1	2	3	4	5	6
Additive genetic variance[+]							
CI121 × NC7	24±3	20±22	20±14	41±21	18±13	40±21	16±14
Jarvis	30±4	12±13	16±19	42±18	38±20	53±22	
Indian Chief	17±4	1±28	31±17	30±19	44±23	---	
Variance among testcross progenies[+]							
Jarvis × Ind. Chief	7±2	18±8	16±8	19±6	9±4	---	---
Ind. Chief × Jarvis	9±2	22±9	10±6	28±8	23±8	---	---

$^+ \times 10^{-4}$, for yield in pounds per plant.

estimates of the quantity $\sigma_A^2 + \sigma_{AE}^2$ (additive + additive-environment interaction variance) for successive selection cycles seemed to be distributed around the more precise estimate for the original population and revealed no trend associated with selection. Variances among testcross progenies after the first cycle of interpopulation selection were greater than estimates in the original population cross. Although no obvious trend was observed in the magnitude of the sequential estimates after the first cycle, data suggested that genetic variance increased after the initial cycle of selection.

Silva and Lonnquist (1968) detected a decrease in genetic variances for yield and days to flower after one cycle of S_1 family and of full-sib family selection. The two methods of selection apparently produced different changes in magnitude of genetic variances. Hallauer (1970) estimated additive genetic variance in the original two populations and after four cycles of RRS. Selection was not effective in increasing yield of Corn Borer Synthetic No. 1 and additive genetic variance estimates showed no difference between C0 ($\hat{\sigma}_A^2 = 143 \pm 35$) and C4 ($\hat{\sigma}_A^2 = 133 \pm 35$). Selection was effective in increasing yield and estimates of additive genetic variance showed a decreasing trend from C0 ($\hat{\sigma}_A^2 = 184 \pm 48$) to C4 ($\hat{\sigma}_A^2 = 130 \pm 35$) in the Iowa Stiff Stalk Synthetic population. Also, the population cross showed an increase in yield and a decrease in σ_A^2 estimates from C0 ($\hat{\sigma}_A^2 = 216 \pm 46$) to C4 ($\hat{\sigma}_A^2 = 96 \pm 30$). Summaries presented in Tables 5.7 and 5.8, however, show no evidence of change in genetic variability with selection.

Hallauer (1971) studied changes in several traits after four cycles of RRS for yield in maize. Additive genetic variance decreased and dominance variance increased for kernel depth in all three populations. For all other traits estimates of σ_A^2 were significant but showed small differences among the three pairs of populations. The change in average silking date for C0 and C4 of both parental populations seemed to have influenced estimates of σ_A^2 for silking date and plant and ear heights.

Genetic variance estimates in the mass selection program in the Hays Golden open-pollinated variety have shown that variability was consistently reduced in the selected populations (Gardner 1969b; Harris et al. 1972). Gardner (1977) summarized results relative to effect of selection on the variability of Hays Golden (control and irradiated) variety, as shown in Table 7.20.

In a number of selection experiments the genetic coefficient of variation

Table 7.20. Relative genetic variance estimates from families derived from Hays Golden (HG) and from control and irradiated mass selected populations of maize (adapted from Gardner 1976).

Population	Generation						
	6	10	15	9[†]			
				(a)	(b)	(c)	(d)
Hays Golden (0)	100	100	100	100	100	100	100
HG – control	136	30	60	40	22	30	77
HG – irradiated	263	89	49	67	47	14	57

[†](a) S_1 lines per se; (b), (c) S_1 testcrossed with NG × NGG and H49 × WF9;
(d) $S_6 \times S_6$ random hybrids.

Table 7.21. Genetic coefficient of variation estimates within populations under half-sib and S_1 family selection

Cycle	Modified ear-to-row selection			Dent Composite[¶]	S_1 selection BSK(S)[#]	S_2 selection BS13[#]
	Paulista Dent[+]	Hays Golden[‡]	Pinamex[§]			
0	15.3	11.3	10.6	7.4	16.2	14.6
1	9.3	4.2	6.1	7.3	15.7	22.3
2	9.1	4.1	5.0	---	10.8	12.9
3	7.1	5.0	3.4	---	15.4	----
4	---	4.8	6.5	---	8.8	---
5	---	---	---	---	13.4	---
6	---	---	---	---	20.6	---
7	---	---	---	---	39.7	---

[+]Paterniani (1967); [‡]Webel and Lonnquist (1967); [§]Paterniani (1969); [¶]Miranda et al. (1972); Lima et al. (1974); [#]Unpublished data from Iowa State University.

has been used to measure the relative amount of genetic variability among entries (half-sibs, full-sibs, selfed, or testcross families). A summary of results of several selection experiments relative to the genetic coefficient of
variation is presented in Table 7.21 for half-sib and S_1 family selection (intrapopulation) and in Table 7.22 for interpopulation improvement and recurrent
selection with unrelated testers.

Table 7.22. Genetic coefficient of variation estimates in the population
cross under reciprocal recurrent selection and half-sib
selection with unrelated testers.

Cycle	Half-sib selection with tester[+]			Reciprocal RS[+]		Modified RRS-2[‡]	
	BSSS[(1)]	BS12(HI)[(2)]	BSK(HI)[(3)]	BSCB1[(4)]	BSSS[(5)]	Flint Composite[(6)]	Dent Composite[(7)]
0	10.2	11.6	4.0	10.8	7.9	11.7	9.5
1	3.8	7.8	5.2	7.7	6.2	5.7	6.4
2	4.8	2.8	2.4	3.2	3.7	12.3	7.7
3	---	4.5	3.4	3.5	2.8	---	---
4	2.6	2.4	3.4	3.6	2.1	---	---
5	3.8	5.0	5.0	4.9	5.3	---	---
6	4.5	9.7	7.7	8.0	5.5	---	---
7	---	---	8.5	5.6	4.6	---	---
8	---	---	---	3.9	8.6	---	---

[+]Unpublished data from Iowa State University; [‡]Paterniani (1971, 1974a, 1974b); testers: (1) Ia13; (2) B14; (3) Ia4652, B14, B73; (4) BSSS; (5) BSCB1; (6) Dent Composite; (7) Flint Composite.

A common feature of Tables 7.21 and 7.22 is that the genetic coefficients of variation generally decrease sharply after the first cycle of selection and remain either unchanged or change very little in subsequent cycles. Selection, of course, is expected to be more effective in the first cycle because a greater amount of genetic variability is available. In subsequent cycles an increase in precision of experiments is required to make selection as effective as in the first cycle. Effective selection increases the mean of the selected trait, which leads to a decrease in the genetic coefficient of variation even if genetic variability remains constant. Because estimates of genetic coefficients of variation are obtained in different years, effects of environment and of genotype-environment interaction may have some influence on the magnitude of genetic variability. In general, empirical results to date have shown that selection seems to have decreased genetic variability, although in most instances only a limited number of cycles of selection have been completed (see Chap. 5). It depends to a great extent on the number of families saved for reproduction in each cycle.

Another important feature of recurrent selection methods is their effect on the potential of improved populations as sources of inbred lines for hybrid development. Theoretical studies have suggested that the increase in frequencies of favorable alleles also increases the opportunity for the development of outstanding hybrids; some empirical results have corroborated this statement. Horner et al. (1973) suggested that commercial maize hybrids could be developed rapidly from a recurrent selection program using as tester a seed parent already in commercial use. Using this procedure they obtained a top-cross hybrid (Florida 200A) that was released for commercial production.

Hallauer (1973) reported that full-sib progenies (S_0 x S_0) after one cycle of reciprocal full-sib selection were at a higher yield level than crosses from the unimproved population. The two parent populations, Iowa Two-ear Synthetic and Pioneer Two-ear Composite, had low frequencies of prolific plants at normal plant densities. After one cycle of selection the increase in yield was 14.8% and 18.7%, respectively. Also, 46% of the full-sib families after selection exceeded the mean of the checks and 16% were one or more standard deviations above the mean of the checks. Only 1% of the full-sib families from crosses between the original populations exceeded the mean of the same checks.

Suwantaradon and Eberhart (1974) observed that hybrids developed from two improved populations outyielded significantly the cross between the parent varieties. The parent varieties BSK and BSSS were improved through five cycles of S_1 family selection and RRS with BSCB1 as tester, respectively. The cross between improved varieties yielded at least 90% as much as the better current hybrids and the best hybrid developed from them outyielded the varietal cross by 18%.

Gardner (1972) developed S_2 lines from the parent variety Hays Golden (HG) from the population after 12 cycles of mass selection (C12), from the irradiated population after 13 cycles of mass selection (I13) for high grain yield, and from the prolific (P7) population mass selected for number of ears per plant. Inbred lines from the selected populations were superior to lines from the parent variety when crossed with Oh43 (inbred line) and the single cross of related lines (N7A x N7B). Lines from C12, I13, and P7 exceeded lines from HG by 11.4%, 10.0%, and 10.6%, respectively, when crossed to Oh43 and by 10.9%, 8.0%, and 7.7% when crossed to N7A x N7B.

Harris et al. (1972) evaluated S_1 lines themselves and in testcross performance after nine cycles of mass selection for yield in Hays Golden. It was shown that the S_1 lines from the selected populations were superior in mean yield to lines of the parent variety either as S_1 lines themselves or in test-

crosses. It was concluded that selection eliminated radiation-induced deleterious mutants from the irradiated population while increasing frequencies of
favorable yield genes common to both selected populations. This apparently
produced similar germplasm reservoirs more suitable than the parent variety
for the development of superior inbred lines.

Martin and Gardner (1976) compared single-cross, three-way, and double-
cross hybrids from inbred lines of Hays Golden and from the control and irradiated derived populations selected after nine cycles of mass selection. Hybrids developed from the mass selected control and irradiated populations were
9.3% and 7.4% higher yielding and 7.4% and 17.6% more prolific, respectively,
than hybrids developed from the original Hays Golden variety.

7.5 FACTORS AFFECTING EFFICIENCY OF SELECTION

Different factors affecting efficiency of selection were treated in detail by Eberhart (1970). He developed the prediction formula that has general
application to the selection methods used in maize breeding and showed how the
variables in the prediction formula can be manipulated to increase efficiency
of selection. Prediction equations for different selection methods are given
in Chap. 6. In this section we give some examples showing how the different
variables in the prediction formula can be changed and how they affect predicted genetic gain.

A simple example to illustrate the use of the prediction formula is given
in Tables 7.23 and 7.24. Three traits were measured for 144 S_1 progenies of
Iowa Stiff Stalk Synthetic in three replications in one experiment. Ten
plants within each plot were inoculated for European corn borer (*Ostrinia nubilalis* Hübner) and for a stalk rot organism (*Diplodia zea* Pass.). Approximately three weeks after inoculation a mean plot rating was made for resistance to first brood European corn borer leaf feeding. Also, about three weeks
after inoculation with *Diplodia zea,* ten plants within each plot were measured
to determine the amount of pressure required to penetrate the rind of the
stalk; later the same ten plants were sliced with a knife to determine the
level of *Diplodia zea* infection. The rind puncture and *Diplodia zea* ratings
were made between the second and third internodes above ground level. All
analyses were made on plot means. The analyses of variance for each trait and
covariances between the three traits are given in Table 7.23. Differences
among entries were significant for the three traits, and a negative genetic
correlation existed between rind puncture and level of *Diplodia zea* infection.

Twenty S_1 progenies were selected for recombination to form the next cycle of selection based on measurements of the three traits. For illustration
we will assume selection was for each trait. Our problem is to determine if
effective selection can be made on three replications in one environment or if
more or fewer replications should be used. Expected gains for each trait for
different numbers of replications are shown in Table 7.24. Because data were
collected in only one environment, relative efficiency of increasing replications for each trait is directly related to level of heritability of the
trait. Rind puncture reading had the largest heritability (85.6%); the effect
of increasing replications was less than for corn borer reading, which had the
lowest heritability. Expected genetic gain for rind puncture reading increased only 6% by increasing replications from two to five, whereas an 18%
increase in genetic gain is predicted for corn borer rating. It seems that
two to three replications would be adequate for each trait, but the effect of
increasing replications would be greater for corn borer rating. Doubling the
number of replications (which doubles land area and labor) increases expected
genetic gain for rind puncture reading by only 6%.

Table 7.23. Analysis of variance and covariance for 144 BSSS S_1 lines for first brood European corn borer, stalk rind puncture, and stalk rot rating compared in one experiment.

Source	df	Mean squares				Mean cross-products		
		Rind puncture	Corn borer	Stalk rot		Rind puncture × Corn borer	Rind puncture × Stalk rot	Corn borer × Stalk rot
Replications	2	18.86	3.02	0.48		4.49	-1.57	-1.20
Entries	143	14.62	5.67	2.19		1.05	-3.97	0.37
Error	286	2.10	2.55	0.60		0.13	-0.25	-0.03
Total	431							
$\hat{\sigma}^2$		2.10 ± .17	2.55 ± .13	0.60 ± .05	σ	0.13	-0.25	-0.03
$\hat{\sigma}^2_g$		4.17 ± .17	1.04 ± .23	0.53 ± .01	σ_g	0.31	-1.24	0.13
$\hat{h}^2$, %[†]		85.6	55.0	72.6				
CV, %		10.8	28.0	28.6	r_g	0.15	-0.83	-0.04
$\bar{x}$		13.4	5.7	2.7	r_p	0.12	-0.70	0.11

[†]Heritability calculated on progeny mean basis as $\hat{\sigma}^2_g / (\hat{\sigma}^2/3 + \hat{\sigma}^2_g) \times 100$.

Table 7.24. Effect of replications on predicted gain for three traits
             of maize.

| Number of replications | Rind puncture reading | | | Relative efficiency |
	$\hat{\sigma}^2_p$ [+]	$\hat{h}^2$ [‡]	ΔG [§]	
5	4.59	90.8	1.61	1.02
4	4.70	88.7	1.60	1.02
3	4.87	85.6	1.57	1.00
2	5.22	79.9	1.51	.96
1	6.27	66.5	1.38	.88
	Corn borer rating			
5	1.55	67.1	0.69	1.10
4	1.68	62.0	0.66	1.05
3	1.89	54.9	0.63	1.00
2	2.32	44.9	0.58	0.92
1	3.59	29.0	0.46	0.73
	Stalk rot rating			
5	0.65	81.8	0.54	1.06
4	0.68	78.2	0.53	1.04
3	0.73	72.9	0.51	1.00
2	0.82	64.2	0.48	0.94
1	1.12	47.3	0.42	0.82

[+] $\hat{\sigma}^2_p$ calculated from the components of variance given in Table 7.23.

[‡] $\hat{h}^2 = \hat{\sigma}^2_g / (\hat{\sigma}^2 / r + \hat{\sigma}^2_g) \times 100$.

[§] $\Delta G = k\hat{\sigma}^2_g / 2\hat{\sigma}_p$, where k is 1.6591, $\hat{\sigma}^2_g$ is the genetic variation among S_1 progenies, 2 is the number of years per cycle, and $\hat{\sigma}_p$ is the square root of the phenotypic variation.

Because data were collected in only one environment, we have biased estimates of heritability. It is not possible from these data to determine how effective two replications in two different environments would be for increasing genetic gain. If the genotype-environment interaction component is relatively large compared to the genotypic component, distribution of replications in different environments would be warranted.

Rogers et al. (1977) and Russell et al. (1978) have computed expected gains from selection for corn rootworms (*Diabrotica* spp.) and second brood European corn borer, respectively, from data collected from different environments. Both studies included S_1 progenies that were evaluated for resistance to the particular maize pest under study. Rootworm resistance was dependent on natural infestation, whereas artificial infestation techniques were used for the second brood European corn borer resistance.

Rogers et al. (1977) studied four synthetic varieties undergoing recurrent selection for rootworm tolerance. Four root traits (root lodging, root damage, root size, and secondary roots) were studied in experiments conducted in two environments in each of two years, each environment including two replications. Except for root damage, most genotypic and genotype-environment components of variance were significantly different from zero. Expected gains from selection for rootworm tolerance were determined for selection based on each of the four root traits and selection indexes including the four root

traits. Indirect selection using root size, secondary roots, and root damage
was not as effective in reducing root lodging as direct selection for root
lodging itself. The only situation in which direct selection would not be as
effective in reducing root lodging would be in environments in which conditions that promote root lodging (e.g., effects of wind and rain) do not occur.
Index selection also would be more effective in nonroot-lodging environments.

Because root lodging resistance seemed to be the best trait for selecting
for rootworm tolerance, Rogers et al. (1977) calculated expected genetic gain
for different combinations for number of replications and environments. Figure 7.3 shows expected gain for three populations. (The fourth population,
BSLR, was not included because root lodging occurred in only one environment;
hence no estimate of genotype-environment interaction was available.) Expected progress among the three populations differed because of the magnitudes of
genotypic (smallest for BSSS) and genotype-environment (smallest for BS1) components of variance. Heritability estimates of root lodging were 45.2%,
41.8%, and 84.6% for BSSS, BSER, and BS1, respectively. Substantial gains can
be expected in all instances by increasing replications from one to two, but
additional locations have a greater impact than additional replications. The
value of adding more than two replications decreases rapidly. For these three
populations, it seems that an effective compromise of two replications at

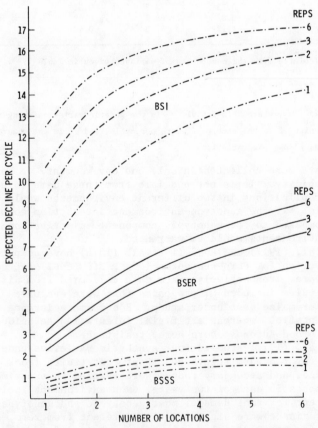

Fig. 7.3. Expected decline percentage of root lodging per cycle for different
combinations of replications and locations in BSSS, BSER, and BS1.

three to four environments would be most desirable. For a trait such as root
lodging that is very dependent on environmental conditions for its expression,
response to selection would be enhanced by increasing the number of locations
as permitted by resources available.

In addition to replications and environments, Russell et al. (1978) had
individual plant data. Russell et al. included 100 S_1 progenies from each of
two synthetic populations BS9Cl and BS16 in their study. BS9Cl had undergone
one cycle of recurrent selection for resistance to European corn borer, whereas BS16 had not undergone any previous selection for resistance. The S_1 progenies from BS9Cl were evaluated at one location in 1975 and 1976, whereas the
S_1 progenies from BS16 were evaluated only in 1976. Components of variance
estimated from the analyses of variance are given in Table 7.25. Estimates of
components of variance were similar, but because the S_1 progenies of BS9Cl
were evaluated in each of two years the BS9Cl estimates are used to illustrate
expected genetic responses from selection.

Selection for resistance to infestation by the second brood of the European corn borer usually is determined by family means obtained by averaging
over plots, replications, and years, where data are obtained on ten plants in
each plot for each of three replications for each year. Figure 7.4 illustrates the effects of number of plants and replications on expected genetic
gain per cycle. Increase in expected genetic gain is minimal when number of
plants is increased beyond 10 in all instances. Comparatively large reduc-
tions in genetic gain are expected when four or fewer replications with 5 or
fewer plants per plot are used. For example, with three replications, a decrease from 10 to 5 plants per plot decreases expected gain only 8%, whereas a
decrease from 5 to 1 plant per plot decreases expected gain 35%. As the number of replications is increased, effect of plot size diminishes. Relation of
expected genetic gain to number of replications is similar to that for number
of plants per plot. Increasing replications from one to two is slightly
greater than increasing replications from four to eight. An increase in plot
size diminishes the effects of added replications. Figure 7.4 shows that reducing plot size by 50% to 5 plants and increasing replications by 33% to 4
does not alter expected gain, but the expensive and time-consuming procedures
of infesting second brood egg masses and counting stalk cavities are reduced
by nearly 30%.

Information on second brood European corn borer resistance is not available until after flowering. If off-season nurseries are available and one

Table 7.25. Variance component estimates obtained for S_1 progenies of
BS9Cl and BS16 evaluated for second-brood European corn
borer resistance (Russell et al. 1978).

| Population | Estimates of variance components[+] | | | | |
	$\hat{\sigma}^2_w$	$\hat{\sigma}^2$	$\hat{\sigma}^2_{ge}$	$\hat{\sigma}^2_g$	$\hat{h}^2$[‡]
BS9Cl	120.3	19.1	15.1	38.3	78.1
BS16C0	117.8	14.5	--	31.6	86.7

[+]$\hat{\sigma}^2_w$, $\hat{\sigma}^2$, $\hat{\sigma}^2_{ge}$, and $\hat{\sigma}^2_g$ refer to the within-plot variation, experimental
error, genotype-environment interaction, and genotypic components of
variance, respectively; $\hat{\sigma}^2_g$ is the variation among S_1 progenies and
equals the additive genetic variance for p = q = 0.5 or no dominance.

[‡]Heritability calculated as $\hat{\sigma}^2_g/(\hat{\sigma}^2/6 + \hat{\sigma}^2_{ge}/2 + \hat{\sigma}^2_g) \times 100$.

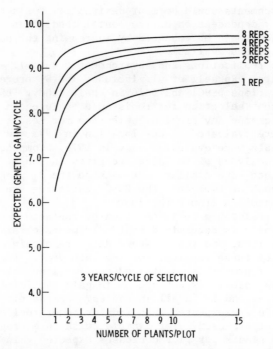

Fig. 7.4.   Expected genetic gain per cycle for second brood European corn bor-
            er resistance for different combinations of plants per plot and
            replications in BS9C1 with S_1 selection.

year's data are sufficient to identify the superior S_1 progenies, one cycle of
S_1 progeny evaluation can be completed in two years. If additional years of
data are required, it would increase the duration of each cycle from two to
three years. Because the BS9C1 progenies were evaluated in two years, the relation of numbers of replications and years to expected gain was examined on a
per cycle and per year basis. The effect of increasing the years of evaluation is positive when expected gain is expressed on a per cycle basis (Fig.
7.5). Results of Fig. 7.5 indicate that reduction in phenotypic variance from
increasing number of years of evaluation is not great enough to compensate for
increased duration of a cycle of selection. It seems, therefore, that the
rate of genetic gain from use of S_1 progenies is maximized by evaluating 5
plants per plot in four replications in one year. If genotype-environment interaction is important in the classification of S_1 progenies for second brood
corn borer resistance, use of more than one location within each year would be
preferable if facilities permit.

Genetic gain expected for different cyclical selection methods for Iowa
Stiff Stalk Synthetic (BSSS) is given to illustrate how genetic gain varies
among selection methods expressed on a per year and per cycle basis. Genetic
gain is given for situations similar to those of the U.S. Corn Belt; only one
season is available for testing but winter nursery facilities can be used for
recombination and development of progenies for testing. Testing usually is
not conducted in winter nurseries because environments are quite different.
Table 7.26 illustrates some possible combinations of methods for use of the
two seasons for different selection methods. For mass and half-sib I selection, it is not possible to use winter nurseries because one cycle of selection and recombination is completed in each summer season. For other methods

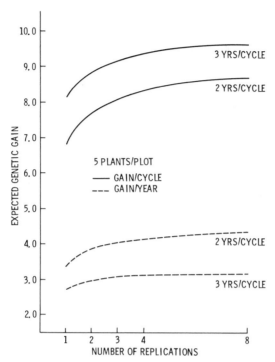

Fig. 7.5. Expected genetic gain relative to numbers of replications and years
           on a per cycle and per year basis for S_1 selection (Russell et al.
           1978).

of selection, winter seasons can be used to great advantage to reduce the number of years to complete each cycle of selection. Use or nonuse of winter
seasons is dictated by availability of data for making selections, sequence of
selection programs, and funds available. If the harvest is delayed for any
reason, the winter planting may be too late to have seed available for planting in the next summer season. Winter seasons are useful to reduce cycle intervals, but researchers must be flexible in their use because unseasonable
conditions may indicate making adjustments in planned selection programs. For
example, full-sib progenies also can be produced at time of recombination to
complete one cycle in 2 years.

     Variance component estimates used to calculate expected gain for differ-
ent selection methods are shown in Table 7.27. Three traits that have different heritability estimates were chosen: 38.9%, 68.4%, and 92.0% on a progeny
mean basis for yield, ear length, and ear height, respectively, and 8.4%,
19.9%, and 59.8% on an individual plant basis. Ear height has a much greater
heritability than yield, and mass selection would be expected to be much more
effective for ear height than for yield. It seems that some type of progeny
evaluation would be considerably better than mass selection for yield improvement, but we must consider the number of years required to complete each cycle
of selection. Hence gain per cycle may be greater by some type of progeny
evaluation but total gain may be greater by mass selection because one cycle
of selection can be completed each summer session.

     Expected gain for different selection methods for the three traits of
Iowa Stiff Stalk Synthetic are given in Table 7.28. Expected gain is presented on a per year and per cycle basis to illustrate the effect of years on pre-

Table 7.26. Schedule of breeding procedures for different methods of selection in which one season is available for selection and one nonselection season is available for recombination or development of progenies for selection; e.g., U.S. Corn Belt.

| Season | | Method of Selection | | | | | | |
	Mass	Half-sib I†	Half-sib II†	Half-sib III†	Full-sib	S_1	S_2	Inbred
Winter	---	---	---	---	---	Produce S_1 progenies	Produce S_1 progenies	Produce S_1 progenies
Summer	Selection	Selection	Produce HS progenies	Produce HS progenies, self	Produce FS progenies	Test and selection	Produce S_2 progenies	Produce S_2 progenies
Winter	---	---	---	---	---	Recombine	---	Produce S_3 progenies
Summer	Selection	Selection	Test and selection	Test and selection	Test and selection	Recombine	Test and selection	Produce S_4 progenies
Winter	---	---	Recombine	Recombine	Recombine	Produce S_1 progenies	Recombine	Produce S_5 progenies
Summer	Selection	Selection	Produce HS progenies	Produce HS progenies, self	Produce FS progenies	Test and selection	Recombine	Produce S_6 progenies
Winter	---	---	---	---	---	Recombine	Produce S_1 progenies	Produce S_6 progenies
Summer	Selection	Selection	Test and selection	Test and selection	Test and selection	Recombine	Produce S_2 progenies	Test and selection
Winter	---	---	Recombine	Recombine	Recombine	Produce S_1 progenies	---	Recombine

†Half-sib I is modified ear-to-row selection; half-sib II is half-sib selection with recombination of remnant half-sib seed; and half-sib III is half-sib selection with recombination of selfed seed.

Table 7.27. Component of variance estimates for three traits of Iowa Stiff Stalk Synthetic obtained from 800 full-sib progenies evaluated in six environments (Silva 1974).

Trait	Estimates of components of variance						$\hat{h}^2$ (%)	
	$\hat{\sigma}^2_w$	$\hat{\sigma}^2$	$\hat{\sigma}^2_{DE}$	$\hat{\sigma}^2_{AE}$	$\hat{\sigma}^2_D$	$\hat{\sigma}^2_A$	Mean[+]	Plant
Yield, g/plant	1301 ± 18	185 ± 7	75 ± 12	92 ± 10	193 ± 21	169 ± 24	38.9	8.4
Ear length, cm	3.98 ± 0.06	0.64 ± 0.02	0.26 ± 0.04	0.22 ± 0.03	0.45 ± 0.06	1.38 ± 0.11	68.4	19.9
Ear height, cm	77.9 ± 1.3	14.0 ± 0.6	7.8 ± 0.9	8.0 ± 0.9	9.5 ± 1.4	174 ± 5.1	92.0	59.8

[+]Heritability estimates calculated on a progeny mean basis as

$$h^2 = \hat{\sigma}^2_A / (\hat{\sigma}^2/8 + \hat{\sigma}^2_{DE}/4 + \hat{\sigma}^2_{AE}/4 + \hat{\sigma}^2_D + \hat{\sigma}^2_A) \times 100$$

Table 7.28. Expected gain per year and per cycle for eight methods of selection in Iowa Stiff Stalk Synthetic for three traits.

Method of selection	Coefficient[‡]	Seasons per cycle	Yield (q/ha)		Ear length (cm)		Ear height (cm)	
			Per year	Per cycle	Per year	Per cycle	Per year	Per cycle
Mass	1/2	1	0.54	0.54	0.46	0.46	8.92	8.92
Mass	1	1	--	--	--	--	17.84	17.84
Half-sib I[§]	1/8	1	1.75	1.75	0.46	0.46	5.63	5.63
Half-sib II[§]	1/4	2	1.75	3.50	0.46	0.92	5.63	11.26
Half-sib III[§]	1/2	3	2.34	7.01	0.61	1.83	7.50	22.50
Full-sib	1/2	2	2.25	4.51	0.63	1.26	7.91	15.82
S_1 progeny	1	2	3.61	7.22	0.96	1.92	11.33	22.66
S_1 progeny[¶]	1	3	2.41	7.22	0.64	1.92	7.55	22.65
S_2 progeny	3/2	3	3.28	9.85	0.82	2.46	5.24	15.72
S_2 progeny[¶]	3/2	4	2.46	9.85	0.62	2.46	3.93	15.72
Inbred line	2	5	2.34	11.73	0.56	2.81	6.48	32.38

[†]Expected gain calculated assuming four environments, each including two replications.

[‡]See Table 6.12.

[§]See Table 7.21.

[¶]Additional season of recombination is used before developing the S_1 progenies.

dicted gain for one cycle.   In most instances the greater the number of years per cycle the greater the expected gain; expected gain per year, however, is reduced when the cycle interval is increased.   Effectiveness of different selection methods is illustrated by comparing expected gain for yield and ear height.   Expected gain for yield per year by mass selection is considerably less than for other selection methods for yield.   Heritability estimates for yield show that the estimate on progeny means is 4.6 times greater than for individual plants.   Heritability estimates for ear height are greater and the progeny mean estimate (92.0%) is 1.5 times greater than the individual plant estimate (59.5%).   Consequently, expected gain of mass selection for ear height per year is greater than all other selection methods except S_1 progeny (two years).   If selection for ear height can be made before flowering, undesirable male plants can be detasseled and expected gain from mass selection is increased (the coefficient is increased from 0.5 to 1) and is greater than all other selection methods on a per year basis.   (Note: Usually selection is for lower rather than higher ear heights in maize breeding.)   Expected gain from mass selection for yield improvement and increased ear length also could be increased by controlling the male parents, but present techniques do not provide for a direct identification of the superior yielding and longer eared plants at flowering time.   Expected gains listed in Table 7.28 apply only when testing is conducted in four environments with two replications in each environment.

Expected gains for different selection methods in Table 7.28 are for a specific situation such as the U.S. Corn Belt.   In tropical and semitropical environments, the frost-free period may extend throughout the year.   Consequently the seasons may not differ very much during the year, or they may be quite different because of some specific weather factor.   It may be possible to get three crop seasons per year as far as temperatures and light durations are concerned; but rainfall may be sparse in one season, thus making yield trials impossible.   In that case it may be possible to irrigate a small area for making recombinations.   Hence, efficiency of selection in tropical and semitropical environments may be enhanced by effective use of different seasons available.

S. A. Eberhart (personal communication, 1971) calculated the expected gain in yield for different selection methods for situations having different numbers and types of seasons.   Estimates of parameters used to calculate expected gain were obtained from four half-sib selection trials in Kenya, East Africa (Table 7.29).   Three situations (A, B, and C) were used to predict gain for five different combinations of seasons that represent different maize growing areas.   Additive genetic variance was constant for situations A, B, and C but the other parameters were varied to show their effects on predicted gain.   The relative changes in predicted gains for situations A, B, and C (Table 7.28) apply only when yield trials are conducted with two replications at four environments within the same season.   The relative changes in predicted gains will not be the same for all methods if numbers of replications and locations are changed.   When the seasons are similar, however, relative gains among selection methods will not change whether there are one, two, or three seasons per year.

Gain per cycle for different selection methods (Table 7.30) again depends on heritability of the trait and type of progeny evaluation used.   Mass selection has the smallest expected gain per cycle for situations A and B but mass selection only requires one growing season to complete a cycle.   With a greater heritability value (50%) in situation C, expected gain per cycle of mass selection exceeds half-sib I and II selection.

Table 7.29. Estimates of genetic parameters for yield for comparing
selection methods.

| Situation | Component of variance estimates[+] | | | | | | | |
	$\hat{\sigma}^2_A$	$\hat{\sigma}^2_D$	$\hat{\sigma}^2_{AL}$	$\hat{\sigma}^2_{DL}$	$\hat{\sigma}^2$	$\hat{\sigma}^2_{me}$	$\hat{\sigma}^2_w$	$\hat{h}^2$
A[‡]	60	30	68	34	98	52	967	5.0
B	60	0	68	0	98	52	967	5.1
C	60	0	0	0	3	0	60	50.0

[+] $\hat{\sigma}^2$, $\hat{\sigma}^2_{me}$, and $\hat{\sigma}^2_w$ refer to the experimental error, microenvironmental interaction, and within-plot variance, respectively.

[‡] Estimates were obtained from four half-sib experiments with 21 plants per plot, two replications, and four environments from analysis pooled over three years. Estimates of dominance variance and interactions were assumed to be half of their corresponding additive genetic variance parameters. The B and C situations are arbitrary to show the effect of changing parameters with additive genetic variance held constant.

Genetic gain per year depends not only on heritability of the trait and type of progeny evaluation but also on types of seasons available each year. For low heritability (situations A and B) and similar seasons, full-sib, S_1, and S_2 selection give the greatest predicted gain per year. If one has either one season or two similar seasons per year and high heritability (situation C), mass selection is clearly superior to other methods. With no dominance variance, low heritability (situation B), and similar seasons, full-sib selection gives as much gain as S_1 and S_2 progeny selection but yield trials are required every second season for full-sib selection. Half-sib I has the greatest predicted gain with two seasons, particularly when heritability is high.

With two nonsimilar seasons per year, full-sib selection has greater predicted gain per year than other selection methods, particularly for situations B and C that have no dominance variance. Two nonsimilar seasons represent conditions used in calculating predicted gain in Table 7.28 for Iowa Stiff Stalk Synthetic. In Table 7.28, however, the predicted gain for full-sib selection was less than for S_1 and S_2 progeny selection because dominance variance was slightly greater than additive genetic variance for yield. The effect of dominance variance on predicted gain for full-sib selection relative to S_1 and S_2 progeny selection for two nonsimilar seasons also is shown in Table 7.30 for situation A.

With three nonsimilar seasons in two years, S_1 progeny selection is predicted to give more gain than S_2 progeny selection. Predicted gain for three nonsimilar seasons per year is greater for half-sib III and S_1 progeny selection than for other selection methods; half-sib III and S_1 progeny selection are about the same for situation C but S_1 progeny selection has greater predicted gain for situations A and B, which have low heritability values.

Choice of a selection method depends on the trait under selection, type of population in which selection is initiated, and objectives of the selection program in relation to the total breeding program. If an exotic population is introduced into the breeding program, mass selection probably will be effective in adapting the population to local conditions (Hallauer and Sears 1972).

Table 7.30. Predicted gain per cycle and per year for three sets of genetic parameters (A, B, and C)[+] and for five combinations of growing seasons.

Selection method	Seasons /cycle	Gain/cycle			Predicted gain per year														
					2 similar seasons			2 nonsimilar seasons			3 nonsimilar seasons in 2 yr			3 nonsimilar seasons/yr			1 season/yr		
		A	B	C	A	B	C	A	B	C	A	B	C	A	B	C	A	B	C
Mass	1	1.5	1.6	6.8	3.0	3.2	13.6	1.5	1.6	6.8	1.5	1.6	6.8	1.5	1.6	6.8	1.5	1.6	6.8
Half-sib I[#]	1	2.3	2.3	3.4	4.7	4.6	6.8	2.3	2.3	3.4	2.3	2.3	3.3	2.3	2.3	3.4	2.3	2.3	3.4
Half-sib II[#]	2	4.7	4.7	6.7	4.7	4.7	6.7	4.7	4.7	6.7	2.3	2.3	3.4	4.7	4.7	6.7	2.3	2.3	3.4
Half-sib III[#]	3	9.4	9.4	13.4	6.2	6.2	8.9	4.7	4.7	6.7	4.7	4.7	6.7	9.4	9.4	13.4	3.1	3.1	4.5
Full-sib	2	6.8	7.3	9.5	6.8	7.4	9.5	6.8	7.4	9.5	3.4	3.7	4.8	6.8	7.4	9.5	3.4	3.7	4.8
S_1	3	10.6	11.1	13.5	7.0	7.4	9.0	5.3	5.6	6.8	5.3	5.6	6.8	10.6	11.1	13.5	3.5	3.7	4.5
S_2	4	13.6	13.9	16.6	6.8	7.0	8.3	6.8	7.0	8.3	4.5	4.6	5.5	6.8	7.0	8.3	3.4	3.5	4.1

[+]A, B, and C as defined in Table 7.27.

[#]See Table 7.26.

As shown in Table 7.28, highly heritable traits (ear height, in this instance) can be effectively changed by mass selection, which also is important for adapting day-length sensitive germplasm to the U.S. Corn Belt. For traits such as days to flower, prolificacy, and ear height, effectiveness of mass selection can be enhanced by detasseling the male plants before anther dehiscence because parental control is increased.

Half-sib I is a modification that seems promising: (1) the method is simple and progeny data are available for yield evaluation and (2) one cycle can be completed each year (or season). If the selection program is an adjunct to an applied breeding program, half-sib III, S_1, and S_2 progeny selection may be favored because preliminary yield test data can be obtained from the evaluation trials. For half-sib III, testcross information is available and S_1 lines used for recombination can be included in the applied breeding program for additional inbreeding and testing.

If genetic variation is primarily additive S_1 and S_2 selection should be effective, and both generate new lines for the applied breeding program. The S_2 selection requires an additional season to complete a cycle and predicted gain per year may be less than for S_1 selection. The slightly smaller predicted gain, however, may be offset by selection advantages for an applied breeding program. For instance, if 500 S_1 progenies are saved from selected S_0 plants of some population, the S_1 progenies can be screened for European corn borer, stalk rot, fungal leaf disease resistance, and phenotypic plant traits in a two-replication test in one environment. Selected S_1 plants within and among the S_1 progenies can be advanced to the S_2 generation. Further selection at harvest may reduce the number of S_2 progenies to 100 for testing. If 20 to 25 S_2 progenies are selected for recombination, intense selection pressure has been given for highly heritable traits before expensive yield tests are conducted. This illustration is one example of how the selection method can be adapted for an applied breeding program that has the primary objective of developing new inbred lines. Innovations by breeders can be made in all selection methods to fit their breeding programs and objectives.

Predicted gains given are for only one cycle. If selection is effective in changing gene frequency of the original population, estimates of genetic parameters of the original populations may be invalid for use in predicting gain in future cycles. For progeny evaluation trials, however, estimates of genetic parameters can be obtained in each cycle and can be used to predict gain expected in the next cycle. Pooling estimates from successive cycles of selection will provide better estimates of parameters used in predicting gain provided the change in gene frequency is not great. If significant changes in gene frequency are suspected, estimates from the most recent cycle of selection should be used even though they may have greater errors than pooled estimates from several cycles.

REFERENCES

Acosta, A. F., and P. L. Crane. 1972. Further selection for lower ear height in maize. *Crop Sci.* 12:165-67.

Anderson, R. L., and T. A. Bancroft. 1952. *Statistical Theory in Research*. McGraw-Hill, New York.

Arboleda-Rivera, F., and W. A. Compton. 1974. Differential response of maize to mass selection in diverse selection environments. *Theor. Appl. Genet.* 44:77-81.

Ariyanayagan, R. P.; C. L. Moore; and V. R. Carangal. 1974. Selection for leaf angle in maize and its effect on grain yield and other characters. *Crop Sci.* 14:551-56.

Bojanowski, J. 1967. Recurrent selection for smut resistance in corn. *Züchter* 37:151-55.

Burton, J. W.; L. H. Penny; A. R. Hallauer; and S. A. Eberhart. 1971. Evaluation of synthetic populations developed from a maize variety (BSK) to two methods of recurrent selection. *Crop Sci.* 11:361-65.

Carangal, V. R.; S. M. Ali; A. F. Koble; E. H. Rinke; and J. C. Sentz. 1971. Comparison of S_1 with testcross evaluation for recurrent selection in maize. *Crop Sci.* 11:658-61.

Collier, J. W. 1959. Three cycles of reciprocal recurrent selection. *Proc. Annu. Hybrid Corn Ind. Res. Conf.* 14:12-23.

Compton, W. A. 1977. Full-sib family selection in Krug under irrigation in Lincoln, Nebraska: Four cycles of progress. North Cent. Corn Breed. Res. Comm. (NCR-2) Rep. Mimeogr. Library, Iowa State Univ., Ames. Pp. 1-10.

Compton, W. A., and K. Bahadur. 1977. Ten cycles of progress from modified ear to row selection in corn (*Zea mays* L.). *Crop Sci.* 17:378-80.

Comstock, R. E.; H. F. Robinson; and P. H. Harvey. 1949. A breeding procedure designed to make maximum use of both general and specific combining ability. *Agron. J.* 41:360-67.

Cortez-Mendoza, H., and A. R. Hallauer. 1979. Divergent mass selection for ear length in maize. *Crop Sci.* 19:175-78.

Darrah, L. L. 1975. Maize genetics. Rec. Res. Annu. Rep., 1971. East African Agric. For. Res. Org. Annu. Rep.

Darrah, L. L.; S. A. Eberhart; and L. H. Penny. 1972. A maize breeding methods study in Kenya. *Crop Sci.* 12:605-8.

————. 1978. Six years of maize selection in Kitale II, Ecuador 573, and Kitale Composite A by use of the comprehensive breeding system. *Euphytica* 27:191-204.

Douglas, A. G.; J. W. Collier; M. F. El-Ebrashy; and J. S. Rogers. 1961. An evaluation of three cycles of reciprocal recurrent selection in a corn improvement program. *Crop Sci.* 1:157-61.

Duclos, L. A., and P. L. Crane. 1968. Comparative performance of top crosses and S_1 progeny for improving populations of corn (*Zea mays* L.). *Crop Sci.* 8:191-94.

Dudley, J. W. 1976. Seventy-six generations of selection for oil and protein percentage in maize. In *Proc. Int. Conf. Quant. Genet.*, E. Pollak, O. Kempthorne, and T. B. Bailey, Jr., eds., pp. 459-73. Iowa State Univ. Press, Ames.

Dudley, J. W., and D. E. Alexander. 1969. Performance of advanced generations of autotetraploid maize (*Zea mays* L.) synthetics. *Crop Sci.* 9:613-15.

Dudley, J. W., and R. J. Lambert. 1969. Genetic variability after 65 generations of selection in Illinois high oil, low oil, high protein, and low protein strains of *Zea mays* L. *Crop Sci.* 9:179-81.

Dudley, J. W.; R. J. Lambert; and D. E. Alexander. 1974. Seventy generations of selection for oil and protein concentration in the maize kernel. In *Seventy Generations of Selection for Oil and Protein in Maize*, J. W. Dudley, ed., pp. 181-212. Crop Sci. Soc. Am., Madison, Wis.

Eberhart, S. A. 1964. Least squares method for comparing progress among recurrent selection methods. *Crop Sci.* 4:230-31.

————. 1970. Factors affecting efficiencies of breeding methods. *African Soils* 15:669-80.

Eberhart, S. A., and M. N. Harrison. 1973. Progress from half-sib selection in Kitale Station Maize. *East African Agric. For. Res. J.* 39:12-16.

Eberhart, S. A.; S. Debela; and A. R. Hallauer. 1973. Reciprocal recurrent

selection in the BSSS and BSCB1 maize varieties and half-sib selection in
BSSS. *Crop Sci.* 13:451-56.

Eberhart, S. A.; M. N. Harrison; and F. Ogada. 1967. A comprehensive breeding system. *Züchter* 37:169-74.

Gardner, C. O. 1961. An evaluation of effects of mass selection and seed irradiation with thermal neutrons on yield of corn. *Crop Sci.* 1:241-45.

⸻. 1968. Mutation studies involving quantitative traits. *Gamma Field
Symp.* 7:57-77.

⸻. 1969a. The role of mass selection and mutagenic treatment in modern
corn breeding. *Proc. Annu. Corn Sorghum Ind. Res. Conf.* 24:15-21.

⸻. 1969b. Genetic variation in irradiated and control populations of
corn after ten cycles of mass selection for high grain yield. In *Induced
Mutation in Plants*, pp. 469-77. Int. At. Energy Agency, Vienna.

⸻. 1972. Performance of hybrids involving selected S_2 lines from Hays
Golden and mass selected populations of corn. *Agron. Abstr.* P. 7.

⸻. 1973. Evaluation of mass selection and of seed irradiation with mass
selection for population improvement in maize. *Genetics* 74:s88-s89.

⸻. 1976. Quantitative genetic studies and population improvement in
maize and sorghum. In *Proc. Int. Conf. Quant. Genet.*, E. Pollack, O.
Kempthorne, and T. B. Bailey, Jr., eds., pp. 475-89. Iowa State Univ.
Press, Ames.

Genter, C. F. 1973. Comparison of S_1 and testcross evaluation after two cycles of recurrent selection in maize. *Crop Sci.* 13:524-27.

⸻. 1976a. Recurrent selection for yield in the F_2 of a maize single
cross. *Crop Sci.* 16:350-52.

⸻. 1976b. Mass selection in a composite of intercrosses of Mexican
races of maize. *Crop Sci.* 16:556-58.

Genter, C. F., and M. W. Alexander. 1966. Development and selection of productive S_1 inbred lines of corn (*Zea mays* L.). *Crop Sci.* 6:429-31.

Genter, C. F., and S. A. Eberhart. 1974. Performance of original and advanced maize populations and their diallel crosses. *Crop Sci.* 14:881-85.

Gevers, H. O. 1974. Reciprocal recurrent selection in maize under two systems of parent selection. *Proc. Fifth Genet. Congr.* Repub. South Africa.

Goulas, C. K., and J. H. Lonnquist. 1976. Combined half-sib and S_1 family
selection in a maize composite population. *Crop Sci.* 16:461-64.

Hakim, R. M.; J. C. Sentz; and V. R. Carangal. 1969. Mass and family selection for yield in a tropical variety of maize. *Agron. Abstr.* P. 7.

Hallauer, A. R. 1968. Effect of mass selection for divergent ear length on
yield in maize. *Agron. Abstr.* P. 9.

⸻. 1970. Genetic variability for yield after four cycles of reciprocal
recurrent selection in maize. *Crop Sci.* 10:482-85.

⸻. 1971. Change in genetic variance for seven plant and ear traits after four cycles of reciprocal recurrent selection for yield in maize.
Iowa State J. Sci. 45:575-93.

⸻. 1973. Hybrid development and population improvement in maize by reciprocal full-sib selection. *Egyptian J. Genet. Cytol.* 2:84-101.

⸻. 1975a. Inbred and hybrid development from improved maize populations. *Proc. 8me Congr. Int. Sect. Mais-Sorgho. EUCARPIA.* Pp. 44-78.

⸻. 1975b. Relation of gene action and type of tester in maize breeding
procedures. *Proc. Annu. Corn Sorghum Res. Conf.* 30:150-65.

⸻. 1977. Four cycles of reciprocal full-sib selection. North Cent.
Corn Breed. Res. Comm. (NCR-2) Rep. Mimeogr. Library, Iowa State Univ.,
Ames. Pp. 11-13.

Hallauer, A. R., and J. H. Sears. 1969. Mass selection for yield in two varieties of maize. *Crop Sci.* 9:47-50.

————. 1972. Integrating exotic germplasm into Corn Belt maize breeding programs. *Crop Sci.* 12:203-6.

Hallauer, A. R., and J. A. Wright. 1967. Genetic variances in the open-pollinated variety of maize, Iowa Ideal. *Züchter* 37:178-85.

Hanson, W. D. 1971. Selection for differential productivity among juvenile maize plants: Associated net photosynthetic rate and leaf area changes. *Crop Sci.* 11:334-39.

————. 1973. Changes in efficiencies and number of chloroplasts associated with divergent selections for juvenile productivity in *Zea mays* L. *Crop Sci.* 13:386-87.

Harland, S. C. 1946. A new method of maize improvement. *Trop. Agric.* 23: 114.

Harris, R. E.; C. O. Gardner; and W. A. Compton. 1972. Effect of mass selection and irradiation in corn measured by random S_1 lines and their testcrosses. *Crop Sci.* 12:594-98.

Hartley, C. P. 1909. Progress in methods of producing higher yielding strains of corn. *USDA Yearbook.* Pp. 309-20.

Hopkins, C. G. 1899. Improvement in the chemical composition of the corn kernel. Illinois Agric. Exp. Stn. Bull. 55:205-40.

Horner, E. S.; W. H. Chapman; H. W. Lundy; and M. C. Lutrick. 1973. Comparison of three methods of recurrent selection in maize. *Crop Sci.* 13:485-89.

Horner, E. S.; W. H. Chapman; M. C. Lutrick; and H. W. Lundy. 1969. Comparison of selection based on yield of topcross progenies and of S_2 progenies in maize (*Zea mays* L.). *Crop Sci.* 9:539-43.

Horner, E. S.; H. W. Lundy; M. C. Lutrick; and R. W. Wallace. 1963. Relative effectiveness of recurrent selection for specific and for general combining ability in corn. *Crop Sci.* 3:63-66.

Horner, E. S.; M. C. Lutrick; W. H. Chapman; and F. G. Martin. 1976. Effect of recurrent selection for combining ability with a single cross tester in maize. *Crop Sci.* 16:5-8.

Hull, F. H. 1945. Recurrent selection for specific combining ability in corn. *Agron. J.* 37:134-45.

Jenkins, M. T. 1940. The segregation of genes affecting yield of grain in maize. *Agron. J.* 32:55-63.

Jenkins, M. T.; A. L. Robert; and W. R. Findley, Jr. 1954. Recurrent selection as a method for concentrating genes for resistance to *Helminthosporium turcicum* leaf blight in corn. *Agron. J.* 46:89-94.

Jinahyon, S., and C. L. Moore. 1973. Recurrent selection techniques for maize improvement in Thailand. *Agron. Abstr.* P. 7.

Jinahyon, S., and W. A. Russell. 1969a. Evaluation of recurrent selection for stalk-rot resistance in an open-pollinated variety of maize. *Iowa State J. Sci.* 43:229-37.

————. 1969b. Effects of recurrent selection for stalk-rot resistance on other agronomic characters in an open-pollinated variety of maize. *Iowa State J. Sci.* 43:239-51.

Johnson, E. C. 1963. Mass selection for yield in a tropical corn variety. *Am. Soc. Agron. Abstr.* P. 82.

Johnson, E. C., and A. Salazar. 1967. Performance of crosses of corn varieties following mass selection. *Agron. Abstr.* P. 12.

Josephson, L. M., and H. C. Kincer. 1973. Selection for low ear placement in corn. *Agron. Abstr.* P. 8.

————. 1976. Mass selection for yield in corn. *Agron. Abstr.* P. 54.

Josephson, L. M.; H. C. Kincer; and B. G. Harville. 1976. Selection studies for low ear placement in corn. *Proc. Annu. Corn Sorghum Res. Conf.* 31:

85-97.

Jugenheimer, R. W., and P. S. Bhatnagar. 1962. Breeding for extremes in lodging resistance in corn by reciprocal recurrent selection. EUCARPIA 2:56-60.

Kiesselbach, T. A. 1922. Corn investigations. Nebraska Agric. Exp. Stn. Res. Bull. 20:5-151.

Kincer, H. C., and L. M. Josephson. 1976. Mass selection for prolificacy in corn. *Agron. Abstr.* P. 55.

Koble, A. F., and E. H. Rinke. 1963. Comparative S_1 line and topcross performance in maize. *Agron. Abstr.* P. 83.

Leng, E. R. 1962. Results of long term selection from chemical composition in maize and their significance in evaluating breeding systems. *Z. Pflanzenzücht.* 47:67-91.

Lima, M.; E. Paterniani; and J. B. Miranda, Fo. 1974. Avaliacão de progênies de meios irmãos no segundo ciclo de selecão em dois compostos de milho. Rel. Cient. Inst. Genét. (ESALQ-USP) 8:78-85.

Lonnquist, J. H. 1949. The development and performance of synthetic varieties of maize. *Agron. J.* 41:153-56.

————. 1951. Recurrent selection as a means of modifying combining ability in corn. *Agron. J.* 43:311-15.

————. 1952. Recurrent selection. *Proc. Annu. Hybrid Corn Ind. Res. Conf.* 7:20-32.

————. 1961. Progress from recurrent selection procedures for the improvement of corn populations. Nebraska Agric. Exp. Stn. Res. Bull. 197.

————. 1963. Gene action and corn yields. *Proc. Annu. Hybrid Corn Ind. Res. Conf.* 18:37-44.

————. 1964. A modification of the ear-to-row procedures for the improvement of maize populations. *Crop Sci.* 4:227-28.

————. 1967. Mass selection for prolificacy in maize. *Züchter* 37:185-88.

————. 1968. Further evidence on testcross versus line performance in maize. *Crop Sci.* 8:50-53.

Lonnquist, J. H., and C. O. Gardner. 1961. Heterosis in intervarietal crosses in maize and its implication in breeding procedures. *Crop Sci.* 1:179-83.

Lonnquist, J. H., and M. F. Lindsey. 1964. Topcross versus S_1 line performance in corn (*Zea mays* L.). *Crop Sci.* 4:580-84.

Lonnquist, J. H., and D. P. McGill. 1956. Performance of corn synthetics in advanced generation of synthesis and after two cycles of recurrent selection. *Agron. J.* 48:249-53.

Lonnquist, J. H., and M. D. Rumbaugh. 1958. Relative importance of test sequence for general and specific combining ability in corn breeding. *Agron. J.* 50:541-44.

Lonnquist, J. H.; O. Cota A.; and C. O. Gardner. 1966. Effect of mass selection and thermal neutron irradiation on genetic variances in a variety of corn (*Zea mays* L.). *Crop Sci.* 6:330-32.

Martin, P. R., and C. O. Gardner. 1976. Comparison of hybrids derived from maize populations after nine cycles of mass selection for yield. *Agron. Abstr.* P. 56.

Mathema, B. B. 1971. Evaluation of progress in adapted × exotic maize population undergoing adaptive mass selection in Nebraska. Master's thesis, Univ. Nebraska, Lincoln.

Miranda, L. T.; L. E. C. Miranda; C. V. Pommer; and E. Sawazaki. 1977. Oito ciclos de selecão entre e dentro de familias de meios irmãos no milho IAC-1. *Bragantia* 36:187-96.

Miranda, J. B., Fo.; R. Vencovsky; and E. Paterniani. 1972. Variância genética aditica da producão de grãos em dois compostos de milho e sua implicacão no melhoramento. Rel. Cient. Inst. Genét. (ESALQ–USP) 5:67–73.

Mock, J. J. 1973. Manipulation of crossing over with intrinsic and extrinsic factors. *Egyptian J. Genet. Cytol.* 2:158–75.

Mock, J. J., and A. A. Bakri. 1976. Recurrent selection for cold tolerance in maize. *Crop Sci.* 16:230–33.

Moll, R. H. 1959. Quantitative genetics in corn and the implications to breeding methodology. *Proc. Annu. Hybrid Corn Ind. Res. Conf.* 14:139–44.

Moll, R. H., and H. F. Robinson. 1966. Observed and expected response in four selection experiments in maize. *Crop Sci.* 6:319–24.

———. 1967. Quantitative genetics investigations of yield of maize. *Züchter* 37:192–99.

Moll, R. H., and C. W. Stuber. 1971. Comparison of response to alternative selection procedures initiated in two populations of maize (*Zea mays* L.). *Crop Sci.* 11:706–11.

Moll, R. H.; C. W. Stuber; and W. D. Hanson. 1975. Correlated response and response to selection index involving yield and ear height of maize. *Crop Sci.* 15:243–48.

Moll, R. H.; C. C. Cockerham; C. W. Stuber; and W. Williams. 1978. Selection responses, genetic-environmental interactions, and heterosis with recurrent selection for yield in maize. *Crop Sci.* 18:641–45.

Noll, C. F. 1916. Experiments with corn. Pennsylvania Agric. Exp. Stn. Bull. 139.

Obilana, T. 1974. Mass selection for yield and earliness in a Nigerian maize composite. Abstr. Annu. Conf. Genet. Soc. Nigeria.

Osuna-Ayalla, J. 1976. Stratified mass selection for production in two maize populations. *Agron. Abstr.* P. 45.

Padgett, C. H.; W. A. Compton; and J. H. Lonnquist. 1968. Divergent mass selection in corn (*Zea mays* L.) for seed size. *Agron. Abstr.* P. 16.

Palma, M. C., and M. Burbano. 1976. Mejoramiento de rendimiento y calidad de maiz (*Zea mays* L.) opaco-2 modificado por seleccion masal. *Inf. Maiz* 16:14.

Paterniani, E. 1967. Selection among and within half-sib families in a Brazilian population of maize (*Zea mays* L.). *Crop Sci.* 7:212–16.

———. 1968. Avaliacão do método de selecão entre e dentro de familias de meios irmãos no melhoramento do milho (*Zea mays* L.). Tese de Cátedra, ESALQ-USP, Piracicaba, Brazil.

———. 1969. Melhoramento de populacões de milho. *Ciênc. Cult.* 21:3–10.

———. 1971. Seleção recorrente reciproca com plantas prolificas. Rel. Cient. Inst. Genét. (ESALQ–USP) 5:129–32.

———. 1974a. Selecão entre e dentro de familias de meios irmãos no milho Piranão. Rel. Cient. Inst. Genét. (ESALQ–USP) 8:174–79.

———. 1974b. Selecão no milho semi-dentado ESALQ-HV-1. Rel. Cient. Inst. Genét. (ESALQ–USP) 8:180–86.

———. 1974c. Selecão recorrent reciproca utilizando progênies de meios irmãos obtidas de plantas prolificas. Rel. Cient. Inst. Genét. (ESALQ–USP) 8:187–91.

Paterniani, E., and R. Vencovsky. 1977. Reciprocal recurrent selection in maize (*Zea mays* L.) based on testcrosses of half-sib families. *Maydica* 22:141–52.

———. 1978. Reciprocal recurrent selection based on half-sib progenies and prolific plants in maize (*Zea mays* L.). *Maydica* 23:209–19.

Penny, L. H. 1959. Improving combining ability by recurrent selection. *Proc. Annu. Hybrid Corn Ind. Res. Conf.* 14:7–11.

———. 1968. Selection induced differences among strains of a synthetic variety of maize. *Crop Sci.* 8:167-68.

Penny, L. H., and S. A. Eberhart. 1971. Twenty years of reciprocal recurrent selection with two synthetic varieties of maize (*Zea mays* L.). *Crop Sci.* 11:900-903.

Penny, L. H.; W. A. Russell; and G. F. Sprague. 1962. Types of gene action in yield heterosis in maize. *Crop Sci.* 2:341-44.

Penny, L. H.; G. E. Scott; and W. D. Guthrie. 1967. Recurrent selection for European corn borer resistance in maize. *Crop Sci.* 7:407-8.

Robinson, H. F., and R. E. Comstock. 1955. Analysis of genetic variability in corn with reference to probable effects of selection. *Symp. Quant. Biol.* 20:127-36.

Rodriguez, C. G.; F. Arboleda-Rivera; and J. E. Vargas. 1976. Efecto de la seleccion masal estratificada ambiental por prolificidad y rendimiento en el comportamiento de algunos caracteres de una poblacion de maiz (*Zea mays* L.). *Inf. Maiz* 16:14-15.

Rogers, R. R.; W. A. Russell; and J. C. Owens. 1977. Expected gains from selection in maize for resistance to corn rootworms. *Maydica* 22:27-36.

Russell, W. A.; S. A. Eberhart; and U. A. Vega. 1973. Recurrent selection for specific combining ability for yield in two maize populations. *Crop Sci.* 13:257-61.

Russell, W. K.; W. A. Russell; W. D. Guthrie; A. R. Hallauer; and J. C. Robbins. 1978. Allocation of resources in breeding for resistance in maize to second brood of the European corn borer. *Maydica* 23:11-20.

Scott, G. E., and E. E. Rosenkranz. 1974. Effectiveness of recurrent selection for corn stunt resistance in maize variety. *Crop Sci.* 14:758-60.

Segovia, R. T. 1976. Seis ciclos de seleção entre e dentro de familias de meios irmãos no milho (*Zea mays* L.) Centralmex. Tese de Doutoramento, ESALQ-USP, Piracicaba, Brazil.

Sevilla, R. 1975. Ocho ciclos de seleccion mazorca-hilera en una variedad peruana de maiz. *Inf. Maiz* 5:11.

Silva, J. C. 1974. Genetic and environmental variances and covariances estimated in the maize (*Zea mays* L.) variety, Iowa Stiff Stalk Synthetic. Ph.D. dissertation, Iowa State Univ., Ames.

Silva, W. J. da, and J. H. Lonnquist. 1968. Genetic variances in populations developed from full-sib and S_1 testcross progeny selection in an open pollinated variety of maize. *Crop Sci.* 8:201-4.

Smith, L. H. 1909. The effect of selection upon certain physical characters in the corn plant. Illinois Agric. Exp. Stn. Bull. 132:48-62.

Smith, L. H., and A. M. Brunson. 1925. An experiment in selecting corn for yield by the method of ear-to-row breeding plot. Illinois Agric. Exp. Stn. Bull. 127.

Sprague, G. F. 1966. Quantitative genetics in plant improvement. In *Plant Breeding*, K. J. Frey, ed., pp. 315-43. Iowa State Univ. Press, Ames.

Sprague, G. F., and B. Brimhall. 1950. Relative effectiveness of two systems for selection for oil content of the corn kernel. *Agron. J.* 42:83-88.

Sprague, G. F., and W. A. Russell. 1957. Some evidence on type of gene action involved in yield heterosis in maize. *Proc. Int. Genet. Symp.* Pp. 522-26.

Sprague, G. F.; P. A. Miller; and B. Brimhall. 1952. Additional studies of the relative effectiveness of two systems of selection for oil content of the corn kernel. *Agron. J.* 44:329-31.

Sprague, G. F.; W. A. Russell; and L. H. Penny. 1959. Recurrent selection for specific combining ability and types of gene action involved in yield heterosis in corn. *Agron. J.* 51:392-94.

Steel, R. G. D., and J. H. Torrie. 1960. *Principles and Procedures of Sta-*

tistics. McGraw-Hill, New York.

Suwantaradon, K., and S. A. Eberhart. 1974. Developing hybrids from two improved maize populations. *Theor. Appl. Genet.* 44:206-10.

Thomas, W. I., and D. G. Grissom. 1961. Cycle evaluation of reciprocal recurrent selection for popping volume, grain yield, and resistance to root lodging in popcorn. *Crop Sci.* 1:197-200.

Thompson, D. L. 1963. Stalk strength of corn as measured by crushing strength and rind thickness. *Crop Sci.* 3:323-29.

———. 1972. Recurrent selection for lodging susceptibility and resistance in corn. *Crop Sci.* 12:631-34.

Thompson, D. L., and P. H. Harvey. 1960. Progress from recurrent and reciprocal recurrent selection for yield of corn. *Agron. Abstr.* Pp. 54-55.

Torregroza, M. 1973. Response of a highland maize synthetic to eleven cycles of divergent mass selection for ears per plant. *Agron. Abstr.* P. 16.

Torregroza, M., and D. D. Harpstead. 1967. Effects of mass selection for ears per plant in maize. *Agron. Abstr.* P. 20.

Torregroza, M.; E. Arias; C. Diaz; and F. Arboleda. 1972. Evaluation of reciprocal recurrent selection in germplasm sources of highland Latin American maizes. *Agron. Abstr.* P. 20.

Torregroza, M.; C. Diaz; E. Arias; J. A. Rivera; and C. Ramirez. 1976. Efecto de la seleccion masal en generaciones avanzades de hibridos varietales de maiz. *Inf. Maiz* 16:12-13.

Troyer, A. F. 1975. Extremely early corns and selection for early flowering. *Proc. 8me Congr. Int. Sect. Mais-Sorgho.* EUCARPIA. Pp. 161-68.

———. 1976. Selection for early flowering in corn: III (18 F_2 population). *Agron. Abstr.* P. 65.

Troyer, A. F., and W. L. Brown. 1972. Selection for early flowering in corn. *Crop Sci.* 12:301-4.

———. 1976. Selection for early flowering in corn: Seven late synthetics. *Crop Sci.* 16:767-72.

Troyer, A. F.; G. R. Herrick; and R. F. Baker. 1965. Ear-to-row selection in corn for agronomic traits: Six cycles compared. *Agron. Abstr.* P. 21.

Vencovsky, R.; J. R. Zinsly; and N. A. Vello. 1970. Efeito da seleção masal estratificada em duas populacões de milho e no cruzamento entre elas. Abstr. 8 Reun. Bras. Milho.

Vera, G. A., and P. L. Crane. 1970. Effects of selection for lower ear height in synthetic populations of maize. *Crop Sci.* 10:286-88.

Walejko, R. N., and W. A. Russell. 1977. Evaluation of recurrent selection for specific combining ability in two open pollinated maize cultivars. *Crop Sci.* 17:647-51.

Webel, O. D., and J. H. Lonnquist. 1967. An evaluation of modified ear-to-row selection in a population of corn (*Zea mays* L.). *Crop Sci.* 7:651-55.

Widstrom, N. W.; W. J. Wiser; and L. F. Bauman. 1970. Recurrent selection in maize for earworm resistance. *Crop Sci.* 10:674-76.

Williams, C. G. 1907. Corn breeding and registration. Ohio Agric. Exp. Stn. Bull. 66.

Williams, C. G., and F. A. Welton. 1915. Corn experiments. Ohio Agric. Exp. Stn. Bull. 282.

Winter, F. L. 1929. The mean and variability as affected by continuous selection for composition in corn. *J. Agric. Res.* 39:451-76.

Woodworth, C. M.; E. R. Leng; and R. W. Jugenheimer. 1952. Fifth generation of selection for oil and protein content in corn. *Agron. J.* 44:60-65.

Zuber, M. S.; M. L. Fairchild; A. J. Keaster; V. L. Fergason; G. F. Krause; E. Hildebrand; and P. J. Loesch, Jr. 1971. Evaluation of 10 generations of mass selection for corn earworm resistance. *Crop Sci.* 11:16-18.

8

Testers and Combining Ability

Progeny test was defined by Allard (1960) as "a test of the value of a genotype based on the performance of its offspring produced in some definite system of mating." It was used as early as 1850 by Vilmorin in France, and it proved to be a highly effective procedure for the improvement of sugar content of sugar beets (*Beta vulgaris*). This method of line selection with progeny testing was known as the *Vilmorin method* or *Vilmorin isolation principle* and was introduced in several plant breeding programs in the latter part of the nineteenth century. The progeny test in maize was first used in 1896 by Hopkins, starting the well-known program for selection of maize oil and protein content. The method became known as the *ear-to-row* procedure.

Davis (1927) suggested the use of the topcross procedure, which is a type of progeny test, to evaluate the combining ability of inbred lines in a hybrid maize breeding program. After Jenkins and Brunson (1932) reported on the effective use of the topcross procedure, it was widely adopted in breeding programs. Other types of progeny tests, such as full-sib and selfed (S_1 or S_2) families, were subsequently introduced.

After Sprague and Tatum (1942) introduced the concepts of general combining ability and specific combining ability, new approaches for the use of progeny test or topcross were suggested. Thus recurrent selection methods (Jenkins 1940; Hull 1945; Comstock et al. 1949; Lonnquist 1949) were introduced, widening the horizon for population improvement in maize. In selection for general combining ability (GCA), a broad base heterogeneous population is used as tester. It can be either the parental population or any broad genetic base (synthetic or open-pollinated variety), unrelated population. In all instances genotypes are tested with a representative sample of genotypes in the tester; i.e., each plant in the base population is crossed to a random sample of gametes from the tester. Each testcross, therefore, is a type of half-sib family. When the tester has a narrow genetic base (inbred line or single cross), selection among testcrosses is said to be for specific combining ability (SCA). Reciprocal recurrent selection (RRS) is similar to selection for GCA because two populations are selected simultaneously, with the testcrosses for each one using the opposite population as the tester.

The use of testcross (or topcross) in maize breeding has one of the following objectives: (1) evaluation of combining ability of inbred lines in a hybrid breeding program, or (2) evaluation of breeding values of genotypes (plants) for population improvement. In each instance a problem of choice of tester is essentially the same, i.e., to find a tester that provides the best discrimination among genotypes according to the purposes of selection. For inbred line evaluation, a *desirable tester* was defined by Matzinger (1953) as one that combines the greatest simplicity in use with the maximum information

on performance to be expected from tested lines when used in other combinations or grown in other environments. It was recognized, however, that no
single tester can completely fulfill these requirements. Rawlings and Thompson (1962) defined a *good tester* as one that classifies correctly relative
performance of lines and discriminates efficiently among lines under test.
For improvement of breeding populations, Allison and Curnow (1966) defined the
best tester as one that maximizes the expected mean yield of the population
produced from random mating of selected genotypes. Hallauer (1975) pointed
out that in general a *suitable tester* should include simplicity in use, provide information that correctly classifies the relative merit of lines, and
maximize genetic gain.

After Jones (1918) suggested the double-cross hybrid procedure, selection
among hybrids was based on the cross performance of a set of inbred lines.
With relatively few inbred lines developed in hybrid programs in the period
1920-30, the $n(n - 1)/2$ combinations possible from a set of n inbred lines
could be tested for the evaluation of their cross performance. As the number
of lines increased, crossing and testing inbred lines in all combinations became impossible. Evaluation of inbred lines themselves had little value because of the inconsistency of correlation between characters of the inbreds
and their performance in F_1 crosses. The topcross test introduced by Davis
(1927) made possible the screening of inbred lines based first on GCA with a
broad base tester. This procedure was shown to be effective by Jenkins and
Brunson (1932) and was widely used subsequently. Johnson and Hayes (1936) also reported that inbred lines giving high yields in topcrosses were more likely to produce better single crosses.

Early testing of inbred lines was suggested by Jenkins (1935) and Sprague
(1939, 1946). Early testing differed in two main respects from the usual procedure for testing inbred lines: (1) S_0 plants were outcrossed to a tester at
the time of first selfing and combining ability and general performance of the
topcross progeny were determined, and (2) the first discarding allowed a
greater concentration of efforts on the families of greatest promise during
the S_1 and S_2 generations where greater opportunities for selection within
lines existed (Sprague 1946). Several subsequent reports showed that the combining ability of lines had a reasonably good stability through the generations of inbreeding, as indicated in reviews presented by Loeffel (1964,
1971). In general, development of inbred lines is a process of sequential selection because some lines are discarded early by their poor performance or by
results based on early testing and others are eliminated by later tests for
general and for specific combining ability.

In either early or late testing, the most questionable point is the
choice of a tester to evaluate combining ability. In some instances when the
objective is the replacement of a line in a specific combination, SCA is of
prime importance and the most appropriate tester is the opposite inbred line
parent of a single cross or the opposite single-cross parent of the double
cross (Matzinger 1953). Often the choice of the best tester is still obscure.

Green (1948a) compared the relative value of two testers, a low yielding,
lodging-susceptible variety and a high yielding, lodging-resistant, double-
cross hybrid. Because the two testers ranked individual segregates of F_2
plants differently and because there was a difference between testers in average topcross yields, the existence of striking differences in their genetic
structures was suggested. It was not possible to determine from available data which tester gave the better estimate of average combining ability; it was
suggested that the best estimate would be obtained using the average performance of topcrosses with both testers rather than topcrosses of either tester
alone. It also was suggested that a synthetic variety composed of the lines

in current use could be employed to measure combining ability of new inbred material. Keller (1949) also found a very low correlation for agronomic characters of related tester (parental population) crosses and the same characters in the unrelated tester (single cross R4 × Hy), indicating that testers did not rank the lines in similar order. No evidence was presented for the superiority of one over the other.

Other recurrent selection experiments also have indicated differences between testers in ranking the genotypes of a population (Lonnquist and Rumbaugh 1958; Horner et al. 1963; Lonnquist and Lindsey 1964; Horner et al. 1969). In selecting for GCA, the only apparent requirement was that the tester parent be genetically heterogeneous, i.e., have a broad genetic base. The GCA of a line, however, is not a fixed property of a line but depends on the genetic composition of the population with which it has been crossed (Kempthorne and Curnow 1961).

8.1 THEORY

The relative value of testers was theoretically examined by Hull (1945). His conclusions were based on variance among and within testcrosses considering one locus with two alleles distributed as 1AA:2Aa:1aa and with complete dominance and arbitrary values for genotypic effects. Evidence for his conclusions can be summarized as shown in Table 8.1.

If inbred lines are drawn from a population, their frequencies will be (1/2)AA and (1/2)aa. When they are testcrossed with the parental population the mean of testcrosses will be 1 and 1/2, respectively. Thus the general mean is 3/4. The variance among testcrosses is 1/16 and consequently the variance within testcrosses is 3/16 - 1/16 = 2/16.

Hull (1945, 1946, 1952) thus stated that theoretically the most efficient tester would be one that is homozygous recessive at all loci and that homozygosity for dominance alleles at any locus should be avoided. Support for this conclusion also was made from the method of constant-parent regression, i.e., regression of performance of offspring on the performance of variable parents for a particular constant parent. Such a regression was shown to be largest with gene frequency 0 for the character in the constant parent. Regression would be zero either for gene frequencies 1 for complete dominance or for gene frequency at equilibrium for overdominant gene action. Some empirical results showed that the regression of F_1 on parental lines was either essentially zero or negative with strong testers and that weaker lines of several samples as

Table 8.1. Genotypic values and testcross means at one locus level in a population with gene frequency of p = 0.5.

Genotypes	Frequency	Genotypic values	Testcross means	
			AA tester	aa tester
AA	1/4	1	1/4	1
Aa	1/4	1	1/4	1/2
Aa	1/4	1	1/4	1/2
aa	1/4	0	1/4	0
Mean		3/4	1	1/2
Variance		3/16	0	2/16

testers gave a regression of the order of 0.6 and higher. Results reported by
Green (1948a) were in accordance with Hull's hypothesis because the lodging-
susceptible tester showed a much greater range for standability than the lodging-resistant tester. On the other hand, Keller (1949) found no difference in
the line × tester interaction variance using two groups (high yielding and low
yielding) of testers. Under Hull's hypothesis it was expected that the low
yielding group would give a higher interaction variance.

8.1.1 Variance among Testcrosses

Rawlings and Thompson (1962) added a significant contribution to the
study of testers. They used six inbred lines, grouped according to their GCA
(low, intermediate, and high), crossed with ten heterozygous strains (five
paired crosses of high yielding selections HH and five paired crosses of low
yielding selections LL) obtained from a divergent recurrent selection program
for yield in Synthetic A. Comparisons were made considering the heterozygous
strains as testers for the six inbred lines (case I) and, conversely, considering the inbred lines as testers for the heterozygous strains (case II). In
case I, with few exceptions, ranking of the lines corresponded to that expected on the basis of the assumed levels of GCA for yield, for both high yielding
and low yielding heterozygous strains as testers. In case II, the correct
relative classification was given on the average by each of the inbred lines
as testers. Considerable overlapping of the ranges of the two groups occurred, but in both instances the testers seemed to give a reasonably correct
ranking of the genotypes.

The authors also pointed out that a second requirement in a good tester
is precision in discriminating among genotypes under test; that is, the best
tester would be the one that would give most precise classification among entries for a given amount of testing, thus allowing the testing of a greater number of entries for a given degree of precision of estimates. They used the F-
test for among-entries variance σ_E^2 to compare relative efficiency of testers
for discriminating among genotypes. In case I, the F-test for the hypothesis
$\sigma_E^2 = 0$ showed that four testers (one of HH group and three of LL group) gave
significant differences among inbred lines. In case II, inbred lines used as
testers (two high, one intermediate, and one low in GCA) measured the differences among the ten intercrosses as significant. Using the Shuman-Bradley
(1957) test of significance for relative sensitiveness of two methods of measurement, they found that only one of the possible comparisons for sensitivity
(that involving the inbred lines NC44 and NC216) was significant. On the other hand, average sensitivity for the LL group in case I was 28% greater than
average sensitivity of the HH group of heterozygous strains used as testers.
In case II, average sensitivity of intermediate and low combining testers was
98% and 90%, respectively, greater than average sensitivity of high combining
testers. Overall results were in favor of the theory that low performing
testers, presumably with low frequency of favorable alleles at important loci,
are the most effective.

Theory supporting Rawlings and Thompson's (1962) results was basically an
extension of Hull's hypothesis. They assumed a model with no epistasis and
examined genetic variability among testers for different levels of dominance
and different gene frequencies in the testers. After adaptation to notation
used in Chap. 2, genetic variance among testcross progenies for one locus and
for a particular tester can be expressed as

$$\sigma_t^2 = (1/2)p(1 - p)(1 + F)[a + (1 - 2r)d]^2$$

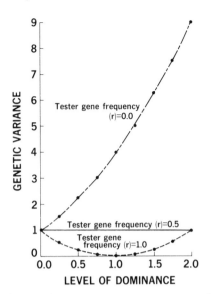

Fig. 8.1. Relative genetic variance among testcrosses for three tester gene
           frequencies and different levels of dominance.

which gives total genetic variance after summation over all loci. For a given
population with fixed values of p, F, a, and d, it is clear that variance
among testcrosses will depend on gene frequency r in the tester. With no dominance (d = 0), genetic variance among testcrosses is constant for any gene
frequency in the tester, i.e., $\sigma_t^2 = (1/2)p(1 - p)(1 + F)a^2$. For any level of
positive dominance (d > 0), genetic variance for tester gene frequency r = 0
will always be greater. For gene frequency r = 0.5 in the tester, σ_t^2 has a
constant value for any level of dominance. For gene frequency r = 1.0 in the
tester, σ_t^2 is zero for complete dominance and equals the value for r = 0.5 for
levels of dominance either 0 or 2. Such relations are presented graphically
in Fig. 8.1 from results of Rawlings and Thompson (1962).

Figure 8.1 shows that a low gene frequency in the tester will give greater variances in the range of partial to complete dominance of genes. High
gene frequency in the tester may give greater variances if overdominance is of
considerable importance. Several reports have shown evidence that genes controlling yield and other important traits in maize are mostly additive and
dominance ranges from partial to complete (see Chap. 5). Therefore, the hypothesis under which a low yielding tester is better than a high yielding
tester is more likely to be true. Although Rawlings and Thompson's results
showed a general trend in favor of their hypothesis, they were considered not
conclusive by the authors because of lack of significance in most comparisons
relative to efficiency of detecting differences among testers. It was recognized that epistasis might completely confound results expected on the basis
of only an additive dominance model. However, Chap. 5 shows that epistasis
seems to be of minor importance in grain yield and other maize characters.

8.1.2 Expected Changes in Gene Frequency
       Comstock (1964) used the same model to analyze the tester problem by com-
paring expected changes in gene frequency Δp after selection among testcrosses

for different testers. It was shown that Δp is proportional to a + (1 - 2p)d (according to notation in Chap. 2) when the parental population is used as tester. The expected change Δp is proportional to a + (1 - 2r)d = a + (1 - 2p)d + 2(p - r)d when an unrelated population is used as tester (see Table 6.15). Thus it is clear that change in gene frequency Δp is expected to be greater using an unrelated tester whenever the quantity p - r is on the average positive and of sufficient magnitude that Δp is significantly larger than when using the parental population as the tester. This situation would be more likely to occur with a poor tester; with low gene frequency at important loci a poor performance tester is used rather than a good performance tester.

Lopez-Perez (1979) conducted a comprehensive study that included 50 unselected S_1 and 50 unselected S_8 lines crossed with five testers that were selected for their expected differences in gene frequencies for yield. The source population for the unselected sample of lines and four of the five testers were Iowa Stiff Stalk Synthetic (BSSS). The samples of 50 lines were from a larger group of 250 S_1 and 247 S_8 lines developed by single-seed descent from Iowa Stiff Stalk Synthetic (Hallauer and Sears 1973). The 50 S_8 lines were direct descendants of the 50 S_1 lines. The five testers were BSSS, the parental source population; BS13(S)C1, an improved strain of BSSS (Hallauer and Smith 1979); BSSS-222, an S_8 line derived from BSSS and identified as one of the poorest yielding lines per se; B73, a high performance line derived from BS13(HT) after five cycles of half-sib recurrent selection; and Mo17, an unrelated line developed by pedigree selection from 187-2 × C103. The 500 testcrosses (50 S_1 lines and 50 S_8 lines crossed to five testers) were evaluated in five experiments to estimate and compare the variability among the testcrosses for each tester.

Estimates of components of variance of S_1 and S_8 lines and their interactions with the five testers are recorded in Table 8.2 for six traits. Estimated components of variance among S_1 and S_8 lines agree with expectations, assuming gene frequency was 0.5 or no dominance effects (see Chap. 2); i.e., variation among S_8 lines was about double that among S_1 lines. Estimates of components of variance of S_1 lines and their interactions with testers were nearly equal, whereas interactions of S_8 lines with testers were less than the line component of variance. Hence, additive effects seem more important than nonadditive effects in causing differences among the S_1 and S_8 lines. Analyses of variance indicated that differences among lines and their interactions with testers were highly significant ($P \leq 0.01$) in all instances except for percentage of root lodging for S_1 lines, which was significant at the 5% level.

Table 8.2. Estimates of components of variance and standard errors for S_1 and S_8 lines and their interactions with testers.

Traits	Lines		Lines × Testers	
	S_1	S_8	S_1	S_8
Yield	8.9 ± 2.5	23.3 ± 5.9	8.7 ± 2.1	13.2 ± 2.7
Grain moisture	0.5 ± 0.1	0.8 ± 0.2	0.2 ± 0.1	0.2 ± 0.1
Days to midsilk	0.7 ± 0.2	1.0 ± 0.3	0.4 ± 0.1	0.4 ± 0.2
Root lodging	1.2 ± 0.8	5.1 ± 1.7	1.1 ± 1.1	2.6 ± 1.2
Stalk lodging	11.0 ± 3.0	24.5 ± 6.7	5.6 ± 1.9	9.9 ± 2.4
Dropped ears	0.4 ± 0.1	0.9 ± 0.3	0.4 ± 0.2	0.8 ± 0.3

Table 8.3. Estimates of components of variance and standard errors for yield (q/ha) for S_1 and S_8 line testcrosses ($\hat{\sigma}_T^2$) and their interactions with environments ($\hat{\sigma}_{TE}^2$) for five testers.

| | Components of Variance | | | |
| | $\hat{\sigma}_T^2$ | | $\hat{\sigma}_{TE}^2$ | |
	S_1	S_8	S_1	S_8
BSSS	18.4 ± 6.5	41.7 ± 12.0	10.5 ± 6.2	21.4 ± 7.4
BS13(S)C1	10.6 ± 4.5	34.1 ± 10.2	3.7 ± 5.5	16.0 ± 6.8
BSSS-222	21.7 ± 6.4	38.7 ± 11.1	-6.9 ± 4.4	18.0 ± 7.1
B73	3.8 ± 3.5	26.4 ± 8.0	12.9 ± 6.5	6.2 ± 5.8
Mo17	25.9 ± 8.5	29.8 ± 9.5	17.2 ± 7.0	21.8 ± 7.5
Average	16.1 ± 5.9	34.1 ± 10.2	7.5 ± 5.9	16.7 ± 6.9

Analyses of variance showed that variation among testcrosses was highly significant for all testers except for B73 S_1 testcrosses. Estimates of components of variance for S_1 and S_8 testcrosses for each tester show that variation among S_8 testcrosses was, on the average, twice that among S_1 testcrosses (Table 8.3). For the four related testers, estimates compare favorably with those expected relative to choice of testers; i.e., BS13(S)C1 (the improved strain of BSSS) and B73 (a good performance line) had less variability among testcrosses than BSSS (the original population) and BSSS-222 (a poor performance line). Hence estimates indicate less variability among testcrosses for testers expected to have a greater frequency of favorable alleles. Mo17 was unrelated to the lines under test, and variability among S_1 Mo17 testcrosses was similar to that for BSSS and BSSS-222, the two testers expected to have a lower frequency of favorable alleles. At the S_8 level, estimates for Mo17 and B73 were similar but less than those for the other three testers. The average variation among S_1 and S_8 testcrosses had the same trend as the variation among S_1 and S_8 lines (Table 8.2); this also indicates additive effects were primarily responsible for differences among testcrosses.

Estimates of the tester-environment interactions were, on the average, about one-half the variation among testcrosses (Table 8.3). At the S_1 level, only testcrosses involving B73 and Mo17 had interactions with environments significantly different from zero, but the B73 testcrosses were the only group that had a nonsignificant interaction at the S_8 level. Testcross-environment interactions at the S_8 level averaged about twice those at the S_1 level: 7.5 for S_1 vs. 16.7 for S_8. There were no trends among testers in their responses to different environments relative to choice of testers. Estimates of components of variance for testcross-environment interactions, however, were about one-half the estimates of components of variance for testcrosses.

Matzinger (1953) reported that as the heterogeneity of the testers increased, the component of variance of the line-tester interaction decreased. Comparisons of estimated components of variance for line-tester interactions support his conclusions for related line testers (Table 8.4). Estimates of the tester-line interaction for the two inbred lines were about twice those for the two broad genetic base testers. Also, tester-line interactions of S_8 lines was about three times greater than for the S_1 lines for both groups of testers. The tester-line interaction for the two elite line testers (B73 and

Table 8.4. Tester-by-line interaction components of
            variance for three groups of testers.

| Testers | Tester-by-line | |
	S_1	S_8
BSSS and BS13(S)C1	4.2 ± 3.0	12.3 ± 4.8
BSSS-222 and B73	7.3 ± 3.6	20.8 ± 6.6
B73 and Mo17	8.7 ± 4.4	6.8 ± 4.0
Matzinger (1953)		
Inbred	--	17.2
Single cross	--	11.9
Double cross	--	6.5

Mo17) was similar for the S_1 and S_8 lines. For the 50 unselected lines included in testcrosses, tester-line interactions were greater for narrow genetic base testers, but interaction components of variance were smaller than
testcross components of variance (Table 8.3). It seems that narrow genetic
base testers can be effectively used to identify lines having good GCA. The
data for an unselected group of lines support those reported by Sprague and
Tatum (1942). Additional data of the study were reported by Lopez-Perez
(1979) and Hallauer and Lopez-Perez (1979); all the evidence was in agreement
with Hull's (1945) hypothesis that the most efficient tester is one having a
low frequency of favorable alleles.

8.1.3 Expected Gain from Selection

Allison and Curnow (1966) reached practically the same conclusions in a
study of the effect of different testers on expected progress after selection
among testcrosses. It was shown that expected change in the population mean
$\Delta_G = (k/\sigma)p(1 - p)[a + (1 - 2p)d][a + (1 - 2r)d]$, according to notation used
in Chap. 2. This expression can be rewritten as

$$\Delta_G = (k/2\sigma)[\sigma_A^2 - 4p(1 - p)\alpha(r - p)d]$$

where $\alpha = [a + (1 - 2p)d]$ is the average effect of substitution of the favorable
allele and $\sigma_A^2 = 2p(1 - p)\alpha^2$ is the additive genetic variance in the noninbred
parental population. Assuming that $\sigma_A^2 > 0$, changes in population means using
various testers are as shown in Table 8.5.

It is clear that under no dominance there is no difference among testers.
With partial to complete positive dominance (d > 0), the best tester is homozygous recessive. Underdominance is not expected to be common, but if it occurs the best tester would be the homozygous dominant. The change in mean Δ_G
= 0 either for p = 0.0 or for p = 1.0 for any degree of dominance. However,
with overdominance there is an optimum gene frequency p_0 ($0 < p_0 < 1$), for
which $\Delta p = 0$. This point of equilibrium occurs when $p_0 = 0.5(1 + a/d)$, so an
increase in the population mean occurs when $p < p_0$ and selection is for high
yielding crosses or when $p > p_0$ and selection is for low yielding crosses. In
all instances the best tester is the recessive homozygote. The probability of
an inbred line being homozygous recessive for all loci is infinitesimally

Table 8.5. Expected genetic gain at one locus level in
a population for three types of testers.

Tester	Gene frequency	Expected change in population mean[†]
Homozygous dominant	$r = 1$	$\frac{k}{2\sigma}[\sigma_A^2 - 4pq^2\,\alpha d]$
Homozygous recessive	$r = 0$	$\frac{k}{2\sigma}[\sigma_A^2 + 4p^2q\,\alpha d]$
Parental population	$r = p$	$\frac{k}{2\sigma}[\sigma_A^2]$

[†]Adapted from Allison and Curnow (1966), where σ_A^2 is the
additive genetic variance at the locus and $\alpha = a + (q - p)d$ is
the average effect of substitution of the favorable allele.

small. Therefore the problem of tester choice is to find an inbred line homozygous recessive at the most important loci or a variety with low gene frequency at the most important loci. Despite the gene frequency in the tester,
the importance of a locus is related to its contribution to the phenotype--a
small increase in gene frequency would result in a relatively large contribution toward changing the phenotypic mean. However, important loci in the
tester are those whose contributions to increase the mean are of such magnitude to be significantly influenced by the gene frequency in the tester. This
problem cannot be studied in individual genes because they act together in the
control of a quantitative trait. From the theoretical point of view some comments can enhance understanding of the problem.

The formula for the expected change in phenotypic mean can be rewritten
as

$$\Delta_G = (k/\sigma)p(1 - p)[a^2 - 2pad - (1 - 2p)d^2] + (2k/\sigma)p(1 - p)[ad + (1 - 2p)d^2]s$$

Thus, the component of Δ_G that depends on gene frequency in the tester is
$(2k/\sigma)p(1 - p)[ad + (1 - 2p)d^2]s$, and the most important locus is that which
exhibits a maximum value for this component. Assuming that all genes have the
same effect, the maximum coefficient for $s = 1 - r$ will depend on gene frequency and degree of dominance. Some values for these coefficients are shown
in Table 8.6.

For example, at a degree of dominance of 0.5, for a given gene frequency
in the tester and a given level of additive effect, maximum change in the mean
as affected by the tester would be due to an initial gene frequency $p = 0.392$.
Under the same conditions, a gene at frequency, say, 0.8 or 0.2 would be less
important in the tester; i.e., the change in mean could be greater but the
portion of these changes that depends on tester gene frequency would be smaller. For no dominance ($d = 0$) all values equal zero in Table 8.6. For positive dominance ($d > 0$) the relative importance of a gene in the tester increases with the degree of dominance for gene frequencies $p \leq 0.5$, but there
is a maximum for gene frequencies $p > 0.5$ in the population.

Allison and Curnow's (1966) general results led to the conclusion that if
there is no overdominance any tester leads to an improvement in the population
mean except for the case of complete dominance and genes fixed at frequency $p
= 1$ at all loci in the tester. However, with overdominance the choice of a
more suitable tester may lead to a decrease in mean yield unless the direction
of selection is changed. Low gene frequency for the most important loci was
shown to be desirable in the tester. However, because the choice of a tester

Table 8.6. Coefficients for s (frequency of recessive allele in the tester) in Δ_G as a function of gene frequency in the population (p) and degree of dominance (d/a) for a fixed value of $(2ka^2/\sigma)$.

Gene frequency (p)	Degree of dominance (d/a)							
	0	0.2	0.4	0.5	0.6	0.8	1.0	1.2
0	0	0	0	0	0	0	0	0
0.1	0	0.021	0.048	0.063	0.080	0.118	0.162	0.212
0.2		.036	.079	.104	.131	.189	.256	.330
0.3		.045	.097	.126	.156	.222	.294	.373
0.4		.050	.104	.132	.161	.223	.288	.357
0.5		.050	.100	.125	.150	.200	.250	.300
0.6		.046	.088	.108	.127	.161	.192	.219
0.7		.039	.071	.084	.096	.114	.126	.131
0.8		.028	.049	.056	.061	.067	.064	.054
0.9	0	0.015	0.024	0.027	0.028	0.026	0.018	0.004
1	0	0	0	0	0	0	0	0
p (at maximum)	--	0.451	0.410	0.392	0.377	0.352	0.333	0.319

by its performance is to a great extent a matter of chance, the choice of the parental variety as tester assures effective selection; only in this case would a high frequency of the recessive allele in the tester always be associated with a high frequency of the recessive allele in the material being tested. An unrelated, low performance variety would be a good tester only if the low performance is due to a low frequency of genes at important loci.

Cress (1966) emphasized that the choice of a tester to maximize gain from selection for a heterogeneous population depends on the average performance of the testcrosses; i.e., the tester with the highest average cross performance is chosen. Nevertheless, "unless the selected individuals are to be used immediately in hybrid combination with the tester, this emphasis on heterotic response is misplaced. The heterotic response reveals little concerning the genetic potential and nothing concerning the expected rate of progress from selection."

Evidence presented by Rawlings and Thompson (1962), Comstock (1964), and Allison and Curnow (1966) all lead to similar conclusions with respect to the choice of the better tester; i.e., either an inbred line homozygous recessive or a population with low gene frequency at important loci is the most effective tester both for discriminating among inbred lines in a hybrid breeding program and for population improvement in a recurrent selection scheme. All available evidence also is in agreement with Hull's hypothesis.

8.1.4 Reciprocal Recurrent Selection

In a reciprocal recurrent selection program there is no opportunity for the choice of testers because each population is the natural tester of its counterpart. However, results reported by Darrah et al. (1972) and Horner et al. (1973) showed that the genetic variance among testcrosses was about twice as large when inbred testers were used as when a noninbred population tester was used in intrapopulation recurrent selection programs. These findings led Russell and Eberhart (1975) to propose that progress in RRS would be greater if inbred lines extracted from the parental populations were used as testers instead of the populations themselves, as in the original procedure proposed by Comstock et al. (1949). Comstock (1977) further analyzed, theoretically, the possible implications of such a modification. Theory based on a model that assumes no multiple alleles, no epistasis, and linkage equilibrium provides the expressions shown in Table 8.7 for variance among testcrosses and expected change in gene frequency at the one locus level.

If inbred lines are extracted from the tester population, there is a probability r that they are homozygous for the better allele AA and a probability 1 − r that they are homozygous for the poorer allele aa. Thus expected genetic variance among testcrosses using an inbred line as tester is

$$V_I = rV_1 + (1 - r)V_0 = (1/2)p(1 - p)[a^2 + 2(1 - 2r)ad + d^2]$$

and it is clear that $V_I = 2Vp$ when $a^2 + 2(1 - 2r)ad + d^2 = 2[a + (1 - 2r)d]^2$; i.e., when $r = (1 + c) \pm \sqrt{2c^2 - (1/4)c}$, where $c = d/a$ is a measure of the degree of dominance (equivalent to a in Comstock's notation). A solution for r is possible only for $c \geq 0.5$, as shown by Comstock (1977). On the other hand, the expected change in gene frequency using inbred lines from the tester population is

$$E(\Delta p) = r\Delta_1 + (1 - r)\Delta_2 = [k/(2\sigma)]p(1 - p)[a + (1 - 2r)d]$$

which equals expected change in gene frequency when using the opposite noninbred population as tester. Such a comparison does not take into account differences in $k/(2\sigma)$ values; k is a function of selection intensity and may be taken as a constant and σ is the phenotypic standard deviation of the testcross means, which is not expected to vary greatly mainly if environment is the most important source of variation.

Because inbred lines are extracted at random, it may occur that a particular sample gives greater changes in gene frequency because of an excess of recessive alleles over the expected number. In terms of expectation there is no a priori reason to expect better results by use of inbred lines rather than the population as tester. Comstock (1977) presented some evidence favoring use of a population as tester in an RRS program. First, phenotypic variance among testcrosses is smaller when the population is used as tester and genotype-environment interaction is also likely to be smaller. Second, variance of expected change in gene frequency $V(\Delta p)$ is smaller for the population as

Table 8.7. Genetic variance and expected change in gene frequency for three different testers (adapted from Comstock 1977).

Tester	Genetic variance	Expected change in gene frequency [$E(\Delta p)$]
Opposite population	$Vp = (1/2)p(1-p)[a+(1-2r)d]^2$	$\frac{k}{2\sigma}p(1-p)[a+(1-2r)d]$
Inbred homozygous for better allele	$V_1 = (1/2)p(1-p)[a-d]^2$	$\frac{k}{2\sigma}p(1-p)[a-d] = \Delta_1$
Inbred homozygous for poorer allele	$V_0 = (1/2)p(1-p)[a+d]^2$	$\frac{k}{2\sigma}p(1-p)[a+d] = \Delta_2$

tester and so probability of fixation of poorer alleles would be slightly less.

Comstock (1979) further extended the comparisons for multiple alleles to support his conclusions that, on the average, rates of change in allele frequencies in RRS programs will not be more rapid when inbred lines extracted from the RRS populations are used as testers rather than the populations themselves. Comstock (1979) also emphasized that when comparing testers to be employed in RRS, the critical parameter is expectation of allele frequency change per unit of time. The suggestion by Russell and Eberhart (1975) would be a consequence of the erroneous assumption that "because of the greater variance among the testcrosses with the inbred tester, gain from selection would be greater."

8.1.5 Combining Ability in Testcrosses

The main difference between general vs. specific combining ability has been attributed to the tester genetic basis, i.e., whether broad or narrow genetic base tester is used. Such differences are essentially a matter of differences in gene frequency. In the broad base tester gene frequencies for different loci may vary from 0 to 1, whereas in the narrow base tester gene frequencies are limited to a few values. For an inbred line, allele frequency at a locus may be 0 or 1 and in the single cross of homozygous lines allele frequencies may be 0, 0.5, or 1. In either case (broad or narrow base tester) selection can lead to change in the population mean as a result of selection for additive effects of genes. Therefore, when considering genotypes in a population or in other genetic materials that are to be evaluated in testcrosses, it seems difficult to distinguish for general vs. specific combining ability and the expression *combining ability* should be used with a broader meaning.

The combining ability of genotypes in a population or in any other genetic materials can be estimated as shown in Table 8.8.

Genetically, combining ability is measured as

$$c_i = \overline{T}_i - \overline{T}_{.} = (p_i - \overline{p}_{.})[a + (1 - 2r)d]$$

where r is the frequency of the favorable allele in the tester. Combining ability c_i is independent from dominance effects only for r = 0.5 or for d = 0 (no dominance). When gene frequency is fixed in the tester (r = 1.0), $c_i = (p_i - \overline{p}_{.})(a - d)$; and for complete dominance (d = a), $c_i = 0$ for every i. In this case the tester is not able to discriminate among lines or populations being tested.

Variance among testcrosses σ_t^2 (same as σ_f^2 in Table 6.3) is an estimate of variance for combining ability σ_c^2. Genetically, we have $\sigma_c^2 = \sigma_p^2[a + (1 - 2r)d]^2$ for any material evaluated in testcrosses (R. Vencovsky, personal communication, 1978). It can be seen that σ_c^2 depends on the square of the average effect of gene substitution in the tester and on the variance of average gene frequencies of the genetic materials under test. For the particular case of genotypes within a population, gene frequencies (for a favorable allele B) within genotypes are 1, 0.5, and 0 in BB, Bb, and bb, respectively. Because gene frequency has a binomial distribution, $\sigma_p^2 = p(1 - p)/2$; and $\sigma_c^2 = [p(1 - p)/2][a + (1 - 2r)d]^2$, which is the variance among testcrosses (see Chap. 2). If the genotypes represent a random sample of inbred lines (F = 1), the genotypic array is p(BB):(1 - p)(bb) with gene frequency 1 and 0 within genotypes, respectively. Thus $\sigma_c^2 = p(1 - p)[a + (1 - 2r)d]^2$ for a random sample

Table 8.8. Combining ability of genotypes in a population or a set of populations (Y) in topcross with a population as tester.

Testcrosses	Plants in a population			Set of populations[+]			Mean of testcrosses	Combining ability (c_i)
	Genotype	Frequency	f_g	Population	Frequency	f_y		
1	BB	p^2	1	P_1	$1/n$	p_1	$\bar{T}_1$	$\bar{T}_1 - \bar{T}.$
2	Bb	$2pq$	0.5	P_2	$1/n$	p_2	$\bar{T}_2$	$\bar{T}_2 - \bar{T}.$
3	bb	q^2	0	P_3	$1/n$	p_3	$\bar{T}_3$	$\bar{T}_3 - \bar{T}.$
·				·				
·				·				
n	--	--	--	P_n	$1/n$	p_n	$\bar{T}_n$	$\bar{T}_n - \bar{T}.$
Average	--	--	p	--	--	$\bar{p}.$	$\bar{T}.$	$\bar{T}.$

[+] Populations with genotypes in Hardy-Weinberg proportion; f_g and f_y is the average gene frequency within genotypes and within populations, respectively.

of inbred lines. In general, for any level of inbreeding F in the base population the variance for combining ability is

$$\sigma^2_c = [p(1 - p)/2](1 + F)[a + (1 - 2r)d]^2$$

The variance σ^2_c must be extended to summation over all loci, and when inbred lines are used as testers the variance will depend on the balance of the
number of loci with frequencies 1 or 0. An excess of important loci with frequency 0 will always be advantageous relative to increasing the variance and
changing the population mean through selection.

All the theoretical aspects so far presented deal with simple models at
one locus level, so it is not possible to predict the relative magnitude of
variances and expected changes in gene frequencies when results have to be extended to all loci controlling the character. Although genes controlling a
quantitative trait cannot be handled individually, it is reasonable to suppose
that they differ from one another with respect to magnitude and type of gene
action and that complex interactions may exist among them. It is also reasonable to suppose that genes occur with different frequencies within the popula-
tion--frequencies varying from 0 to 1 according to some definite distribution;
it is more likely to suppose that in heterogeneous populations most genes are
at intermediate frequencies and few genes have frequencies near the fixation
(0 or 1). Therefore, additional theory for the study of problems (such as
that presented in this section) considering the whole population of genes and
their properties would be straightforward and should provide additional important tools for the use of quantitative genetic theory in maize breeding. Additional experimental results would be required to support the current and future theoretical studies and contribute to the formulation of more adequate
models.

8.2 COROLLARY STUDIES

Corollary studies related to the choice of testers have been conducted to
determine: (1) relation of line performance to hybrid performance, (2) effectiveness of visual selection for developing lines that have good combining
ability, (3) genetic diversity of the lines being tested, (4) appropriate
stage of inbred line development for testing for combining ability, and (5)
relative importance of general and specific combining abilities. Any one
study, however, does not provide information on all the above five items.
Data are often conflicting and interpretations differ. In some instances relative merits of the art and science of maize breeding are in conflict.

Development of new lines has been found to be considerably easier than
determination of their worth in hybrids; all lines are not equally good in hybrids and methods are needed to screen for elite lines for potential use in
hybrids. During inbred line development it is a common practice to discard
lines that have obvious morphological deformities, are difficult to maintain
because of either poor pollen production or ear shoot development, and are obviously susceptible to diseases and insects, have poor plant vigor, etc. Phenotypic elimination of lines is effective for many traits for use in hybrids,
but effective phenotypic elimination of lines that have poor combining ability
has not been resolved. Some traits, such as disease and insect resistance,
are used in discarding lines because they are not acceptable for commercial
use although they may have good yield potential in hybrids in a pest-free environment.

8.3 CORRELATIONS BETWEEN LINES AND HYBRIDS

Relations of plant and ear traits and yield of inbred lines to the same traits in hybrids have been studied from the time of initiation of inbred and hybrid development programs to the present. Because yield trials are expensive to conduct, any information on inbred lines that is indicative of their performance in hybrids is desirable in order to eliminate the need for making crosses and conducting yield trials. Hence, it is desirable to investigate possible methods to reduce testing of inbred lines in hybrids and to determine if expression of traits in inbred lines is transmissible to their hybrids.

Correlation studies between inbred line traits and either the same or different traits in their hybrids have been used to determine the effectiveness of selection on hybrid performance. A brief summary of some correlation studies is given in Table 8.9. The correlations are independent of those in Table 5.17, which were obtained from progenies developed by use of mating designs. The correlations in Table 8.9 were, with one exception, obtained from what was considered the best and most vigorous inbred lines that had survived selection and their crosses--in most instances single crosses. Consequently, sample size usually was small and the correlations cannot be interpreted relative to some population. Simple phenotypic correlations were calculated in most instances.

Kiesselbach (1922), Richey (1924), and Richey and Mayer (1925) reported evidence that certain inbred lines were better than others in transmitting high yielding ability to their crosses. Although there was a general tendency for some inbred lines to be better than others in crosses, Richey and Mayer (1925) emphasized that the lack of any definitive correlation between yields of parent inbred lines and their crosses indicates that selection for combining ability in the final analysis must be based on the performance of the lines in crosses rather than on the inbred lines. Richey (1924) also emphasized that the best chances of success seem to be in the use of large numbers and in avoiding preconceived notions as to what are the best lines.

Jorgenson and Brewbaker (1927), Nilsson-Leissner (1927), Jenkins (1929), Johnson and Hayes (1936), and Hayes and Johnson (1939) studied the relation of traits of inbred lines to the same traits in their crosses and to yield of their crosses. In most instances simple correlations were obtained that were considered of sufficient magnitude to be useful in selection, but Johnson and Hayes (1936) concluded that yield of inbred lines was not significantly correlated with combining ability of topcrosses. Relatively large multiple correlations between yield of crosses with yield and other traits of inbred lines were obtained by Jorgenson and Brewbaker (1927), Nilsson-Leissner (1927), and Hayes and Johnson (1939); in all instances R values indicated that more than 50% of the total variation in yield of the crosses was accounted for by inbred line traits ($R = 0.82$ and 0.67 for flint and dent; $R = 0.61$; and $R = 0.67$, respectively).

Jenkins (1929), in a comprehensive study for the correlation of inbred traits with the same traits in their hybrids, determined the correlations between: (1) various traits of parental inbred lines and the same traits of the crosses, and (2) traits of parental inbred lines and means of the same traits of all crosses that included the inbred line common in all crosses. Positive correlations were found for all 19 traits studied but they were small in most instances. In the first case, none of the characters of the parental inbreds was closely related to its F_1 cross; correlations ranged from -0.10 to 0.24. Correlations between yield of parental inbred lines and yield of their F_1 crosses were 0.14 for case 1 and 0.20 for case 2 (Table 8.9). Multiple corre-

Table 8.9. Summary of six studies reporting the relation between traits of inbred lines and the same traits of their crosses.

	Nilsson-Leissner (1927)	Jorgenson & Brewbaker (1927)	Johnson & Hayes (1936)	Hayes & Johnson (1939)	Jenkins (1929)[†]		Gama & Hallauer (1977)[‡]		
					1	2	1	2	Avg.
Yield	.46 ± .05	.50 ± .08	-.02	.25	.14 ± .02	.20 ± .03	.09	.11	.22
Ear length	.73 ± .02	.58 ± .08	.28	.46	.30 ± .02	.43 ± .03	.21**	-.06	.37
Ear diameter	.92 ± .01	.63 ± .06	--	--	.35 ± .02	.40 ± .03	.18*	-.02	.41
Kernel row no.	.95 ± .01	.79 ± .04	--	--	.47 ± .02	.67 ± .02	--	--	.72
Kernel depth	--	--	--	--	--	--	.25**	.05	.15
Plant height	.75 ± .02	.48 ± .08	.28	.37	.32 ± .02	.45 ± .03	.39**	.07	.39
Days to flower	--	.29 ± .09	--	.57	.24 ± .02	.34 ± .03	.28**	.28**	.33
Second ears	.48 ± .08	--	-.26[‡]	--	.26 ± .02	.36 ± .03	--	--	.21

[†]The left-hand correlations are between one parental inbred line and the mean of all its crosses; the right-hand correlations are between the mean of the two parental inbreds and their specific cross.
[‡]Ears from tillers included.

lations for different combinations of inbred line traits and yield of their F_1 crosses ranged from 0.20 to 0.42. In the second case, correlations were generally larger, ranging from 0.25 to 0.67. Because of the positive correlations between traits of inbred lines and their crosses, Jenkins concluded that the correlations were sufficiently large in some instances to have predictive value. Gama and Hallauer (1977) reported a study similar to that of Jenkins for unselected inbred lines and their single-cross hybrids developed from Iowa Stiff Stalk Synthetic. Genetic correlations were determined for eight traits and they were too small in all instances for predictive purposes; correlations for yield, for example, were only 0.09 and 0.11 for cases 1 and 2, respectively. Multiple correlations of plant traits, ear traits, and plant and ear traits of inbred lines with yield of their crosses were also small.

Kyle and Stoneberg (1925), Hayes (1926), Jenkins (1929), Kovacs (1970), Obilana and Hallauer (1974), and Bartual and Hallauer (1976) determined correlations among traits of inbred lines; and Kempton (1926), Jenkins (1929), El-Lakany and Russell (1971), and Silva (1974) determined correlations among traits of hybrids. Correlations by Hayes (1926) were for eight traits of inbred lines in different generations of inbreeding. Correlations of ear length and number of ears with yield were significantly positive. Jenkins (1929) and Kovacs (1970) determined correlations for several traits with yield of inbred lines in the same generation of inbreeding. Jenkins (1929) reported that yields of inbred lines were positively and significantly correlated with plant height, number of ears per plant, ear length, ear diameter, and shelling percentage; but they were negatively and significantly correlated with date of silking, ear shrinkage, chlorophyll grade, and ear shape index.

Kovacs (1970) reported data for 5 ear traits and yield for a selected group of inbred lines in a breeding study. Average correlations for ear traits of four groups of inbred lines with yield were 0.68 (ear length), 0.56 (number of kernel rows), 0.74 (number of kernels), 0.72 (kernel length), and 0.59 (thousand-seed weight), although there were large differences in the correlations among the four groups of inbred lines that were considered good and poor for yielding ability. Genetic correlations among 11 traits and the same traits with yield reported by Obilana and Hallauer (1974) and Bartual and Hallauer (1976) were for two sets of unselected inbred lines derived from Iowa Stiff Stalk Synthetic. Except for some of the correlations among ear components and ear components with yield, most correlations were small; kernel depth (0.82 and 0.76) and kernels per row (0.86) had the greatest correlations with yield.

Additional correlation studies were given by Love (1912), Collins (1916), Love and Wentz (1917), Etheridge (1921), Kempton (1924, 1926), and Richey and Willier (1925). The studies calculated the correlations primarily to determine if some ear and plant traits could be used for yield improvement. Detailed data were taken on several different traits and positive and significant correlations usually were reported. Kempton (1926) reported significant correlations between length of central spike of tassel and total ear length of 0.272 (F_2 plants) and 0.340 (F_1 plants), both of which exceeded three times their standard errors. Similarly, significant correlations were obtained for number of tassel branches and number of rows on the ear (0.185 and 0.167), but further analysis by partial correlations showed no genetic correlation between number of rows on the ear with either season or grain yield. All correlations were too small to adequately predict future selection progress and correlations would be restricted to the F_2 population of a wide cross of the Mexican variety Jala and the popcorn variety Tom Thumb.

Love and Wentz (1917) summarized the literature and presented results from their studies for the traits listed on the early maize show cards (see

Table 1.1) and yield. All correlations were small and not consistent among
the different studies. Love and Wentz concluded: "(1) The judge at a corn
show or a farmer in selecting his seed corn cannot pick the high yielding seed
ears when judging from outward characters of the ears. It is evident that the
points emphasized on a score card are of no value for seed ear purposes and
are entirely for show purposes. (2) The only basis left for selecting high
yielding seed corn is the ear-to-row progeny test." These remarkable conclusions are equally valid today, although experimental techniques have been refined.

Although several reported correlations were relatively large, nearly all
the authors stated in their conclusions that comparative yield trials of the
hybrids are needed. It is necessary, however, that selection be practiced
during inbred line development. The ultimate use of inbred lines is in hybrids, and it is necessary that they possess certain standards of vigor and
productiveness as a line before testing. It may be an advantage to the breeder for some relationships if the inbred line traits are not correlated with
yield of their crosses because selection against many traits can be emphasized
without fear of discarding all the potentially high yielding genotypes. If
the correlation is small for traits that are not of economic importance (e.g.,
plant color), we would have a normal distribution on inbred lines with acceptable vigor and productiveness for testing in hybrids. For other traits
(e.g., stalk lodging) we would benefit from a correlation of the trait between
inbred lines and their hybrids because weak stalk inbred lines could be discarded before testing. Jenkins (1929), for example, reported a correlation of
0.88 ± 0.03 for percentage of erect plants at harvest between inbred lines and
their crosses, indicating that 77% of the variation in the crosses for erect
plants could be associated with the inbred lines used in the crosses. For
commercial usage, this correlation would be desirable in the inbred lines and
their crosses. In summary it seems that effective selection can be made for
certain traits, but the ultimate use of inbred lines in crosses must be determined from yield evaluations of the crosses.

8.4 VISUAL SELECTION

Studies determining the effectiveness of visual selection for inbred line
development were reported by Jenkins (1935), Sprague and Miller (1952), Wellhausen and Wortman (1954), Osler et al. (1958), and Russell and Teich (1967).
Except for recent studies for unselected inbred lines, all studies included
inbred lines that had undergone visual selection and comparisons were not possible between visual selection and no selection. Because some of the correlation studies showed a good relation between vigor and productiveness of the
inbred lines and their crosses, these investigators tested the effectiveness
of visual selection compared to testcross evaluation.

Jenkins (1935) compared seven pairs of lines from Iodent and five pairs
of lines from Lancaster in topcross evaluation trials for yield in generations
S_2 to S_8, omitting S_7. Each line in each generation was represented by a selected and a discarded sib strain. Yield tests of the Iodent lines showed
that for the first two generations of inbreeding average yield of selected
strains was significantly better than for discarded strains. For all generations of Lancaster lines and after the second generation of Iodent lines, topcross yields of the selected and discarded sib strains were not significantly
different but the small differences consistently favored selected strains.
Sprague and Miller (1952) also found no effect of visual selection on the combining ability of the selections during four generations of inbreeding for two
sets of progenies.

Brown (1967) tested topcrosses of 20 lines visually selected as the most desirable phenotypes and 20 lines visually selected as the poorest phenotypes of 1160 unselected lines from four open-pollinated varieties: Krug, Reid, Lancaster, and Midland. Yields were identical for the two groups of topcrosses (54.16 q/ha for selected vs. 54.15 q/ha for unselected). Brown (1967) concluded that "while it cannot be said that the type of selection practiced had any adverse effect on yield, neither did it contribute to any improvement in yield." (Topcrosses of visually selected lines, however, were superior to those of unselected lines for root and stalk lodging.) On the contrary, Wellhausen and Wortman (1954) and Osler et al. (1958) reported that visual selection resulted in small positive gains in combining ability of the selections.

Russell and Teich (1967) conducted a comprehensive study comparing inbred line performance and combining ability for yield of inbred lines developed by (1) visual selection within and among ear-to-row progenies during successive inbreeding generations and (2) selection within and among ear-to-row progenies based on testcross performance during successive inbreeding generations. The inbred lines were developed from an F_2 population of M14 × C103 under two plant densities: Low plant densities were 29,652 plants/ha for visual selection and 39,536 plants/ha for testcross selection; high plant density was 59,304 plants/ha for both selection schemes. Testcross evaluations were made in the F_2, F_3, and F_4 generations, whereas visual selection was practiced in the F_2 through the F_6 generations. Inbred selections for both selection schemes and for both plant densities, M14 and C103 inbred lines, and M14 × C103 were crossed to WF9 × I205 (tester for the testcross selection) and Ia4810 (double-cross hybrid chosen to give a measure of GCA). The inbred lines themselves also were evaluated. For simplification, the groups of materials were designated as follows:

group 0: inbred lines selected by testcross performance at both rates
groups 1, 2: inbred lines selected by testcross performance in low (1) and
          high (2) rates
groups 3, 4: inbred lines visually selected in low (3) and high (4) rates
groups 5, 6, 7: testcrosses of M14 (5), C103 (6), and M14 × C103 (7)

A comparison of group means showed that selection as measured by combining ability was effective for groups 0, 1, 2, and 4 but not for group 3 (visual selection at low density) when compared with (M14 × C103) × tester. Group 0 lines were superior to groups 1, 2, 3, and 4 lines, and the lines in group 0 showed less interaction with environments. Although Russell and Teich (1967) concluded that visual selection of inbred line performance in high stands was as effective as selection by extensive testcrossing, it seems that more extensive evaluation of the lines in group 0 resulted in higher yielding genotypes with greater stability over environments. Because testcross evaluations were at two plant densities, additional information on environmental response was available for the lines in group 0. Inbred line yields of groups 3 and 4 were superior to yields of groups 1 and 2, and yields of inbred lines selected at high density were superior to those selected at low density.

Visual selection is always practiced by the maize breeder during each generation of inbreeding for inbred line development. How effective visual selection is for the expression of yield in crosses remains questionable; the evidence shows at least that it is not detrimental in crosses for yield.

Attempting to measure effectiveness of visual selection in inbred line development is difficult because of the phenomenon of sampling. In some sets of selected materials, a relation may be obtained between some traits of the inbred lines and expression for yield in crosses, whereas in others there may

be no relation. The differences may be due to sampling and the original assemblage of genes in the population. For an unselected set of inbred lines,
Gama and Hallauer (1977) failed to detect any correlations of predictive value.

It seems that the art of plant breeding, or visual selection, has a place
in maize breeding and certainly does no harm. Because vigor and productiveness of inbred lines are important for the production of commercial single
crosses, it is imperative that inbred lines whose traits do not impair the final products (high performance hybrids) be easily reproducible to plant on
sizable areas. Visual selection in modifying traits of the lines themselves
is generally accepted, but how this influences combining ability is not determined.

8.5 GENETIC DIVERSITY

Importance of genetic diversity of inbred lines used in crosses is generally accepted. In the past, general experience usually but not always showed
that crosses of unrelated genotypes contributed to greater yields. Two inbred
lines when crossed to produce an exceptionally good yielding single cross were
said to "nick" well. It usually required extensive evaluation trials to determine the unique combination of two parental inbred lines; hence the development of the inbred-variety cross technique for selecting inbred lines that
have the greatest potential. Although the genetic basis of the importance of
genetic diversity was not clear, general experience showed that the better
double crosses involved inbred lines derived from two or more varieties (Hayes
1963).

Empirical data on the effects of genetic diversity of parental inbred
lines on hybrid performance were given by Hayes and Johnson (1939), Wu (1939),
Eckhardt and Bryan (1940a, 1940b), Johnson and Hayes (1940), Cowan (1943), and
Griffing (1953). In all instances parental inbred lines of related origin
consistently produced lower yielding crosses than those that had one or no
parent lines in common. There were some exceptions, but the greater frequency
of high yielding crosses involved lines from different varieties. Most of the
varieties were of U.S. Corn Belt origin, but Griffing (1953) showed that parental lines including germplasm from outside the U.S. Corn Belt also contributed to the crosses of diverse origin.

The importance of maintaining genetic diversity of the parent inbred
lines was emphasized by the North Central Corn Breeding Research Committee
(NCR-2) in 1939 when it developed the A and B grouping of inbred lines. The
designation of inbred lines in the A and B grouping was somewhat arbitrary,
but it was an attempt to minimize gene exchange between the two gene pools.
Strict enforcement of this arbitrary grouping was impossible, but it was recognized that opportunities for the production of high yielding hybrids would
be enhanced by crossing and testing inbred lines from the two groups rather
than within the two groups. Of course, genetic erosion of the two groups is
continuing because of pedigree selection in F_2 populations of the best single-
cross hybrids; this is the cause of the increasing interest for introducing
exotic germplasm in the breeding programs. In the U.S. Corn Belt at the present it seems that inbred lines of Reid origin crossed with those of Lancaster
origin produce higher yielding crosses on the average. Similarly, crosses of
lines originating from flint and dent varieties are commonly used in Europe
and other areas of the world. Unlike correlation and visual selection, however, genetic diversity of the lines used in crosses is generally recognized to
be important.

8.6 TESTING STAGE

The stage of testing inbred lines for combining ability also has received attention. Inbreeding five to seven generations accompanied by visual selection usually was, and often still is, completed before attention was given to the performance of the inbred lines in crosses. Not until the publication of Jenkins's study (1935) was empirical evidence given comparing the testcrosses of lines in different generations of inbreeding. Richey and Mayer (1925) concluded from their study that comparison of crosses indicates no general advantage for crosses made after five generations of inbreeding over analogous crosses made after three generations of inbreeding. This suggested to them that there is little inherent relation between the yield of a cross and the number of generations of inbreeding of its parent lines before crossing.

Because correlation studies of inbred line traits with the same traits of their F_1 crosses (and hence visual selection during the inbreeding generations) did not seem to provide a satisfactory index of the yield potential of inbred lines in crosses, the next logical step was to determine if cross performance in the early generations of inbreeding was predictive of its performance at increased levels of inbreeding. It is generally acknowledged that isolation of genotypes from a breeding population (e.g., open-pollinated variety) depends on sufficiency of sampling. Because the potential ceiling of any derived line is determined at the time of the first selfing generation of an S_0 plant, it seemed logical that better sampling techniques could be used if material could be discarded in early generations of inbreeding. Thus greater attention and effort could be expended on selection within progenies of the genotypes saved on the basis of the preliminary early yield test information.

Jenkins (1935), in the study evaluating effectiveness of visual selection, reported data that supported early testing as an effective procedure for saving lines that have greater potential than others. Comparisons of selected and discarded sib strains for the seven lines of the Iodent series and the five lines of the Lancaster series showed that association with generations was not significant for the Iodent series and was not significant after the second generation of inbreeding for the Lancaster series. On the basis of his results, Jenkins concluded that "the inbred lines acquired their individuality as parents of topcrosses very early in the inbreeding process and remained relatively stable thereafter." His conclusion stimulated thinking about the merits of early testing to determine which progenies should be maintained in the nursery for further selection and inbreeding. In some respects, Jenkins's conclusion seemed reasonable because small populations of each progeny usually are included in the breeding nursery; because of low heritability of yield, linkage of genes, and limited effectiveness of visual selection, it seemed that the selection for more desirable gene combinations for yield would be small. The probabilities of success would decrease with each generation of inbreeding.

Sprague (1946) reported the results of a program designed to compare the relative merits of early testing, as suggested by Jenkins (1935) and Sprague (1939). Sprague (1946) selected 167 phenotypically desirable S_0 plants of Iowa Stiff Stalk Synthetic, selfed each S_0 plant, and outcrossed to the double-cross tester Ia13. Yields of the 167 testcrosses were normally distributed and ranged from 38.6 to 63.0 q/ha. The wide range of the selected S_0 plants for yield also is evidence of the poor relation between visual selection and yield of crosses. The difference (6.0 q/ha) necessary for significance at the 5% level showed that 4 of the crosses were significantly poorer than Iowa Stiff Stalk Synthetic and 2 were significantly better than the double cross Ia13.

Two samples based on the testcross information were subsequently studied: (1) S_1 lines representing the best 10% of the 167 testcrosses were self-polli-nated and crossed to Ia13, the tester; and (2) a group of 12 lines was chosen that represented a seriated sampling of the 167 testcrosses in which 20 self-pollinations and testcrosses were attempted in each family. Only six fami-lies ultimately were available for testing in the second group. Distribution of S_1 testcross yields clearly indicated that S_0 plants exhibiting high com-bining ability transmitted this trait to their S_1 progeny. Significant dif-ferences were noted among the 20 testcrosses within each of the six families, which suggested that heterozygosity was similar for S_0 plants having good and poor testcross yields. Distributions from the four best yielding families were not significantly different, but they were significantly different from the two poorest yielding families.

Finally, Sprague (1946) compared three of the selected samples in the S_3 generations in testcrosses with five standard lines. All possible crosses were produced and the three selected lines were superior to the standard lines for yield and resistance to root and stalk lodging. Sprague (1946) concluded from these data that early testing might be a useful tool in a breeding pro-gram and the data seem to support Jenkins (1935), Johnson and Hayes (1940), Cowan (1943), and Green (1948a, 1948b) that combining ability is a heritable trait.

Additional studies support and discourage early testing. Lonnquist (1950) conducted an early testing study for a series of selected plants from a strain of Krug. After the S_0 plant testcross performance data were available, Lonnquist practiced divergent selection for good and poor testcross yields for a group of lines that had the best and poorest S_0 testcross yields. Results from the first four generations of inbreeding showed that testcross combining ability could be modified by a combination of selection and testing. Diver-gent selection for good and poor combining ability for three generations with-in the original lines that had poor combining ability produced S_4 lines that were not significantly different from those in the good combining group se-lected for three generations for poor combining ability. Thus selection in the low combining group was not fruitful and additional selection and testing should be concentrated on the progenies that exhibit the best combining abili-ty in the S_1 generation.

Richey (1945), Singleton and Nelson (1945), and Payne and Hayes (1949) expressed doubt about the value of early testing for combining ability. Rich-ey's conclusions were based on a reanalysis of Jenkins's data indicating that his interpretations for the value of early testing were not warranted. Rich-ey's conclusions are somewhat surprising in view of the conclusions given by Richey and Mayer in 1925. Singleton and Nelson's data were for a group of se-lected lines continued for three generations of inbreeding; they concluded that early testing was ineffective, but there were no significant differences among the 10 lines studied. Payne and Hayes presented data comparing the S_0 and S_1 testcrosses and concluded that testing of S_0 plants was of doubtful value. However, as Sprague (1955) has pointed out, Payne and Hayes's data in-dicated that the S_0 plants with the best combining ability produced a greater percentage of good combining S_1 crosses than the S_0 plants with poor combining ability.

Lopez-Perez (1979) also was able to compare yields of the testcrosses of the same lines at the S_1 and S_8 levels of inbreeding (Hallauer and Lopez-Pe-rez 1979). Genetic correlations were determined between the S_1 and S_8 test-crosses, among testers at the S_1 and S_8 levels, and between S_8 testcrosses and yields of S_7 lines per se (see Table 8.10). There was no correlation between yields of S_7 lines per se and S_8 testcross yields (last column, Table 8.10).

Table 8.10. Genetic correlations between S_1 and S_8 testcrosses and between S_7 lines per se and S_8 testcrosses for five testers for yield.

Testers	Testers					Avg.	S_7 lines[+]
	BSSS	BS13(S)C1	BSSS-222	B73	Mo17		
BSSS	0.20[‡]	0.74	0.25	0.22	0.65	0.46	0.17
BS13(S)C1	0.68	0.17	0.71	0.48	0.91	0.71	0.10
BSSS-222	0.61	0.71	0.42	0.61	0.04	0.40	-0.09
B73	0.41	0.58	0.37	0.56	0.62	0.48	0.07
Mo17	0.64	0.68	0.77	0.78	0.35	0.56	-0.04
Average	0.58	0.66	0.62	0.54	0.72	0.34	0.04

[+]Correlations between S_7 lines per se and their respective S_8 testcrosses.

[‡]Correlations on diagonal are for S_1 vs. S_8 testcrosses; correlations above diagonal are among S_1 testcrosses; and correlations below diagonal are among S_8 testcrosses.

Genetic correlations between S_1 and S_8 testcrosses ranged from 0.22 for BSSS and B73 testcrosses to 0.91 for BS13(S)C1 and Mo17 testcrosses at the S_1 level and from 0.37 for BSSS-222 and B73 testcrosses to 0.78 for B73 and Mo17 testcrosses at the S_8 level. At both levels of inbreeding, the lowest correlation was between a low performance and a good performance tester and the highest correlation was between two good performance testers. Averaged over the four correlations for each tester, BS13(S)C1 at the S_1 level and Mo17 at the S_8 level had the highest correlations. The individual and average correlations were consistently greater at the S_8 level.

Correlations between S_1 and S_8 testcrosses for each tester are listed on the diagonal of Table 8.10. The correlations for each tester were too small to accurately predict S_8 testcross yields on the basis of S_1 testcross yields. Correlations for the narrow genetic base testers were greater than for the two broad genetic base testers. Although correlations between S_1 and S_8 testcrosses were too small for predictive purposes, several S_1 testcrosses were consistently ranked high by each tester and S_1 testcrosses also were good indicators of their S_8 testcrosses (Hallauer and Lopez-Perez 1979). The S_1 testcrosses for BSSS-222 (the poor performance line tester) correctly predicted S_8 testcrosses in 34 of 50 instances. The graphic relations of the two sets of testcrosses BSSS-222 (Fig. 8.2) show only one major exception: Line 46 in testcross with BSSS-222 yielded 59.1 q/ha as an S_1 line and 74.8 q/ha as an S_8 testcross, the greatest yielding S_8 testcross. The odds, however, were in favor of S_1 testcrosses for determining which lines to retain for further selection and testing. Relations were not as good for the other testers, with B73 S_1 testcrosses having the poorest and Mo17 S_1 testcrosses similar to those for BSSS-222. Because additive effects seemed more important for differences among testcrosses for the five testers, average performance of the lines with the five testers would provide the best measure of combining ability. Figure 8.3 shows graphically the relative S_1 and S_8 testcrosses averaged for the five testers. The highest yielding S_1 testcrosses generally were the higher yielding S_8 testcrosses and seem to support the proposed merits of early testing.

Although some disagreement exists for the appropriate stage of testing for combining ability, it seems that some form of early testing is included in

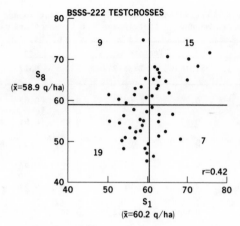

Fig. 8.2. Distributions of BSSS-222 S_1 and S_8 testcrosses.

most breeding programs; perhaps in most instances testcrossing is done at some
intermediate stage of inbreeding. Testcrossing may be called early testing
but may be delayed until the S_2 or S_3 generation of inbreeding. Although early testing did not imply a perfect relation between the initial and later generations of inbreeding, it was designed to separate the population of lines
into good and poor groups for combining ability. If early testing is effective in assigning lines into good and poor groups, greater emphasis for selection and testing can be placed on those in the good group; Lonnquist's (1950)
data show this is an advantage. Yield testing, however, is expensive in money, time, and energy at whatever stage of inbreeding it is used for making the
initial testcrosses. Early testing was designed to reduce effort on progenies
that have poor combining ability, but sampling may be limited in the number of
crosses that can be produced and tested. Alternatively, a larger sample of S_1
progenies of selected S_0 plants can be produced and screened visually for resistance to pests and for morphological traits unacceptable in lines for use
in producing hybrid seed. Selection among and within S_1 progenies and advancing to the S_2 generation by self-pollination can be done relatively easily.

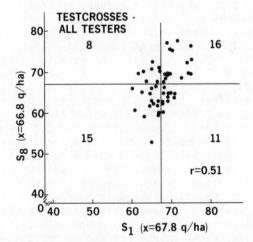

Fig. 8.3. Distributions of S_1 and S_8 testcrosses averaged over five testers.

Additional selection among and within S_2 progenies can be continued and initial testcrosses made in the S_3 generation. If for example, 500 to 600 S_1 progenies can be produced from a breeding population, selection during S_1 and S_2 generations may reduce the number for testcrossing to say 100 in the S_3 generation. In contrast, it may not be possible to produce and test 500 to 600 testcrosses of selected S_0 plants. Although correlations between line and hybrid traits may be low and effectiveness of visual selection for combining ability questioned, combining ability tests at the S_3 generation would be for a group of phenotypically elite lines that can be propagated as lines and useful as parental stocks in hybrids. Lack of a genetic correlation between selected traits of the lines and their combining ability would be an advantage rather than a disadvantage for many traits. Thus one would expect an approximately normal distribution for combining ability of phenotypically desirable lines, provided the lines are from the same population.

Other traits, stalk quality for example, are very important in all types of circumstances in which maize is grown. Jenkins (1929) and Sprague (1946) found correlations of 0.88 and 0.98 for stalk quality between S_0 and S_1 generations, which is excellent, on the assumption that a negative correlation does not exist between stalk quality and combining ability. Eberhart (1974), however, failed to detect any significant correlations between stalk quality and combining ability. Effectiveness in modifying traits of the lines themselves is generally accepted, but visual selection for modifying combining ability is still questionable.

The final analysis of testing determines the effectiveness of testing methods used in identifying lines that have good performance and extensive use in hybrids. Of the 27 lines studied by Jenkins (1934), two lines (L289 and L317) of the Lancaster series were used extensively in hybrids. L289 and L317 ranked in the upper half of the lines tested and would have been selected with early testing. Two other lines (I224 and L304A) included in Jenkins's study were used to a limited extent; I224 had the greatest yields in the Iodent S_1 series and L304A was in the upper 50% of the Lancaster S_1 series. From the data reported by Sprague (1952), three lines were identified by early testing that had some use in commercial hybrids; B10 and B11 had limited use, whereas B14 was used extensively. A fourth line B37 also was later released and used extensively. Recently B73 and B84 were released for use in commercial hybrids. Both lines were initially selected on the basis of early testing in recurrent selection programs. Both B73 and B84 exhibit high yield in crosses with Mo17. B73 is used extensively at the present time in commercial hybrids that have wide adaptability and B84 seems to have the same potential. Available evidence indicates early testing is capable of identifying lines that have eventually had use in hybrids. Of course errors have been made in selecting lines on the basis of early testing, but the early discarding of lines permits savings in time, money, and effort expended on lines tested at more advanced stages of inbreeding. Greater effort in selection and testing can be directed to those lines that have survived early testing.

Development and use of different methods of recurrent selection have emphasized early testing for discriminating among progenies to determine which ones to recombine to form the next cycle of selection. The objective of early testing in recurrent selection is the same as originally proposed for early testing: to identify genotypes that have good and poor combining ability in crosses. Errors will be made in those cases intermediate in combining ability. Evidence given in Chap. 7, however, indicates that long-term cyclical selection programs have been effective for yield improvement.

Early testing, in the purest sense, was used in nearly all instances. Although Payne and Hayes (1949) and Hayes (1963) questioned the value of early testing, Hayes (1963) considered recurrent selection important for the future progress of corn breeding and Hayes and Garber (1919) recommended recurrent selection for use in maize breeding. A ceiling on the assemblage of genes is imposed by the particular S_0 plant selected for self-pollination. Recombination of genes permits some additional selection in later generations, but it is minor compared to original selection of S_0 plants. Thus it seems some form of early testing is desirable to enhance the efficiency of any breeding program.

8.7 GENERAL VS. SPECIFIC COMBINING ABILITY

Combining ability of inbred lines is the ultimate factor determining future usefulness of the lines for hybrids. Combining ability initially was a general concept considered collectively for classifying an inbred line relative to its cross performance. Sprague and Tatum (1942) refined the concept of combining ability, and the two expressions of general (GCA) and specific (SCA) combining ability have had a significant impact on inbred line evaluation and population improvement in maize breeding.

Sprague and Tatum interpreted their results relative to the type of gene action operative for two groups of inbred lines: (1) six tests of single-cross hybrids involving inbred lines that had survived previous selection and testing and (2) two tests of single-cross hybrids involving inbred lines that had not been closely selected for yielding ability. They analyzed the diallel crosses to determine the relative importance of GCA and SCA for the lines included in each set of crosses. Although the diallel had been used before in maize breeding to determine the yield of inbred lines in crosses, Sprague and Tatum apparently were the first to partition the total combining ability of the lines into GCA and SCA. They defined GCA as the average performance of a line in hybrid combinations and SCA as those instances in which certain hybrid combinations are either better or poorer than would be expected on the average performance of the parent inbred lines included. They also emphasized that estimates of GCA and SCA are relative to and dependent on the particular set of inbred lines included in the hybrids under test, an important principle that is often forgotten.

Briefly, Sprague and Tatum (1942) found that GCA was relatively more important than SCA for unselected inbred lines, whereas SCA was more important than GCA for previously selected lines for influencing yield and stalk lodging. Also, they interpreted GCA as an indication of genes having largely additive effects and SCA as indicative of genes having dominance and epistatic effects. Their results supported usage of the topcross test for preliminary evaluation of inbred lines for GCA, although single-cross trials are necessary to determine the most productive specific combinations. Although GCA can be determined from single-cross combinations, it can be more effectively determined with topcross tests--particularly if preliminary information is needed for a large number of lines. The conclusions of Sprague and Tatum (1942) are equally valid in present breeding programs because of interest in production and growing of single crosses.

The notions of GCA and SCA have been used extensively in breeding of maize and other crop species. Types of genetic materials used, methods used, traits studied, and interpretations made have been diverse. Some problems associated with diallel analysis of crosses for estimating GCA and SCA were emphasized in Chap. 4.

Statistical geneticists have made important contributions to clarifying genetic information from an analysis of a diallel set of crosses. Henderson

(1952) and Griffing (1953) defined and applied GCA and SCA concepts to animal and plant experiments; Hull (1946, 1952), Griffing (1950), Hayman (1954, 1958), and Jinks (1954) gave procedures for estimating the genetic parameters for restricted genetic models; Griffing (1956a) showed the relation of the diallel cross mating design to the concept of covariances between relatives in terms of additive and dominance genetic effects; Griffing (1956b) presented the analysis of variance for four situations of the diallel mating design; and Matzinger et al. (1959) showed how GCA and SCA and their interactions with environments can be calculated and interpreted for the diallel mating design. We have not included any experimental results of the diallel mating design, but a cursory review of the literature shows that it is used extensively. In most if not all instances the estimates from the diallel mating cannot be interpreted relative to a reference population.

The concepts of GCA and SCA became useful for characterizing inbred lines in crosses and often were two of the traits included in the description of an inbred line. Later developments in the characterization of genetic variance and types of gene action operative in crosses of inbred lines also often were interpreted relative to GCA and SCA of inbred lines. Different selection methods proposed for recurrent selection also were considered in the context of GCA and SCA and the type of gene action contributing to the heterosis expressed in crosses. If additive gene effects with partial to complete dominance are important, recurrent selection methods that emphasize GCA should be used (Jenkins 1940). If overdominance is of primary importance, recurrent selection methods that emphasize SCA would be appropriate (Hull 1945). Comstock et al. (1949) designed the RRS method to enhance selection for all types of gene action, i.e., both GCA and SCA. Before development of the RRS method, maize breeders often were polarized relative to the importance of GCA and SCA.

Experimental results listed in Chap. 7 indicate that selection based on either broad base or narrow base testers generally has been effective for the improvement of both the population and testcross performance. It was commonly accepted that the use of a narrow base tester would improve the combining ability to the specific tester but would have little value for the improvement of GCA, which is based mainly on additive genetic effects. Recent reports have shown that the use of inbred lines as testers has improved both combining ability for specific testers and GCA as measured by the performance of populations themselves or of testcrosses with unrelated broad base populations (Horner et al. 1973; Russell et al. 1973; Russell and Eberhart 1975; Hoegemeyer and Hallauer 1976; Horner et al. 1976; Walejko and Russell 1977). The results seem logical in view of the summaries in Tables 5.1 and 5.2 for relative types of genetic variance, and Sprague and Tatum's (1942) original conclusion was that GCA is more important for previously unselected material.

In selection experiments, the testcross progenies (usually half-sibs) are the original cross information used to select those progenies for recombination. Horner (1973, 1975) reported a study testing Hull's hypothesis of the effectiveness of recurrent selection for SCA and speculated that about one-third of the gain from use of an inbred tester was specific to the tester and about two-thirds was due to the additive effects transferable to other combinations. Horner (1975) also concluded that maize breeders should be able to change testers as the occasion demands because overdominance and epistasis do not seem of major importance in yield heterosis. Horner's (1973, 1975) conclusions were supported by Russell et al. (1973) and Walejko and Russell (1977) for different populations and inbred testers.

The conclusions for the selection studies were corroborated by Russell and Eberhart (1975) and Hoegemeyer and Hallauer (1976). Russell and Eberhart compared selected lines from two recurrent selection programs that evaluated half-sib progenies. Analysis of variance of the crosses among lines indicated that most of the variation within each set of crosses was attributable to the average performance (or GCA) of the lines included in the crosses. In five of six instances, the GCA mean square was highly significant and relatively large compared with the SCA mean square, indicating additive gene effects were the primary type of gene action operative in the crosses. Hoegemeyer and Hallauer tested single crosses among S_7 lines selected by a full-sib selection scheme that emphasized selection for SCA. Crosses among the elite lines that had survived selection for four generations of selection for SCA also had high GCA with other elite lines. Hoegemeyer and Hallauer also showed that their selection procedure was effective for SCA, but the SCA effects must be small relative to the GCA effects. Although present evidence seems to show that GCA (or additive gene effects) is more important than SCA, the terms still are commonly used in the sense of the maize breeding jargon and will continue to be so. Maize breeders breed and test to identify that unique combination of inbred lines for high productivity; i.e., SCA is expressed although the GCA was probably more important in identifying the lines for the unique combination. Non-additive gene effects seem to be small on the average, but they may be important for that one unique combination.

REFERENCES

Allard, R. W. 1960. *Principles of Plant Breeding*. Wiley, New York.

Allison, J. C. S., and R. W. Curnow. 1966. On the choice of tester parent for the breeding of synthetic varieties of maize (*Zea mays* L.). *Crop Sci.* 6:541–44.

Bartual, R., and A. R. Hallauer. 1976. Variability among unselected maize inbred lines developed by full-sibbing. *Maydica* 21:49–60.

Brown, W. L. 1967. Results of non-selective inbreeding in maize. *Züchter* 37:155–59.

Collins, G. N. 1916. Correlated characters in maize breeding. *J. Agric. Res.* 6:435–54.

Comstock, R. E. 1964. Selection procedures in corn improvement. *Proc. Annu. Corn Sorghum Res. Conf.* 19:87–94.

————. 1977. An evaluation of RRS with inbred line testers. North Cent. Corn Breed. Res. Comm. (NCR-2) Rep. Mimeogr. Library, Iowa State Univ., Ames.

————. 1979. Inbred lines versus the populations as testers in reciprocal recurrent selection. *Crop Sci.* 19:881–86.

Comstock, R. E.; H. F. Robinson; and P. H. Harvey. 1949. A breeding procedure designed to make use of both general and specific combining ability. *Agron. J.* 41:360–67.

Cowan, J. R. 1943. The value of double cross hybrids involving inbreds of similar and diverse genetic origin. *Sci. Agric.* 23:287–96.

Cress, C. E. 1966. Heterosis of the hybrid related to gene frequency differences between two populations. *Genetics* 53:269–74.

Darrah, L. L.; S. A. Eberhart; and L. H. Penny. 1972. A maize breeding study in Kenya. *Crop Sci.* 12:605–8.

Davis, R. L. 1927. Report of the plant breeder. Rep. Puerto Rico Agric. Exp. Stn. Pp. 14–15.

Eberhart, S. A. 1974. Annu. Rep. Corn Breed. Invest. Mimeogr. Ames, Iowa.

Eckhardt, R. C., and A. A. Bryan. 1940a. Effect of method of combining the four inbred lines of a double cross of maize upon the yield and variability of the resulting double crosses. *J. Am. Soc. Agron.* 32:347-53.

————. 1940b. Effect of the method of combining two early and two late inbred lines of corn upon the yield and variability of the resulting double crosses. *J. Am. Soc. Agron.* 32:645-56.

El-Lakany, M. A., and W. A. Russell. 1971. Relationship of maize characters with yield in testcrosses of inbreds at different plant densities. *Crop Sci.* 11:698-701.

Etheridge, W. C. 1921. Characters connected with the yield of the corn plant. Missouri Agric. Exp. Stn. Res. Bull. 46.

Gama, E. E. G., and A. R. Hallauer. 1977. Relation between inbred and hybrid traits in maize. *Crop Sci.* 17:703-6.

Green, J. M. 1948a. Relative value of two testers for estimating topcross performance in segregating maize populations. *J. Am. Soc. Agron.* 40:45-57.

————. 1948b. Inheritance of combining ability in maize hybrids. *J. Am. Soc. Agron.* 40:58-63.

Griffing, J. B. 1950. Analysis of quantitative gene action by constant parent regression and related techniques. *Genetics* 35:303-21.

————. 1953. An analysis of tomato yield components in terms of genotypic and environmental effects. Iowa Agric. Exp. Stn. Res. Bull. 397.

Griffing, B. 1956a. A generalized treatment of the use of diallel crosses in quantitative inheritance. *Heredity* 10:31-50.

————. 1956b. Concept of general and specific combining ability in relation to diallel crossing systems. *Australian J. Biol. Sci.* 9:463-93.

Hallauer, A. R. 1975. Relation of gene action and type of tester in maize breeding procedures. *Proc. Annu. Corn Sorghum Res. Conf.* 30:150-65.

Hallauer, A. R., and E. Lopez-Perez. 1979. Comparisons among testers for evaluating lines of corn. *Proc. Annu. Hybrid Corn. Ind. Res. Conf.* 34:57-75.

Hallauer, A. R., and J. H. Sears. 1973. Changes in quantitative traits associated with inbreeding in a synthetic variety of maize. *Crop Sci.* 13:327-30.

Hallauer, A. R., and O. S. Smith. 1979. Registration of maize germplasm (reg. GP 81 and GP 82). *Crop Sci.* 19:755.

Hayes, H. K. 1926. Present day problems of corn breeding. *J. Am. Soc. Agron.* 18:344-63.

————. 1963. *A Professor's Story of Hybrid Corn.* Burgess, Minneapolis.

Hayes, H. K., and R. J. Garber. 1919. Synthetic production of high protein corn in relation to breeding. *J. Am. Soc. Agron.* 11:308-18.

Hayes, H. K., and I. J. Johnson. 1939. The breeding of improved selfed lines of corn. *J. Am. Soc. Agron.* 31:710-24.

Hayman, B. I. 1954a. The analysis of variance of diallel tables. *Biometrics* 10:235-44.

————. 1954b. The theory and analysis of diallel crosses. *Genetics* 39:789-909.

Henderson, C. R. 1952. Specific and general combining ability. In *Heterosis*, J. W. Gowen, ed., pp. 350-70. Iowa State Univ. Press, Ames.

Hoegemeyer, T. C., and A. R. Hallauer. 1976. Selection among and within full-sib families to develop single-crosses of maize. *Crop Sci.* 16:76-81.

Horner, E. S. 1963. A comparison of S_1 line and S_1 plant evaluation for combining ability in corn. *Crop Sci.* 3:519-22.

Horner, E. S.; W. H. Chapman; M. C. Lutrick; and H. W. Lundy. 1969. Comparison of selection based on yield of topcross progenies and S_2 progenies in maize (*Zea mays* L.). *Crop Sci.* 9:539-43.

Horner, E. S.; W. H. Lundy; M. C. Lutrick; and W. H. Chapman. 1973. Comparison of three methods of recurrent selection in maize. *Crop Sci.* 13:485-
89.

Horner, E. S.; M. C. Lutrick; W. H. Chapman; and F. G. Martin. 1976. Effect
of recurrent selection for combining ability with a single-cross tester
in maize. *Crop Sci.* 16:5-8.

Hull, H. F. 1945. Recurrent selection for specific combining ability in
corn. *J. Am. Soc. Agron.* 37:134-45.

————. 1946. Maize genetics cooperation newsletter 20.

————. 1952. Recurrent selection and overdominance. In *Heterosis*, J. W.
Gowen, ed., pp. 451-73. Iowa State Univ. Press, Ames.

Jenkins, M. T. 1929. Correlation studies with inbred and crossbred strains
of maize. *J. Agric. Res.* 39:677-721.

————. 1935. The effect of inbreeding and of selection within inbred lines
of maize upon the hybrids made after successive generations of selfing.
Iowa State J. Sci. 3:429-50.

————. 1940. The segregation of genes affecting yield of grain in maize.
J. Am. Soc. Agron. 32:55-63.

Jenkins, M. T., and A. M. Brunson. 1932. Methods of testing inbred lines of
maize in crossbred combinations. *J. Am. Soc. Agron.* 24:523-30.

Jinks, J. L. 1954. Analysis of continuous variation in a diallel cross of
Nicotiana rustica varieties. *Genetics* 39:767-88.

Johnson, I. J., and H. K. Hayes. 1936. The combining ability of inbred lines
of Golden Bantam sweet corn. *J. Am. Soc. Agron.* 28:246-52.

————. 1940. The value of hybrid combinations of inbred lines of corn selected from single crosses by the pedigree method of breeding. *J. Am.
Soc. Agron.* 32:479-85.

Jones, D. F. 1918. The effect of inbreeding and crossbreeding upon development. Connecticut Agric. Exp. Stn. Bull. 207:5-100.

Jorgenson, L., and H. E. Brewbaker. 1927. A comparison of selfed lines of
corn and first generation crosses between them. *J. Am. Soc. Agron.* 19:
819-30.

Keller, K. R. 1949. A comparison involving the number of and relationships
between testers in evaluating inbred lines of maize. *Agron. J.* 41:323-
31.

Kempthorne, O., and R. W. Curnow. 1961. The partial diallel cross. *Biometrics* 17:229-50.

Kempton, J. H. 1924. Correlation among quantitative characters in maize. *J.
Agric. Res.* 28:1095-1102.

————. 1926. Correlated characters in a maize hybrid. *J. Agric. Res.* 32:
39-50.

Kiesselbach, T. A. 1922. Corn investigations. Nebraska Agric. Exp. Stn.
Res. Bull. 20:5-151.

Kovacs, I. 1970. Some methodological problems of the production of inbred
lines. In *Some Methodological Achievements of the Hungarian Hybrid Maize
Breeding*, I. Kovacs, ed., pp. 54-72. Akademiai Kiado, Budapest.

Kyle, C. W., and H. F. Stoneberg. 1925. Associations between number of kernel rows, productiveness, and deleterious characters in corn. *J. Agric.
Res.* 31:83-99.

Loeffel, F. A. 1964. S_1 crosses compared with crosses of homozygous lines.
Proc. Annu. Corn Sorghum Res. Conf. 19:95-104.

————. 1971. Development and utilization of parental lines. *Proc. Annu.
Corn Sorghum Res. Conf.* 26:209-17.

Lonnquist, J. H. 1949. The development and performance of synthetic varieties of corn. *Agron. J.* 41:153-56.

————. 1950. The effect of selecting for combining ability within segregating lines of corn. *Agron. J.* 42:503–8.

Lonnquist, J. H., and M. F. Lindsey. 1964. Topcross versus S_1 line performance in corn (*Zea mays* L.). *Crop Sci.* 4:580–84.

Lonnquist, J. H., and M. D. Rumbaugh. 1958. Relative importance of test sequence for general and specific combining ability in corn breeding. *Agron. J.* 50:541–44.

Lopez-Perez, E. 1979. Comparisons among five different testers for the evaluation of unselected lines of maize (*Zea mays* L.). Ph.D. dissertation, Iowa State Univ., Ames.

Love, H. H. 1912. The relation of certain ear characters to yield in corn. *Am. Breeders' Assoc. Rep.* 7:29–40.

Love, H. H., and J. B. Wentz. 1917. Correlations between ear characters and yield in corn. *J. Am. Soc. Agron.* 9:315–22.

Matzinger, D. F. 1953. Comparison of three types of testers for the evaluation of inbred lines of corn. *Agron. J.* 45:493–95.

Matzinger, D. F.; G. F. Sprague; and C. C. Cockerham. 1959. Diallel crosses of maize in experiments repeated over locations and years. *Agron. J.* 51:346–50.

Nilsson-Leissner, G. 1927. Relation of selfed strains of corn to F_1 crosses between them. *J. Am. Soc. Agron.* 19:440–54.

Obilana, A. T., and A. R. Hallauer. 1974. Estimation of variability of quantitative traits in BSSS by using unselected maize inbred lines. *Crop Sci.* 14:99–103.

Osler, R. D.; E. J. Wellhausen; and G. Palacios. 1958. Effect of visual selection during inbreeding upon combining ability in corn. *Agron. J.* 50:45–48.

Payne, K. T., and H. K. Hayes. 1949. A comparison of combining ability in F_3 and F_2 lines of corn. *Agron. J.* 41:383–88.

Rawlings, J. O., and D. L. Thompson. 1962. Performance level as criterion for the choice of maize testers. *Crop Sci.* 2:217–20.

Richey, F. D. 1924. Effects of selection on the yield of a cross between varieties of corn. USDA Bull. 1209.

————. 1945. Isolating better foundation inbreds for use in corn hybrids. *Genetics* 30:455–71.

Richey, F. D., and L. S. Mayer. 1925. The productiveness of successive generations of self-fertilized lines of corn and of crosses between them. USDA Bull. 1354.

Richey, F. D., and J. G. Willier. 1925. A statistical study of the relation between seed-ear characters and productiveness in corn. USDA Bull. 1321.

Russell, W. A., and S. A. Eberhart. 1975. Hybrid performance of selected maize lines from reciprocal recurrent selection and testcross selection programs. *Crop Sci.* 15:1–4.

Russell, W. A., and A. H. Teich. 1967. Selection in *Zea mays* L. by inbred line appearance and testcross performance in low and high plant densities. Iowa Agric. Home Econ. Exp. Stn. Res. Bull. 542.

Russell, W. A.; S. A. Eberhart; and U. A. Vega. 1973. Recurrent selection for specific combining ability for yield in two maize populations. *Crop Sci.* 13:257–61.

Shuman, D. E. W., and R. A. Bradley. 1957. The comparison of the sensitiveness of similar experiments. *Theor. Ann. Math. Stat.* 28:902–20.

Silva, J. C. 1974. Genetic and environmental variances and covariances estimated in the maize (*Zea mays* L.) variety, Iowa Stiff Stalk Synthetic. Ph.D. dissertation, Iowa State Univ., Ames.

Singleton, W. R., and O. E. Nelson. 1945. The improvement of naturally cross-pollinated plants by selection in self-fertilized lines. IV. Combining ability of successive generations of inbred sweet corn. Connecticut Agric. Exp. Stn. Bull. 490:458-98.

Sprague, G. F. 1939. An estimation of the number of top-crossed plants required for adequate representation of a corn variety. *J. Am. Soc. Agron.* 38:11-16.

———. 1946. Early testing of inbred lines of corn. *J. Am. Soc. Agron.* 38:108-17.

———. 1952. Early testing and recurrent selection. In *Heterosis*, J. W. Gowen, ed., pp. 400-417. Iowa State Univ. Press, Ames.

———. 1955. Corn breeding. In *Corn and Corn Improvement*, G. F. Sprague, ed., pp. 221-92. Academic Press, New York.

Sprague, G. F., and P. A. Miller. 1952. The influence of visual selection during inbreeding on combining ability in corn. *Agron. J.* 44:258-62.

Sprague, G. F., and L. A. Tatum. 1942. General vs. specific combining ability in single crosses of corn. *J. Am. Soc. Agron.* 34:923-32.

Walejko, R. N., and W. A. Russell. 1977. Evaluation of recurrent selection for specific combining ability in two open-pollinated maize cultivars. *Crop Sci.* 17:647-51.

Wellhausen, E. J., and S. Wortman. 1954. Combining ability in S_1 and derived S_3 lines of corn. *Agron. J.* 46:86-89.

Wu, S. K. 1939. The relationship between the origin of selfed lines of corn and their value in hybrid combinations. *J. Am. Soc. Agron.* 31:131-40.

9

Inbreeding

Maize is a cross-pollinated species; consequently, a maize population or variety includes a mixture of genotypes, each unique, that depends on the particular combination of the two gametes that united to form the zygote. Maize is monoecious with the male and female organs located in two different and separate inflorescences; staminate flowers (male) are in the tassel located at the top of the plant and the pistillate flowers (female) are usually located at the sixth or seventh node from the top of the plant. The structures that contain the male (tassel) and female flowers (ears) are prominent and facilitate hand pollination. Although the tassel and ear of maize are distinctive and easily recognizable from other crop species, variations in the tassel and ear occur in races, populations, varieties, hybrids, and inbred lines. Tassels may differ in length, number and spacings of tassel branches, compactness, color, ease of pollen shed, amount of pollen shed, etc. Ears vary in relative placement on the plant, length, diameter, kernel-row number, ear number, kernel color, etc.

Pollen dispersal is by wind currents, a system that favors cross-pollination. The protandrous nature of maize also promotes cross-pollination, but a small amount of self-fertilization may occur when there is some overlap of pollen shed with silk receptivity. In a few examples protogyny occurs (Weatherwax 1955). Kiesselbach (1922) reported only 0.7% self-pollination occurred in Nebraska White Prize plants grown in a field of Hogue's Yellow Dent. He concluded from his data that the amount of self-fertilization that occurs under ordinary field conditions for maize is negligible.

Departures from randomness of mating also have been reported; they occur because of differences in maturity of individual plants in a population and environmental conditions that exist at flowering time. Gutierrez and Sprague (1959) reported that maize exhibited marked departures from randomness of mating. They also concluded that dates of silking and pollen shedding, number of plants shedding pollen, length of pollen shedding period, plant height, and possibly selective fertilization (which also was demonstrated by Jones 1924) may have been factors contributing to departures from randomness of mating.

Some of these factors would tend to develop subpopulations within a presumably random mating population--subpopulations because early flowers would tend to mate with early flowers and late flowers would tend to mate with late flowers. Although some physiologically isolated subpopulations would exist within an open-pollinated maize population, there also would tend to be cross-pollination among subpopulations sufficient to prevent easily distinguishable subpopulations.

Although complete randomness of mating may not be realized in an open-pollinated population of maize, the genotypic array within a maize population is extensive and complex. Each individual within a population arises from the union of two different gametes; hence each plant is in reality a different F_1 hybrid. An array of all the plants (hypothetical F_1 individuals) in a population would approximate a normal distribution for quantitatively inherited traits; i.e., there would be a range in distribution for each trait but the population would be characterized by the plants within two standard deviations of the mean. A maize population, therefore, would include individual plants that are as productive as hybrids produced from crosses of inbred lines. The problem is one of identification and perpetuation. Not until studies were conducted to reduce the complex genotypes of individual plants of maize populations to pure genotypes (inbred lines) that were reproducible and usable as parental stocks for the production of hybrids was the modern concept of maize breeding developed.

Inbreeding occurs, either naturally or artificially, in matings between individuals that are more closely related than by random chance, i.e., closer than the average relationship within the particular population defined as non-inbred. We are concerned with effects of inbreeding that occur from controlled matings, but inbreeding can occur naturally in small populations because of the limited number of matings that can occur. Occurrence of inbreeding as a result of population size rather than by controlled matings is of particular concern in maize breeding programs that include recombination of selected individuals in various recurrent selection procedures, because a limited number of individuals (often 10 to 20) is selected for recombination and random drift in gene frequencies.

Inbreeding is determined by the percentage of homozygosis of the zygote; i.e., inbreeding is a description of the zygotes and not the gametes. Gametes from a homozygous individual, such as an inbred line of maize, are no more or less inbred than gametes from a completely heterozygous individual, such as an F_1 hybrid, between two relatively homozygous inbred lines. This relation indicates why individual plants in a cross-pollinated population of maize can be considered as unique F_1 hybrids--they each were formed from the union of two gametes that are not identifiable and repeatable. Hence inbreeding, primarily by self-fertilization, is important in the modern concept of maize breeding.

9.1 EARLY REPORTS OF INBREEDING AND OUTBREEDING

The early history of inbreeding and outbreeding was reviewed and summarized by East and Jones (1918) and Jones (1918). Kölreuter (1776), Knight (1799), Gärtner (1849), and Focke (1881) conducted extensive experiments with many plant species and were prominent hybridizers before the rediscovery of Mendel's laws of segregation and recombination. None of them, however, realized that inbreeding and outbreeding were opposite expressions of the same phenomenon.

Darwin (1877) also recorded many observations of effects of inbreeding and outbreeding in plant species, but his interpretations also were made before Mendelism was understood. Darwin thought that inbreeding was not a normal process of mating and that the evil effects of inbreeding accumulated until eventually the species propagated by inbreeding was doomed to extinction. However, he made one important interpretation of effects of inbreeding. These effects usually had been attributed to the inbreeding process itself rather than to homozygosity. Darwin observed that it made no difference in vigor whether plants in an inbred lot were selfed or crossed among themselves; this he attributed to the members of an inbred lot becoming germinally alike. This interpretation was supported by crosses of his selfed lines with other lines

inbred to a less degree that did not express as great an increase in vigor as
the crosses of the same lines with a fresh stock from different regions.
Crosses between two inbred lines, however, did give a noticeable increase in
vigor, often exceeding the original variety. Differences in the performance
of inbred lines and their crosses were attributed to germinal similarity and
diversity. Darwin's interpretations were similar to our concept of segregation and recombination, and many of the problems that bothered Darwin would
have been clarified if knowledge of Mendelism had been available to him.

Although the phenomenon of inbreeding and crossing was studied extensively in the plant kingdom, not until the studies reported by Shull (1908, 1909,
1910) in maize was the fundamental similarity of inbreeding and crossing
clearly understood; and not until Shull's (1908) interpretations became available was the exact genetic basis of inbreeding understood. East (1908, 1909)
observed similar phenomena. But East (1909) also was apprehensive about the
practical utility of inbreeding to develop parental stocks to produce crosses
for the improvement of maize production.

Although Darwin had conducted some inbreeding studies with maize, the beginnings of the modern concept of maize breeding methods (inbred lines and hybrids) are accredited to Shull. Long-term inbreeding studies in maize were
few and most of the results were not encouraging. Shamel (1905), Collins
(1909), and Davenport and Holden (cited by Shull 1952) reported on the effects
of inbreeding in maize; in most instances the injurious effects of inbreeding
were recognized but the advantages of its use in crosses were not. The studies of Shull and East were conducted independently at the same time at locations in close proximity. A historical perspective of the studies by Shull
and East was given by Shull (1952, pp. 17-18). The inbreeding studies by East
and Shull were reported the same year (1908), but the correct interpretation
of the phenomenon of inbreeding and crossing was given by Shull.

Shull's remarkable conclusions were based on a limited amount of information, but they formed the basic principles of maize breeding as we know them
today. Shull's studies in maize were designed not to study the effects of inbreeding and crossing but rather the effects of the phenomenon on the inheritance of number of kernel rows per ear. Based on the inheritance of kernel
rows per ear, plant height, and yield of the inbred lines developed by selfing
and their performance in crosses among the inbred lines, Shull (1908, p. 299)
concluded: "The obvious conclusion to be reached is that an ordinary corn-
field is a series of very complex hybrids produced by the combination of numerous elementary species. Self-fertilization soon eliminates the hybrid elements and reduces the strain to its elementary components. In the comparison
between a self-fertilized strain and a cross-fertilized strain of the same origin, we are not dealing then, with the effects of cross- and self-fertilization *as such*, but with the relative vigor of biotypes and their hybrids
[*sic*]." Thus he concluded that self-fertilization was not the direct effect
of the observed differences but an indirect effect of separating the hybrid
combinations into their elementary species or biotypes. The differences among
inbred lines were heritable and were similar to the pure lines described by
Johannsen.

Shull's conclusion that self-fertilization separated hybrids into pure
forms also was an important contribution to what is now termed pedigree selection; i.e., specific crosses are made between inbred lines and self-fertilization is imposed on the F_1 hybrid in an attempt to derive inbred lines that are
superior to either parent because of genetic segregation and recombination,
i.e., transgressive segregants. Shull (1908) emphasized that "there is no in-
telligent attempt in these methods to determine the relative value of the sev-

eral biotypes *in hybrid combination*, but only in the pure state. In the present state of our knowledge it is impossible to predict from a study of two pure strains what will be the relative vigor of their hybrid offspring." The insight of these conclusions is interesting because maize breeders are still faced with determining the relative merits of inbred lines in hybrids (see Chap. 8).

Shull (1909) discussed further the results of his studies and outlined the two essential features for the production of hybrid maize. He referred to the process as the pure line method in maize breeding, which he considered under two heads: "(1) finding the best pure-lines and (2) the practical use of the pure-lines in the production of seed corn." He discussed the necessity of making as many self-fertilizations as possible and continuing self-fertilization until the lines were in the homozygous state. He correctly hypothesized that lines developed by self-fertilization would not be useful for all situations and that the only method for determining relative merit of the lines was to make all possible crosses among the lines and evaluate them in F_1 hybrids. He concluded that after one has determined which pair of pure strains (pure lines) produce the desired results, the method of producing the seed is relatively simple but it may be somewhat costly.

Shull (1910) reinforced his arguments for the advantages of the pure line method of maize breeding. The salient feature of his discussion was the comparison of the pure line method with varietal hybridization (broad breeding) methods discussed by East (1909) and Collins (1909). Shull reasoned correctly that only by reducing the complex mixtures of genotypes in a variety to their pure forms and determining the best pair in F_1 hybrids would maximum uniformity and performance be attained. Because varieties were complex mixtures of genotypes, the modal level of heterozygosity of the variety cross would be less than the cross of two pure lines. Although Shull agreed with East that the pure line method is "more correct theoretically but less practical" than the variety cross method, he was anxious that the pure line method be seriously tested before comparisons were made between the two methods. Developments since then have proven the tenets of Shull correct and they are still used today. East and Hayes (1912) and Jones (1918) studied the procedures, and the suggestion of the use of the double cross by Jones provided the impetus for extensive experimentation of the pure line method of maize breeding. Shull (1908, 1909, 1910) outlined the subsequent course of maize breeding.

9.2 INBREEDING SYSTEMS

Inbreeding in maize usually is by self-fertilization because of the rapid approach to homozygosity. Because maize has the male and female organs in separate inflorescences, it is easy to manipulate self- and cross-fertilizations artificially. Techniques have been developed in which inexperienced individuals can self and cross selected plants and produce 300 to 500 seeds per pollination (Russell and Hallauer 1980). Self-fertilization is the most extreme form of inbreeding, but other methods are available. Other less restrictive forms of inbreeding have been suggested for producing more vigorous inbred lines (Macaulay 1928; Lindstrom 1939; Harvey and Rigney 1947; Kinman 1952; Stringfield 1974). Theoretically, less restrictive forms would permit less rapid fixation of deleterious genes than the selfing method. The zygotic array of the progeny by selfing is determined by the gametic array of the plant that is selfed. Future selection during inbreeding would be fixed within the limits of the genotype of the original S_0 plant selfed. If, for example, some form of sibmating is used to effect inbreeding, fixation of deleterious alleles would be slower and opportunities for selection would be enhanced. The advantages and disadvantages of intense inbreeding (i.e., selfing vs. a milder form) have to be considered; i.e., do the advantages of selection during inbreeding outweigh the advantages of rapid approach to homozygosity by selfing?

Some of the common systems of inbreeding are listed in Table 9.1. Earlier methods for measuring expected average level of inbreeding were suggested, but formulas for measuring inbreeding from matings between different kinds of relatives were first derived by Wright (1921, 1922a). Wright's coefficient of inbreeding (F) is the degree of correlation between uniting gametes and measures directly the probable increase in homozygosis. Although Wright (1922a, 1922b) was the first to derive such formulas, others have obtained the same results with different methods. The method derived by Malécot (1948) has been the one most widely used; it is easier to visualize and gives the same results as the path coefficient method of Wright (1922a). Malécot defines the coefficient of inbreeding as the probability that two genes at the same locus are identical by descent. Two allelic genes at a locus may be the same either because they are alike in state or identical by descent. Alleles that are alike in state have the same function, but so far as known they do not occur together because of ancestry unless they originated from some unknown remote ancestor. Hence, the coefficient of inbreeding, or level of inbreeding, of relatives is relative to a specified base population.

Theory, use, and illustrations for determining the coefficient of inbreeding of relatives for different systems of inbreeding are given by Kempthorne (1957), Pirchner (1969), and Li (1976). The coefficient of inbreeding, therefore, is computed to determine the level of homozygosity in a specified generation. The approach to homozygosity varies among the different systems of inbreeding. Self-fertilization approaches homozygosity faster than any of

Table 9.1. Coefficients of inbreeding (expected homozygosity) in different generations for different systems of inbreeding.

Generation	Half-sib	Full-sib[+]	Selfing	Backcrossing[‡] F of parent = 0.0	Backcrossing[‡] F of parent = 1.0
0	0.000	0.000	0.000	0.000	0.000
1	0.000	0.000	0.500	0.250	0.500
2	0.125	0.250	0.750	0.375	0.750
3	0.219	0.375	0.875	0.438	0.875
4	0.305	0.500	0.938	0.469	0.938
5	0.381	0.594	0.969	0.484	0.969
6	0.448	0.672	0.984	0.492	0.984
7	0.509	0.734	0.992	0.496	0.992
10	0.654	0.859	0.999	0.500	0.999
	$(1/8)(1+6F' + F'')$[§]	$(1/4)(1+2F' + F'')$	$(1/2)(1+F')$	$(1/4)(1+2F')$	$(1/2)(1+F')$

[+]Also the same as crossing the offspring to the younger parent.

[‡]Involves the crossing of one parent to its offspring.

[§]Recurrent relation for calculating expected homozygosity in the system of inbreeding where F is Wright's coefficient of inbreeding, F' refers to the previous generation, and F" refers to the second generation removed.

the other systems, but in plant species that are dioecious or have self-incompatibility systems (and in higher animal species having separate sexes) it is not possible to self-fertilize; hence, full-sib mating is the closest form of inbreeding possible. Maize, however, is amenable to all systems of inbreeding, and the system used depends on either the objectives of the breeding program or the fundamental studies under investigation.

The advantages of self-fertilization to develop inbred lines that are relatively homozygous are obvious in Table 9.1. It requires three generations of full-sib matings and six generations of half-sib matings to have the same theoretical level of inbreeding, or homozygosity, as one generation of self-fertilization. Even if the off season (e.g., winter nurseries for the U.S. Corn Belt) is available and used, the number of seasons required to attain the same theoretical level of homozygosity will be greater. Half-sib and full-sib systems of inbreeding also will usually require one additional season to develop the families for continued matings. If one wants to use the full-sib method of inbreeding in a particular population, it is necessary to make controlled pollinations to develop full-sib families. Seed harvested from plant-to-plant crosses (full-sibs) will be noninbred and cross-pollination of the full-sib individuals in the next season will yield seed that has an expected homozygosity of 25%; in the same two seasons, the expected homozygosity by selfing is 75%.

Half-sib families can be obtained two ways, depending on the population to be used to initiate inbreeding by half-sibbing: (1) Half-sib seed can be obtained from an isolation field by harvesting and shelling each ear separately. Assuming no natural self-fertilization, female gametes of each plant were pollinated by a random sample of male gametes from the population; hence the seeds are half-sibs having a common female parent. (2) In the nursery half-sib progenies are established by controlled pollinations by bulking pollen hoped to be representative of the population and applying the pollen to the silks of a separate sample of plants. In subsequent generations, inbreeding is established by bulk pollination of a plant within the half-sib progeny. In either instance the theoretical level of inbreeding after two generations is 12.5% compared with 75% after two generations of self-fertilization.

Backcrossing usually is not considered one of the common systems of inbreeding, but backcrossing is used extensively either to transfer a specific gene from one genotype to another or to improve an inbred line for some quantitative trait. The merits of the two will not be discussed, but the transfer of a simple, dominant gene (e.g., *Ht* gene for *Helminthosporium turcicum* resistance) for a specific trait is usually more useful. Improvement of an inbred line for a quantitative trait for say European corn borer resistance or components of yield is less successful because of the dilution effects in successive backcrosses (Geadelmann and Peterson 1978). If the parent used for backcrossing is homozygous (F = 1) and a simple dominant gene is being transferred by backcrossing, the rate of inbreeding is the same as selfing and the genotype is theoretically rapidly recovered. The theoretical level of inbreeding by backcrossing to a noninbred parent (F = 0) never exceeds 50%. If complete homozygosity is desired from a backcrossing program that includes a noninbred parent, some other form of inbreeding is necessary. Homozygosity is eventually achieved by all systems of inbreeding except backcrossing to the noninbred parent. For maize, we have a choice of inbreeding system, but in many plant species and higher animal species the choice is restricted.

Macaulay (1928) suggested an inbreeding system, less severe than self-fertilization, that he called plot-inbreeding. He suggested planting seed from selected ears in plots containing 200 to 250 plants, each plot isolated from the others. Within each plot only large uniform ears on selected plants

are retained each generation for continued selection. The suggested procedure
seems to be a form of half-sib inbreeding. Although no data were presented,
claims were made that more vigorous lines were obtained by plot-inbreeding
than by self-fertilization and that the best lines developed by self-fertilization must be inferior to the best lines from plot-inbreeding because: (1)
Fifty percent of all heterozygous factors are reduced to homozygosity in one
generation by self-fertilization, whereas it takes several generations in
plot-inbreeding. Thus there is more chance for selection under plot-inbreeding. (2) Lines developed from self-fertilized plants can only inherit favorable factors that are present in the parent plant, whereas plot-inbreeding provides for the gradual accumulation into a single strain all the favorable factors possessed by the individuals in a plot. Macaulay also believed that it
was possible to develop homozygous strains as vigorous as F_1 hybrids (which
are discussed further in Chap. 10). Availability of isolation would limit the
use of this system even if all claims were fully justified. It would require
many more generations to develop the lines, and their merit in hybrids would
have to be established by conventional procedures.

Lindstrom (1939) and Harvey and Rigney (1947) compared inbred lines developed by full-sib and selfing systems and combinations of full-sibbing and
selfing. In both instances the relative number of lines was limited, and they
arrived at opposite conclusions. Lindstrom suggested that mild inbreeding
should be used at the beginning to prevent rapid fixation of deleterious
traits and provide a broader base for selection under diverse environmental
conditions. He further theorized that the same results may be attained by using greater numbers in a self-fertilization program, which is an important
point. Harvey and Rigney, on the other hand, suggested that it may be better
to inbreed intensely at the beginning to select out deleterious factors than,
by mild inbreeding, to retain deleterious factors that segregate out in later
generations. Their conclusions were based on the high incidence of barrenness
that occurred after full-sibbing for three generations and then selfing for
four generations; similar results were reported by Good (1976).

Kinman (1952) also suggested developing lines for use in hybrids by composite sibbing. On the average, however, testcross hybrids involving sibbed
lines were not different in performance or variability from those of S_3 lines
(3 generations of selfing) developed from the same source. Sibbed lines were
more variable than the related selfed lines (S_4) in maturity, number of tassel
branches, plant height, and yield. Loeffel (1971) tested a series of inbred
lines in hybrids after one to four generations of inbreeding to compare (1)
selfing with sibbing, (2) selfing followed by sibbing, and (3) intensity of
inbreeding. He found no appreciable gains in maturity, yield, or other traits
that could be attributed to either the level of inbreeding or the choice between selfing and sibbing methods of inbreeding.

The most recent discussion for use of milder forms of inbreeding for developing inbred lines for use in hybrids was given by Stringfield (1974). He
discussed the negative aspects of inbreeding by self-fertilization as follows:
(1) continuous self-fertilization seems too violent, (2) homozygous lines bear
many marks of lapses in selection, (3) homozygous lines are essentially in-
flexible, and (4) homozygous lines have disadvantages in production of seed
and may limit the final crop. All aspects are related to the rapid fixation
of genes in the homozygous condition by self-fertilization. Stringfield was
of the opinion that better and more useful inbred lines could be developed
from a milder form of inbreeding that permitted effective selection (the
breeder's art) concurrently with inbreeding. He proposed assortative matings
in a population by chain crossing selected plants and making selections from
choice plants that were pollinated by choice plants. The procedure is a modi-

fication of the plot-inbreeding system suggested by Maculay (1928). Stringfield called it broad-line development and developed and tested a few lines.
The broad-line concept has not been extensively used.

Suggestions for use of systems of inbreeding in maize other than self-
fertilization have been made, but they have not been popular. Often the suggestions are philosophical, emphasizing the rigidity of self-fertilization,
particularly if the art of plant breeding is important in line development.
Because of rapid fixation of genes by self-fertilization, opportunities for
visual selection would be limited within progenies for particular plant types.
For example, when $F = 0.5$, the total genetic variance assuming only additive
effects between means of families will be twice as great as that within families (Chap. 2). After two generations of self-fertilization, $F = 0.75$ and total genetic variance among S_2 families is six times greater than within S_2
families. Inbreeding increases effectiveness of selection among families but
decreases effectiveness of selection among individuals within a family. Visual selection (Chap. 8) is important for some traits, but visual selection
among progenies rather than within progenies is much more effective. The usual attitude is to use visual selection where possible and inbreed to homozygosity as quickly as possible, i.e., by self-fertilization. The differences
certainly will be greater among progenies than within progenies. Recycling of
desirable factors can be accomplished by intercrossing desirable genotypes
that also have had some testing for combining ability and starting another cycle of inbreeding and selection (Fig. 1.4).

The reasons for use of milder systems of inbreeding have some validity,
but the time required to attain a level of homozygosity acceptable to current
standards of uniformity for use as parental stocks in hybrids discourages
their use. Applied breeders particularly are under pressure to develop new
lines for use in hybrids, and they resort to self-fertilization because of its
rapid approach to homozygosity. Recycling of new lines and additional pedigree selection is initiated by self-fertilization. Hence the general philoso-
phy is to screen greater numbers of lines developed by self-fertilization than
to patiently select fewer lines by inbreeding systems that have a slower fixation of genes. Right or wrong, this system is currently used and will undoubtedly continue to be used in the future. Self-fertilization by pedigree
selection certainly contributes to narrowing the genetic base of parental
stocks used to produce hybrids. It is not the system of inbreeding but the
parental source material that contributes to the problem.

9.3 INBREEDING BECAUSE OF SMALL POPULATION SIZE

Maize breeders conducting recurrent selection programs are continually
faced with decisions concerning the number of progenies of a population for
evaluation and the number of progenies to select for recombination to form the
next cycle. The number of individuals sampled for evaluation and recombination is recurring in each cycle. In most instances greater numbers dictate
compromises--compromises that often are not desirable but are accepted to
maintain continuous recycling of material. The populations included are usually closed populations because the researchers are interested in determining
the long-term effects of the specific selection program (Chap. 7). Eventually
all members of a closed population become related to one another; the time at
which this happens depends on the size of the population. Inbreeding is
avoided as much as possible in recombining and random mating of the resynthe-
sized populations--inbreeding is less than if mating were strictly at random.
But because of the number of individuals that can be reasonably included, inbreeding is inevitable. Once the locus becomes fixed (or inbred) within a

closed population, the only recourse is to outcross to an unrelated population. In small populations, which are common in recurrent selection programs,
the closed population cannot be retained without some inbreeding.

The amount of inevitable inbreeding that occurs in a wholly random mating
population is determined by the number N of unrelated individuals in the population, whose gametes unite at random in every generation. In an isolation
planting pollinated by wind movement of male gametes, a small amount of self-
fertilization can occur, depending on population size. Controlled hand-polli-
nations in a breeding nursery essentially eliminates self-fertilization, but
inbreeding inevitably occurs because of limited population size. Li (1976)
showed that the probability that a gamete will unite with a gamete from the
same individual is $1/N$ and from a different individual is $(N - 1)/N$, including
the possibility of self-fertilization. Because the probability of a gene
identical by descent from full-sib matings is $1/4$ (Chap. 3) and the probability of full-sibs mating together is $2/N$, the proportion of individuals identical by descent becomes $1/(2N)$. Hence $F = 1/(2N)$. If $N = 1$ (self-fertilization), $F = 0.50$; and if $N = 2$ (crosses made between two individuals), $F =$
0.25, coefficient of inbreeding by full-sibbing. In small populations, probabilities of matings among related individuals are high. It has been common
in some recurrent selection studies to recombine ten individuals; the average
expected level of inbreeding would be 5% $(1/20)$.

In the next cycle or generation we have two proportions that contribute
to the level of inbreeding: (1) genes at $1/(2N)$ of the loci identical by descent and (2) proportion of homozygous loci because of inbreeding (F') in the
previous generation. The total inbreeding, as shown by Li (1976) is

$$F = 1/(2N) + [1 - 1/(2N)]F'$$

where $1/(2N)$ is the increment for population size and $[1 - 1/(2N)]F'$ is the
inbreeding already present in the population. This formula is exact and ap-
plies to populations whether they are large or small. Li also gives an example for monoecious individuals in which self-fertilization is prevented. The
number of possible matings among N individuals, therefore, is $N(N - 1)$. This
type of mating is common in recurrent selection studies among the selected individuals used to form the next cycle population. The recurrence relation for
monoecious individuals that does not include self-fertilization is

$$F = [1/(2N)][1 + 2(N - 1)F' + F'']$$

where $1/(2N)$ is the portion due to population size and F' and F'' the increments of inbreeding present in previous cycles. This expression reflects the
number of matings made with the restriction preventing self-fertilization.
For example, expected total inbreeding in the third cycle from recombining ten
individuals, using the first formula, is 0.1426, whereas expected total inbreeding using the latter formula is 0.1402. The difference in expected total
inbreeding reflects the correction for number of matings. Exclusion of self-
fertilization shifts the increment of new inbreeding back to the grandparental
(F'') generation.

If the increment of inbreeding is ΔF, the relation

$$\Delta F = (F' - F)/(1 - F)$$

shows the increase in homozygosity relative to the heterozygosity that is
present. The important feature is the heterozygosity present because once the
loci are fixed in a closed population they remain fixed unless outcrossing occurs. The new inbreeding ΔF measures the porportionate increase in the rate
of inbreeding (Falconer 1960). Inbreeding ΔF and population size N, therefore, have the following relation:

$\Delta F = 1/(2N)$

A maize population of N diploid monoecious individuals in Hardy-Weinberg equilibrium would produce 2N gametes. If we assume the conditions of equal males and females mating at random, the offspring population of N individuals will have the expected distribution of $(pA + qa)^{2N}$, which is the familiar binomial expansion. Gene frequency is expected to be the same, but random sampling from a population can cause random deviations from the expected. If we let $p' - p$ denote the random deviation of gene frequency in the progeny population from the preceeding population, the variance of p becomes

$$\sigma^2_{\Delta p} = [p(1 - p)]/(2N)$$

Since $\Delta F = 1/(2N)$, the increment of the inbreeding coefficient rather than population size can be used to estimate the variance of gene frequency changes:

$$\sigma^2_{\Delta p} = [p(1 - p)]\Delta F$$

Li (1974) used these relations to determine the effective size of a breeding population, based on the results of Wright (1931). The rate of decrease in heterozygosis for N dioecious individuals composed of unequal numbers of males m and females f was $1/(8N_m) + 1/(8N_f)$; for equal numbers of males and females the relation was approximately $1/(2N)$. If we consider an ideal maize population of N individuals having equal numbers of males and females mating at random, it was shown that the variance of gene frequency was $[p(1 - p)]/2N$ and the rate of inbreeding was $1/(2N)$. The effective size of a breeding group of individuals is the actual breeding size compared with the ideal population. Effective size N_e of a population when equal numbers of males and females contribute the same number of gametes to the next generation is 2N, or twice the number of N individuals recombined. With unequal numbers of females and males, the rate of decrease of heterozygosis is $1/(8N_m) + 1/(8N_f)$, which becomes $[(N_m + N_f)]/(8N_mN_f)$ or $1/(2N_e)$. Rearranging, the effective size of the breeding population is $N_e = 4N_mN_f/(N_m + N_f)$, which reduces to 2N for equal numbers of males and females. Males and females are equally used in most maize populations, but if unequal males and females are used, effective size depends much more on the sex that is fewer in number. For example, for 10 males and 100 females, $N_e = 4(10)(100)/110 = 36.4$. For comparison, for 10 males and 10 females equally mated, $N_e = 4(10)(10)/20 = 20$, which is the same as $N_e = 2N$.

In some recurrent selection programs, progenies that have some level of inbreeding are recombined. The effect of inbreeding is to reduce effective size, as shown by Sprague and Eberhart (1977): $N_e = 2N/(1 + F_p)$, where F_p is the coefficient of inbreeding of the parental plants of the lines being recombined. If the recurrent selection is based on S_2 performance itself, one has a choice of recombining remnant S_1 or S_2 seed. If S_1 seed is used, $F_p = 0$ because the gametic array of the S_1 lines is equivalent to an S_0 plant; N_e would be 40 if 20 S_1 lines are recombined. If remnant S_2 seed is used, N_e is reduced to 26.7. Because of the reduction in N_e it would be an advantage to use S_1 rather than S_2 seed for recombination. Another common technique is to recombine elite inbred lines to form synthetic varieties. For this case, N_e would be equal to the number of lines included, e.g., 16 lines recombined would be the effective size of the population. The intercrossed lines would be random mated; but F for $1/(2N)$ is 0.031 unless some outcrossing, intentional or unintentional, is permitted.

Importance of effective size and associated effects of inbreeding with recurrent selection programs was emphasized by Sprague and Eberhart (1977). It is relatively easy to intercross and random mate sizable populations of maize, but choices have to be made relative to selection intensity and numbers of progenies tested (Robertson 1960; Baker and Curnow 1969; Rawlings 1970). Although the number of individuals selected for recombination can cause some problems in handling, usually the testing phase of the program imposes restraints on the number of individuals recombined. The formulas for inbreeding and effective size are for ideal situations, which rarely occur. Individuals intermated in recurrent selection programs probably do not contribute equally to the next generations, but their contributions are not known unless pedigrees are maintained throughout the selection program. Table 9.2, adapted from Sprague and Eberhart (1977), shows expected levels of inbreeding of different cycles of recurrent selection for different effective size of populations.

It is obvious from Table 9.2 that effective size of population has a pronounced effect on expected level of inbreeding and that cumulative effects of inbreeding with cycles of selection is dramatic. If $N = 10$, expected level of inbreeding is 0.22 after five cycles of selection, compared with 0.12 if $N = 20$; expected level of inbreeding is 47.8% less for $N = 20$ than for $N = 10$ after five cycles of selection. After 40 cycles of selection for $N = 10$ and $N = 20$, expected level of inbreeding is only 26.8% less for $N = 20$. Use of S_2 lines for recombination also shows that expected levels of inbreeding are increased considerably. Use of individuals (or lines) that have some inbreeding reduces effective size of the population; e.g., for $N = 10$, N_e is reduced from 20 to 13.3. Hence expected level of inbreeding in the first cycle is increased by 60%, from 0.05 to 0.08.

Because inbreeding in successive cycles depends on inbreeding in previous cycles, expected level of inbreeding from use of inbred individuals will always be greater than that from use of noninbred parents. Additional recombination by random mating will not change expected level of fixation; loci that are fixed will remain so unless mutation or outcrossing occurs. To reduce expected level of inbreeding, one alternative would be to use remnant S_1 seed of the progenitors of the S_2 progenies that were tested in a recurrent selection program. In all instances, expected level of inbreeding becomes an important factor in long-term selection programs. Inbreeding can be reduced by use of either a larger number of individuals (which either reduces selection intensity or increases number of progenies tested) or use of individuals that are not inbred for recombination. It is not unreasonable to expect an individual conducting a recurrent selection based on half-sib, full-sib, S_1, or S_2 progeny evaluation to complete ten cycles of selection. If noninbred individuals are recombined, expected level of inbreeding would be reduced from 0.39 to 0.18 if $N = 25$ rather than 10.

9.4 ESTIMATES OF INBREEDING DEPRESSION

Although a tremendous number of self-pollinations have been made in maize, estimates of the inbreeding depression for different traits are surprisingly few. From the earliest studies in maize (Shamel 1905; East 1908; Shull 1908) effects of inbreeding were obvious: With increasing homozygosity, (1) vigor and productiveness were reduced and traits became fixed and (2) differences among lines increased whereas variability within lines decreased. Effects of inbreeding were interpreted on the basis of Mendelian genetics (Shull 1908) because of fixation of alleles with increased homozygosity.

Table 9.2. Expected levels of inbreeding for the different cycle populations with varying effective size of population.

Cycle of selection	Effective population size (N_e)[†]									
	20		30		40		50		60	
	F = 0	F = 0.5	F = 0	F = 0.5	F = 0	F = 0.5	F = 0	F = 0.5	F = 0	F = 0.5
	$20^†$	$13.3^†$	30	20	40	26.7	50	33.3	60	40
1	0.05	0.08	0.03	0.05	0.02	0.04	0.02	0.03	0.02	0.02
2	.10	.14	.06	.10	.05	.07	.04	.06	.03	.05
3	.14	.20	.09	.14	.07	.10	.06	.08	.05	.07
4	.18	.25	.12	.18	.10	.14	.08	.11	.06	.09
5	.22	.31	.15	.22	.12	.17	.10	.14	.08	.12
6	.25	.35	.18	.26	.14	.20	.11	.16	.09	.14
7	.29	.40	.20	.29	.16	.23	.13	.19	.11	.16
8	.32	.44	.23	.32	.18	.26	.15	.21	.12	.18
9	.35	.48	.26	.36	.20	.28	.16	.23	.14	.20
10	.39	.52	.28	.39	.22	.31	.18	.26	.15	.22
20	.62	.76	.48	.62	.39	.52	.33	.45	.28	.39
40	0.86	0.94	0.73	0.86	0.63	0.77	0.55	0.69	0.48	0.63

[†] $N_e = 2[N/(1 + F)]$, where N is the number of lines recombined and F is the inbreeding of the lines recombined; e.g., F = 0.5 for S_2 lines.

Means and frequency distributions (Jones 1918; Shull 1952) usually were given, but rates of inbreeding depression were not given. In most instances too few lines were included and the observations were made in different years for different generations. Estimates would have been biased by environmental effects in the different years and the errors of estimation large. For the limited number of lines included, detailed data were collected and observations recorded for each line in each generation; information then was interpreted, correctly, relative to the expected level of homozygosity by self-fertilization on the basis of Mendelian segregation of genetic factors.

The classic study of the effects of self-fertilization in maize was reported by Jones (1918, 1939). The study originally started with 12 self-fertilizations in the open-pollinated variety Chester's Leaming in 1904 and the first detailed report was given by East and Hayes (1912). Jones (1918) reported on the first 11 generations of self-fertilization for four inbred lines (two of which were substrains of the original cross) for yield and plant height. Measurements were made in each year of self-fertilization. A gradual reduction in yield and plant height with self-fertilization occurred, but environmental effects for estimating rate of inbreeding depression were obvious. For example, yield of line 1-6-1-3 was 49.2 q/ha in generation 8 vs. 17.3 and 15.9 q/ha in generations 5 and 9, respectively.

Self-fertilization for three lines was continued for 30 generations and the results were given by Jones (1939). Because of the disturbing environmental effects in each generation, he grouped the measurements for each 5-generation segment to average the seasonal fluctuations (Table 9.3). A decreasing trend for yield occurred throughout the 30 generations of self-fertilization, but ear height showed no reduction after 5 generations. Linear regressions for the decrease in means with increased homozygosity were similar for each line for plant height and yield. On the average, plant height was reduced 29% and yield reduced 79% after 30 generations of self-fertilization. Effects of self-fertilization for the two traits were quite different. After 5 generations plant height was reduced 30% with little change thereafter; yield continued to decrease with self-fertilization and was reduced 75% by generation 20.

Table 9.3. Effect of 30 generations of self-fertilization on plant height and yield for three lines of maize (adapted from Jones 1939).

| Generations selfed no. | Plant height (cm) | | | | | Yield (q/ha) | | | | |
| | Inbred lines | | | | % change | Inbred lines | | | | % change |
	1-6	1-7	1-9	Avg.		1-6	1-7	1-9	Avg.	
0	297.2	297.2	297.2	297.2	--	50.6	50.6	50.6	50.6	--
1-5	221.0	205.7	195.6	207.5	30	40.0	31.9	25.6	32.5	36
6-10	246.4	213.4	208.3	222.8	25	28.1	22.5	21.2	23.9	53
11-15	246.4	213.4	210.8	223.5	25	23.8	21.2	16.2	20.4	60
16-20	223.5	215.9	190.5	210.0	29	13.8	15.0	8.8	12.5	75
21-25	205.7	190.5	180.3	192.3	35	12.5	13.1	8.1	11.2	78
26-30	233.7	203.2	195.6	210.8	29	15.0	11.2	5.6	10.6	79
b[†]	-8.7	-11.1	-12.6	-10.8		-6.3	-5.8	-6.5	-6.2	

[†]Linear regression of trait on seven groups of self-fertilization.

Jones (1939) discusses possible reasons for the differences in inbreeding depression for the two traits. Because the level of heterozygosity is reduced 50% on the average with each self-fertilization, the more rapid attainment of stability of means for plant height may be due to less complexity in its inheritance than for yield. If fewer genetic factors influence the expression for plant height, homozygosity would be approached more rapidly than for a more complex trait, particularly grain yield. Grain yield is the ultimate expression of the productivity of a genotype that includes expression of physiological mechanisms throughout the growing season. Average estimates of heritability on either an individual plant or a progeny basis (Tables 5.1 and 5.4) show that the estimate of heritability for plant height is about three times greater than for yield. Jones (1939) produced and tested crosses of the substrains of the 1-6 and 1-7 lines (Table 9.3) and found no evidence that the F_1 significantly differed from either parent.

An illustration of the approach to homozygosity is given in Fig. 9.1 for different numbers of genetic factors. Loss of heterozygosity on the average is less as the number of genetic factors increases. Expected relative heterozygosity is given for the same locus in all individuals in the population, or put in another way, for all loci within the same individual. As the number of loci affecting a trait is increased, loss of heterozygosity for the population is not as rapid as for say one factor, where the heterozygosity is reduced 50% after one generation of self-fertilization. Number of factors affecting most traits considered quantitative in inheritance is unknown, but Jones's data for ear height certainly suggest fewer factors affecting plant height than grain yield. From Fig. 9.1 and Allard (1960) the number of factors affecting yield must be greater than 15 because the approach to homozygosity is slower than shown for 15 factors.

Recent estimates of inbreeding depression in maize have shown the relative rates for different plant and ear traits. In comparison to the study reported by Jones (1939) and others, the recent studies included a greater number of progenies compared in the same environments by use of some experimental

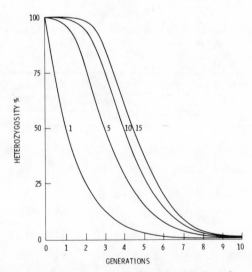

Fig. 9.1. Percentage of heterozygous individuals in the population for each generation of self-fertilization when 1, 5, 10, and 15 allelic pairs are considered.

design. Remnant seeds of different generations of inbreeding and the original
population were compared in the same experiments to obtain estimates of environmental effects and experimental error. Estimates of rates of inbreeding
depression were calculated by regressing the means for different generations
of inbreeding on expected level of homozygosity.

Estimates of inbreeding depression for 17 traits of maize are given in
Table 9.4 for 11 populations, but 5 of these are for advanced cycles of 6 populations. The precision of average estimates for the different traits is not
the same because different numbers of estimates were available, ranging from 1
for ear leaf width to 19 for yield. Estimates of inbreeding depression for
yield are given on a per area basis (q/ha) and per plant basis (g/plant). For
yield, on the average we can expect 0.5 q/ha (1.2 g/plant) decrease for each
1% increase in homozygosity. On a per plant basis, estimates of inbreeding
depression for yield were similar for all populations, ranging from -0.82
(Synthetic O.P. by full-sibbing) to -1.44 (BSSSCO) g/plant.

Although estimates of inbreeding depression for yield are similar, they
were obtained in different populations evaluated in different environments
with theoretical levels of homozygosity attained by different systems of inbreeding. Sing et al. (1967) obtained their levels of inbreeding from the
pedigrees of double-double crosses; Genter (1971) and Harris et al. (1972) obtained their estimates from S_1 lines; and Hallauer and Sears (1973), Cornelius
and Dudley (1974), and Good and Hallauer (1977) obtained their estimates by
comparing different generations of inbreeding attained by either self-fertilization or sibbing. In spite of differences in populations, environments, and
procedures for attaining expected levels of homozygosity, estimates of the
linear decrease in yield per plant for each 1% increase in inbreeding were
nearly the same. Differences in yield loss with inbreeding on a per area basis are greater among the studies than on a per plant basis because different
plant stand densities were used in the different studies. If linear rates of
decrease per plant were adjusted to a constant stand, the rates of decrease in
yield with increasing homozygosity would be more similar. Response in yield
with inbreeding corroborates observations reported from earliest studies in
maize: Yield, or productivity, decreased in all instances.

Estimates of inbreeding depression were negative for all other traits except days to flower and percentage of barrenness. These expressions of inbreeding also agree with previous reported effects; vigor and plant size are
reduced, flowering is later, and barrenness increases. Although observations
of the inbreeding effects have been common, Table 9.4 shows the average expected rates of inbreeding depression, which were not available until recently. As for yield, the average estimates of inbreeding depression for each
trait are similar. Days to flower, for example, had positive estimates in all
instances, ranging from +0.02 to +0.09 for an average of +0.05. Estimates obtained by selfing or sibbing also were similar. Inbreeding is expected to reduce yield and components of yield, reduce plant size, and increase time to
flowering and incidence of barrenness; all estimates confirm previous observations.

Expected changes from inbreeding of the traits listed in Table 9.4 are
shown in Table 9.5 for five levels of homozygosity. Average estimates for
each 1% increase in level of homozygosity (Table 9.4) were used to calculate
the other four levels of homozygosity. The calculated means in Table 9.5 assume that linear regression is a valid mathematical description of the average
performance of a population for different levels of inbreeding. Assuming a
linear relation, we can expect on the average that grain yield is reduced
115.2 g/plant from the noninbred generation to 100% homozygosity, or on a per
unit basis we can expect a 51 q/ha decrease in yield. The population of in-

Table 9.4. Inbreeding depression showing the change in phenotypic means per 1% increase of the coefficient of inbreeding for different maize populations.

Source	Population	Yield q/ha	Yield g/plant	Height, cm Plant	Height, cm Ear	Days to flower	Ear number	Kernel weight g
Sing et al. (1967)	Indian Chief	-.2018	-.8438	-.2400	-.2000	+.0225	-.0004	
	Jarvis	-.2360	-.9865	-.3300	-.1800	+.0410	-.0004	
Genter (1971)	BSSSCO	-.7820	-1.4385					
	BSSS(HT)C7	-.5560	-1.0228					
	VaCBSCO	-.7760	-1.4275					
	VaCBS(S)C4	-.6510	-1.1975					
Harris et al. (1972)	NHG	-.7210	-1.3956	-.2742	-.3888	+.0680		
	NHG(M)C9	-.7172	-1.3883	-.2114	-.2342	+.0752		
	NHG(I)C9	-.6232	-1.2063	-.1570	-.2074	+.0852		
Hallauer & Sears (1973)	BSSSCO	-.4490	-1.1124	-.4800	-.3000	+.0460	---	---
Cornelius & Dudley (1974)	Synthetic O.P.							
	Selfing	-.3855	-.9510	-.5805	-.2900	---	---	-.0427
	Sibbing	-.3328	-.8175	-.5280	-.2480	---	---	-.0441
	Combined	-.3797	-.9355	-.5760	-.2860	---	---	-.0429
Good & Hallauer (1977)	BSSSCO							
	Selfing	-.4628	-1.1528	-.4904	-.3342	+.0535	+.0003	-.1040
	Sibbing	-.4511	-1.1176	-.5268	-.3507	+.0422	+.0004	-.0872
	Sibbing & selfing	-.4588	-1.1366	-.5381	-.3416	+.0459	+.0002	-.1053
	Combined	-.4628	-1.1466	-.4942	-.3210	+.0519	-.0002	-.1006
Good & Hallauer (1977)	BSSS(R)C7							
	Selfing	-.4700	-1.1645	-.5020	-.4130	+.0300	-.0004	-.2030
	Full-sibbing	-.5820	-1.4420	-.6690	-.5920	+.0380	-.0009	-.3170
	n	19	19	15	15	12	8	9
	Average	-.5104	-1.1517	-.4398	-.3124	+.0499	-.0004	-.1163
Levings et al. (1967)	Autotetraploids		-1.8426	-.6189				
Rice & Dudley (1974)	Autotetraploids							
	Selfing		-1.5360	-.8580	-.4740			
			-1.3210	-.5740	-.2350			
	Full-sibbing		-1.1140	-.5580	-.2950			
			-1.1730	-.3770	-.1610			

bred lines on the average is expected to be 44 cm shorter, have 31 cm lower ear placement, and be about 5 days later in maturity. Shull (1908) studied kernel-row number, and from Table 9.5 we can expect to have 1.8 less kernel-row numbers in the population of inbred lines. Incidence of barrenness increases with increased homozygosity, but all other traits show a decrease (Table 9.5).

Percentage of decrease by self-fertilization for an unselected set of inbred lines developed from Iowa Stiff Stalk Synthetic was reported by Good and Hallauer (1977). The relative decreases from the original noninbred population to the S_8 generation (eight generations of self-fertilization) are shown in Table 9.6. The relative decrease from the S_0 to the S_8 generation was greatest for yield, which was reduced 68%. There was a steady decline in yield throughout the range of inbreeding, which agrees with Jones's (1939) data. Plant height, the other trait studied by Jones, was reduced 22% by the S_3 generation with only 2% additional reduction for the five additional generations of self-fertilization, which also agrees with Jones. Ear height had a greater relative reduction with inbreeding, with a decreasing trend in all generations; the change was only 5% after the S_3 generation vs. 30% for the

Oil, %	Cob diam. mm	Leaf width mm	Average barren-ness	Stand	Grain moist. %	Shell-ing, %	Ear, mm Diam.	Length	Kernel row no.	Kernel depth mm
					-.0528	-.0726				
					-.0288	-.0822				
					+.0632	-.0614				
---	-.0350	-.1340			---	---	-.1008	-.4400	-.0180	-.0650
-.0039					-.0036					
-.0027					-.0023					
-.0038					-.0038					
---	-.0240	---	+.0012	-.0994	---	---	-.0870	-.4250	---	-.0680
---	-.0220	---	+.0011	-.1022	---	---	-.0850	-.4180	---	-.0660
---	-.0260	---	+.0010	-.0883	---	---	-.0870	-.4340	---	-.0680
---	-.0250	---	+.0010	-.0974	---	---	-.0860	-.4160	---	-.0660
---	-.0430	---	+.0007	-.0520	---	---	-.0900	-.1800	---	-.0430
---	-.0740	---	+.0006	-.0360	---	---	-.1130	-.1500	---	-.0740
3	7	1	6	6	6	3	7	7	1	7
-.0034	-.0355	-.1340	+.0009	-.0792	-.0046	-.0720	-.0926	-.3518	-.0180	-.0642
								-.0760		

first three generations of self-fertilization. Days to flower increased 25% from S_0 to S_8 and increased throughout the range of self-fertilization, but the greatest one-generation increase occurred from S_0 to S_1. Of yield components, kernel depth exhibited the greatest decrease with inbreeding (37%) and occurred throughout the generations of self-fertilization. Although none of the components of yield was affected by inbreeding to the same extent as yield itself, the cumulative effects of inbreeding for all traits affected yield.

Good and Hallauer (1977) conducted a comprehensive study estimating and comparing rates of inbreeding depression for unselected lines developed by three methods of inbreeding: self-fertilization, full-sibbing, and full-sibbing followed by self-fertilization. Each of the three series of lines was isolated from Iowa Stiff Stalk Synthetic with no intentional selection during the generations of inbreeding. Originally 250 S_0 plant were self-fertilized and 243 pairs of S_0 plants were sib-mated. Two separate samples of S_0 plants were used to develop the self and sib series of lines. The sib series was sib-mated for five generations at which time two progeny rows were grown; sib-mating was continued in one row and self-fertilization was initiated in the second row. Bulk entries of each of the three series of inbred lines for the

Table 9.5. Inbreeding depression expected from the estimates obtained
            from maize populations (Table 9.4) for five levels of
            homozygosity.

Trait	\multicolumn{5}{c}{Level of homozygosity, %}
	1	25	50	75	100
Yield, q/ha	- .5104	-12.76	-25.52	-38.28	- 51.04
g/plant	-1.1517	-28.79	-57.58	-86.38	-115.17
Plant height, cm	- .4398	-11.00	-21.99	-32.98	- 43.98
Ear height, cm	- .3124	- 7.81	-15.62	-23.43	- 31.24
Days to flower, no.	+ .0499	+ 1.25	+ 2.50	+ 3.74	+ 4.99
Ear number	- .0004	- 0.01	- 0.02	- 0.03	- 0.04
Grain moisture, %	- .0046	- 0.12	- 0.23	- 0.34	- 0.46
Shelling, %	- .0720	- 1.80	- 3.60	- 5.40	- 7.20
Ear diameter, mm	- .0926	- 2.32	- 4.63	- 6.94	- 9.26
Ear length, mm	- .3518	- 8.80	-17.59	-26.38	- 35.18
Kernel-row number	- .0180	- 0.45	- 0.09	- 1.35	- 1.80
Kernel depth, mm	- .0642	- 1.60	- 3.21	- 4.82	- 6.42
Kernel weight, g	- .1163	- 2.91	- 5.82	- 8.72	- 11.63
Oil, %	- .0034	- 0.08	- 0.17	- 0.26	- 0.34
Cob diameter, mm	- .0355	- 0.89	- 1.78	- 2.66	- 3.55
Leaf width, mm	- .1340	- 3.35	- 6.70	-10.05	- 13.40
Average barrenness	+ .0009	+ 0.02	+ 0.04	+ 0.07	+ 0.09
Stand, no.	- .0792	- 1.98	- 3.96	- 5.94	- 7.92

generations of inbreeding were evaluated in five-replicate experiments at nine
Iowa environments. Rates of inbreeding depression were determined for each
series of lines (Good and Hallauer 1977) and are included in Table 9.4.

Although some statistically significant differences occurred among the
linear regression coefficients, the differences were small (Table 9.4). Linear regressions among the three series of lines were significantly different
for plant height, cob diameter, yield, 300-kernel weight, stand, and barrenness but, for example, linear regression estimates for yield (-1.15, -1.12,
and -1.13 for self-fertilization, full-sibbing, and full-sibbing and self-fertilization, respectively) did not have much practical significance. Relative
rates of inbreeding depression (Fig. 9.2) emphasize the similarity of the decrease in yield by the three methods of inbreeding. There is little indication that inbreeding by full-sibbing rather than selfing will result in more
vigorous lines, although the rate of inbreeding depression for yield is
slightly less for the two instances that self-fertilization and full-sibbing
were directly compared (Cornelius and Dudley 1974; Good and Hallauer 1977, Table 9.4). Good and Hallauer found that the linear model accounted for more
than 99% of the variation for yield.

Good and Hallauer also compared the means of the traits for the three
methods of inbreeding at comparable levels of homozygosity. The theoretical
levels of homozygosity were not exactly the same, but they were similar enough
for comparative purposes (Table 9.7). In very few instances were the means at
similar levels of homozygosity significantly different. Probably the most important difference was for yield at the greatest level of homozygosity: Inbred lines developed by full-sibbing for ten generations and selfing for three

Table 9.6. Observed means by selfing in Iowa Stiff Stalk Synthetic (BSSS) and percentage changes in traits relative to original noninbred population.

Generation	Yield, g X̄	%	Plant height cm X̄	%	Ear height cm X̄	%	Days to flower X̄	%	Ear number X̄	%	Ear diam. mm X̄	%	Ear length, mm X̄	%	Kernel depth, mm X̄	%	Kernel weight, g X̄	%	Cob diam. mm X̄	%	Barren-ness X̄	%	Stand X̄	%
Single crosses[+]	172.3	102	192.1	99	95.7	97	26.2	98	.12	120	48	100	185	103	18	95	78.7	102	30	100	.01	25	41.9	95
Double crosses[+]	165.5	98	195.4	101	97.9	99	26.9	101	.08	80	48	100	184	102	18	95	77.0	100	30	100	.01	25	44.5	100
BSSS S$_0$	168.6	100	194.1	100	98.9	100	26.7	100	.10	100	48	100	180	100	19	100	77.3	100	·30	100	.04	100	44.3	100
S$_1$	112.1	66	175.5	90	84.1	85	30.4	114	.12	120	45	94	166	92	16	84	71.3	92	29	97	.04	100	39.1	88
S$_2$	87.0	52	162.0	83	73.9	75	30.3	113	.13	130	43	90	155	86	15	79	70.3	91	28	93	.07	175	37.7	85
S$_3$	75.1	44	152.4	78	69.6	70	31.1	116	.13	130	42	88	150	83	14	74	69.6	90	28	93	.10	250	35.3	80
S$_4$	64.9	38	150.5	78	67.5	68	32.1	120	.13	130	41	85	143	79	13	68	69.2	90	28	93	.13	325	33.0	74
S$_6$	61.2	36	148.1	76	65.7	66	32.3	121	.12	120	41	85	143	79	13	68	67.5	87	28	93	.12	300	33.9	76
S$_8$	54.4	32	147.9	76	64.3	65	33.5	125	.13	130	40	83	140	78	12	63	65.9	85	28	93	.14	350	35.0	79
(SE)X̄	2.8		2.0		1.6		0.6		.01		0.2		1.4		0.2		0.8		0.2		.01		0.8	
Change S$_0$-S$_8$, %	-68		-24		-35		+25		+30		-17		-22		-37		-15		-7		+250		-21	

[+] Bulk of crosses among S$_7$ lines included as checks.

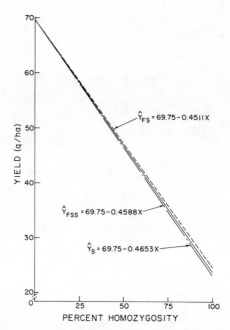

Fig. 9.2. Inbreeding depression of yield in Iowa Stiff Stalk Synthetic by
selfing (S), full-sibbing (FS), and full-sibbing followed by three
generations of selfing (FSS).

generations had an average yield that was significantly greater than the lines
developed by self-fertilization (25.7 vs. 22.0 q/ha). It is doubtful that the
yield advantage is great enough to warrant the five additional generations of
inbreeding required to attain the same level of homozygosity.

The data in Table 9.7 substantiate the theoretical levels of homozygosity
for the three methods of inbreeding. It seems that for unselected lines the
effects of homozygosity are the same whatever method of inbreeding is used to
attain it. Empirical data agree with theoretical expectations for the different methods. Except for belief in the effectiveness of visual selection for
milder forms of inbreeding, it does not seem that other forms can be recommended. Effects of inbreeding are the same whatever method is used to attain
a given level; e.g., 50% homozygosity can be attained by one generation of
self-fertilization or three generations of full-sibbing, but in only one instance (ear diameter) were the generation means significantly different. At
expected homozygosity levels of 99%, the generations for the three methods of
inbreeding significantly differed in five instances. In all except one (barrenness), the means of the lines developed by ten generations of sib-mating and
three generations of self-fertilization were greater than those obtained by
eight generations of self-fertilization. Although significant, the differences for the five traits were relatively small in all instances.

The comprehensive inbreeding study conducted by Good and Hallauer (1977)
included 22 levels of homozygosity for the three series of lines developed by
self-fertilization, full-sibbing, and combinations of self-fertilization and
full-sibbing. Additionally, a level of homozygosity for 12.5% was established
from half-sibs. Linear and quadratic regression models were fitted for the
means of each trait for the 22 levels of homozygosity. Linear regression coefficients for the linear model were significantly different from zero for all

Table 9.7. Comparisons between generation means with comparable levels of expected homozygosity for lines developed by three methods of inbreeding.

Generation[+]	Expected homozygosity, %	Yield q/ha	Yield g/plant	Plant height cm	Ear height cm	Days to flower[§]	Ear no.	Ear diam. mm	Ear length mm	Kernel depth mm	Kernel weight g	Cob diam. mm	Average barrenness, %	Stand no.
S_0	0.0	68.0	168.6	194.1	98.9	26.7	1.10	48	180	19	77.3	30	0.04	44.3
S_1	50.0	45.2	112.1	175.5	84.1	30.4	1.12	45[a]	166	16	71.3	29	0.04	39.1
FS_3	50.0	45.9	113.5	170.2	81.4	30.6	1.08	46[b]	164	16	72.7	29	0.04	38.2
FS_6	73.4	37.7	93.5	159.3	73.5	29.6	1.13	44[b]	157	15	70.9	29[b]	0.06	38.7
S_2	75.0	35.1	87.0	162.0	73.9	30.3	1.13	43[a]	155	15	70.3	28[a]	0.07	37.7
FS_8-S_1	82.6	33.8	83.8	153.6	69.3	30.6	1.12	43	152	14[a]	69.8	28	0.08	35.8
FS_5-S_1	83.6	31.9	79.0	152.8	69.4	30.5	1.12	43	150	15[b]	68.4	28	0.08	38.3
S_3	87.5	30.3[b]#	75.1[b]	152.4	69.6	31.0	1.13	42	150[b]	14	69.2	28	0.10	35.3
FS_9	85.9	31.2[b]	77.4[b]	150.2	66.8	30.1	1.13	42	151[b]	14	70.0	28	0.02	36.9
FS_{10}	88.6	26.7[a]	66.1[a]	148.4	67.6	30.7	1.13	41	145[a]	13	69.8	28	0.13	35.3
S_4	93.8	26.2	64.9	150.5	67.5	32.1	1.13	41	143	13	69.2	28	0.13[b]	33.0
FS_5-S_2	91.8	27.4	67.9	147.9	67.1	31.7	1.12	41	146	13	67.5	28	0.08[a]	36.6
S_6	98.4	24.7	61.2	148.1	65.7	32.3	1.12	41[b]	143	13	67.5	28	0.12	33.9
FS_5-S_4	98.0	26.1	64.6	144.2	65.7	31.1	1.12	41[b]	144	13	68.3	28	0.12	34.6
$FS_{10}-S_2$	97.2	24.9	61.7	146.1	63.6	31.5	1.15	40[a]	141	13	69.4	28	0.12	33.3
S_8	99.6	22.0[a]	54.4[a]	147.9	64.3	33.5	1.13	40	140	12[a]	65.9[a]	28[b]	0.14[b]	35.0
FS_5-S_6	99.5	22.8[ab]	56.2[ab]	142.6	65.3	32.2	1.12	40	138	12[a]	67.5[ab]	27[a]	0.09[a]	33.8
$FS_{10}-S_3$	98.6	25.7[b]	63.7[b]	144.8	67.0	31.6	1.13	40	142	13[b]	69.6[b]	28[b]	0.12[ab]	33.0

[+] S_i, FS_i, and FS_i-S_i indicate number of generations of self-fertilization, full-sibbing, and full-sibbing followed by self-fertilization, respectively.

Generation means with different letters are significantly (P ≤ 0.05) different. All comparisons without superscripts are not significantly different.

[§] Days after July 1.

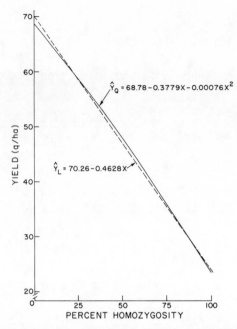

Fig. 9.3.   Inbreeding depression of yield in Iowa Stiff Stalk Synthetic for 22
            entries from three series of inbreeding.

12 traits; linear regressions were positive for days to flower, second ears
per plant, and average barrenness per plant and were negative for all other
traits. Fitting the quadratic model resulted in nonsignificant linear regressions for ear diameter, second ears per plant, and average barrenness per
plant and in nonsignificant quadratic regressions for days to flower, plant
and ear heights, cob diameter, stand, and second ears per plant. Although
quadratic regressions were significant for 6 of the 12 traits, amount of total
variation among the levels of homozygosity accounted for by quadratic regression was small. Linear prediction for yield, for example, accounted for 99.1%
of total variation. The quadratic parameter for yield, which was highly significant, accounted for only 0.5% more of the total variation for yield. Predicted linear and quadratic models and observed means for yield are shown in
Fig. 9.3. There is a slight quadratic trend but the inbreeding depression for
yield is essentially linear with increased homozygosity.
     Wright (1922b) determined for independent loci that the mean performance
of a trait is proportional to the decrease in heterozygosis (or increase in
homozygosis), regardless of the number of alleles or level of dominance at
each locus; i.e., a linear decrease would be expected with a decrease in heterozygosis. A curvilinear response is interpreted as evidence of interallelic
interactions or epistasis. All the inbreeding studies relating mean performance to level of heterozygosity seemed to be adequately described by a genetic
model that includes additivity of unlinked locus effects. Linear regression
coefficients generally accounted for most of the variation among the generations of inbreeding; in most instances the proportion of the total sums of
squares among generations explained by linear regression or percentage of homozygosity F exceeded 90%, and usually it was greater than 98% to 99%. Dominance (intralocus interactions) is necessary to exhibit inbreeding depression;
linear regression measures the net dominance deviation. The absence of any

appreciable deviation from regression does not mean that interlocus interactions (or epistasis) do not exist; it only reflects the net interlocus interactions, which can be caused by cancellation effects of interlocus interactions. Because there are generally no significant deviations from the linear model, the effects of epistasis on inbreeding depression do not seem to be important. The linear regression model seems to be appropriate for describing average performance of different generations of inbreeding regardless of the system of inbreeding used in diploid and autotetraploid maize.

9.5 FREQUENCY OF USEFUL LINES

Lindstrom (1939) summarized a survey of the maize breeders in the USDA and in 24 U.S. Agriculture Experiment Stations to determine the number of inbred lines that had been isolated. He reported that 27,641 lines had been selfed for one to three years and that only 677 (2.4%) were presumably useful lines. His assessment of the quality of the inbred lines was rather pessimistic because every one of the useful lines was seriously defective for one or more traits, including poor yields. Lindstrom concluded that probably 100,000 lines had been tested for at least three years but very few were useful. He questioned the use of self-fertilization as the proper inbreeding method for isolating inbred lines. Whatever inbreeding system is used, it is unlikely that the relative number of useful lines would have changed. If 100,000 inbred lines were tested from say 1920 to 1939, one hesitates to even guess the number of lines that have been tested since 1939.

Public agencies were the principal developers of inbred lines until 1939. Since World War II and particularly during the 1960s and 1970s there has been a tremendous expansion in the commercial sector in number and size of breeding programs primarily designed to isolate and test new inbred lines. The cooperative state-federal maize breeding program at Ames, Iowa, includes on the average about 360 testcrosses each year, which usually involve the initial testing of S_2 or S_3 lines for combining ability for yield, lodging, and grain moisture at harvest. These lines have survived previous selection for various agronomic traits in the breeding nursery and in some instances have had some previous testing in recurrent selection programs. If we consider 360 testcrosses as an average for each year of the 40 years since 1939, 14,400 lines have merited testing. If we extrapolate from the program to include 25 public and 25 private breeding programs, we have an average of 18,000 lines tested each year or 720,000 since 1939. The numbers seem large, but they may be conservative because more than 25 companies also are involved in maize breeding and several of the larger companies have more than one breeding station--some have more than ten breeding stations to select lines for use in particular environments. The number of lines that have been inbred, selected, and advanced to testcrossing may well be in the neighborhood of 1 million. Recent surveys reported by Sprague (1971) and Zuber (1975) list the inbred lines and the extent of their use in producing hybrid seed. The 38 inbred lines developed by public agencies were used in 0.1% or more of the seed requirements for the 1976 U.S. crop year, as reported by Zuber. The relative proportion of the 38 useful inbred lines relative to the 360,000 tested by public agencies is only 0.01%, which means that only 1 in 10,000 S_2 or S_3 lines tested were eventually used to any extent in commercial hybrids.

There are several pitfalls in calculating the estimates. In some ways the estimate of 0.01% seems too small. Again using estimates for the Iowa location, four inbred lines (B37, B73, B14A, and B57--excluding B68, which is a

recovered B14A) were included in the survey reported by Zuber (1975). Relative to the estimated 14,400 testcrosses for the Iowa location, the percentage
of useful lines is 0.027, an estimate about three times greater than for all
stations. This estimate seems extremely small, but it seems realistic compared to the number of lines inbred to the S_2 or S_3 generation and the number
of lines that are eventually used to any extent in commercial hybrids. The
base of reference may not seem valid, but several of the lines (B14, B37,
Oh43, and C103) were developed during this period of 40 years so that the total span of time seems correct. Also several of the 38 lines included in the
survey are recoveries (A632, A619, Va26, H84, B68, etc.) of lines and may in-
flate the actual number of usable lines developed. The present usage of lines
is primarily in single crosses rather than double crosses. Use of single
crosses would require fewer usable lines than double crosses, and the few
elite lines would be used extensively, as shown by Zuber; hence fewer lines
would be used for different combinations of single crosses than in double
crosses. Weighing the pros and cons for calculating the percentage of usable
lines from inbreeding and testcrossing programs, 0.01% may be a good estimate,
but it is considerably lower than the 2.4% reported by Lindstrom (1939).

9.6 TYPES OF HYBRIDS PRODUCED FROM INBRED LINES
    Shull's (1908, 1909, 1910) original concept was the production and grow-
ing of single-cross hybrids, but the costs of seed production seemed to limit
its usefulness. This limitation was overcome with Jones's (1918) suggestion
that double-cross hybrids can be produced from two single-cross hybrids to reduce the costs of seed production; consequently, double-cross hybrids became
rapidly accepted in the United States (Fig. 1.2). Single-cross hybrids, however, have been gradually replacing double-cross hybrids in the U.S. Corn Belt
and other areas of the world since 1960. Some estimates indicate more than
80% of the U.S. Corn Belt is currently planted to single-cross hybrids. In
addition three-way crosses and modified single crosses have been used in some
instances, mainly because some problems of seed production are alleviated by
use of the single cross as a seed parent. Another modification used in some
instances is commonly referred to as a modified single cross, in which either
one or both parents are crosses of related lines; i.e., parent lines have a
common parent in their ancestry. Usually the seed parent is produced from a
cross of two closely related lines to enhance the quantity of seed produced
because of some hybrid vigor from the cross. The degree of ancestry of the
parents included in a modified single cross is not exact and varies among modified single crosses. For example, one may consider (H84 × H93)Va26 a modified single cross. Both H84[(B37 × GE440)HtHt] and H93[(B37 × GE440)B37[4]Ht-
Ht] have B37 in their ancestry in different proportions, some hybrid vigor is
expressed in the cross of H84 X H93, and both H84 and H93 combine well with
Va26. The parents of this example may be more or less related to other examples referred to as modified single crosses.
    Different types of maize hybrids have advantages and disadvantages. Ini-
tially, the low vigor and productiveness of the inbred lines available dis-
couraged the production of single-cross hybrid seed at an acceptable cost.
Recycling of lines and selection have produced lines that are more productive
than the earlier derived lines, but increased usage of fertilizers and herbicides and improved husbandry practices also have contributed to the feasibility of producing single-cross hybrid seed. Schnell (1973) discussed three main
aspects in his comparisons of different types of hybrids: (1) uniformity, (2)

yield, and (3) stability. A fourth aspect may be considered: the relative
simplicity of selecting and testing the three types of hybrids.

Uniformity. Uniformity has been an important factor for the acceptance of
single crosses. If good practices are used in the production of a single-
cross hybrid, single crosses are uniform genotypically and phenotypically. A
field of a single-cross hybrid is very attractive to the producer because of
its uniformity of appearance, maturity, and harvesting characteristics. A
single-cross is genetically homogeneous, which may be a disadvantage. If
large areas are planted to the same single-cross hybrid, uniformity will be
evident not only within a given field but also in the large area. This situation, for a particular maturity zone, existed when the 1970 *Helminthosporium
maydis* outbreak occurred in the United States. It was caused by the use of T-
cytoplasm in the production of the hybrids. It is difficult to fathom that a
situation similar to the use of T-cytoplasm would occur over as an extensive
area for genotypic uniformity. The extreme uniformity of single crosses, however, does cause some concern about the host-pathogen relationship within each
field and area that could develop as genetic diversity among and within single
crosses is reduced.

Yield. One of the advantages often given for the use of single-cross hybrids
is that greater yields are possible. Intuitively, this seems correct because
it is easier to identify two inbred lines that are superior yielding in crosses than to identify three or four. Superior performing single crosses are
identified in the first stages of testing; and single-cross data are used to
predict the performance of three-way and double-cross hybrids, usually based
on the average performance of the nonparental single crosses (Chap. 10). A
summary of comparisons among types of hybrids given in Table 9.8 includes number of hybrids included in comparisons, average yield, and relative yields,
with single-cross hybrids the base of comparison. There is some variation in
advantage of one type of hybrid relative to another, but the differences generally are not large. Weighted and unweighted averages are very similar in
Table 9.8.

        Average comparisons, however, can be misleading in some instances, and
two examples illustrate some of the comparisons. Schnell (1973) examined in
some detail the data reported by Weatherspoon (1970), who used nine unrelated
inbred lines to produce the 36 possible single crosses and balanced sets of 36
three-way and double crosses. Schnell's summary of Weatherspoon's data (Table 9.9) shows that the single crosses had greater average yield, greater
standard deviation, and greater range from low to high crosses than did double
crosses, with three-way crosses intermediate. The best single cross was 13.8
and 8.6 q/ha greater than the best double and three-way crosses, respectively.
Expected greatest yield also was obtained for each type of cross predicted on
the basis of the largest deviate expected in a sample of 36 crosses. In all
instances, it seems that the best single crosses have a striking advantage
over the best three-way and double crosses. In predicting the best cross for
each type, a constant k value was used. Schnell emphasized, however, that all
possible crosses among the nine inbred lines were produced and tested but only
a balanced set of the possible three-way and double crosses were included.
The single-cross information was used to predict the 252 possible three-way
and 378 double crosses. Greatest yields of the predicted three-way and double
crosses (Table 9.9) are only 1.5 and 2.4 q/ha, respectively, less than the
single crosses. Predictions may be biased with regard to the average and individual crosses, but Otsuka et al. (1972) indicated that yields of the better

Table 9.8.   Summary of the average and relative yields (q/ha) for single, three-way, and double-cross hybrids of maize.

Source	Single crosses			Three-way crosses			Double crosses		
	Yield	No.	%	Yield	No.	%	Yield	No.	%
Doxtator and Johnson (1936)	47.0	6	100	51.4	2	109.4	44.9	3	95.5
Stringfield (1950)	53.1	6	100	53.1	12	100	51.2	3	96.4
Jones (1958)	44.4	317	100	--	--	--	45.0	483	101.4
Jugenheimer (1958)	65.4	6	100	62.3	12	95.3	61.3	3	93.8
Sprague et al. (1962)	68.1	60	100	66.6	60	97.8	--	--	--
Sprague and Thomas (1967)	75.3	15	100	73.8	60	98.0	--	--	--
Eberhart and Hallauer (1968)	71.0	6	100	70.7	12	99.6	70.4	3	99.2
	70.4	6	100	71.2	12	101.1	71.1	3	101.0
Eberhart and Russell (1969)	71.0	45	100	--	--	--	69.6	45	98.0
Weatherspoon (1970)	65.1	36	100	62.0	36	95.2	60.3	36	92.6
Wright et al. (1971)	47.8	150	100	48.4	600	101.2	--	--	--
Stuber et al. (1973) - U[†]	59.4	84	100	60.4	168	101.7	59.2	42	99.7
S	60.6	84	100	61.4	168	101.3	61.1	42	100.8
Lopez-Perez (1977) - S[‡]	92.2		100	87.4		94.8	89.9		97.5
FS	89.0		100	89.3		100.3	88.0		98.9
Unweighted	65.3	15	100	66.0	13	100.4	64.3	12	98.6
Weighted	53.5	821	100	55.5	1142	103.8	49.8	663	93.1
	64.0[§]	354	100	63.3[§]	542	98.9	62.5[§]	180	97.7

[†]Hybrids produced among selected and unselected lines.

[‡]Hybrid bulks from lines developed by selfing (S) and full-sibbing (FS) systems of inbreeding.

[§]Averages omitting Jones (1958) and Wright et al. (1971) for single crosses, Wright et al. (1971) for three-way crosses, and Jones (1958) for double crosses.

three-way and double crosses were underestimated from the nonparental single crosses.

Jones (1958) made a comparison among 317 single and 483 double crosses available from field trials grown in Iowa in 1951 (Table 9.10). Single crosses were made by crossing in all combinations a series of highly selected inbred lines, and double crosses were produced from a series of single crosses. Mean yields of the two types of hybrids were not different, but there were differences in the range and distribution of the two types. No double crosses were represented in the two extreme low and high classes, and single crosses had a distinctly bimodal distribution. Single crosses also had a greater standard deviation than double crosses. Although only two types of hybrids were included, comparisons for single and double crosses were similar to those reported by Weatherspoon (1970); i.e., single crosses had a greater variation in yield and a greater standard deviation.

The differences in yield among the three types of crosses do not seem to be as great as expected. Cockerham (1961) showed an advantage in selecting among single crosses compared with three-way and double crosses for lines developed from a specified population regardless of the type of gene action expressed in the crosses. Stuber et al. (1973) presented data that conformed with the genetic theory. If only additive genetic effects are important, the expected selection advance is twice as great among single crosses as among

Table 9.9.   Yield (q/ha) distributions of balanced sets of 36 single,
three-way, and double crosses (Schnell 1973).

Crosses	Average	Standard deviation	Extremes		Expected greatest cross[†]
			Least	Greatest	
Single	65.1	8.8	43.6	81.5	83.7
Three-way	62.0	6.2	47.7	72.9	75.1
Double	60.3	3.8	54.0	67.7	68.3
Predicted					
Three-way	65.1	6.4	47.4	80.0	
Double	65.1	4.8	52.5	79.1	

[†]$\overline{X} + k\hat{\sigma}_x$, where k = 2.12, the greatest expected deviate in a sample of 36.

double crosses. As the relative importance of nonadditive effects (dominance and epistasis) increases, the relative advantages of selecting among single crosses are even greater.

It is shown in Chap. 5 that the estimation of epistatic variance for maize populations is not satisfying. But comparisons among means of genetic populations formed from specific pairs of genotypes invariably show evidence for significant net epistatic effects. If epistatic effects contribute to the heterosis expressed in a cross of two inbred lines, single crosses are superior to three-way and double crosses because the unique epistatic combinations of the single cross are disrupted in the gametes of the single-cross parents used to produce the three-way and double crosses. If unique combinations of epistatic effects are important in single crosses, a greater range of yields is expected for single crosses than for three-way and double crosses, as demonstrated by Jones (1958) and Weatherspoon (1970).

Sprague et al. (1962) and Sprague and Thomas (1967) made comparisons of single and three-way crosses for two sets of inbred lines. Sprague et al. used a highly selected set of six inbred lines from different sources and found that 18 of 20 single-cross sets yielded more than the corresponding three-way sets. The highly selected lines were identified on the basis of their performance in crosses; selection and testing apparently identified lines that contributed unique combinations of epistatic effects to their crosses. Sprague and Thomas used a set of unselected inbred lines developed from the Midland variety; they found 11 of 19 instances where the single crosses were higher yielding than the three-way crosses. The approximate equivalence of the single- and three-way cross yields produced from their unselected lines conforms with the expectation that any selection involved was neutral with respect to types of gene action. The data of Lopez-Perez (1977)

Table 9.10. Frequency distributions for 317 single and 483 double crosses for yield (Jones 1958).

Crosses	Class centers (q/ha)														$\overline{X}$	SD
	23.8	26.9	30.0	33.1	36.2	39.4	42.5	45.6	48.8	51.9	55.0	58.1	61.2	64.4		
Single	6	16	30	27	5	7	16	53	63	50	28	13	1	2	44.4	9.5
Double	--	--	4	18	32	55	89	89	119	52	24	1	--	--	45.0	5.6

show similar yields of the three types of hybrids produced from unselected lines.

The contrasts of the results for specific studies and the averages of Table 9.8 are not as serious as they appear. It seems epistasis is important in single crosses, but sampling would dampen the results. For a fixed set of lines, the number of possible combinations of three-way and double crosses is much greater than for single crosses. If all possible three-way and double crosses were tested extensively, it may be possible to identify three-way and double crosses that also have unique combinations of epistatic effects. The unique combinations may be different from those expressed in single crosses because of genetic recombination of single-cross parents. Because the number of possible crosses (e.g., for nine lines, we have 36 single-, 252 three-way, and 378 double crosses) increases dramatically for a fixed set of inbred lines, prediction methods of Jenkins (1934) have been used to identify the particular three-way and double crosses for testing. Otsuka et al. (1972) and Stuber et al. (1973) found that use of nonparental single crosses underestimated three-way and double cross performances, but they concluded that the underestimates seem small and the biases due to epistasis and genotype-environment interactions were similar; the same conclusions are reported by Eberhart and Hallauer (1968). None of the studies recommended use of more complex procedures to predict three-way and double crosses.

Stability. Single crosses are heterozygous, but they are homogeneous with respect to the genotype because each plant is the same. The lack of genetic variability within a single cross has concerned maize breeders and producers on both an individual field and an area basis. Although single crosses may be superior to three-way and double crosses, the consistency of performance over environments (or stability) was of concern. It seemed that external environmental factors (weather, soil, and pests) would have a greater effect on the genotypically uniform single crosses than on the genotypically variable three-way and double crosses. Jones (1958) examined the comparison of single and double crosses and concluded that the greater genetic homeostasis of double crosses was more important than the problems associated with producing single crosses. Federer and Sprague (1947) and Sprague and Federer (1951) estimated the genotype-environment component of variance for three types of hybrids and found that the interaction component was greater for single crosses than for double crosses, indicating a more sporadic performance of single crosses in different environments. The mixture of related genotypes of double crosses, therefore, showed smaller hybrid-environment interactions than the single crosses and presumably had greater stability over environments than single crosses. No comparisons were made for individual hybrids within types. Gama and Hallauer (1980) found no differences in relative stability of single-cross hybrids produced from selected and unselected inbred lines.

Eberhart and Russell (1969) used their stability analysis (Eberhart and Russell 1966) for a diallel set of 45 single crosses and a corresponding balanced set of double crosses derived from 10 selected inbred lines. The 90 hybrids were grown at 21 U.S. Corn Belt environments and the traditional analysis of variance showed nearly three times greater variation among single crosses (490.0) than among double crosses (170.2). The hybrid-environment interaction was slightly greater for the single crosses (55.5 vs. 37.6). The stability analysis identified two single crosses as being as stable as any of the double crosses, and these two single crosses outyielded four commercial single crosses by 11% and three commercial double crosses by 13%. Comparisons of the highest yielding single and double crosses for the 21 environments showed that the differences among the stability parameters were not distinctly

different. On the average double-cross hybrids were slightly more stable with
respect to stability parameters, but single crosses were slightly higher
yielding in all types of environments (poor, average, and good); none of the
differences, however, was significant. Eberhart and Russell (1969) concluded
that single crosses as stable as double crosses over environments could be
identified, but that extensive testing over a wider range of environments
would be needed to identify good yielding single crosses for commercial use.
It is doubtful that the resources needed to identify stable single crosses
would be any greater than needed for three-way and double crosses. To provide
valid data for predicting three-way and double crosses, it is necessary to
have adequate yield testing of single crosses. Single-cross data are used to
predict the combinations of three-way and double crosses, which will then need
to be adequately tested to determine the best performing and most stable
three-way and double crosses. The two-stage testing required for three-way
and double crosses fulfills the wider testing requirement needed to identify
stable single crosses. Hence it does not seem that additional resources are
required to identify superior, stable single crosses.

Simplicity. All aspects of comparisons among single, three-way, and double
crosses do not show any striking advantages for a type except for the simplicity of studying and testing single crosses. Good yielding single crosses with
desired agronomic traits are usually easier to identify than the more complex
hybrids, but this does not imply that equally good yielding three-way and double crosses could not be identified. As shown in Table 9.8, good yielding
three-way and double crosses can be identified if sampling of the possible
three-way and double crosses is included; this is prohibitive in most instances for a given set of lines because the number of possible combinations
increases rapidly.

The advantage of simplicity for single crosses also is evident in the mechanical procedures of breeding and production. Greater breeding effort can
be allotted to line development if resources are not needed to produce the
three-way and double-cross seed for testing. Testing is simpler because only
one-stage testing is required rather than two-stage for the more complex hybrids. Production of single crosses is simpler because they require only
three isolation fields (two for foundation seed stocks and one to produce hybrid seed), whereas perhaps seven isolation fields (four for the foundation
seed stocks, two for parental single crosses, and one to produce double-cross
hybrid seed) are needed to produce double crosses. The difference in number
of isolation fields is usually not as great for the two types of hybrids because foundation seed stock organizations are available to produce the parental stocks. Production of single-cross seed is certainly more expensive per
unit area than double-cross seed, but this disadvantage also must be considered with respect to maintenance of two seed stocks for single crosses vs. six
for double crosses. The other advantages of single crosses and the ratio of
seed costs to other production costs have encouraged the use of single-cross
hybrids.

9.7 HETEROZYGOSITY-PERFORMANCE RELATION

Recent studies estimating the rates of inbreeding depression by the regular inbreeding systems supplement studies relating the level of heterozygosity
and performance of some quantitative trait as evidence for the presence (or
absence) of epistatic effects. The basis of this relation stems from Wright's
(1922b) investigations of heterosis and inbreeding depression in guinea pigs.
These experiments were the basis for Wright's general conclusion that the

change in vigor is directly proportional to the change in heterozygosity in
the population. Wright summarized his findings by stating, "A random-bred
stock derived from n inbred families will have (1/n)th less superiority over
its inbred ancestry than the first cross or a random-bred stock from which the
inbred families might have been derived without selection." The results of
Wright can be summarized mathematically as

$$F_2 = F_1 - (F_1 - \bar{P})/n$$

where n is the number of inbred parents and $\bar{P}$ is the mean of all the parents.
Application of the heterozygosity-performance relation has been reported in
maize. The effects of inbreeding were discussed by Gilmore (1969).

Kiesselbach (1933) reported data that he interpreted to support Wright's
conclusion. Yield data from F_1 hybrids with 2, 4, 8, and 16 inbred parents
and their F_2 generations seemed to indicate that yield was highly correlated
with heterozygosity. Yield data of the inbred parents were not available, but
Kiesselbach summarized his findings that the decrease in yield of any hybrid
in the F_2 generation is equal to half the difference in yield between the F_1
and the mean yield of the open-pollinated varieties. The reduced yields of
the F_2 generations were attributed to a reduction in the number of favorable
growth factors as a consequence of close breeding. Kiesselbach further concluded that for any orthodox hybrid in which the parents were composed of
equal numbers of parent inbreds, the reduction in yield tended to be inversely
proportional to the number of lines involved.

Neal (1935) used Wright's formula to predict the performance of advanced
generations in maize hybrids. He used ten single, four three-way, and ten
double-cross hybrids; their F_2 generations; and F_3 generations of six of the
single crosses to test the validity of Wright's relation in maize. Neal reported that the single, three-way, and double crosses lost 29.5%, 23.4%, and
15.8% vigor, respectively, in the F_2 generation relative to the F_1, which was
in close agreement with the predicted losses of 31.0%, 21.0%, and 15.3% for
the three respective types of hybrids. Neal concluded that vigor decreased
according to Wright's formula. Wright's formula is based on the assumption of
arithmetic gene action. Powers (1941) reanalyzed Neal's data using arithmetic
and geometric models; best agreement between observed and predicted values was
obtained with the arithmetic model.

Kinman and Sprague (1945) also studied the type of model involved in loss
of vigor with advanced generations. They compared the observed and expected
performance of 10 maize inbred lines, the 45 possible single crosses, and
their F_2 generations by use of arithmetic and geometric models. They also
concluded that the arithmetic model was a closer approximation to the observed
data than was the geometric model. They emphasized, however, that the calculation of an arithmetic value does not imply that all the genetic factors involved in the expression of a particular trait operate in an additive fashion;
i.e., cancellation of interlocus interactions may prevent their detection.
The studies of Neal (1935) and Kinman and Sprague (1945) verified the hypothesis that performance was a linear function of the percent heterozygosity based
on the use of inbred lines and on the F_1 and F_2 generations corresponding to
0%, 100%, and 50% heterozygosity, respectively. With these three levels of
heterozygosity, there was no indication that the observed results deviated
from a linear model.

Subsequent studies have been reported that included additional generations with levels of heterozygosity intermediate to the three levels of 0%,
50%, and 100%. Stringfield (1950) used combinations of four inbred lines to
produce seven genetic populations that represented four levels of heterozygos-

ity (0%, 50%, 75%, and 100%). He concluded that the heterozygosity-performance relation was curvilinear for maturity and height of ear-bearing node and less so for grain yield. The curvilinear relation indicated that as 100% heterozygosity was approached the added increments of heterozygosity had a lesser effect. He hypothesized that the added genes may be duplicating the functions of those already present, which agrees with the diminishing rate of returns hypothesis suggested by Rasmusson (1934).

In a similar study, Sentz et al. (1954) included five levels of heterozygosity in two maize populations. The heterozygosity-performance relation for seven traits over four environments showed a significant deviation from linearity for all traits except maturity and plant height in one population and ear diameter in the second. A curvilinear relation was noted between the 25% and 75% levels of heterozygosity for all traits except ear number. Within an individual environment, the responses were similar in both populations for yield and maturity; but the responses differed among environments, indicating interactions with environments. Sentz et al. concluded that the curvilinear response to levels of heterozygosity was evidence for epistatic gene action, but they also acknowledged that the genetic base of their materials was restricted.

Martin and Hallauer (1976) examined the relation between four levels of heterozygosity and five traits for four groups of inbred lines. The lines in each group were assigned as follows: type I, lines from open-pollinated varieties (first cycle); type II, lines from pedigree crosses or improved populations (second cycle); type III, good lines based on good vigor and general combining ability; and type IV, poor lines based on poor vigor and combining ability. Each of the four groups of lines included seven lines, the 21 possible F_1, F_2, and F_3 generations among the seven lines, the 21 backcrosses to each of the lines, and the 21 backcrosses selfed for each of the inbred lines. Data were collected in two replicate experiments conducted in five environments. The relation between level of heterozygosity and mean performance was determined for each environment and combined across the five environments. Sums of squares for levels of heterozygosity were partitioned into those due to linear, quadratic, and lack-of-fit components. Some instances of detectable epistatic effects were noted in all types for each trait in each environment. The frequency of detectable epistatic effects, however, was much less than that of significant linear effects. Yield, for example, combined across the four types and the five environments had 84 instances of significant (P $\leq$ 0.01) linear, 3 quadratic, and 1 lack-of-fit mean squares. If the linear model assumes that individual loci contribute their effects independently of all other loci and that the quadratic and lack-of-fit mean squares indicate evidence of epistatic effects, the occurrence of net epistatic effects for yield was very low in the 84 crosses. Occurrence of significant epistatic effects was greater in the individual environments than in the combined analysis across the five environments.

In addition to testing mean squares for the presence of epistatic effects, Martin and Hallauer (1976) determined the relative proportions of levels within crosses sums of squares that were due to the linear, quadratic, and lack-of-fit sources of variation. For yield, the linear model accounted for 99.0%, 97.8%, 98.0%, and 97.9% of the total variation for the four types of lines; the quadratic was 1.5% or less in all types. In 15 of 20 instances for all traits for all types, the linear model accounted for 90% or more of the total variation. Figure 9.4 shows the relation of yield to level of heterozygosity, and a linear trend is evident for each of the four types of lines. Although at least one instance of significant epistatic effect was detected in

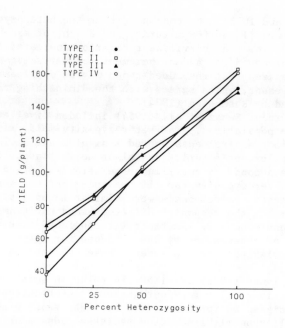

Fig. 9.4. Relation between level of heterozygosity and yield averaged over
           five environments for four types of inbred lines (Martin and Hal-
           lauer 1976).

each type for each trait, the relative contribution of net epistasis to the
mean seems small.

All the previous studies on heterozygosity-performance relation included
crosses among selected inbred lines as their experimental material. Some data
also are available for crosses generated from open-pollinated varieties. Pollak et al. (1957) used three open-pollinated varieties and their F_1, F_2, and
backcross populations to establish three levels of expected heterozygosity.
Based on expectations, the backcrosses would equal the F_2 generations and the
F_2 would be midway between the F_1 and the mean of the two parent varieties.
The F_2 generations, for yield, were intermediate to the means of the parents
and the F_1 for the three crosses, with the backcrosses similar to the F_2; all
comparisons except one were within one standard error of the means. Pollak et
al. concluded that the evidence does not suggest the presence of important epistatic combinations of genes conditioning yield. Robinson and Cockerham
(1961) related yield and ear height to level of heterozygosity for two open-
pollinated varieties, their F_1 and F_2 generations, and the selfs of each.
There was no significant deviation from their proposed additive with dominance
genetic model.

Moll et al. (1965) included open-pollinated varieties and their F_1 and F_2
generations for different levels of genetic divergence, based on geographical
origin of the parental varieties. Per plant means of the parents (159.8 g),
the F_1 generation (196.3 g), and the F_2 generation (179.3 g) show that the F_2
mean deviates only 0.7% from the average (178.0 g) of the parents' and the F_1
means; hence the presence of net epistatic effects on the average was not important for these variety crosses. For the F_1 and F_2 generations of the varieties included by Moll et al. (1965), the F_2 averaged 8.9% lower yields than
the F_1 crosses. Pollak et al. (1957) for their materials found the F_2 deviat-

ed from the midparent by only 1.2% and the F_2 mean, on the average, was 6.0% lower than the F_1 mean.

Shehata and Dhawan (1975) determined the inbreeding depression for three sets of diallel crosses, each set including 10 parents that were either open-pollinated or synthetic varieties. Average inbreeding for each of the three diallel series of crosses was 10.2%, 11.4%, and 11.2%, comparing the F_2 with the F_1 generations. The deviations of the F_2 generations from the midparents of the parental varieties and their F_1 variety crosses were small in all instances and are interpreted as evidence that net epistatic effect in yield is small relative to the additive effects among loci.

The decreases from inbreeding F_1 generations to produce F_2 generations generally were less for the varieties than for specific crosses of a pair of inbred lines (Table 9.11). Although the data are limited, the frequency of significant epistatic effects was greater for crosses of inbred lines than for varieties. This difference seems reasonable because of the nature of the material included in the crosses. For varieties this would involve an array of genotypes included in the crosses whereas crosses of inbred lines would involve the expression for two specific genotypes. Martin and Hallauer (1976) found differences in frequencies of detectable epistasis among four types of inbred lines based on origin of the lines included in the four types. The effects of inbreeding depression also seem greater for F_1 generations of inbred lines than for variety crosses. The results from studies of F_1 and F_2 generations produced from inbred lines for two different eras show a large reduction from the F_1 to the F_2 generations: Kiesselbach (1922) reported the F_2 generations were 46.7% lower yielding than the F_1 generations; Martin and Hallauer (1976), for 84 crosses among 28 lines, found the F_2 generations were 30.1% lower yielding than the F_1 generations. Some of the differences can be attributed to the improvements made in the yield capacity of the inbred lines used in the crosses. Kiesselbach presented data for one year and found the F_1 hybrids were 404.7% higher yielding than the inbred lines; Martin and Hallauer's study showed the hybrids were only 184.2% higher yielding than the inbred lines. For an unselected group of lines developed from Iowa Stiff Stalk Synthetic, Lopez-Perez (1977) found that the F_2 generations yielded 35.5% less than the F_1 generation single crosses.

Data estimating the rates of inbreeding depression and comparing the means of different generations with level of heterozygosity indicate that net epistatic effects are small relative to the additive effects among loci regardless of the methods and materials used in the comparisons. The greatest proportion of the total variation usually was accounted for by the linear model, which, particularly for yield, was greater than 90% and usually greater than 98%. As for the experiments with generation means (Chap. 5), some instances of epistatic effects are detected in special crosses but their relative frequency is low and the total variation attributed to epistasis is small. It should be reemphasized that only net epistasis is detected, which implies not that epistasis is absent (which is unrealistic) but that it must be of a cancelling nature from the comparisons of generations. Genetic models based on additivity of locus effects seem to be an adequate mathematical description of the gene action operative in the inbreeding effects in maize.

The effects of inbreeding are known, their rates for different traits determined, and various systems of inbreeding discussed. The present stage of hybrid maize in the U.S. Corn Belt requires use of hybrids produced from inbred lines. Although the effects of inbreeding on phenotype are usually undesirable, use of inbred lines to produce the high yielding, uniform hybrids ensures extensive use of inbreeding in the foreseeable future, which for the

Table 9.11. Some comparisons of parents and F_1 and F_2 generations for crosses among inbred lines and varieties of maize for yield.

Source	Type of cross[+]	Parents (P)	F_1	$\frac{(F_1-P)}{P} \times 100$	F_2	$\frac{(F_1-F_2)}{F_1} \times 100$
			Inbred parents			
Kiesselbach (1922)	SC	6.4	32.3	404.7	15.7	51.4
		---	32.2	---	18.4	42.8
		---	32.6	---	17.4	46.6
Kiesselbach (1930)	SC	15.0	35.6	137.3	24.0	32.6
		---	33.4	---	22.6	32.3
Richey et al. (1934)	DC	---	49.1	---	41.5	15.5
Neal (1935)	SC	14.8	39.2	164.9	27.6	29.6
	3W	14.9	40.1	169.1	30.8	23.2
	DC	15.6	40.1	157.0	33.8	15.7
Kinman and Sprague (1945)	SC	17.6	49.9	183.5	31.8	36.3
Stringfield (1950)	SC	25.9	53.1	104.9	38.9	36.6
Martin and Hallauer (1976)	SC	21.5	61.1	184.2	42.7	30.1
Lopez-Perez (1977)	SC	---	90.6	---	58.4	35.5
	3W	---	87.4	---	55.9	36.0
	DC	---	89.9	---	56.0	37.7
			Variety parents			
Pollak et al. (1957)[‡]	VC	174.0	202.7	16.5	190.6	6.0
Robinson and Cockerham (1961)[‡]	VC	236.0	267.8	13.5	181.6	32.2
Moll et al. (1965)[‡]	VC	159.8	196.3	22.8	179.3	8.9
Shehata and Dhawan (1975)	VC	82.2	101.8	23.8	91.4	10.2
	VC	75.4	89.0	18.0	78.1	11.4
	VC	81.9	93.1	13.7	83.5	11.2

[+] Crosses are designated as SC, 3W, DC, and VC for single crosses, 3-way crosses, double crosses, and variety crosses, respectively.

[‡] Grams per plant; all other yields are expressed as quintals per hectare.

most part will be self-fertilization. Kiesselbach (1922) stated that "no elemental strain of corn has yet been found . . . , which is as vigorous or productive as the original variety from which it was derived by repeated self-fertilization." This statement is equally true now. Improvements have been made in general productiveness and vigor of inbred lines since the suggestion of Shull (1908), but they are still poor relative to the original source population from which they were extracted. Recurrent selection procedures are effective for improving the general level of our breeding populations, but inbreeding will reduce the complex genotypes of the populations to their homozygous form and, consequently, vigor and productiveness will be reduced. As we continue to upgrade the general performance of our breeding populations, we can expect to obtain greater yielding and more vigorous inbred lines (see Genter 1971), but inbreeding depression will continue to be evident when self-fertilization is imposed on the complex array of genotypes within a population. Because of the expression of heterosis (Chap. 10) observed among crosses of pure line genotypes, self-fertilization is an important feature of nearly all maize breeding programs. Reducing the complex genotype of a breeding population to its pure line (inbred lines) components permits the identification and reproduction of genotypes that can be used to repeatably produce a specific elite hybrid. This is the primary objective of inbreeding in maize.

REFERENCES

Allard, R. W. 1960. *Principles of Plant Breeding*. Wiley, New York.

Baker, L. H., and R. N. Curnow. 1969. Choice of population size and use of variation between replicate populations in plant breeding selection programs. *Crop Sci.* 9:555-60.

Cockerham, C. C. 1961. Implications of genetic variances in a hybrid breeding program. *Crop Sci.* 1:47-52.

Collins, G. N. 1909. The importance of broad breeding in corn. USDA Bull. 141 (IV).

Cornelius, P. L., and J. W. Dudley. 1974. Effects of inbreeding by selfing and full-sibbing in a maize population. *Crop Sci.* 14:815-19.

Darwin, C. 1859. *The Origin of Species*. World Famous Books, Merrill & Baker, New York.

Darwin, C. 1877. *The Effects of Cross- and Self-Fertilization in the Vegetable Kingdom*. Appleton, London.

Doxtator, C. M., and I. J. Johnson. 1936. Prediction of double cross yields in corn. *J. Am. Soc. Agron.* 28:460-62.

East, E. M. 1908. Inbreeding in corn. Connecticut Agric. Exp. Stn. Rep. 1907. Pp. 419-28.

————. 1909. The distinction between development and heredity in inbreeding. *Am. Nat.* 43:173-81.

East, E. M., and H. K. Hayes. 1912. Heterozygosis in evolution and in plant breeding. USDA Bur. Plant Ind. Bull. 243.

East, E. M., and D. F. Jones. 1918. *Inbreeding and Outbreeding*. Lippincott, Philadelphia.

Eberhart, S. A., and A. R. Hallauer. 1968. Genetic effects for yield in single-, three-way, and double-cross maize hybrids. *Crop Sci.* 8:377-79.

Eberhart, S. A., and W. A. Russell. 1966. Stability parameters for comparing varieties. *Crop Sci.* 6:36-40.

————. 1969. Yield and stability for a 10-line diallel of single-cross and double-cross maize hybrids. *Crop Sci.* 9:357-61.

Falconer, D. S. 1960. *Quantitative Genetics*. Ronald Press, New York.

Federer, W. T., and G. F. Sprague. 1947. A comparison of variance components in corn yield trials. I. Error, tester × line, and line components in top-cross experiments. *J. Am. Soc. Agron.* 39:453-63.

Focke, W. O. 1881. *Die Pflanzen-Mischlinge*. Berlin.

Gama, E. E. G., and A. R. Hallauer. 1980. Stability of hybrids produced from selected and unselected lines of maize. *Crop Sci.* 20:623-26.

Gärtner, C. F. 1849. *Versuche und Beobachtungen über die Bastarderzengung in Pflanyenreich*. Stuttgart.

Geadelmann, J. L., and R. H. Peterson. 1978. Effects of two yield component selection procedures on maize. *Crop Sci.* 18:387-90.

Genter, C. F. 1971. Yield of S_1 lines from original and advanced synthetic varieties of maize. *Crop Sci.* 11:821-24.

Gilmore, E. C. 1969. Effect of inbreeding of parental lines on predicted yields of synthetics. *Crop Sci.* 9:102-04.

Good, R. L. 1976. Inbreeding depression in Iowa Stiff Stalk Synthetic (*Zea mays* L.) by selfing and full-sibbing. Ph.D. dissertation, Iowa State Univ., Ames.

Good, R. L., and A. R. Hallauer. 1977. Inbreeding depression in maize by selfing and full-sibbing. *Crop Sci.* 17:935-40.

Gutierrez, M. G., and G. F. Sprague. 1959. Randomness of mating in isolated polycross plantings in maize. *Genetics* 44:1075-82.

Hallauer, A. R., and J. H. Sears. 1973. Changes in quantitative traits associated with inbreeding in a synthetic variety of maize. *Crop Sci.* 13:327-30.

Harris, R. E.; C. O. Gardner; and W. A. Compton. 1972. Effects of mass selection and irradiation in corn measured by random S_1 lines and their testcrosses. *Crop Sci.* 12:594–98.

Harvey, P. H., and J. A. Rigney. 1947. Inbreeding studies with prolific corn varieties. Dept. Agron., North Carolina State Univ., Raleigh.

Jenkins, M. T. 1934. Methods of estimating performance of double-crosses in corn. *J. Am. Soc. Agron.* 26:199–204.

Jones, D. F. 1918. The effects of inbreeding and crossbreeding upon development. Connecticut Agric. Exp. Stn. Bull. 207:5–100.

————. 1924. Selective fertilization among the gametes from the same individuals. *Proc. Nat. Acad. Sci.* 10:218–21.

————. 1939. Continued inbreeding in maize. *Genetics* 24:462–73.

————. 1958. Heterosis and homeostasis in evolution and in applied genetics. *Am. Nat.* 92:321–28.

Jugenheimer, R. W. 1958. Hybrid maize breeding and seed production. FAO, Rome.

Kempthorne, O. 1957. *An Introduction to Genetic Statistics*. Wiley, New York.

Kiesselbach, T. A. 1922. Corn investigations. Nebraska Agric. Exp. Stn. Res. Bull. 20:5–151.

————. 1930. The use of advanced generation hybrids as parents of double cross seed corn. *J. Am. Soc. Agron.* 22:614–26.

————. 1933. The possibilities of modern corn breeding. *Proc. World Grain Exhib. Conf.* (Canada) 2:92–112.

Kinman, M. L. 1952. Composite sibbing versus selfing in development of corn inbred lines. *Agron. J.* 44:209–41.

Kinman, M. L., and G. F. Sprague. 1945. Relation between number of parental lines and theoretical performance of synthetic varieties of corn. *J. Am. Soc. Agron.* 37:341–51.

Knight, T. A. 1799. An account of some experiments on the fecundation of vegetables. *Philos. Trans. R. Soc. London* 89:195.

Kölreuter, J. G. 1776. *Dritte Fortsetzung der vorläufigen Nachricht von einigen das Geschlecht der Pflanzen betreftender Versuchen and Beobachtunger*. Leipzig.

Levings, C. S., III; J. W. Dudley; and D. E. Alexander. 1967. Inbreeding and crossing in autotetraploid maize. *Crop Sci.* 7:72–73.

Li, C. C. 1976. *Population Genetics*. Boxwood Press, Pacific Grove, CA.

Lindstrom, E. W. 1939. Analysis of modern maize breeding principles and methods. *Proc. Seventh Int. Genet. Congr.* 7:191–96.

Loeffel, F. A. 1971. Development and utilization of parental lines. *Proc. Annu. Corn Sorghum Res. Conf.* 26:209–17.

Lopez-Perez, E. 1977. Comparisons among maize hybrids made from unselected lines developed by selfing and full-sibbing. Master's thesis, Iowa State Univ., Ames.

Macaulay, T. B. 1928. The improvement of corn by selection and plot inbreeding. *J. Hered.* 19:57–72.

Malécot, G. 1948. *Les Mathématiques de l'Hérédité*. Masson et Cie, Paris.

Martin, J. M., and A. R. Hallauer. 1976. Relation between heterozygosis and yield for four types of maize inbred lines. *Egyptian J. Genet. Cytol.* 5:119–35.

Moll, R. H.; J. H. Lonnquist; J. V. Fortuna; and E. C. Johnson. 1965. The relation of heterosis and genetic divergence in maize. *Genetics* 52:139–44.

Neal, N. P. 1935. The decrease in yielding capacity in advanced generations of hybrid corn. *J. Am. Soc. Agron.* 27:666-70.

Otsuka, Y.; S. A. Eberhart; and W. A. Russell. 1972. Comparisons of prediction formulas for maize hybrids. *Crop Sci.* 12:325-31.

Pirchner, F. 1969. *Population Genetics in Animal Breeding*. W. H. Freeman, San Francisco.

Pollak, E.; H. F. Robinson; and R. E. Comstock. 1957. Interpopulation hybrids in open-pollinated varieties of maize. *Am. Nat.* 91:387-91.

Powers, L. 1941. Inheritance of quantitative characters in crosses involving two species of Lycopersicon. *J. Agric. Res.* 63:149-74.

Rasmusson, J. A. 1934. A contribution to the theory of quantitative character inheritance. *Hereditas* 18:245-61.

Rawlings, J. O. 1969. Present status of research on long- and short-term recurrent selection in finite populations: Choice of population size. *Proc. Second Meet. Work. Group Quant. Genet.*, sect. 22. IUFRO, Raleigh, N.C. Pp. 1-15.

Rice, J. S., and J. W. Dudley. 1974. Gene effects responsible for inbreeding depression in autotetraploid maize. *Crop Sci.* 14:390-93.

Richey, F. D.; G. H. Stringfield; and F. F. Sprague. 1934. The loss of yield that may be expected from planting second generation double-crossed corn. *J. Am. Soc. Agron.* 26:196-99.

Robertson, A. 1960. A theory of limits in artificial selection. *Proc. R. Soc.* B153:234-49.

Robinson, H. F., and C. C. Cockerham. 1961. Heterosis and inbreeding depression in population involving two open-pollinated varieties of maize. *Crop Sci.* 1:68-71.

Russell, W. A., and A. R. Hallauer. 1980. Corn. In *Hybridization of Crop Plants*, W. R. Fehr and H. H. Hadley, eds., pp. 299-312. Am. Soc. Agron., Madison, Wis.

Schnell, F. W. 1975. Type of variety and average performance in hybrid maize. *Z. Pflanzenzüchtg.* 74:177-88.

Sentz, J. C.; H. F. Robinson; and R. E. Comstock. 1954. Relation between heterozygosis and performance in maize. *Agron. J.* 46:514-20.

Shamel, A. D. 1905. The effect of inbreeding in plants. *USDA Yearbook.* Pp. 377-92.

Shehata, A. H., and N. L. Dhawan. 1975. Genetic analysis of grain yield in maize as manifested in genetically diverse varietal populations and their crosses. *Egyptian J. Genet. Cytol.* 4:90-116.

Shull, G. H. 1908. The composition of a field of maize. Am. Breeders' Assoc. Rep. 4:296-301.

————. 1909. A pure line method of corn breeding. Am. Breeders' Assoc. Rep. 5:51-59.

————. 1910. Hybridization methods in corn breeding. *Am. Breeders' Mag.* 1:98-107.

————. 1952. Beginnings of the heterosis concept. In *Heterosis*, J. W. Gowen, ed., pp. 14-48. Iowa State Univ. Press, Ames.

Sing, C. F.; R. H. Moll; and W. D. Hanson. 1967. Inbreeding in two populations of *Zea mays* L. *Crop Sci.* 7:631-36.

Sprague, G. F. 1946. The experimental basis for hybrid maize. *Biol. Rev.* 21:101-20.

————. 1971. Genetic vulnerability to disease and insects in corn and sorghum. *Proc. Annu. Corn Sorghum Res. Conf.* 26:96-104.

Sprague, G. F., and S. A. Eberhart. 1977. Corn breeding. In *Corn and Corn Improvement*, G. F. Sprague, ed., pp. 305-62. Am. Soc. Agron., Madison, Wis.

Sprague, G. F., and W. T. Federer. 1951. A comparison of variance components in corn yield trials. II. Error, year × variety, location × variety, and variety components. *Agron. J.* 43:535-41.

Sprague, G. F., and W. T. Thomas. 1967. Further evidence of epistasis in single and three-way cross yields of maize (*Zea mays* L.). *Crop Sci.* 7: 355-56.

Sprague, G. F.; W. A. Russell; L. H. Penny; and T. W. Horner. 1962. Effects of epistasis on grain yield of maize. *Crop Sci.* 2:205-8.

Stringfield, G. H. 1950. Heterozygosis and hybrid vigor in maize. *Agron. J.* 42:145-51.

————. 1974. Developing heterozygous parent stocks for maize hybrids. De-Kalb AgResearch, DeKalb, Ill.

Stuber, C. W.; W. P. Williams; and R. H. Moll. 1973. Epistasis in maize (*Zea mays* L.). III. Significance in predictions of hybrid performance. *Crop Sci.* 13:195-200.

Weatherspoon, J. H. 1970. Comparative yields of single, three-way, and double crosses of maize. *Crop Sci.* 10:157-59.

Weatherwax, Paul. 1955. Structure and development of reproductive organs. In *Corn and Corn Improvement*, G. F. Sprague, ed., pp. 89-121. Academic Press, New York.

Wright, J. A.; A. R. Hallauer; L. H. Penny; and S. A. Eberhart. 1971. Estimating genetic variance in maize by use of single and three-way crosses among unselected inbred lines. *Crop Sci.* 11:690-95.

Wright, S. 1921. Systems of mating. II. The effects of inbreeding on the genetic composition of a population. *Genetics* 6:124-43.

————. 1922a. Coefficients of inbreeding and relationship. *Am. Nat.* 56: 330-38.

————. 1922b. The effects of inbreeding and crossbreeding on guinea pigs. III. Crosses between highly inbred families. USDA Bull. 1121.

————. 1931. Evolution in Mendelian populations. *Genetics* 16:97-159.

Zuber, M. S. 1975. Corn germplasm base in the United States: Is it narrowing, widening, or static? *Proc. Annu. Corn Sorghum Res. Conf.* 30:277-86.

10

Heterosis

The phenomenon of heterosis has been exploited extensively in maize breeding. Heterosis or hybrid vigor and inbreeding depression are complementary and the two phenomena often are observed in the same studies. Maize breeding methods during the twentieth century have been developed to take advantage of the manifestation of heterosis in crosses of inbred lines.

Changes in breeding methods also caused changes in seed multiplication. To take advantage of the heterosis expressed in the first generation cross of inbred lines, new seed supplies are required for each growing season; thus farmers are unable to use seed from their harvested crop for the following growing season. Farmers do not usually have the genetic materials, know-how, and equipment to produce hybrid seed. Production of hybrid seed, consequently, has been reserved for the specialists.

Publicly financed organizations in the United States initially developed inbred lines, produced and tested hybrids, and recommended their use to farmers. Because of the rapid acceptance of hybrid maize (Fig. 1.2), it soon became obvious that participation of other organizations was necessary to ensure an adequate supply of high quality seed for each growing season. The commercial aspects of hybrid maize became obvious because parental lines used for production of hybrids could be controlled, and commercial companies were developed that specialized in the production and sale of hybrid seed. It has become a highly competitive business and commercial companies have expanded their efforts to provide better service. They have breeding programs for developing and modifying parental inbred lines, conduct research on seed quality and seedling vigor, conduct extensive yield trials to determine the best hybrids for each area of adaptation, and have agronomists in the field to assist farmers with problems of growing hybrid maize. The development and growing of hybrid maize can justifiably be called one of plant breeding's greatest accomplishments. The success of hybrid maize has established an ideal that is being used by plant breeders and commercial companies for other crop species.

Zirkle (1952) summarized early work on the occurrence of heterosis of maize, as well as of other plant species. Köelreuter (1766), Knight (1799), and Gärtner (1849, p. 791) investigated and described plant hybridization, but apparently the first correct interpretation of hybridization in maize was provided by a letter written by Cotton Mather in 1716. Darwin (1877, p. 482) was the first to conduct experiments comparing selfed and crossed plants from the same stocks of maize; he found that plant height of crossed plants was greater than selfed plants at the juvenile (19%) and mature (9%) plant stages.

Beal (1880) was aware of Darwin's results and conducted an experiment that is analogous to present-day methods used for maize hybridization on a large-scale basis. Beal's parental stocks, however, were open-pollinated va-

rieties rather than the inbred lines commonly used today. Beal collected two
stocks of maize that were similar but had been grown about 160 km apart for an
unknown number of years. The two stocks were grown in the same field; one was
detasseled and the other was used as the male parent. The hybrid seed harvested from the detasseled rows was planted and was found to be 51% superior
in yield to the original varieties. Beal suggested that use of this procedure
was one method to increase yields of maize, but later developments overshad-
owed widespread usage of variety hybrids in the United States. Beal's results
were confirmed by Sanborn (1890), McClure (1892), Morrow and Gardner (1893),
and Webber (1900, 1901).

        Because of the superiority of variety hybrids, it has been questioned why
they did not become more important until the development of the inbred line-
hybrid concept in the 1920s. Three reasons are: (1) The time was not right
to realize the commercial potential for producing hybrid seed. Variety crosses were not significantly different in phenotypic appearance from varieties
and the heterosis was not as great as observed for hybrids produced among inbred lines. (2) Research on inbreeding depression and heterosis among inbred
lines was striking enough to divert attention from variety hybrids to the potential use of inbred lines to produce hybrids. (3) Although comparisons
showed that crosses were superior, the heterosis concept was not known, and
the principal objective of these studies was to control the parentage of the
seed being produced.

        The beginning of the heterosis concept in maize started with the studies
reported by Shull (1908), "The composition of a field of maize." Shull (1952)
summarized his studies and apparently was the first to correctly interpret the
phenomena of inbreeding depression and hybrid vigor. Hybrid vigor and heterosis are nearly synonymous; the word *heterosis* was coined by Shull (1914) to
provide a term to describe the phenomenon but it did not include a description
of the genetic mechanism involved in its expression.

10.1 EMPIRICAL EVIDENCE

        The manifestation of heterosis in crosses of maize varieties by Beal
(1880) alerted other researchers to possible benefits of variety crosses.
Several studies evaluating open-pollinated varieties and their crosses in the
early part of the twentieth century were summarized by Richey (1922). Most of
the studies involved crosses of a series of open-pollinated varieties with an
open-pollinated variety that was either high yielding or popular with local
farmers. Hence a group of open-pollinated varieties were crossed to a common
tester variety to determine performance of the variety crosses and heterosis
expressed in the variety crosses relative to either the average performance of
the two open-pollinated parental varieties or the high parent. Because of
Jones's (1918) suggestion that single crosses be used as parent stocks for
production of double-cross hybrids, procedures for inbred line and hybrid development received strong emphasis during the 1920s; consequently, little information is available in the literature after 1920 for variety crosses. Performance of variety crosses received renewed interest in the 1950s because of
the development of quantitative genetics and recurrent selection procedures
for the improvement of breeding populations. Differences of opinion of relative importance of dominance vs. overdominance in expression of heterosis in
hybrids, the apparent yield plateau in double-cross hybrid yields in the
1950s, and potential of recurrent selection procedures for improvement of
breeding populations stimulated development of different maize breeding procedures. The apparent yield plateau suggested that adequate genetic variation
was not present in open-pollinated varieties for development of superior dou-

ble-cross hybrids; quantitative genetic studies were dependent on a reference population for making inferences; and reciprocal recurrent selection was suggested on the premise of using two base populations and selecting for both general and specific combining abilities. In all instances the use of open-pollinated varieties became important because they were the source populations used to develop the original group of inbred lines for production and growing of double-cross hybrids.

The method of evaluation and the choice of varieties included for evaluation of heterosis also changed. Instead of crossing a group of varieties to a common tester variety, the diallel mating design was used to determine general performance of a variety in comparison with other varieties and specific performance of a particular pair of varieties. The latter information was important in the choice of two varieties for initiating reciprocal recurrent selection (RRS). Open-pollinated varieties were included in many of the diallel series of crosses, but synthetic varieties, composites, and varieties improved by selection also were often included. In most instances a measure of heterosis was desired among the variety crosses, but in some instances genetic information was obtained by selfing either the parental varieties or the variety crosses.

A summary of the manifestation of heterosis in crosses of maize varieties is given in Tables 10.1 and 10.2. Information on heterosis of crosses ranges from that of Morrow and Gardner (1893) to recent information evaluating effectiveness of recurrent selection. Because yield is the most important economic trait of maize, only the heterosis information on yield is given. Table 10.1 includes 611 varieties and 1394 variety crosses that were evaluated for yield heterosis. Note that many varieties and also some crosses are repeated in the numbers given in Table 10.1. Heterosis relative to the average of the two parent varieties (midparent) and the high-parent variety are given for each reported study and averaged over all studies. Average midparent heterosis for the 1394 crosses weighted for the number of crosses in each study was 19.5%. Average midparent heterosis was evident in nearly all studies; the only exception was for the 6 varieties and 10 variety crosses reported by Noll (1916), which was -0.5%. Midparent heterosis in Table 10.1 is the average for each study. Variety crosses that were either above or below the midparent also are shown in Table 10.1. Except for the study by Noll, which had 7 of the 10 variety crosses below the midparent, a majority of variety crosses exceeded the midparent.

High-parent heterosis and frequency of variety crosses that exceeded the high parent varied considerably among the reported studies. High-parent heterosis for variety crosses evaluated before 1932 was generally quite small. Average high-parent heterosis ranged from -9.9% for the one variety cross reported by Garber and North (1931) to 43.0% for 10 flint variety crosses reported by Troyer and Hallauer (1968). Average high-parent heterosis for the 1394 variety crosses was 8.2%.

Table 10.2 gives a summary of midparent and high-parent heterosis for six different categories of variety crosses. Compared with the summary of heterosis prepared by Richey (1922), percentage of variety crosses that exceeded the midparent and high parent were amazingly similar. Table 10.1 also includes some of those reported by Richey, but only 244 of the 1394 crosses were available at the time of Richey's summary. The percentages of the variety crosses that exceeded the high parent by 0% to 5%, 6% to 15%, and 16% or more also were very similar. The most obvious difference of the comparisons in Table 10.2 is between average high-parent heterosis for variety crosses tested before 1932 (0.0%) and those tested after 1955 (10.0%). Perhaps the small and inconsistent expression of high-parent heterosis was an important factor in the limit-

Table 10.1. Summary of comparisons between parental varieties of maize and first generation variety crosses.

Source	Parent varieties	Variety crosses	Average heterosis (%)		Variety crosses					
					Midparent (no.)		Exceeds high parent (%)			
			Midparent	High parent	Above	Below	Total	0-5	6-15	16+
Morrow and Gardner (1893)	7	5	14.9	7.9	5	0	4	2	0	2
Hayes and East (1911)	3	2	41.2	5.0	2	0	0	0	0	0
Hartley et al. (1912)	81	75	9.8	-1.9	62	13	36	12	14	10
Hayes (1914)	39	35	10.7	-0.1	28	7	18	8	7	3
Williams and Welton (1915)	13	17	7.4	1.7	13	4	7	6	1	0
Noll (1916)	6	10	-0.5	-6.8	3	7	3	1	2	0
Hutcheson and Wolfe (1917)	5	4	4.4	-9.0	3	1	1	0	1	0
Jones et al. (1917)	55	50	7.9	1.1	43	7	34	21	12	1
Hayes and Olson (1919)	13	12	14.3	11.5	12	0	11	3	4	4
Griffe (1922)	5	6	14.1	6.9	6	0	6	3	3	0
Kiesselbach (1922)	14	13	-3.6	-8.8	5	8	0	0	0	0
Waldron (1924)	13	12	9.1	4.8	11	1	9	5	2	2
Garber et al. (1926)	5	8	37.2	-2.1	8	0	3	0	1	2
Garber and North (1931) - 0	2	1	45.4	13.6	1	0	1	0	1	0
- I	2	1	32.9	-9.9	1	0	0	0	0	0
Robinson et al. (1956)	6	15	19.9	11.5	15	0	12	1	5	6
Torregroza (1959)	10	7	17.0	--	7	0	--	--	--	--
Lonnquist and Gardner (1961)	12	66	8.5	2.8	61	5	49	30	19	0
Moll et al. (1962)	6	14	20.2	8.6	13	1	12	5	4	3
Paterniani and Lonnquist (1963)	12	63	33.0	14.0	61	2	47	2	22	23

Table 10.1. (Cont.)

Source	Parent varieties	Variety crosses	Average heterosis (%) Midparent	Average heterosis (%) High parent	Midparent (no.) Above	Midparent (no.) Below	Exceeds high parent (%) Total	Exceeds high parent (%) 0-5	Exceeds high parent (%) 5-16	Exceeds high parent (%) 16+
Timothy (1963)	8	28	31.0	22.0	--	--	--	--	--	--
	8	6	50.3	15.1	6	0	4	1	0	3
Moll et al. (1965)	8	28	22.8	8.7	28	0	16	1	5	10
Wellhausen (1965)	15	18	59.2	35.9	18	0	18	2	2	14
Hallauer and Eberhart (1966)	9	36	11.0	6.0	36	0	31	12	18	1
Paterniani (1967)	9	26	40.5	30.1	26	0	26	0	2	24
Castro et al. (1968)	5	10	45.5	2.9	10	0	5	1	3	1
Paterniani (1968)	10	45	34.8	5.6	43	2	25	4	5	16
Crum (1968)	10	45	21.3	7.0	43	2	32	7	7	18
Hallauer and Sears (1968)	9	36	9.8	4.2	35	1	28	20	6	2
Troyer and Hallauer (1968)	10	45	72.0	43.0	45	0	43	1	6	36
Silva (1969)	3	3	18.2	0.3	3	0	2	2	0	0
Paterniani (1970)	16	55	2.6	-5.6	33	22	14	7	7	0
	6	12	18.1	6.9	12	0	10	2	7	1
Moll and Stuber (1971)‡ - 0	2	1	18.9	17.8	1	0	1	0	0	1
- I	2	2	23.0	21.0	2	0	2	0	0	2
Eberhart (1971) - Corn Belt	9	36	14.2	4.5	32	4	23	6	11	6
- Southern	6	15	21.4	11.2	15	0	12	1	8	3
Tavares (1972)	6	3	13.6	0.4	3	0	2	1	1	0
Hallauer (1972)	9	36	14.0	8.1	32	4	33	11	17	5
El-Rouby and Galal (1972)	7	21	6.4	-0.8	20	1	10	8	2	0

Table 10.1. (Cont.)

Source	Parent varieties	Variety crosses	Average heterosis (%) Midparent	Average heterosis (%) High parent	Midparent (no.) Above	Midparent (no.) Below	Exceeds high parent (%) Total	Exceeds high parent (%) 0-5	Exceeds high parent (%) 0-16	Exceeds high parent (%) 16[+]
Crum (1972)	10	45	18.4	11.2	43	2	32	7	13	12
Barriga and Vencovsky (1973)	5	10	13.7	2.5	10	0	5	3	1	1
Valois (1973)‡ - 0	2	1	9.8	9.5	1	0	1	0	1	0
- I	2	1	11.6	8.1	1	0	1	0	1	0
- III	2	1	4.2	0.8	1	0	1	1	0	0
Vencovsky et al. (1973)	10	45	21.6	12.0	45	0	38	9	14	15
Miranda (1974a)	9	36	15.2	5.8	32	4	24	5	9	10
Genter and Eberhart (1974) - 0	6	15	15.7	10.8	15	0	13	2	7	4
- 0×I	13	42	14.9	9.0	42	0	37	11	17	9
- I	7	21	14.2	8.8	21	0	19	5	11	3
Shehata and Dhawan (1975) - set 1	10	45	24.0	11.8	26	19	19	--	--	--
- set 2	10	45	19.0	8.0	22	23	15	--	--	--
- set 3	10	45	19.0	9.2	31	14	20	--	--	--
Hallauer and Malithano (1976)‡ - 0	7	21	18.8	5.8	21	0	17	9	4	4
- I	10	45	20.6	12.2	45	0	39	8	14	17
Paterniani (1977)	12	36	9.7	2.8	30	6	21	8	9	4
Obilana et al. (1979) - 0	2	1	10.8	8.1	1	0	1	0	1	0
- I	2	1	16.5	13.0	1	0	1	0	1	0
Paterniani and Goodman (1977)	6	15	18.7	7.9	15	0	11	2	5	4
Totals and means§	611	1394	19.5	8.2	1206	160	905	256	313	282

†Heterosis of variety crosses was 16% or greater.
‡Data for original (0) and improved (I) variety crosses also are included in Tables 7.14 and 7.15.
§Weighted by the number of crosses.

Table 10.2. Summary of comparisons between parental varieties of maize and the first generation crosses for different eras and original and improved varieties.

| Source | Parent varieties | Variety crosses | Average heterosis (%) | | Variety crosses | | | | | | | |
| | | | Midparent | High parent | Midparent (%) | | Exceeds high parent (%) | | | | | |
					Above	Below	Total	0-5	6-15	16[+]		
Richey (1922)	---	244	---	---	86.5	13.5	67.8	25.8	25.8	16.2
Table 10.1	611	1394	19.5	8.2	88.3	11.7	66.6	20.9	25.6	23.0
Before 1932	263	251	9.9	0.0[‡]	80.9	19.1	53.0	24.3	19.1	9.6
After 1955	348	1143	21.6	10.0	90.0	10.0	69.7	20.0	27.2	26.5
Original	21	40	17.9	8.3	100.0	0.0	85.0	27.5	35.0	22.5
Improved	25	71	18.8	11.1	100.0	0.0	87.3	18.3	38.0	31.0

[†]Greater than 16%; [‡]exact value is −0.028%.

ed use of variety crosses before development of double crosses from use of inbred lines. Because high-parent heterosis was not always observed (variety
crosses exceeded high parent in only 53.0% of the crosses, Table 10.2) and often less than 5% (24.3% of the crosses, Table 10.2), it is not surprising that
variety crosses were not widely accepted. The superiority of variety crosses
generally was not as great as reported by Beal (1880). Often (as mentioned
previously) varieties were crossed to a common tester. A highly productive
variety used as tester also would have reduced expression of high-parent heterosis. Table 10.2 shows that 17.2% of variety crosses tested before 1932
yielded less than the midparent. Since 1955, 10.0% of variety crosses yielded
less than the midparent; the diallel mating design was used in nearly all instances for the latter comparisons. Comparisons of the varieties and variety
crosses for the populations included in the recurrent selection programs show
that all the variety crosses, both original and improved, exceeded the midparent. This may be expected because in most instances the varieties included
were based on previous variety cross information and specifically selected for
heterosis manifested in variety crosses. Midparent and high-parent heterosis,
however, were very similar for the original (8.3%) and improved variety crosses (11.1%).

As expected, there was considerable variation among individual variety
crosses within each of the studies reported in Table 10.1. In the earlier
studies, Richey (1922) pointed out that greatest heterosis was manifested in
variety crosses of extreme types; i.e., heterosis in crosses between flint or
flour varieties and dent varieties was greater than crosses among dent varieties. No critical data are available but it seems that greatest expression of
heterosis is not dependent on differences in only endosperm types. Paterniani
and Lonnquist (1963) included different endosperm type varieties and found
that manifested heterosis was as great among variety crosses produced from varieties having the same endosperm type as among those produced from varieties
having different endosperm types. Heterosis manifested in variety crosses was
dependent on more than just endosperm type. Greatest expression of heterosis
was reported by Troyer and Hallauer (1968) for crosses produced among early
flint varieties.

Two statements in summaries involving variety crosses illustrate the status of the potential of variety crosses. Hayes and Olson (1919) stated, "The
use of first-generation crosses between pure varieties is a means of increasing yield of corn although all such crosses are not equally productive, some
being of no value." Richey (1922) in his summary of variety crosses stated,
"In such more or less haphazard crossing, therefore, the chances seem about
equal of obtaining a cross that is or is not better than the better parent."
The conclusions of these authors show that advantages of variety crosses were
not always obvious. All the data were collected before experimental plot
techniques and statistical analysis were developed, both of which would tend
to mask the true differences. These conclusions were made at the time vigorous attention was being given to the inbred line–hybrid breeding programs.

The manifestation of heterosis usually depends on genetic divergence of
the two parental varieties. Genetic divergence among varieties usually is unknown, and the only recourse is to determine level of genetic divergence empirically by means of variety crosses. Genetic divergence of the parental varieties is inferred from the heterotic patterns manifested in the series of
variety crosses. If heterosis manifested from the cross of two parental varieties is relatively large, it is concluded that the two parental varieties are
more genetically diverse than two varieties that manifest little or no heterosis in their variety crosses.

Establishment of heterotic patterns among varieties has important implications for selecting inbred lines as potential seed stocks in hybrids. One
of the first decisions maize breeders usually make in determining the matrix
of crosses to produce among a set of elite inbred lines is the parental origin
of the lines. If the origins of the inbred lines are known the logical crosses can be produced, based on the heterotic pattern of the parental source populations. For example, in the U.S. Corn Belt, elite inbred lines of Reid
background tend to be crossed and tested in combination with elite lines of
Lancaster background. Although we have discussed importance of genetic divergence of elite inbred lines (Chap. 8), the same procedures are used for early
testing of inbred lines to obtain preliminary information on their combining
abilities. If origins of the lines are unknown, crosses usually are made to
elite inbred lines of known origin. Subsequent yield test information provides the information necessary for classification into known heterotic patterns.

Examples for determining heterotic pattern of varieties and importance of
genetic diversity in the manifestation of heterosis were reported by Moll et
al. (1962), Moll et al. (1965), and Tsotsis (1972). The objective of Tsotsis's study was to determine the heterotic pattern among a group of open-pollinated varieties in order to establish two broad genetic base breeding populations. Because applied maize breeders want to maximize heterotic response
in crosses among elite inbred lines, the task will be enhanced if genetic differences are established in the basic breeding populations. From evaluation
of variety crosses Tsotsis was able to assign varieties to the two gene pools,
based on the heterotic patterns of the variety crosses. The logical sequence
is then to cross and test lines between the two gene pools.

Diallel cross analysis for a fixed set of open-pollinated varieties provides the basis for a preliminary analysis of the heterotic pattern among variety crosses (see Chap. 4). Preliminary inferences are taken from the significance of effects in the analysis of variance. Thus, average heterosis $\bar{h}$
is indicative of the superiority of variety crosses over midparent values.
Variety heterosis, when significant as a source of variation, indicates that
the heterotic pattern of at least one of the varieties differs from the others
when crossed with the remaining varieties. Specific heterosis results from
specific crosses; and when significant it means that at least one cross differs from the others due to nonadditive effects and differences in gene frequency of varieties. A summary of several diallel cross analyses for variety
crosses relative to the significance of effects is shown in Table 10.3. Only
4 out of 15 or 26.7% of the studies detected significance for specific effects
for yield, suggesting that in most instances the choice of parental varieties
for heterosis exploitation can be based on the yield of varieties themselves
and on the average of their crosses. Only occasionally is the superiority of
a given cross increased by positive and significant specific effects.

Moll et al. (1962) and Moll et al. (1965) studied crosses produced among
varieties with different levels of genetic diversity. The level of genetic
diversity was inferred from the geographical origin of the varieties. Two
open-pollinated varieties from the U.S. Corn Belt, for instance, were assumed
to have less genetic divergence than a variety from the U.S. Corn Belt compared with one from southeastern United States. Moll et al. (1962) crossed in
diallel series 6 open-pollinated varieties: 2 from the U.S. Corn Belt, 2 from
southeastern United States, and 2 that originated in the Caribbean region. All
parental varieties and 14 of the possible 15 variety crosses were tested in
North Carolina. Relative heterosis of the variety crosses agreed with the
original classification of the genetic divergence of the parental varieties.

Table 10.3. Level of significance of variety and heterotic
component effects for yield in several variety
diallel crosses.

Source	Number of varieties	Effects[†]			
		v_j	$\bar{h}$	h_j	$s_{jj'}$
Gardner (1965)	4	*	*	ns	ns
Gardner and Eberhart (1966)	6	*	*	ns	ns
Hallauer and Eberhart (1966)	9	*	*	*	ns
Gardner and Paterniani (1967)	6	*	*	ns	ns
Hallauer and Sears (1968)	9	*	*	*	*
Troyer and Hallauer (1968)	10	*	*	*	*
Castro et al. (1968)	5	*	*	ns	ns
Vencovsky (1969)	12	*	*	*	ns
Eberhart (1971) - Corn Belt	9	*	*	*	*
- Southern	6	*	*	ns	ns
Hallauer (1972)	9	*	*	*	*
Barriga and Vencovsky (1973)	5	*	*	*	ns
Miranda (1974a) - set 1[‡]	7	*	*	ns	ns
- set 2[‡]	9	*	*	*	ns
Genter and Eberhart (1974)[§]	13	*	*	*	ns

[†]See Sec. 10.5 Components of Heterosis in Intervarietal Diallel
Crosses for the meaning of effects.

*Significant, $P < 0.05$; ns, nonsignificant.

[‡]Set 1: seven short plant varieties; set 2: set 1 plus
two high yielding varieties.

[§]Includes 6 original (0) and 7 advanced (I) populations; significance levels refer to each effect on the average of 0 and I; only
h_j showed a significant interaction with (0 vs. I).

Moll et al. (1965) extended the previous study to include 2 races from
Mexico. The 8 parental varieties and 28 variety crosses were tested in each
region of origin of the parental varieties. They found that heterosis manifested in crosses of varieties hypothesized to be the most genetically diverse
was less than heterosis expressed between varieties considered to be less genetically diverse. These results suggested that the concept of genetic divergence for maximum expression of heterosis has limits. Apparently crosses between extremely divergent parents create a situation where the harmonious
functioning of alleles is disrupted; consequently the physiological functions
are not as efficient as in situations where the alleles have had similar selection pressures.

Heterosis observed in variety crosses is the average expression of heterosis of the genotypes formed by crossing a sample of genotypes from each of
the two parental varieties. Individual F_1 plants in the variety crosses will
vary in relative heterosis, depending on the two parental gametes sampled to

produce the F_1 plant. Some of the F_1 variety cross plants would be expected
to show considerably more heterosis than others. If sampling is adequate the
range of heterosis of individual plants in the variety crosses will approximate a normal distribution. The primary goal is to identify the parental
plants in each variety that maximize heterosis when crossed; i.e., the performance of a specific hybrid is greater than the modal performance of a variety cross. This concept was the basis of Shull's (1908) suggestion for developing pure lines and using them to produce the crosses.

Not all crosses among a set of lines will exceed the variety cross. If a
set of unselected inbred lines is developed from a variety and a large number
of crosses is made among the unselected inbred lines, the average of all
crosses is expected to regenerate the original variety (Good and Hallauer
1977). Since Shull's suggestion, the objective of maize breeding has been to

Table 10.4. Heterosis for four groups of inbred lines for yield and four
components of yield (Martin and Hallauer 1976).

Generation		Yield (g/plant)	Ear length (cm)	Ear diameter (cm)	Kernel-row number	300 kernel wt. (g)
				Trait		
Group I (first cycle - 7 lines and 21 crosses)						
Midparent		48.62	12.60	3.82	14.16	63.63
F_2, BC_1, BC_2		100.97	15.87	4.30	15.24	70.61
F_1		152.65	18.37	4.57	15.84	76.07
	h^+	213.96	45.8	19.6	11.9	19.6
	I^+	-33.85	-13.6	-5.9	-3.8	-7.2
Group 2 (second cycle - 7 lines and 21 crosses)						
Midparent		63.58	14.68	4.02	14.46	71.75
F_2, BC_1, BC_2		116.36	16.64	4.46	15.45	75.74
F_1		159.59	18.19	4.71	15.78	82.52
	h^+	151.00	23.9	17.2	9.1	15.0
	I^+	-27.08	-8.5	-5.3	-2.1	-8.2
Group 3 (good lines - 7 lines and 21 crosses)						
Midparent		67.04	14.67	4.04	14.73	67.72
F_2, BC_1, BC_2		111.21	16.45	4.40	15.80	68.89
F_1		150.72	18.22	4.62	16.41	72.57
	h^+	124.82	24.2	14.4	11.4	7.6
	I^+	-26.22	-9.7	-4.8	-3.7	-5.1
Group 4 (poor lines - 7 lines and 21 crosses)						
Midparent		38.24	12.73	3.67	13.26	73.46
F_2, BC_1, BC_2		103.59	16.58	4.27	15.12	72.87
F_1		158.47	19.25	4.55	15.63	76.40
	h^+	314.4	51.2	24.0	17.9	4.0
	I^+	-34.64	-13.9	-6.2	-3.3	-4.6

$^+$h (heterosis) = $[(F_1-MP)/MP] \times 100$; I (inbreeding depression) = $[(F_1-F_2)/$
$F_1] \times 100$.

obtain an elite group of inbred lines and to identify the specific set of lines that maximizes expression of heterosis in hybrids. Because the inbreds are nearly homogeneous and homozygous, the superior hybrid identified is reproducible. Consequently, expression of heterosis for yield in say single-cross hybrids of inbred lines would be expected to be much greater.

One example is provided by data of Martin and Hallauer (1976) (Table 10.4). They compared means of four sets of inbred lines and their single-cross hybrids in the same experiment. Heterosis for yield exceeded 150% for each of the four sets of inbred lines. Heterosis manifested in single crosses produced from selected homozygous lines is about tenfold greater than in varietal crosses shown in Tables 10.1 and 10.2. Differences in magnitudes of heterosis observed between the two sets of material have at least two obvious explanations: (1) The base of comparison is much different in most instances because average yield of two inbred lines is less than that of two noninbred varieties, and (2) inbred lines included by Martin and Hallauer were selected because of their performance in hybrids and would not be expected to represent a random sample of gametes from a maize population. The first explanation is common in comparisons that involve means. The greater heterosis among flint variety crosses reported by Troyer and Hallauer (1968) in comparison with dent varieties studied by Lonnquist and Gardner (1961) could be explained in part by differences in yield levels of parents included and environments in which they were tested. The second explanation is a consequence of the breeding procedures used to develop elite inbred lines for use in single-cross hybrids.

We have only discussed observed heterosis for yield. Several studies have been reported attempting to define the origin of heterosis for specific crosses. These studies included different plant organs at different stages of development of parents and crosses. Generally they did not give an adequate explanation of the phenomenon of heterosis. Measurements usually showed that differences existed, but the basic mechanism was not determined. For a detailed account of these studies for maize and other plant species, the reader is referred to the summary given by Sprague (1953).

10.2 GENETIC BASIS

Heterosis, whether between crosses of varieties or inbred lines, is observed in maize, but the genetic basis of observed heterosis is still conjectural. Different theories have been proposed as an explanation, some having and others not having a genetic basis. Although extensive research has been conducted for the past 70 years, it has been difficult to prove or disprove the different theories proposed. Because of the importance of maize and the heterosis manifested in maize crosses, breeders and geneticists have been active in studying the heterosis phenomenon and attempting to develop genetic models for its basis.

Several hypotheses have been proposed and discussed to account for heterosis in maize, but most can be included in either one of the two following categories: (a) physiologic stimulation (or allelic interaction or overdominance) and (b) dominant favorable growth factors. The evidence supporting either of the two hypotheses depends on the available data and the interpretation of these data. Since the hypotheses were proposed and discussed, rapid progress has been made in quantitative genetic theory. Yield, as an example, has a very complex mode of inheritance, and heterosis of yield has been of great interest to both early and recent researchers. Yield is a measure of reproductive capacity and nearly always is treated as a quantitative trait. Information from quantitative genetic studies has, therefore, contributed to the understanding of heterosis in maize. In the past, however (e.g., see

Richey 1946), heterosis and quantitative genetics were often treated as separate entities, which they are not. We will give a cursory review of past developments for the explanation of heterosis. Interested readers should refer to Whaley (1944), Richey (1946), and Sprague (1953) for more extensive reviews.

Shull (1908) presented the first theory of heterosis, designated as the physiologic stimulation or heterozygosity hypothesis. The theory was based on the premise that heterozygosity itself was the cause of heterosis, which is a non-Mendelian explanation. East and Hayes (1912) and Shull (1912) supported this theory. In 1936, East reviewed the evidence for heterosis and concluded that dominant favorable growth factors were inadequate to explain heterosis. He proposed that multiple alleles at a locus are differentiated with respect to their physiological functions. As Sprague (1953) indicated, the ideas proposed are included within the limits of the physiologic stimulation hypothesis, but the proposed model was similar to the concept of overdominance suggested by Hull (1945).

The second hypothesis of heterosis, dominant favorable growth factors, was first presented mathematically by Bruce (1910). Although the hypothesis was briefly presented, the mathematical derivation given by Bruce is amenable to the current status of the dominance hypothesis and the occurrence and magnitude of heterosis that can be expected from crosses. The derivations presented by Bruce could include any number of gene pairs, any range of gene frequencies, and any level of dominance. The salient features of Bruce's hypothesis were that heterosis would occur if the parents differed in gene frequency and dominance was present; it was a Mendelian genetic hypothesis. Jones (1917, 1945, 1958), among others, was a strong supporter of the hypothesis that heterosis is manifested by the accumulation of dominant favorable factors at different loci.

Proponents of the overdominance hypothesis attacked the dominance hypothesis because evidence did not substantiate the proposed model.

1. *If the phenomenon of heterosis were due to the accumulation of dominant favorable growth, one should be able to obtain inbred lines as productive as the single-cross hybrids; this has never been accomplished.* However, although critical data are not available, empirical evidence indicates that relative vigor, health, and productivity of inbred lines are increasing in successive recyclings, which is why it is possible to produce single-cross hybrids in the U.S. Corn Belt. Some evidence is given in Table 10.4. Second cycle inbred lines were 30.8% greater yielding than first cycle lines. Also, Collins (1921) showed that if number of factor pairs for a trait exceeds 10, the chance of obtaining a plant homozygous for all 10 factors is remote. Gene number for traits such as vigor, health, and productivity are unknown but they must be extremely great in all instances. Resolution of gene structure at the present indicates that the nucleotide base pairs influencing yield are very great. Genes defined on this basis would preclude any opportunities (at our present state of knowledge) of obtaining inbred lines as vigorous as single-cross hybrids. But advances are being made in improvement of vigor, health, and productivity of our inbred lines and would substantiate the dominance hypothesis.

2. *Absence of skewed distributions in the F_2 populations also is an indication that dominance is not the primary feature of heterosis. If dominance were present, we would expect a skewed distribution from the expansion of the binomial* $(3 + 1)^n$. Collins (1921) showed however that skewness because of dominance is not great if a large number of factor pairs are involved in ex-

pression of the trait. Again, for yield where number of factor pairs must be
extremely large, absence of skewed distributions does not detract from the
dominance hypothesis as the basis of heterosis.

3. *Failure to obtain convincing evidence for occurrence of dominance in
the expression of traits quantitatively inherited also is given as a reason
for not supporting the dominance hypothesis.* On the contrary, most evidence
indicates that partial to complete dominance is the primary mode of inheritance. Or, most evidence indicates overdominance is not the primary mode of
inheritance. Some of the first quantitative genetic studies of F_2 populations
indicated that overdominance (see Table 5.3) may be important. It was found,
however, that estimates were biased because of linkage; thus, pseudooverdominance because of linkage rather than true overdominance existed. Also, results of recurrent selection experiments designed to maximize selection for
overdominant gene action (Hull 1945) show that partial to complete dominance
is more in concert with results of selection than overdominant gene action.
Level of dominance is difficult to estimate for complex traits controlled by a
large unknown number of genes, but most of the cumulative evidence favors partial to complete dominance for the majority of genes involved.

Definitive proof for either of the hypotheses proposed for the genetic
basis of heterosis probably will be difficult to establish. Because of complexity of the inheritance of quantitative traits, all types of gene action,
both inter- and intraallelic, are probably involved. Maize breeders work on
the basis of chromosome segments, and the resolution of factor pairs is limited. Heterosis in maize can be utilized without knowledge of the exact genetic
basis of its occurrence. What is important is knowledge of the predominant
types of gene action operative in designing effective and efficient breeding
schemes for continued progress. For practical purposes, it seems that evidence supports the hypothesis that heterosis results from an accumulation of
dominant favorable growth factors.

Crow (1948, 1952), however, presented arguments that the dominance hypothesis advanced by Bruce (1910) and Jones (1917) was not sufficient to account for observed heterosis among hybrids obtained by crossing inbred lines
extracted from an equilibrium population. Crow's argument was based on mathematical calculations for a given set of assumptions, some of which are not
valid in applied breeding programs. Crow defined an equilibrium population as
one in which gene frequencies were in equilibrium between mutation and selection, and genotypic frequencies were those expected with random mating and
linkage equilibrium. With this definition of an equilibrium population and
assuming 5000 loci and 10^{-5} mean mutation rate as reasonable estimates for
yield, a fitness trait, it was calculated that maximum heterosis manifested in
hybrids over the parent equilibrium population would be 5%; i.e., the best hybrid among inbred lines extracted from the equilibrium population would not
exceed the equilibrium by more than 5%. Because the best hybrids among a set
of inbred lines usually exceed the equilibrium population by more than 5%, the
only plausible explanation of heterosis was the importance of overdominance,
as suggested by East (1936) and Hull (1945).

The critical feature of Crow's theoretical derivation is the definition
of an equilibrium population. To maximize heterosis manifested in crosses of
inbred lines, maize breeders usually cross lines between populations that show
evidence of a definite heterotic pattern, such as lines of Reid Yellow Dent
origin with those of Lancaster Surecrop origin or dent with flint. If the
lines were extracted from different varieties, the assumption of an equilibrium population would apply only if all varieties are genetically identical,
i.e., gene frequencies the same for all source populations from which the

lines were extracted. Evidence given in Tables 10.1 and 10.2, however, demonstrates genetic differences among open-pollinated varieties because of the heterosis manifested among the variety crosses. It is interesting that the 5% derived by Crow is similar to the average high parent of 8.2% given in Table 10.1. If we visualize the maize species *Zea mays* L. as our equilibrium population, the estimate of 5% is similar to the observed high parent heterosis of 8.2%. If the number of gene loci affecting yield is greater than 5000, as it could be, the average high parent heterosis would satisfy Crow's model. Unless the open-pollinated varieties have been selected for a common equilibrium status, differences in average grain yield between parental varieties and selected hybrids involving inbred lines extracted from different source varieties do not provide critical evidence relative to the level of dominance. But heterosis manifested in crosses among open-pollinated varieties (Tables 10.1 and 10.2) provides evidence that genetic differences exist among varieties. Usually, maize breeders produce hybrids between lines of diverse parentage. Because there are genetic differences among source populations or varieties, Crow's arguments do not negate the dominance hypothesis as an explanation of heterosis in hybrids produced among inbred lines, or, for that matter, variety crosses produced among different varieties.

Epistasis also may contribute to the heterosis expressed in crosses. Although studies indicate epistasis does not seem to be a major component of genetic variability (see Chap. 5), epistatic effects have been shown to occur in specific crosses of inbred lines of maize (Bauman 1959; Gorsline 1961; Gamble 1962; Sprague et al. 1962; Sprague and Thomas 1967; Eberhart and Hallauer 1968; Stuber and Moll 1971). Detection of epistatic effects indicates that some specific crosses of lines with unique combinations of genes contribute to heterosis. Cress (1966) has shown that multiple alleles can show negative dominance effects among some of the combinations, and hence could account for the observed results in the absence of epistasis. The curvilinear relation among different generations generated from crosses of inbred lines and varieties also is interpreted as evidence of epistasis (see Chap. 9). Epistasis for quantitative traits surely exists, but it has been difficult to determine that epistatic interactions account for very little of the genetic variability in maize populations beyond that accounted for by additive and dominance variances. Although it has not been quantified, epistasis may be important in the expression of heterosis in a single cross of two inbred lines; i.e., the unique combination of gene interactions is restricted to the cross of the two inbred lines. Of course this would be restricted to the specific cross and may be of relatively small importance in a maize population.

10.3 BIOMETRICAL CONCEPT

Falconer (1960) has shown that heterosis will be expressed when we have the following conditions: (1) presence of some level of dominance and (2) relative difference in gene frequency of the two parents to determine the magnitude of the heterosis expressed in crosses. If either or both of the conditions do not exist, heterosis will not be manifested. The conditions demonstrated by Falconer are essentially the same as those given by Bruce (1910), but they are generalized in modern vernacular to include any number of genes, range in gene frequency of the two parents, and arbitrary level of dominance. The statistical description of heterosis given by Falconer does not support or detract from either of the proposed hypotheses for heterosis.

As previously shown, heterosis is defined as $h = \bar{F}_1 - \overline{MP}$, where $\bar{F}_1$ is the mean of the first generation cross between two parental populations of lines in Hardy-Weinberg equilibrium and $\overline{MP}$ is the midparent value. Table 2.1 shows that at the one locus level the population mean is $\mu_A = (p - q)a + 2pqd$, where p and $q = 1 - p$ refer to gene frequencies and a and d to genotype effects. If another population is considered, with corresponding gene frequencies r and s, its mean is given by $\mu_B = (r - s)a + 2rsd$. Thus, the midparent value $\overline{MP}$ = $(1/2)(p - q + r - s)a + (pq + rs)d$. From Table 2.6, the mean of the first generation cross between two populations is shown to be $\overline{F}_1 = (pr - qs)a + (ps + qr)d$. Therefore, we have

$$h = [(pr - qs) - (1/2)(p - q + r - s)]a + [(ps + qr) - (pq - rs)]d$$

$$= (1/2)[2pr - 2qs - (p - q) + (r - s)]a + (ps + qr - pq - rs)d$$

$$= (1/2)(q^2 - s^2 - 2qs - p^2 + r^2 + 2rp)a + p(s - q)d + r(q - s)d$$

$$= (1/2)[(q - s)^2 - (p - r)^2]a + (s - q)(p - r)d$$

$$= 0 + (1 - r - 1 + p)(p - r)d = (p - r)^2 d$$

From the above derivations, it can be seen that the crossed population mean equals the midparent value when no dominance is assumed.

The F_2 generation mean ($\overline{F}_2$) is obtained considering the genetic structure of the crossed population. At the one locus level, the F_2 generation is in Hardy-Weinberg equilibrium and its gene frequency is the average gene frequency of the two parental populations; i.e., $p' = (p + r)/2$. The array of genotypic values is $p'^2 a : 2p'q'd : q'^2(-a)$ so that the crossed population mean is $\overline{F}_2$ = $(p' - q')a + 2p'q'd = (1/2)[(p - q + r - s)a + (p + r)(q + s)d]$, which equals $\overline{MP}$ and $\overline{F}_1$ under the assumption of no dominance. Thus, the excess of $\overline{F}_2$ over $\overline{MP}$ is $[(1/2)(p + r)(q + s) - (pq + rs)]d = (1/2)(ps + rq - pq - rs)d$ = $(1/2)(p - r)^2 d = (1/2)h$. Therefore, $\overline{F}_2$ is greater than $\overline{MP}$ by an amount that is half the excess expressed in the first cross generation, or in other words, half the heterotic effect is lost from the first to the second generation under the restriction of no epistasis when several loci are considered. The complete heterotic effect would be obtained by summation of effects over all loci.

When considering the cross between two inbred lines, the only difference is that for one line $p = 0$ or $p = 1$, depending on whether the gene is in homozygous recessive or homozygous dominant condition. For the other line, $r = 0$ or $r = 1$ for the same locus. The heterotic response in the first generation cross is due to the loci where $p = 1$ and $r = 0$ or vice versa, so the heterotic effect depends on the number of such contrasting loci and also on the level of dominance at each locus. In any case, heterotic response is expected to occur whenever there is difference in gene frequencies and some degree of directional dominance at one or more loci involved in the control of the character.

10.4 HETEROSIS AND PREDICTION METHODS

One of the contributions of genetics to agriculture experimentation has been predicting results following controlled crosses. Following the rediscovery of Mendelian laws, prediction methods have been used extensively in the study of qualitative traits. In quantitative traits, parameters, such as means rather than proportions of genotypes, are of greater importance. Prediction of means, as well as prediction of results from selection, is one of the important contributions of quantitative genetics to plant and animal breeding. Theory developed for prediction purposes has brought about a better understanding of the nature of gene action and its relation to population means and their components.

In the following sections several cases are considered where prediction of population means has possible uses. The theory involved is restricted to the specific conditions of parental populations that have diploid segregation, no preferential fertilization, Hardy-Weinberg equilibrium, linkage equilibrium, and/or negligible epistasis. The procedures presented are useful when a large number of possible crosses cannot be produced and evaluated in experimental trials.

10.4.1 Populations under Selfing

Theories about prediction involving homozygous (pure) lines were first developed for self-fertilizing crops, but they are useful in maize and other cross-pollinated crops where inbred lines are used. The following formulas are given by Mather (1949) and Mather and Jinks (1971).

Let P_1 and P_2 represent two completely homozygous lines and F_1 the first generation cross between them. For any quantitative trait the mean of the F_2 generation (obtained by selfing or sibmating F_1 plants) can be predicted by

$$\overline{F}_2 = (1/4)(\overline{P}_1 + \overline{P}_2 + 2\overline{F}_1)$$

where $\overline{P}_1$ and $\overline{P}_2$ are the parent-line means and $\overline{F}_1$ the first generation cross mean.

In the same manner other advanced populations means ($\overline{F}_3$, $\overline{F}_4$,...) may be predicted by

$$\overline{F}_3 = (1/8)(3\overline{P}_1 + 3\overline{P}_2 + 2\overline{F}_1), \quad \overline{F}_4 = (1/16)(7\overline{P}_1 + 7\overline{P}_2 + 2\overline{F}_1),...$$

The general formula to predict population means in any generation under selfing without selection is

$$\overline{F}_n = (1/2)[1 - (1/2)^{n-1}](\overline{P}_1 + \overline{P}_2) + (1/2)^{n-1}\overline{F}_1$$

where the nth generation is obtained after $n - 1$ generations under selfing (Mather 1949).

When the F_1 is backcrossed to either parent (BC1 and BC2), its mean is predicted by

$$\overline{BC1} = (1/2)(\overline{P}_1 + \overline{F}_1) \quad \overline{BC2} = (1/2)(\overline{P}_2 + \overline{F}_1)$$

The F_2 mean also is expected to be $\overline{F}_2 = (1/2)(\overline{BC1} + \overline{BC2})$.

10.4.2 Double Crosses and Three-way Crosses

Double-cross hybrids have been extensively used in maize since their use was suggested by Jones (1918). In subsequent years, maize breeders' primary objective has been the development of new superior hybrids. Double-cross hybrids result from crosses between two single crosses that are themselves the result of crosses between two inbred lines. The best results are expected to occur when four different inbred lines are used. If the same inbred line is included in both parental single crosses some inbreeding is expected in the double cross, thus precluding the maximization of heterosis.

A set of n inbred lines can be subdivided into C_n^4 subsets of four inbred lines. Within each subset only three distinct double crosses are possible. Within the subset [A, B, C, D], the following double crosses are possible: (A × B) × (C × D), (A × C) × (B × D), and (A × D) × (B × C). Thus the total number of possible double crosses is

$$N_{dc} = 3C_n^4 = (1/8)[n(n - 1)(n - 2)(n - 3)]$$

For example, for n = 10, 630 distinct double crosses are possible. Such a number of entries cannot be evaluated in experimental trials and the need of predictions is evident. Prediction based on single crosses ($N_{sc} = C_n^2$) would require only 45 single crosses to get the information for 630 double crosses.

Double-cross hybrid breeding programs usually use two genetically divergent populations. Theory and empirical data have demonstrated that superior hybrids are more likely to be obtained if both populations are improved by some method of recurrent selection, especially reciprocal recurrent selection (Hallauer 1973; Suwantaradon and Eberhart 1974; Moll et al. 1977). In such instances the best procedure is to develop one parental single cross from each base population to maximize heterotic effects in the double crosses. If n_1 inbred lines are available from one population and n_2 from another, the number of possible double crosses is

$$N_{dc} = C_{n_1}^2 C_{n_2}^2$$

In this case only $n_1 n_2$ single crosses are needed for prediction. If $n_1 = n_2 = n$, $N_{dc} = (1/4)[n(n-1)]^2$, and n^2 single crosses are needed to predict double crosses.

Three-way crosses have also been successfully used in commercial maize hybrids. They result from crosses between one single cross and one inbred line with the single cross used as the female parent for seed production. Prediction of three-way crosses also is important whenever the number of inbred lines is too great for experimental evaluation.

If a fixed set of n inbred lines is available, the number of possible three-way crosses is

$$N_{tc} = 3C_n^3 = (1/2)[n(n-1)(n-2)]$$

For example, for n = 10 we have $N_{tc} = 360$.

If two sets of n_1 and n_2 inbreds are from two distinct populations, we have

$$N_{tc} = n_1 C_{n_2}^2 = n_1 n_2 (n_2 - 1)/2 \quad \text{when the single cross is from set 2}$$

$$N_{tc} = n_2 C_{n_1}^2 = n_1 n_2 (n_1 - 1)/2 \quad \text{when the single cross is from set 1}$$

Total number of possible three-way crosses is $(1/2)n_1 n_2 (n_1 + n_2 - 2)$. In this case only $n_1 n_2$ single crosses are needed for prediction. If $n_1 = n_2 = n$, then n^2 single crosses are needed. For example, if $n_1 = n_2 = 5$, $N_{tc} = 100$.

Prediction of double-cross performance in maize was first reported by Jenkins (1934) using single-cross data. Jenkins suggested four alternative methods for prediction:

A. Mean performance of six possible single crosses among any set of four inbred lines
B. Average performance of four nonparental single crosses
C. Average performance of four inbred lines over a series of single crosses
D. Average performance of a set of four inbred lines when tested by the topcross procedure

The four methods of prediction developed by Jenkins differ with respect to type of gene action involved. Methods A, C, and D are related only to additive gene action, while method B involves additive as well as nonadditive (dominance and various types of epistasis) effects. Jenkins, using all methods, found a significant correlation between observed and predicted means. However, the correlation was greatest in method B, which agrees with the present knowledge of quantitative genetics theory. The efficiency of method B also was reported by several authors (Doxtator and Johnson 1936; Anderson 1938; Hayes et al. 1943; Hayes et al. 1946).

Prediction of double-cross hybrids using Jenkins's method B is as follows. In each set of four inbred lines (say P_1, P_2, P_3, and P_4), the six possible single crosses are S_{12}, S_{13}, S_{14}, S_{23}, S_{24}, and S_{34}; and the three possible double crosses may be predicted:

$$S_{12} \times S_{34}{:}\overline{D}_{12.34} = (1/4)(\overline{S}_{13} + \overline{S}_{14} + \overline{S}_{23} + \overline{S}_{24})$$
$$S_{13} \times S_{24}{:}\overline{D}_{13.24} = (1/4)(\overline{S}_{12} + \overline{S}_{14} + \overline{S}_{23} + \overline{S}_{34})$$
$$S_{14} \times S_{23}{:}\overline{D}_{14.23} = (1/4)(\overline{S}_{12} + \overline{S}_{13} + \overline{S}_{24} + \overline{S}_{34})$$

A simple model can be used to illustrate the theory of prediction procedure, although a rather complex theory may be involved. Suppose that the four parental lines have the following genotypes: P_1:AABB, P_2:AAbb, P_3:aaBB, and P_4:aabb. Denoting by a (or -a) and d the genotypic effects for homozygotes and heterozygotes as deviations around a mean μ of the two extreme homozygotes, the following effects can be assigned in a diallel table, not considering epistatic effects, where

1	2	3	4
$\mu + a_A + a_B$	$\mu + a_A + d_B$	$\mu + d_A + a_B$	$\mu + d_A + d_B$
	$\mu + a_A - a_B$	$\mu + d_A + d_B$	$\mu + d_A - a_B$
		$\mu - a_A + a_B$	$\mu - a_A + d_B$
			$\mu - a_A - a_B$

Double cross $D_{12.34}$ results from the following cross:

Parental single-crosses	Double-cross Genotypes	Genotypic effects
AABb	AaBB	$\mu + d_A + a_B$
aaBb	AaBb	$\mu + d_A + d_B$
	AaBb	$\mu + d_A + d_B$
	Aabb	$\mu + d_A - a_B$
	Average	$\mu + d_A + (1/2)d_B$

Average genotypic effects of a double cross can be predicted by average genotypic effects of the four genotypes in the upper right-hand corner of the diallel table. Those effects are relative to the nonparental single crosses.

Following the same procedure, three-way crosses of a set of three inbred lines can be predicted by

$$\overline{T}_{12.3} = (1/2)(\overline{S}_{13} + \overline{S}_{23})$$

$$\overline{T}_{13.2} = (1/2)(\overline{S}_{12} + \overline{S}_{23})$$

$$\overline{T}_{23.1} = (1/2)(\overline{S}_{12} + \overline{S}_{13})$$

According to Eberhart (1964) the following formulas can be used to predict double crosses:

1. $\hat{D}^{sa}_{ij.kl} = (1/6)(S_{ij} + S_{ik} + S_{il} + S_{jk} + S_{jl} + S_{kl})$

2. $\hat{D}^{sb}_{ij.kl} = (1/4)(S_{ik} + S_{il} + S_{jk} + S_{jl})$

3. $\hat{D}^{tij}_{ij.kl} = (1/2)(T_{ij.k} + T_{ij.l})$

4. $\hat{D}^{tkl}_{ij.kl} = (1/2)(T_{kl.i} + T_{kl.j})$

5. $\hat{D}^{t}_{ij.kl} = (1/2)(D^{tij}_{ij.kl} + D^{tk.l}_{ij.kl})$

The first two formulas correspond to methods A and B as proposed by Jenkins (1934). The others are based on three-way cross performance and have not been extensively used because from a fixed set of inbred lines there are more possible three-way crosses than single crosses. Such formulas may be useful when a desirable single cross (S_{ij}) is available and two new inbred lines (k and l) must be developed to form the double cross $D_{ij.kl}$.

All of the five formulas mentioned above are unbiased by additive effects and all but formula 1 also by dominance effects. Hence, formulas 2, 3, 4, and 5 can be efficiently used to predict double-crosses if epistatic effects are negligible. When epistatic effects of the dominance type are unimportant relative to other types of epistasis, the following linear relation is suggested to predict the double cross (Eberhart 1964):

6. $\hat{D}^{t-s}_{12.34} = 2\hat{D}^{t}_{12.34} - \hat{D}^{sb}_{12.34}$

$$= (1/2)(T_{12.3} + T_{12.4} + T_{34.1} + T_{34.2}) - (1/4)(S_{13} + S_{14} + S_{23} + S_{24})$$

Eberhart et al. (1964) used formulas 2, 5, and 6 to predict double-cross performance. Although observed values were not available for comparisons, they concluded that differences in predicted values were not significant among the methods. In addition, although epistasis was present it was not of sufficient magnitude relative to experimental error and genotype-environment interaction to give formula 6 a higher efficiency.

The joint number of single crosses and three-way crosses required to predict double crosses is $n_s + n_t = (1/2)n(n - 1)^2$, which only exceeds the number of possible double crosses when the number of inbred lines $n \leq 7$. If two distinct populations are used, $n_s + n_t = (1/2)n_1n_2(n_1 + n_2)$, where n_1 and n_2 are the number of inbred lines from each population. So $n_d = (n_s + n_t)$ when $n_1n_2 - 3n_1 - 3n_2 + 1 = 0$. If $n_1 = n_2$, then $n_d > (n_s + n_t)$ when $n > 5$. Prediction based on single and three-way crosses is justified only when n_d is large; but in this case, $n_s + n_t$ is also large as shown in Table 10.5. Hence prediction based on this procedure has a practical limitation.

Cockerham (1967) presented a unified theory that makes use of both the genetic and experimental conditions to predict double crosses from single-cross hybrid data. The random sample approach was considered and an *optimum predictor* was suggested. Cockerham found that differences in efficiency when comparing the optimum predictor with Jenkins's methods A and B were small, although the optimum predictor was generally more efficient.

Table 10.5. Number of possible double crosses (n_d) and number of entries ($n_s + n_t$) needed for their prediction.

	one set (n lines)			two sets ($n_1 = n_2 = n'$)		
n	n_d	n_s	$n_s + n_t$	n'	n_d	$n_s + n_t = (n')^3$
4	3	6	18	2	1	8
5	15	10	40	3	9	27
6	45	15	75	4	36	64
7	105	21	126	5	100	125
8	210	28	196	6	225	216
9	378	36	288	7	441	343
10	630	45	405	8	784	512
15	4,095	105	1,470	9	1,296	729
20	14,435	190	3,610	10	2,025	1,000
30	82,215	435	12,615	20	36,100	8,000

Otsuka et al. (1972) used the concept of optimum predictor to develop equations for use in double-cross, three-way cross, and single-cross estimation, based on general and specific effects estimated from a diallel analysis of fixed effects. The models were

$$\hat{S}_{ij} = m + \hat{g}_i + \hat{g}_j + \hat{s}_{ij}$$

$$\hat{D}_{ij.kl} = m + (1/2)(\hat{g}_i + \hat{g}_j + \hat{g}_k + \hat{g}_l) + \lambda(1/4)(\hat{s}_{ik} + \hat{s}_{il} + \hat{s}_{jk} + \hat{s}_{jl})$$

$$\hat{T}_{ij.k} = m + (1/2)(\hat{g}_i + \hat{g}_j) + \hat{g}_k + \lambda(1/2)(\hat{s}_{ik} + \hat{s}_{jk})$$

where λ, varying between 0 and 1, was a weighting coefficient for specific effects. When $\lambda = 1$ the formula for double-cross prediction corresponds to Jenkins's method B. The use of *optimum weight* predictor was possible, but it was found that prediction with Jenkins's method B ($\lambda = 1$) was nearly as efficient. The authors also found that Jenkins's method A was slightly superior to any other predictor in some instances, suggesting that information from parental single crosses may be effective for predicting hybrids from a fixed set of highly selected lines.

Accuracy of prediction depends more on number of replications and environments than on small differences in prediction methods. Otsuka et al. (1972) noted, for example, that 45 single crosses from 10 lines would require 450 field plots (two replications and five environments) to predict 630 possible double crosses. On the other hand, 2520 plots (two locations and two replications) would have been required to have a similar precision on double cross performance itself.

10.4.3 Synthetic Varieties

Synthetic varieties have been widely used for commercial and breeding purposes since they were first suggested by Hayes and Garber (1919). *Synthetics* were defined by Lonnquist (1961) as "open-pollinated populations derived

from the intercrossing of selfed plants or lines and subsequently maintained
by routine mass selection procedures from isolated plantings." When open-pollinated varieties instead of inbred lines are intercrossed, resulting populations are usually called *composites* or *composite varieties*. Synthetic varieties and composites are very similar in structure, but the distinction may be
useful for practical purposes. Some breeders use synthetic variety in its
broadest sense, including any open-pollinated population derived from artificial selection; to avoid confusion, the term is used herein according to Hayes
and Garber's (1919) concept (defined by Lonnquist, 1961).

Yield or any quantitative traits of synthetic varieties are predicted according to a formula based on Wright's (1922) statement: "A random-bred stock
derived from n inbred families will have (1/n)th less superiority over its inbred ancestry than the first cross or a random-bred stock from which the inbred families might have been derived without selection." The formula developed from this principle is commonly called *Wright's formula* and was cited by
Kinman and Sprague (1945) as

$$\hat{\overline{Y}}_2 = \hat{\overline{Y}}_1 - (\hat{\overline{Y}}_1 - \hat{\overline{Y}}_0)/n$$

where $\overline{Y}_2$ = mean of a synthetic variety obtained by intercrossing all possible
single crosses among a set of n inbred lines, $\overline{Y}_1$ = average performance of all
single crosses among n inbred lines, and Y_0 = average performance of n parental inbred lines. The closeness of observed and predicted means using
Wright's formula has been reported (e.g., Neal 1935).

A general formula was given by Busbice (1970) to predict synthetic varieties. The general formula is

$$\overline{Y}_t = \overline{Y}_0 + [(F_0 - F_t)/(F_0 - F_1)](\overline{Y}_1 - \overline{Y}_0) \qquad F_0 \neq F_1$$

where F_t (i = 0, 1, 2, ..., t) is Wright's coefficient of inbreeding in the ith
generation. Values of F_1 and F_t can be computed from F_0, using special formulas where the coefficient of parentage of parents (r_0), the ploidy number
(2k), the frequency of selfing (s), and the number of parents (n) are consid-
ered. Terms $\overline{Y}_0$, $\overline{Y}_1$, and $\overline{Y}_t$ are the means in generations 0 (parents), 1 (all
possible single crosses among n parents), and t. If the parents are complete-
ly homozygous ($F_0 = 1$), unrelated, diploid (2k = 2), and no selfing occurs (s
= 0), the general formula reduces to Wright's formula.

Gilmore (1969) concluded that Wright's formula can be used when lines are
at any level of inbreeding. He showed that Wright's formula is valid for S_1,
S_2,..., S_n lines; parental lines need not be limited to those developed from
selfing; and gene frequency must be limited to 1 or 0.5. It is only required
that parental lines be in Hardy-Weinberg equilibrium for each locus. As pointed
out by Mochizuki (1970), Wright's formula is used to predict composite populations because parental varieties are in genetic equilibrium.

10.4.4 Composite Populations

Composite populations result from random mating all possible intervarietal crosses among a fixed set of heterogeneous varieties (populations or
races). Composites have been widely used as breeding (base) populations because greater genetic variability is expected to be available if populations
of diverse origins (diverse populations, including exotic germplasm) are combined. Heterosis among intervarietal crosses also has been found to be relatively high, and when a composite is formed the mean yield of the new population is expected to be greater than the average of the parental varieties.

Selection among varieties used as parents may give a higher composite mean. In some instances poorer varieties can be discarded and the breeder can start the program at a higher level of productivity. Choosing varieties to be used as parents may be based on a variety-cross diallel evaluated in a series of environments (Table 10.3). Composite means thus can be predicted from the diallel data.

The number of possible different composites N_{co} increases greatly with an increase in n, the number of parental varieties (Vencovsky and Miranda 1972):

$$N_{co} = 2^n - (n + 1)$$

For n = 10, it is possible to have 1013 different composites for the cases in which each parental variety contributes equal proportions of germplasm. Actually the number of possible composites is infinite if we include those types where unequal contributions of parental varieties is assumed. The formula above also gives the number of possible synthetic varieties from a set of n inbred lines.

Prediction of composite means was first suggested by Eberhart et al. (1967), where an expression similar to Wright's formula was suggested:

$$\hat{Y}^r_{co} = \overline{Y}_c - (\overline{Y}_c - \overline{Y}_v)/n$$

where $\hat{Y}^r_{co}$ represents predicted mean of a quantitative trait for a composite obtained from random mating, $\overline{Y}_c$ is the average of all possible intervarietal crosses among n parental varieties, $\overline{Y}_v$ is the mean of n parental varieties, and n is the number of varieties.

The predicted means are obtained from a diallel table as follows. Taking the complete diallel table as reference, each predicted mean is obtained by averaging all n^2 values of a partial diallel table containing only the values from parental varieties and their crosses relative to the composite under consideration. An example is shown below of a reference diallel table from which data are used to predict the composite means A ($\hat{Y}_{123}$) and B ($\hat{Y}_{456}$)

Reference diallel table					
	1	2	3	4	5	6
1	60	71	67	69	63	62
2		58	62	62	58	57
3			46	60	55	54
4				48	55	54
5					47	51
6						41

Composite A:	(123)		Composite B:	(456)	
60	71	67	48	55	54
71	58	62	55	47	51
67	62	46	54	51	41
Mean:	Y_{123} = 62.7		Mean:	Y_{456} = 50.7	

Mochizuki (1970) showed a detailed derivation of the above formula, presenting also its alternative forms, as follows:

$$Y_{co}^{r} = \mu + (1/n)(a_1 + a_2 + \ldots + a_n) + (1/n)(d_1 + d_2 + \ldots + d_n)$$
$$+ (2/n^2)(h_{12} + h_{13} + \ldots + h_{n-1,n})$$

$$= \mu + (1/n) \sum_{j=1}^{n} (a_j + d_j) + (2/n^2) \sum_{j<j'} h_{jj'}$$

where μ is the parent varieties mean, a_j and d_j are deviations due to homozygotes and heterozygotes relative to the jth variety, and $h_{jj'}$ is the total heterosis relative to cross between varieties j and j'.

In terms of observed means it follows that

$$Y_{co}^{r} = \frac{2}{n(n-1)} \sum_{j<j'} Y_{jj'} - \frac{1}{n}\left[\frac{2}{n(n-1)} \sum_{j<j'} Y_{jj'} - \frac{1}{n} \sum_j \Sigma Y_{jj'} \right]$$

$$= \overline{Y}_c - \frac{\overline{Y}_c - \overline{Y}_v}{n}$$

When the objective of a recurrent selection program is hybrid development, the recommended procedure is the use of two base populations with some genetic divergence between them. Starting from a set of n varieties, a pair of composites can be formed so that a good complementation, or a good specific combining ability, exists between them. Such a pair of base populations may be efficiently used for a reciprocal recurrent selection or for the development of inbred lines and hybrids. The number of possible pairs of composites is given by (Vencovsky and Miranda 1972):

$$N_{pc} = (1/2)[3^n - 2^n(n+2) + n(n+1) + 1]$$

where varieties in each composite contribute equally to predict each pair of composites; i.e.,

$$\hat{Y}_{A \times B} = [1/(mn)](Y_{11} + Y_{12} + \ldots Y_{1n} + Y_{21} + Y_{22} + \ldots + Y_{2n}$$
$$+ \ldots + Y_{m1} + Y_{m2} + \ldots + Y_{mn})$$

where Y_{ij} is a cross of the ith variety (i = 1, 2,..., m) in composite A and the jth variety (j = 1, 2,..., n) in composite B.

For example, from the reference diallel table the composite cross (123) × (456) mean is predicted by

$$\hat{Y}_{A \times B} = (1/9)(69 + 63 + \ldots + 54) = 60.0$$

and the composite cross is predicted by

$$\hat{Y}_{A' \times B'} = (1/9)(71 + 69 + \ldots + 51) = 60.2$$

as shown in the following example:

Composite cross: (123) × (456)			Composite cross: (135) × (246)		
69	63	62	71	69	62
62	58	57	62	60	54
60	55	54	58	55	51
Mean: $\hat{Y}_{A \times B} = 60$			$\hat{Y}_{A' \times B'} = 60.2$		

Note that the prediction formulas for composites and pairs of composites are valid when parental varieties are in genetic equilibrium and when they contribute equally in each composite. In the case of unequal contribution of parental varieties, a general formula is given by Vencovsky (1970). Denoting by p_{ij} the frequency of the (favorable) allele at the ith locus in the jth variety, in the composite population the gene frequency is $\bar{p}_{i.} = \Sigma_j f_j p_{ij}$, where f_j is the proportion of jth variety germplasm in the composite and $\Sigma_j f_j = 1$ (Wahlund 1928).

At genetic equilibrium the composite mean is

$$\hat{Y}^r_{co} = \mu' + \Sigma_i (2\bar{p}_{i.} - 1)a_i + 2\Sigma_i (\bar{p}_{i.} - \bar{p}_{i.}^2)d_i$$

Introducing parameters defined by Gardner and Eberhart (1966), this formula may be expressed as

$$\hat{Y}^r_{co} = \mu' + \Sigma_j f_j (\hat{a}_j + \hat{d}_j) + 2 \Sigma_{j<j'} f_j f_{j'} \hat{h}_{jj'}$$

which also can be expressed in terms of observed means as

$$\hat{Y}^r_{co} = \Sigma_j f_j^2 Y_{jj} + 2 \Sigma_{j<j'} f_j f_{j'} Y_{jj'}$$

where Y_{jj} and $Y_{jj'}$ are taken from a diallel table.

In the same way, a cross between two composites can be predicted as

$$\hat{Y}_{A \times B} = \Sigma_{jj'} f_j f_{j'} Y_{jj'}$$

where j and j' refer to varieties that enter in composites A and B, respectively.

Prediction of composite means has been used by several breeders both for equal contribution of parental varieties (Hallauer and Eberhart 1966; Darrah et al. 1972; Vencovsky et al. 1973; Miranda 1974a) and for unequal contribution (Miranda). Prediction of pairs of composites as a first step for a reciprocal recurrent selection program also has been used (Vencovsky et al.).

10.4.5 Generalization of Prediction Methods

When epistasis is taken as negligible, several formulas are available for prediction of double crosses, three-way crosses, composites, and synthetics. A general formula was suggested by Vencovsky (1973) to predict means in any noninbred population obtained through controlled crossing if the parents are in genetic equilibrium:

$$\bar{Y} = (\Sigma_i f_i P_i)(\Sigma_j f_j P_j)$$

In the above expression, it is assumed that the resulting population comes from two original sources of gametes: one source of female gametes and another source of male gametes. Hence the expression within the first parentheses represents a male (or female) gamete source and the second a female (or male) gamete source; f_i and f_j represent the proportion of parental germplasm (original) in the resulting population, so $\Sigma_i f_i = \Sigma_j f_j = 1$. Terms P_i and P_j denote symbolically the parental (original) types. After parents have been identified and their proportions assigned, the expression must be algebraically extended and terms P_i^2 or P_j^2 replaced by $\bar{Y}_i$ or $\bar{Y}_j$ (parent means) and cross products P_iP_j and P_jP_i (assuming no differences between reciprocal crosses) by $\overline{Y}_{ij}$ (hybrid means).

A numerical example may illustrate the application of the general formula. Let the diallel table below represent means of a quantitative character, where any parental type is in Hardy-Weinberg equilibrium:

1	2	3	4
1.6	1.8	2.0	1.7
	1.4	1.6	1.8
		1.7	1.9
			1.5

(a) Three-way cross prediction:

$$\hat{T}_{12.3} = [(1/2)P_1 + (1/2)P_2](P_3) = (1/2)P_1P_3 + (1/2)P_2P_3$$

and after transformation,

$$\hat{T}_{12.3} = (1/2)\overline{Y}_{13} + (1/2)\overline{Y}_{23} = (1/2)(2.0) + (1/2)(1.6) = 1.80$$

A three-way cross like $\hat{T}_{12.3}$ results from a cross between a single cross (S_{12}) and an inbred line (P_3). If the single cross is used as the female, the original sources of female gametes are the inbred lines P_1 and P_2 with equal participation ($f_1 = f_2 = 1/2$) and are represented in the first term. In the same way the original source of male gametes is the inbred line P_3, which is represented in the second term.

(b) Double-cross prediction:

$$\hat{D}_{12.34} = [(1/2)P_1 + (1/2)P_2][(1/2)P_3 + (1/2)P_4]$$

$$= (1/4)(\overline{Y}_{13} + \overline{Y}_{14} + \overline{Y}_{23} + \overline{Y}_{24}) = 1.78$$

(c) Prediction of an F_2 generation from the cross S_{12}:

$$\overline{F}_2 = [(1/2)P_1 + (1/2)P_2][(1/2)P_1 + (1/2)P_2]$$

$$= (1/4)(\overline{Y}_1 + \overline{Y}_2 + 2\overline{Y}_{12}) = 1.65$$

Note that an F_2 generation is obtained either by selfing F_1 plants when parents are inbred lines or by crossing among themselves. For both instances, the female and male gametes come from the same original source (P_1 and P_2).

(d) Prediction of a synthetic variety or a composite with equal contributions, i.e., $f_1 = f_2 = f_3 = f_4 = 1/4$:

$$\overline{Y}_{syn} = [(1/4)P_1 + (1/4)P_2 + (1/4)P_3 + (1/4)P_4]^2$$

$$= (1/16)(\overline{Y}_1 + \overline{Y}_2 + \overline{Y}_3 + \overline{Y}_4)$$

$$+ (1/8)(\overline{Y}_{12} + \overline{Y}_{13} + \overline{Y}_{14} + \overline{Y}_{23} + \overline{Y}_{24} + \overline{Y}_{34}) = 1.74$$

The gametes that form a synthetic variety come from the same original source, which is represented by the parental inbred lines. The same situation holds for the composites.

(e) Prediction of a synthetic variety with unequal contributions, i.e., $f_1 = 1/2$, $f_2 = 1/6$, $f_3 = 1/6$, and $f_4 = 1/6$:

$$\hat{Y}_{syn} = [(1/2)P_1 + (1/6)P_2 + (1/6)P_3 + (1/6)P_4]^2$$
$$= (1/4)\overline{Y}_1 + (1/36)(\overline{Y}_2 + \overline{Y}_3 + \overline{Y}_4) + (1/16)(\overline{Y}_{12} + \overline{Y}_{13} + \overline{Y}_{14})$$
$$+ (1/18)(\overline{Y}_{23} + \overline{Y}_{24} + \overline{Y}_{34}) = 1.74$$

(f) Prediction of a backcross, $Y_{12.2}$:

$$Y_{12.2} = [(1/2)P_1 + (1/2)P_2](P_2) = (1/2)\overline{Y}_{12} + (1/2)\overline{Y}_2 = 1.6$$

10.4.6 Heterosis as a Component of Means

Heterosis or hybrid vigor is regarded as the superiority of a hybrid over its parents. Its quantitative measure usually is in relation to the average of the parents:

$$h = \overline{F}_1 - (\overline{P}_1 + \overline{P}_2)/2 \quad \text{or} \quad \overline{F}_1 = (\overline{P}_1 + \overline{P}_2)/2 + h$$

which is applied for both cross- and self-pollinated species. The mean of any derived population from an F_1 cross may be associated with changes in heterosis, e.g., for the F_2 generation, obtained by random mating in the F_1 generation:

$$\overline{F}_2 = (\overline{P}_1 + \overline{P}_2)/2 + (1/2)h$$

This expression also is valid for selfed F_1 when parents are homozygous lines.

It can be seen that half the heterotic effect is lost after the first generation of selfing. With continuous selfing, the heterosis in each generation is expected to be half that existing in the preceding generation. So the general formula to predict an F_n generation is

$$\overline{F}_n = (\overline{P}_1 + \overline{P}_2)/2 + (1/2)^{n-1}h$$

where F_n is a population after $n - 1$ generations under selfing (Mather and Jinks 1971).

The decrease in heterosis by half in each generation is explained by the decrease in the frequency of heterozygous loci. For one locus we have:

Generation	Genotypes			Frequency of heterozygotes, %
F_1		Aa		100
F_2	$(1/4)$AA	$(1/2)$Aa	$(1/4)$aa	50
F_3	$(3/8)$AA	$(1/4)$Aa	$(3/8)$aa	25
.		.		.
F_n	$(1/2) - (1/2)^n$AA	$(1/2)^{n-1}$Aa	$(1/2) - (1/2)^n$aa	$100(1/2)^{n-1}$

When several generations of selfing are evaluated with the parents, several estimates for heterosis are available. When the data have a good fit to the theoretical model, the best estimate for heterosis as well as for parent means is least squares. A general least squares formula may be useful for estimating heterosis, e.g.,

$$\hat{h} = \frac{(n + 2)\left[\sum_{k=1}^{n}\left(\frac{1}{2}\right)^k \overline{F}_k\right] - \left(\frac{2^n - 1}{2^n}\right)G}{2\left[\frac{(n + 2)}{3}\frac{(4^n - 1)}{4^n} - \frac{(2^n - 1)^2}{2^n}\right]}$$

where $G = \overline{P}_1 + \overline{P}_2 + \overline{F}_1 + \overline{F}_2 + \ldots + \overline{F}_n$ and

$$\sum_{k=1}^{n} (1/2)^k \overline{F}_k = (1/2)\overline{F}_1 + (1/4)\overline{F}_2 + (1/8)\overline{F}_3 + \ldots$$

Other types of crosses besides selfing change the original heterotic effect. For backcrosses we have

$$\overline{BC1} = (3\overline{P}_1 + \overline{P}_2)/4 + (1/2)h$$

In a synthetic variety (or composite), means also can be expressed as functions of the average heterosis of all crosses among n parent inbred lines (or n varieties in the case of composites):

$$Y_{syn} = Y_{co} = \overline{P} + [(n - 1)/n]\overline{h}$$

where $\overline{P}$ is the parent means (inbred lines or varieties). Note that a portion of heterotic effects can be retained in composites. Thus the use of composites may be considered as a method of direct utilization of heterosis expressed in intervarietal crosses.

Table 10.6 shows the number of possible composites and pairs of composites when a given number n of varieties is available. It also indicates the

Table 10.6. Number of composites, number of pairs of composites, and average heterosis retained from use of n varieties to synthesize the composites.

n	Number of composites[+]	Number of pairs of composites	Average heterosis retained[‡] (%)
2	1	0	50.0
3	4	0	66.7
4	11	3	75.0
5	26	25	80.0
6	57	130	83.3
7	120	546	85.7
8	247	2,037	87.5
9	502	7,071	88.9
10	1,013	23,436	90.0
20	1,048,555	1,731,858,075	95.0
30	1,073,741,793	102,928,386,178,606	96.7

[+]For equal contribution of each variety to the composite germplasm.

[‡]When all (n) varieties are used to synthesize one composite.

percentage of average heterosis that is theoretically retained when all varieties are used.

In the synthesis of composites it is not always convenient to have equal contributions of all varieties. Sometimes varieties are included because they have specifically defined traits, such as yield, plant height, disease resistance, lodging resistance, earliness. If some of the traits are more important than others, it may be desired to have some varieties contributing a greater proportion to the composite germplasm. A similar situation was discussed by Miranda and Vencovsky (1973), where a high yield variety and seven short plant varieties were available to form a composite. It was desired to introduce genes for shorter plants into the high yield variety. Because emphasis was mainly for yield, the first variety contributed 50% and the seven short plant varieties the other 50% of the new composite germplasm. In this situation the possible number of composites is $2^n - 1$, and the new composite can be predicted by (Miranda 1974a):

$$Y_{co}^r = (1/2)(\overline{Y}_o + \overline{Y}_v) + (1/2)\overline{h}_o + [(n - 1)/(4n)]\overline{h}_v$$

where $\overline{Y}_o$ = observed mean of the base population, $\overline{Y}_v$ = average for the remaining n varieties (for the given example, n = 7), $\overline{h}_o$ = average heterosis of all crosses between the base population and other varieties, and $\overline{h}_v$ = average heterosis of crosses among the short plant varieties. In this example (n = 7) it can be seen that the new composite can theoretically retain 50% of $\overline{h}_o$ and 21.4% of $\overline{h}_v$.

A general formula for situations like the above can be derived by assuming that k sets of varieties (grouped according to defined characteristics) are available. If each set contributes a proportion f_i (i = 1, 2,..., k) to the new composite germplasm, each variety of the ith set enters with a proportion f_i/n_i, where n_i is the number of varieties in the ith set. Thus the new composite mean may be predicted by

$$\hat{Y}_{co}^r = \sum_{i=1}^{k} f_i\overline{Y}_i + \sum_{i=1}^{k} f_i^2[(n_i - 1)/n_i]\overline{h}_i + 2 \sum_{i<i'} f_if_{i'}\overline{h}_{ii'}$$

where $\overline{Y}_i$ = average of all varieties in the ith set, $\overline{h}_i$ = average heterosis of all crosses among varieties of the ith set, and $h_{ii'}$ = average heterosis of all crosses between varieties of the ith set and varieties of the i'th set (Miranda 1974b).

10.5 COMPONENTS OF HETEROSIS IN INTERVARIETAL DIALLEL CROSSES

A better understanding of the intervarietal heterosis expressed as components of means was given by Gardner and Eberhart (1966). They expressed an intervarietal cross mean by

$$Y_{jj'} = \mu + (1/2)(v_j + v_{j'}) + \overline{h} + h_j + h_{j'} + s_{jj'}$$

where μ = mean of n parental varieties; v_j = variety effect, where the jth variety enters as one of the parents (j = 1, 2,..., n); $\overline{h}$ = average heterosis of all crosses; h_j = variety heterosis, which is a constant contribution of the jth variety to heterosis of all crosses where it enters as one of the parents; and $s_{jj'}$ = specific heterosis for the cross between varieties j and j' and is a deviation from the expected mean based on $\overline{h}$ and h_j effects, i.e., $s_{jj'}$ = $h_{jj'} - \overline{h} - h_j - h_{j'}$. All parameters are estimated from an intervarietal diallel. The Gardner-Eberhart (1966) model partitions the total heterosis $h_{jj'}$ into three components,

$$h_{jj'} = \bar{h} + h_j + h_{j'} + s_{jj'}$$

The model given above is the complete model. Other reduced models can be used, where some effects are omitted. Reduced models were suggested by Eberhart and Gardner (1966) when some effects are found to be unimportant from the analysis of variance. In such instances reduced models give a greater precision in the estimates. The models available are:

model 1: $Y_{jj'} = \mu + (1/2)(v_j + v_{j'})$
model 2: $Y_{jj'} = \mu + (1/2)(v_j + v_{j'}) + \bar{h}$
model 3: $Y_{jj'} = \mu + (1/2)(v_j + v_{j'}) + \bar{h} + h_j + h_{j'}$
model 4: $Y_{jj'} = \mu + (1/2)(v_j + v_{j'}) + \bar{h} + h_j + h_{j'} + s_{jj'}$

Least squares estimates of the parameters included in the models are

Parameter estimate	Model
$\hat{\mu}$	1, 2, 3, 4
$\hat{v}_j$	1, 2, 3, 4
$\hat{\bar{h}}$	2, 3, 4
$\hat{h}_j$	3, 4
$\hat{s}_{jj'}$	4

Eberhart and Gardner (1966) presented a general model for genetic effects, where heterosis of intervarietal crosses was defined according to gene frequencies and dominance effects. Thus, if p_{ij} and q_{ij} are frequencies of two alleles at the ith locus in variety j and $p_{ij'}$ and $q_{ij'}$ are the frequencies of the same alleles in variety j', the intervarietal heterosis is defined by

$$h_{jj'} = \sum_i (p_{ij} - p_{ij'})(q_{ij'} - q_{ij})\delta_i = \sum_i (p_{ij} - p_{ij'})^2 \delta_i$$

which can be extended for multiple alleles.

An alternative expression was given by Casas and Wellhausen (1968) for diallel crosses as follows:

$$h_{jj'} = z_j + z_{j'} - 2w_{jj'}$$

where $z_j = \sum_i (p_{ji} - \bar{p}_{.i})^2 d_i$

$z_{j'} = \sum_i (p_{j'i} - \bar{p}_{.i})^2 d_i$

$$w_{jj'} = \sum_i (p_{ji} - \overline{p}_{.i})(p_{j'i} - \overline{p}_{.i})d_i$$

Vencovsky (1970) used this model (based on z_j and $w_{jj'}$) to permit genetic interpretation of $\overline{h}$, h_j, and $s_{jj'}$. The author found the least squares estimates of z_j and $w_{jj'}$ to be

$$\hat{z}_j = (1/n)\{Y_{(j).} - [(n-2)/2]Y_{jj}\} - (1/n^2)(Y_H + Y_v/2)$$

$$\hat{w}_{jj'} = (1/n^2)(Y_{jj} + Y_{j'j'} + Y_{(j).} + Y_{(j)'.} - (1/n^2)(Y_H + Y_v/2) - (1/2)Y_{jj'}$$

A first conclusion from these expressions is that if a given variety k has $\hat{z}_k = 0$ (its gene frequency has no deviation from the average), it necessarily has $\hat{w}_{jk} = 0$, and consequently $\hat{h}_{jk} = z_j$ ($j = 1, 2, \ldots, n$; $j \neq k$). The term z_j is the expected heterosis in a cross where variety j is one of the parents and the other parent is a pool of all other varieties that form the complete diallel.

Components of total heterosis, according to Gardner and Eberhart's (1966) model, were then demonstrated to be as follows.

Average heterosis:

$$\overline{h} = 2[n/(n-1)]\overline{z}_. = 2[n/(n-1)]\sum_i \sigma_{ip}^2 d_i$$

where $\overline{z}_. = (1/n)\sum_j \overline{z}_j$ and σ_{ip}^2 is the variance of gene frequency (locus i) over varieties. Thus, average heterosis is zero when $d_i = 0$ for all loci or when $\sigma_{ip}^2 = 0$ for all loci (no difference in gene frequency among varieties). When considering only two varieties, $\overline{h}$ is the heterosis in the variety cross. A recurrent selection program will lead to an increase in heterosis if there is an increase in the variance of gene frequencies. However, if the increase in gene frequency is in the same direction and with the same magnitude in both populations, heterosis will be unchanged.

Variety heterosis:

$$h_j = [n/(n-2)](z_j - \overline{z}_.) = [n/(n-2)][\sum_i (\overline{p}_{ji} - \overline{p}_{.i})^2 d_i - \sum_i \sigma_{ip}^2 d_i]$$

Basically, h_j is a deviation of z_j around $\overline{z}$. The most negative value among the h_j will occur when $z_j = 0$; i.e., $p_{ji} = \overline{p}_{.i}$ (meaning that gene frequency equals the average gene frequency of all parents, assuming a positive dominance). Positive values for h_j will occur in varieties having $z_j > \overline{z}$. Considering many loci, this situation may occur for (1) varieties with many loci at a high gene frequency, (2) varieties with many loci at a low gene frequency, and (3) varieties that show a dispersion of gene frequency (high and low) relative to the average gene frequency at each locus. Such relations are shown hypothetically in Table 10.7.

Specific heterosis:

$$s_{jj'} = [2/(n-2)]\{[n/(n-1)]\overline{z}_. - z_j - z_{j'}\} - 2w_{jj'}$$

Specific combining ability or specific heterosis $s_{jj'}$ depends on the size n of the diallel set, on the average heterosis, and on the heterotic component of general combining ability in addition to the $w_{jj'}$ component. For large n,

Table 10.7. Values for z_j in three varieties that show
different gene frequency. Example with four
loci with $d_i = 1$ and $\overline{p}_{.i} = 0.6$ (Vencovsky
1970).

| Variety | Frequency of the dominant allele | | | | z_j |
	A	B	C	D	
1	0.8	0.8	0.8	0.8	0.16
2	.4	.4	.4	.4	.16
3	.8	.4	.8	.4	0.16
.	.	.	.	.	.
.	.	.	.	.	.
.	.	.	.	.	.
$\overline{P}_{.i}$	0.6	0.6	0.6	0.6	

$s_{jj'} = -2w_{jj'}$, showing that greater values of specific combining ability are
expected for varieties that are more divergent for genes showing dominance effects. Even for large n, $s_{jj'}$ is not a fixed property of a specific cross,
because it depends also on the average gene frequency $\overline{p}_{.i}$ (Vencovsky 1970).
The general combining ability effect is given by

$$g_j = (1/2)(a_j + d_j) + [n/(n - 2)](z_j - \overline{z})$$

where a_j and d_j are defined by Gardner and Eberhart (1966) as being related to
homozygous and heterozygous contributions, respectively. Eberhart and Gardner
(1966) also defined the variety effect as $v_j = a_j + d'_j$ and the relation for
general combining ability as

$$g_j = (1/2)v_j + h_j$$

The general combining ability effect depends not only on the variety effect
but also on dominance effects that arise in intervarietal crosses, i.e., on
the variety heterosis h_j.

REFERENCES

Anderson, D. C. 1938. The relation between single and double cross yields in
   corn. *J. Am. Soc. Agron.* 30:209–11.
Bauman, L. F. 1959. Evidence of non-allelic gene interaction in determining
   yield, ear height, and kernel-row number in corn. *Agron. J.* 51:531–34.
Barriga, P. B., and R. Vencovsky. 1973. Heterose da producão de graõs e de
   outros caracteres agronômicos em cruzamentos intervarietais de milho.
   Ciênc. Cult. 25:880–85.
Beal, W. J. 1880. Rep. Michigan Board Agric. Pp. 287–88.
Bruce, A. B. 1910. The Mendelian theory of heredity and the augmentation of
   vigor. *Science* 32:627–28.
Busbice, T. H. 1970. Predicting yield of synthetic varieties. *Crop Sci.* 10:
   265–69.
Casas, D. E., and E. J. Wellhausen. 1968. Diversidad genetica y heterosis.
   Fitotec. Latinoam. 5 (2):53–61.
Castro, M.; C. O. Gardner; and J. H. Lonnquist. 1968. Cumulative gene ef-
   fects and the nature of heterosis in maize crosses involving genetically

divergent races. *Crop Sci.* 8:97-101.

Cockerham, C. C. 1967. Prediction of double crosses from single crosses. *Züchter* 37:160-69.

Collins, G. N. 1921. Dominance and the vigor of first generation hybrids. *Am. Nat.* 55:116-33.

Cress, C. E. 1966. Heterosis of the hybrid related to gene frequency differences between two populations. *Genetics* 53:269-74.

Crow, J. F. 1948. Alternative hypothesis of hybrid vigor. *Genetics* 33:478-87.

————. 1952. Dominance and overdominance. In *Heterosis*, J. W. Gowen, ed., pp. 282-97. Iowa State Univ. Press, Ames.

Crum, W. S. 1968, 1972. Evaluation of Corn Belt open-pollinated varieties for use in breeding populations. Personal communication.

Darrah, L. L.; S. A. Eberhart; and L. H. Penny. 1972. A maize breeding methods study in Kenya. *Crop Sci.* 12:605-8.

Darwin, C. 1877. *The Effects of Cross and Self Fertilization in the Vegetable Kingdom*. Appleton, New York.

Doxtator, C. W., and I. J. Johnson. 1936. Prediction of double cross yields in corn. *J. Am. Soc. Agron.* 28:460-62.

East, E. M. 1936. Heterosis. *Genetics* 26:375-97.

East, E. M., and H. K. Hayes. 1912. Heterozygosis in evolution and in plant breeding. USDA Bur. Plant Ind. Bull. 243.

Eberhart, S. A. 1964. Theoretical relations among single, three-way, and double cross hybrids. *Biometrics* 20:522-39.

————. 1971. Regional maize diallels with U.S. and semi-exotic varieties. *Crop Sci.* 11:911-14.

Eberhart, S. A., and C. O. Gardner. 1966. A general model for genetic effects. *Biometrics* 22:864-81.

Eberhart, S. A., and A. R. Hallauer. 1968. Genetic effects for yield in single, three-way, and double-cross maize hybrids. *Crop Sci.* 8:377-79.

Eberhart, S. A.; M. N. Harrison; and F. Ogada. 1967. A comprehensive breeding system. *Züchter* 37:169-74.

Eberhart, S. A.; W. A. Russell; and L. H. Penny. 1964. Double cross hybrid prediction when epistasis is present. *Crop Sci.* 4:363-66.

El-Rouby, M. M., and A. R. Galal. 1972. Heterosis and combining ability in variety crosses of maize and their implications in breeding procedures. *Egyptian J. Genet. Cytol.* 1:270-79.

Falconer, D. S. 1960. *Introduction to Quantitative Genetics*. Ronald Press, New York.

Gamble, E. E. 1962. Gene effects in corn (*Zea mays* L.). I. Separation and relative importance of gene effects for yield. *Canadian J. Plant Sci.* 42:339-48.

Garber, R. J., and H. F. A. North. 1931. The relative yield of a first generation cross between two varieties of corn before and after selection. *J. Am. Soc. Agron.* 23:647-51.

Garber, R. J.; T. E. Odland; K. S. Quisenberry; and T. C. McIbvaine. 1926. Varietal experiments and first generation crosses in corn. West Virginia Agric. Exp. Stn. Bull. 199.

Gardner, C. O. 1965. Teoria de genetica estatistica aplicada a las medias de variedades, sus cruces y poblaciones afines. *Fitotec. Latinoam.* 2:11-22.

Gardner, C. O., and S. A. Eberhart. 1966. Analysis and interpretation of the variety cross diallel and related populations. *Biometrics* 22:439-52.

Gardner, C. O., and E. Paterniani. 1967. A genetic model used to evaluate the breeding potential of open pollinated varieties of corn. *Ciênc. Cult.* 19:95-101.

Gärtner, C. F. 1849. *Versuche und Beobachtungen über die Bastarderzengung in Pflanzenreich*. Stuttgart.

Genter, C. F., and S. A. Eberhart. 1974. Performance of original and advanced maize populations and their diallel crosses. *Crop Sci.* 14:881-85.

Gilmore, E. C. 1969. Effect of inbreeding of parental lines on predicted yields of synthetics. *Crop Sci.* 9:102-4.

Good, R. L., and A. R. Hallauer. 1977. Inbreeding depression in Iowa Stiff Stalk Synthetic (*Zea mays* L.) by selfing and full-sibbing. *Crop Sci.* 17: 935-40.

Gorsline, G. W. 1961. Phenotypic epistasis for ten quantitative characters in maize. *Crop Sci.* 1:55-58.

Griffee, F. 1922. First generation corn varietal crosses. *J. Am. Soc. Agron.* 14:18-27.

Hallauer, A. R. 1972. Third phase in the yield evaluation of synthetic varieties of maize. *Crop Sci.* 12:16-18.

————. 1973. Hybrid development and population improvement in maize by reciprocal full-sib selections. *Egyptian J. Genet. Cytol.* 2:84-101.

Hallauer, A. R., and S. A. Eberhart. 1966. Evaluation of synthetic varieties of maize for yield. *Crop Sci.* 6:423-27.

Hallauer, A. R., and D. Malithano. 1976. Evaluation of maize varieties for their potential as breeding populations. *Euphytica* 25:117-27.

Hallauer, A. R., and J. H. Sears. 1968. Second phase in the evaluation of synthetic varieties of maize for yield. *Crop Sci.* 8:448-51.

Hartley, C. P.; E. B. Brown; C. H. Kyle; and L. L. Zook. 1912. Crossbreeding corn. USDA Bur. Plant Ind. Bull. 218.

Hayes, H. K. 1914. Corn improvement in Connecticut. Connecticut Agric. Exp. Stn. Rep. Pp. 353-84.

Hayes, H. K., and E. M. East. 1911. Improvement in corn. Connecticut Agric. Exp. Stn. Bull. 168.

Hayes, H. K., and R. J. Garber. 1919. Synthetic production of high protein corn in relation to breeding. *Agron. J.* 11:309-18.

Hayes, H. K., and P. J. Olson. 1919. First generation crosses between standard Minnesota corn varieties. Minnesota Agric. Exp. Stn. Bull. 183.

Hayes, H. K.; R. P. Murphy; and E. H. Rinke. 1943. A comparison of the actual yield of double crosses of maize with their predicted yield from single crosses. *J. Am. Soc. Agron.* 35:60-65.

Hayes, H. K.; E. H. Rinke; and Y. S. Tsiang. 1946. The relationship between predicted performance of double crosses of corn in one year with predicted and actual performance of double crosses in later years. *Agron. J.* 38:60-67.

Hull, F. H. 1945. Recurrent selection for specific combining ability. *J. Am. Soc. Agron.* 37:134-45.

Hutcheson, T. B., and T. K. Wolfe. 1917. The effect of hybridization on maturity and yield in corn. Virginia Agric. Exp. Stn. Tech. Bull. 18.

Jenkins, M. T. 1934. Methods of estimating the performance of double crosses in corn. *J. Am. Soc. Agron.* 26:199-204.

Jones, D. F. 1917. Dominance of linked factors as a means of accounting for heterosis. *Genetics* 2:466-79.

————. 1918. The effects of inbreeding and crossbreeding upon development. Connecticut Agric. Exp. Stn. Bull. 207:5-100.

————. 1945. Heterosis resulting from degenerative changes. *Genetics* 30: 527-42.

————. 1958. Heterosis and homeostasis in evolution and in applied genetics. *Am. Nat.* 92:321-28.

Jones, D. F.; H. K. Hayes; W. L. Slate, Jr.; and B. G. Southwick. 1917. Increasing the yield of corn by crossing. Connecticut Agric. Exp. Stn. Rep. Pp. 323-47.

Kiesselbach, T. A. 1922. Corn investigations. Nebraska Agric. Exp. Stn. Bull. 20:5-151.

Kinman, M. L., and G. F. Sprague. 1945. Relation between number of parental lines and theoretical performance of synthetic varieties of corn. *Agron. J.* 37:341-51.

Knight, T. A. 1799. An account of some experiments on the fecundation of vegetables. *Philos. Trans. R. Soc. London* 89:195.

Köelreuter, J. G. 1776. *Dritte Forsetzung der vorläufigen Nachricht von einigen das Geschlecht der Pflanzen betreftenden Versuchen und Beobachtunger.* Leipzig.

Lonnquist, J. H. 1961. Progress from recurrent selection procedures for the improvement of corn populations. Nebraska Agric. Exp. Stn. Res. Bull. 197.

Lonnquist, J. H., and C. O. Gardner. 1961. Heterosis in intervarietal crosses in maize and its implications in breeding procedures. *Crop Sci.* 1: 179-83.

McClure, G. W. 1892. Corn crossing. Illinois Agric. Exp. Stn. Bull. 290.

Martin, J. M., and A. R. Hallauer. 1976. Relation between heterozygosis and yield for four types of maize inbred lines. *Egyptian J. Genet. Cytol.* 5: 119-35.

Mather, K. 1949. *Biometrical Genetics.* Methuen, London.

Mather, K., and J. L. Jinks. 1971. *Biometrical Genetics.* Chapman & Hall, London.

Miranda, J. B., Fo. 1974a. Cruzamentos dialélicos e síntese de compostos de milho (*Zea mays* L.) com ênfase na produtividade e no porte da planta. Tese de Doutoramento, ESALQ-USP, Piracicaba, Brazil.

———. 1974b. Predicão de médias de compostos em funcão das médias das variedades parentais e das heteroses dos cruzamentos. Rel. Cient. Inst. Genét. (ESALQ-USP) 8:134-38.

Miranda, J. B., Fo., and R. Vencovsky. 1973. Predicão de médias na formacão de alguns compostos de milho visando a producao de graõs e o porte da planta. Rel. Cient. Inst. Genét. (ESALQ-USP) 7:117-26.

Mochizuki, N. 1970. Theoretical approach for the choice of parents and their number to develop a highly productive synthetic variety in maize. *Japan J. Breed.* 20:105-9.

Moll, R. H., and C. W. Stuber. 1971. Comparisons of response of alternative selection procedures initiated with two populations of maize (*Zea mays* L.). *Crop Sci.* 11:706-11.

Moll, R. H.; A. Bari; and C. W. Stuber. 1977. Frequency distribution of maize yield before and after reciprocal recurrent selection. *Crop Sci.* 17:794-96.

Moll, R. H.; W. S. Salhuana; and H. F. Robinson. 1962. Heterosis and genetic diversity in variety crosses of maize. *Crop Sci.* 2:197-98.

Moll, R. H.; J. H. Lonnquist; J. V. Fortuna; and E. C. Johnson. 1965. The relation of heterosis and genetic divergence in maize. *Genetics* 52:139-44.

Morrow, G. E., and F. D. Gardner. 1893. Field experiments with corn. Illinois Agric. Exp. Stn. Bull. 25:173-203.

Neal, N. 1935. The decrease in yielding capacity in advanced generations of hybrid corn. *Agron. J.* 27:666-70.

Noll, C. F. 1916. Experiments with corn. Pennsylvania Agric. Exp. Stn. Bull. 139.

Obilana, A. T.; A. R. Hallauer; and O. S. Smith. 1979. Predicted and observed response to reciprocal full-sib selection in maize (*Zea mays* L.). *Egyptian J. Genet. Cytol.*

Otsuka, Y.; S. A. Eberhart; and W. A. Russell. 1972. Comparison of prediction for maize hybrids. *Crop Sci.* 12:325-31.

Paterniani, E. 1967. Cruzamentos intervarietais de milho. Rel. Cient. Inst. Genét. (ESALQ-USP) 1:49-50.

———. 1968. Cruzamentos interracias de milho. Rel. Cient. Inst. Genét. (ESALQ-USP) 2:108-10.

———. 1970. Heterose em cruzamentos intervarietais de milho. Rel. Cient. Inst. Genét. (ESALQ-USP) 4:95-100.

———. 1977. Avaliacão de cruzamentos semi-dentados de milho (*Zea mays* L.). Rel. Cient. Inst. Genét. (ESALQ-USP) 11:101-7.

Paterniani, E., and M. M. Goodman. 1977. Races of maize in Brazil and adjacent areas. CIMMYT.

Paterniani, E., and J. H. Lonnquist. 1963. Heterosis in interracial crosses of corn (*Zea mays* L.). *Crop Sci.* 3:504-7.

Richey, F. D. 1922. The experimental basis for the present status of corn breeding. *J. Am. Soc. Agron.* 14:1-17.

———. 1946. Hybrid vigor and corn breeding. *J. Am. Soc. Agron.* 38:833-41.

Robinson, H. F., and C. C. Cockerham. 1961. Heterosis and inbreeding depression in populations involving two open-pollinated varieties of maize. *Crop Sci.* 1:68-71.

Robinson, H. F.; R. E. Comstock; A. Klalil; and P. H. Harvey. 1956. Dominance versus overdominance in heterosis: Evidence from crosses between open-pollinated varieties of maize. *Am. Nat.* 90:127-31.

Sanborn, J. W. 1890. Indian corn. *Rep. Maine Dep. Agric.* 33:54-121.

Shehata, A. H., and M. L. Dhawan. 1975. Genetic analysis of grain yield in maize as manifested in genetically diverse varietal populations and their crosses. *Egyptian J. Genet. Cytol.* 4:90-116.

Shull, A. F. 1912. The influence of inbreeding on vigor in *Hydatina senta*. *Biol. Bull.* 24:1-13.

Shull, G. H. 1908. The composition of a field of maize. Am. Breeders' Assoc. Rep. 4:296.

———. 1914. Duplicate genes for capsule form in *Bursa bursapastoris*. Z. Ind. Abstr. Ver. 12:97-149.

———. 1952. Beginnings of the heterosis concept. In *Heterosis*, J. W. Gowen, ed., pp. 14-48. Iowa State Univ. Press, Ames.

Silva, J. C. 1969. Estimativa dos efeitos gênicos epistáticos em cruzamentos intervarietais de milho e suas geracões avancadas. Master's thesis, ESALQ-USP, Piracicaba, Brazil.

Sprague, G. F. 1953. Heterosis. In *Growth and Differentiation in Plants*, W. E. Loomis, ed., pp. 113-36. Iowa State Univ. Press, Ames.

Sprague, G. F., and W. I. Thomas. 1967. Further evidence of epistasis in single and three-way cross yields in maize. *Crop Sci.* 7:355-56.

Sprague, G. F.; W. A. Russell; L. H. Penny; T. W. Horner; and W. D. Hanson. 1962. Effect of epistasis on grain yield in maize. *Crop Sci.* 2:205-8.

Stuber, C. W., and R. H. Moll. 1971. Epistasis in maize. II. Comparison of selected with unselected populations. *Genetics* 67:137-49.

Suwantaradon, K., and S. A. Eberhart. 1974. Developing hybrids from two improved maize populations. *Theor. Appl. Genet.* 44:206-10.

Tavares, F. C. A. 1972. Componentes da producão relacionados a heterose em hibridos intervarietais de milho (*Zea mays* L.). Master's thesis, ESALQ-USP, Piracicaba, Brazil.

Timothy, D. H. 1963. Genetic diversity, heterosis, and use of exotic stocks of maize in Colombia. In *Statistical Genetics and Plant Breeding*, W. D.

Hanson and H. F. Robinson, eds., pp. 581-93. NAS-NRC Publ. 982.

Torregroza, C. M. 1959. Heterosis in populations derived from Latin American open-pollinated varieties of maize. Master's thesis, Univ. Nebraska, Lincoln.

Troyer, A. F., and A. R. Hallauer. 1968. Analysis of a diallel set of maize. *Crop Sci.* 8:581-84.

Tsotsis, B. 1972. Objectives of industry breeders to make efficient and significant advances in the future. *Proc. Annu. Corn Sorghum Res. Conf.* 27: 93-107.

Valois, A. C. C. 1973. Efeito da selecão massal estratificada em duas populacões de milho (*Zea mays* L.) e na heterose de seus cruzamentos. Master's thesis, ESALQ-USP, Piracicaba, Brazil.

Vencovsky, R. 1969. Análise de cruzamentos dialélicos entre variedades pelo métado de Gardner e Eberhart. Rel. Cient. Inst. Genét. (ESALQ-USP) 3: 99-111.

———. 1970. Alguns aspectos teóricos e aplicodos relativos a cruzamentos dialélicos de variedades. Tese de Livre-Docência, ESALQ-USP, Piracicaba, Brazil.

———. 1973. Synthesis of composite populations (abstr.). *Genetics* 74:284.

Vencovsky, R., and J. B. Miranda, Fo. 1972. Determinacão do número de pos síveis compostos e pares de compostos. Rel. Cient. Inst. Genét. (ESALQ-USP) 6:120-23.

Vencovsky, R.; J. R. Zinsly; N. A. Vello; and C. R. M. Godoi. 1973. Predicão da média de um caráter quantitativo em compostos de variedades e cruzamentos de compostos. *Fitotec. Latinoam.* 8:25-28.

Wahlund, S. 1928. Zuzammensetzung von Populationen und Korrelationsercheinungen von Stadpunkt der Vererbungslehre aus betrachtet. *Hereditas* 11:65-106.

Waldron, L. R. 1924. Effect of first generation hybrids upon yield of corn. North Dakota Agric. Exp. Stn. Bull. 177.

Webber, H. J. 1900. Xenia or the immediate effect of pollen in maize. USDA Div. Veg. Phys. Bull. 22.

———. 1901. Loss of vigor in corn from inbreeding (abstr.). *Science* 13: 257-58.

Wellhausen, E. J. 1965. Exotic germplasm for improvement of Corn Belt maize. *Proc. Annu. Hybrid Corn Ind. Res. Conf.* 20:31-45.

Whaley, W. G. 1944. Heterosis. *Bot. Rev.* 10:461-98.

Williams, C. G., and F. A. Welton. 1915. Corn experiments. Ohio Agric. Exp. Stn. Bull. 282.

Wright, S. 1922. The effects of inbreeding and cross breeding on guinea pigs. USDA Bull. 1121.

Zirkle, C. 1952. Early ideas on inbreeding and crossbreeding. In *Heterosis*, J. W. Gowen, ed., pp. 1-13. Iowa State Univ. Press, Ames.

11

Germplasm

Choice of germplasm, either fortuitous or planned, plays an important role in any breeding program, whether an applied breeding program for inbred line development or population improvement or a selection study comparing breeding methods. There are certainly differences among breeding populations, and the particular choice of germplasm may decide the ultimate success or failure of selection. Maize breeders have always had a wealth of germplasm available for their use. Even within regions having similar environmental aspects, experience and testing soon identified certain populations (or varieties) that were better than others for use as varieties themselves, use in variety crosses, and use as source populations for the development of inbred lines as parent stocks for hybrids. Experience in the United States has shown that some breeding populations (e.g., Lancaster, Midland, Stiff Stalk Synthetic) yield a greater frequency of usable inbred lines than others (e.g., Hickory King, Krug, Corn Borer Synthetic 1). Differences among breeding populations occur because of the original genes and past selection that created an assemblage of genes in the greater frequencies that are desired in modern hybrids. Selection may have been as intense and effective for other populations, but perhaps the original germplasm in the populations did not include the genes necessary to meet the standards of the breeder. If the genes are not present, the efforts of the breeder will be futile regardless of the patience and experimental techniques used. Hence, maize breeders have two important but separate decisions to make in developing their breeding programs: (1) choice of germplasm and (2) choice of breeding procedure.

Maize is an extremely diverse genus, having many morphological and biological differences. Maize is a monoecious plant that has separate male and female inflorescences; and because it is essentially 100% cross-pollinated through wind movement, the opportunities for intercrossing are ample. Maize is grown from 58° north latitude without interruption through the temperate, subtropical, and tropical regions of the world to 40° south latitude. Growing of maize in Canada, northern Europe, and Russia to Australia, South Africa, and Argentina permits selection of maize types for each ecological niche. In the Andean region, maize is grown from sea level to elevations over 3808 m (above Lake Titicaca in Peru) and from areas with less than 25.4 cm of rainfall (such as the Guajira Peninsula of Colombia) to over 1016 cm (Department of Choco on the Pacific coast of Colombia) (Grant et al. 1963). Maize is grown in all states of the United States and in every other important agricultural area of the world. The extent of maize culture, probably greater than any other cultivated crop, is important in the world economy.

Maize breeders have been increasingly cognizant of the importance of *genetic diversity* of germplasm in the twentieth century. The expression of het-

375

erosis, whatever its genetic basis, depends on the differences in gene frequency of the parental stocks, whether varieties or inbred lines, that are
used to produce the crosses. Because of the heterotic responses observed in
crosses of maize, breeders have emphasized crossing of parental stocks that
were derived from different breeding populations. Initially, the concern of
diversity of source populations, which were usually open-pollinated varieties,
was either for crossing to produce variety crosses or to initiate inbreeding
for developing inbred lines. Past selection, either natural or human, in different regions developed germplasm that had distinctive phenotypic features
(e.g., Wallace and Bressman 1925) and different gene frequencies for different
traits as evidenced by the expression of heterosis in crosses (Table 10.1).

The concern of genetic diversity of germplasm has received renewed emphasis in recent years because of (1) the rapid shift from double crosses to
simpler types of hybrids and (2) the *Helminthosporium maydis* outbreak on hybrids produced on T-cytoplasm in the United States in 1970. It was shown in
Chap. 9 that the use of simpler types of hybrids does not seem to have any serious disadvantages compared with double crosses. A series of surveys spon-
sored by the American Seed Trade Association (Sprague 1971; Corn 1972; Zuber
1975) indicated extensive usage of a few publicly developed inbred lines in
hybrids. The first two surveys related primarily to use of inbred lines in
double crosses; the last two surveys were taken after the rapid shift from
double crosses to single crosses. Double crosses, because they are produced
from the union of gene arrays of two single crosses, were more variable and
provided a level of genetic diversity within each field and among fields that
would be expected to be greater than that resulting from the use of single
crosses. It seems, therefore, that genetic diversity of hybrids produced and
grown has been reduced substantially within the past 20 years in the U.S. Corn
Belt. Level of genetic uniformity, however, would not approach the extensive
use of only one cytoplasm in producing hybrid seed. Because maize is an annual crop the problem of uniformity seems to be less serious than for a perennial crop, provided that a source of germplasm and of new breeding materials is
available to shift, in a short period of time, the hybrids of varieties cultivated for grain production. The 1970 *H. maydis* debacle did, however, emphasize the potential seriousness of the problem of reduced genetic diversity of
our germplasm. Subsequent reports (Sprague 1971; Genetic vulnerability of major crops 1972; Recommended actions and policies 1973; Lonnquist 1974; Brown
1975) have discussed the potential seriousness of reduced genetic diversity,
outlined steps that are necessary to correct the situation, and recommended
possible avenues of research to reduce genetic uniformity of breeding materials. Although breeders have at their disposal a nearly unlimited diversity
of maize germplasm, the main problem arises from usage of parental inbred
lines in hybrids. If a single cross of two inbred lines is superior to other
single crosses (e.g., B73 × Mo17), economic considerations and competition
among seed producers force widespread usage of either one or a few specific
hybrids.

We briefly consider some of the aspects of maize germplasm--potential,
possible usage, and problems associated with use of exotic germplasm. The origin of maize is discussed briefly for historical background and the development of the U.S. Corn Belt germplasm.

11.1 ORIGIN OF MAIZE

Maize is a member of the grass family placed in the tribe *Maydeae* (also
called *Tripsaceae* by some authorities, e.g., Hitchcock 1935). The tribe *Maydeae* includes seven genera, two that are native to the Western Hemisphere and

five that are native to Asia. Each genus has the common trait of separate
male and female inflorescences on the same plant. The two genera native to
the Western Hemisphere include *Zea* and *Tripsacum;* in some instances a third
genus, *Euchlaena,* is considered. Currently, the annual weed teosinte previously classified as *Euchlaena mexicana* is included in the same genus as maize
(*Zea mexicana*). A perennial tetraploid (2n = 40) form of teosinte (*Zea perennis*), thought to be derived from *Zea mexicana*, is often classified as a separate species, and *Zea perennis* usually is not considered in the origin of any
of the other living species. The rediscovery of perennial diploid teosinte
(*Zea diploperennis* [Gramineae]) in southern Jalisco, Mexico, may provide valuable clues to the evolution of *Zea* and to the origin of *Zea perennis* (Iltis
1979).

Annual teosinte plants resemble maize plants but commonly are more slend-
er and have several stalks (or tillers) per plant. Teosinte ears are smaller
than maize ears, usually having only five or six seeds per row and two rows
per ear. Each kernel is enclosed by a horneous shell borne on a hardened,
brittle rachis. Annual teosinte and maize each have 10 pairs of chromosomes.
The chromosomes of teosinte resemble those of maize but tend to have more
knoblike structures than those of maize. Maize and teosinte cross relatively
easily and produce fertile offspring.

Tripsacum (*Tripsacum*) includes several perennial species and the resemblance between maize and tripsacum is much less than between maize and teosinte. The relationship between tripsacum and maize seems to have been dis-
covered about the same time as the relationship between maize and teosinte.
Male flowers are located on the upper portions of the tassel and the female
flowers on the lower portion. Each seed is enclosed by a horny covering, but
seeds are not covered by leaves or husks as for maize and teosinte. Tripsacum
species have chromosome numbers that are multiples of 36. Special techniques
are usually required for crossing tripsacum and maize. All maize-tripsacum
crosses so far have been male sterile and the male sterility persists after
several generations of backcrossing to maize.

The five Asiatic genera of *Maydeae* have not been studied as extensively
as the two genera native to the Western Hemisphere. *Coix* is the only Asiatic
genus ever seriously considered as a possible ancestor of maize. Relatively
little is known about the other four Asiatic genera, *Chionachne, Polytoca,
Sclerachne,* and *Trilobachne.* *Chionachne* and *Sclerachne* have 20 chromosomes;
Polytoca has 40; different species of *Coix* have 10, 20, and 40; and the chromosome number is unknown for *Trilobachne.* *Coix* is the only Asiatic genus that
has been successfully crossed with maize. Maize also has been crossed with
sugarcane (*Saccharum officinarum*), but apparently sugarcane crosses with many
grasses without regard to degree of relationship (Goodman 1965a).

The origin of maize has been studied extensively, but the putative parents of cultivated maize as we know it today are still conjectural. Extensive
literature indicates the extent of the study of the origin of cultivated
maize, but the issue is still unsettled and it may be too late to ever definitely establish its origin. Reviews by Weatherwax (1955), Goodman (1965a),
Mangelsdorf (1974), and Galinat (1977) illustrate the differences of opinion
regarding the origin of maize, but there seems to be general agreement that it
originated in the Western Hemisphere.

Historically, four hypotheses regarding the possible origin of maize are:
(1) Maize, teosinte, tripsacum, and perhaps some of the *Andropogoneae* descend-
ed from a common, extinct ancestor native to the highlands of Mexico or Guatemala (Weatherwax 1955). (2) Maize originated from a cross between two species, perhaps *Coix* and *Sorghum,* each with 10 chromosomes (Anderson 1945). (3)
The tripartite hypothesis of Mangelsdorf and Reeves (1939) postulated that (a)

wild maize was a form of pod corn native to the lowlands of South America, (b) teosinte originated from crossing cultivated maize and tripsacum in Central America, and (c) modern varieties of maize arose from crosses between maize and tripsacum or teosinte. (4) Maize was derived from teosinte by direct human selection (Beadle 1939). These four divergent theories of the origin of maize have stimulated research in an attempt to resolve the issue. The hypotheses have been modified in some instances as additional evidence became available (Goodman 1965a; Galinat 1977).

Weatherwax's proposal is the simplest hypothesis and has received support (e.g., Brieger et al. 1958) for that reason. He does not consider the oriental *Maydeae* to be closely related to maize. Weatherwax (1955) described the traits that would be necessary in a wild form of maize and concluded that teosinte was not wild maize. He suggested that wild maize probably became extinct soon after the Indians began growing maize; that the chromosome number in tripsacum probably prevented crossing between tripsacum and either maize or teosinte; and that maize and teosinte, as well as tripsacum, arose in mutually exclusive isolation because of the ease of crossing between the two species. Weatherwax's proposal has been criticized for relying on three separate areas of origin; for assuming the development of modern races of maize from primitive races of maize and wild maize; and because of the presence of the heterochromatic knobs in maize, which he attributes to contamination from teosinte.

Anderson (1945) hypothesized that the origin of maize was in southwestern Asia; hence the suggestion that perhaps maize originated from a cross of *Coix* and *Sorghum*. The hypothesis (*Coix* × *Sorghum* = maize), however, has not been considered seriously. Objections to this hypothesis have arisen as to when and how the genera of *Maydeae* became distributed in southern Asia and Central America because of the discovery of ancient maize pollen and the little genetic evidence that maize originated by doubling a basic chromosome number of 10. Recent crosses of maize and *Coix* may provide additional information on the validity of Anderson's hypothesis.

The tripartite theory of Mangelsdorf and Reeves (1939) has stimulated research that either supported or detracted from the validity of the postulated theory. It has probably received more attention than the others. The first aspect of the tripartite theory was modified: The original form of pod corn as described by Weatherwax was changed to a pod corn type in which each kernel is only partially enclosed by a small husk, and the center of origin was changed from the lowlands of South America to Mexico. Teosinte was considered to have arisen from crosses of *Zea* and *Tripsacum* because teosinte seemed intermediate for several traits. This second aspect has been questioned because (1) maize and tripsacum are so variable that teosinte seems more uniform by comparison; (2) some authorities claim that teosinte is not intermediate to maize and tripsacum; and (3) hybrids of maize and tripsacum are sterile, forming a barrier for the exchange of genetic factors from tripsacum to maize. The third aspect of the tripartite theory has received greater acceptance than the first two, but there has been some disagreement on the relative influence of tripsacum and teosinte on the development of maize. Because teosinte and maize are nearly identical, genetically and cytologically, the general consensus is that teosinte has contributed more to the development of maize than has tripsacum. The validity of the tripartite hypothesis for explaining the origin of maize has been abandoned by its principal proponent. Mangelsdorf (1974), after spending nearly 30 years investigating the origin of maize, concluded that the hypothesis was not adequate. Electron microscope studies of the pollen of maize, teosinte, *Tripsacum*, and maize-*Tripsacum* hybrids by Mangelsdorf's colleagues showed convincingly that teosinte was not a hybrid of maize and *Tripsacum* (Mangelsdorf 1974).

The hypothesis that maize was derived directly by human selection from teosinte was suggested by Beadle (1939). Subsequent studies by Galinat (1970, 1971, 1975), Iltis (1970, 1972), de Wet and Harlan (1971, 1972, 1976), Beadle (1972, 1977), and Kato (1975) have presented additional evidence. Beadle (1977) studied F_2 and backcross generations of crosses between teosinte and primitive maize types (Argentine pop, Chapalote, and Chalco teosinte). From genetic analysis of 16,000 segregants, he found occurrence of parental types in frequencies of about 1 in 500, suggesting a relatively few independently segregating genes. Demonstration trials also showed that teosinte yields were comparable to those of wild wheat in the Near East. Sufficient quantities of seed could be harvested to be used for human food.

Galinat (1977) gave a detailed account that reviewed the evidence currently available for the origin of maize. Comparisons of maize with its relatives (or putative parents) were made on the basis of genetics, cytology, floral structures, fossil evidence, and the relative contributions of maize's relatives to maize improvement. Galinat summarized as follows, "Of the various hypotheses on the origin of maize, essentially only two alternatives now remain as viable options: (1a) Present-day teosinte is the wild ancestor of maize; (1b) A primitive teosinte is the common wild ancestor of both maize and Mexican teosinte; (2) An extinct form of pod corn was the ancestor of maize with teosinte being a mutant form of this pod corn." The four hypotheses for the origin of maize were not always incompatible in their features, and as additional evidence became available it is only natural that modifications and revisions were made in the original hypotheses. The issue is not settled, but it seems that teosinte was important in the evolution of maize and offers greater opportunities than *Tripsacum* as a source of genetic variation for maize breeding. *Tripsacum* has possibilities but the technical problems of introducing its germplasm into maize germplasm are much greater. For future studies, Goodman (1965a) feels that greater emphasis should be given to the Asiatic genera of *Maydeae* and Galinat (1977) feels that the greater genetic variability of the nine or so species of *Tripsacum* offers potential usefulness for maize improvement. Rediscovery of *Zea diploperennis* will provide geneticists and plant breeders potentially valuable germplasm for gene transfer (Iltis 1979). Future research with *Zea diploperennis* may have important implications in studying the origin of maize, but practical uses in maize improvement remain in the future.

11.2 CLASSIFICATION OF MAIZE GERMPLASM

If one considers only the obvious phenotypic differences, the range of variability among races, varieties, hybrids, and inbred lines of maize for plant type, ear type, tassel type, and maturity is striking. The obvious phenotypic differences are relative to the germplasm included in the breeding nurseries, which may vary from adapted germplasm to introduced germplasm; but the differences are always present. Because of the range in latitude and altitude in which maize can be grown, it is little wonder that so many different types of maize have been developed. It seems that maize has been cultivated for 5000 years; consequently its use in satisfying food, fuel, and fiber as well as cultural needs of Indian settlements created a vast array of germplasm. Development of the cultures of the different groups of peoples, their migrations, discovery of the Western Hemisphere, and the subsequent movement of Europeans into it also were important factors in creation of a diversity of maize germplasm. Because of the cross-pollination accompanied by a continuous interchange of genes among populations, additional pools of genetic variation have been created by the movement of peoples. Subsequent selection, both natural and artificial, developed germplasm that was often quite different in phenotype and genotype from the original parental germplasm.

The vast array of maize germplasm was obvious to students, fanciers, taxonomists, botanists, and breeders of maize, but no natural classification was attempted until the 1940s. Sturtevant (1899), one of the first to make the attempt, classified the maize germplasm known to him into six main groups, five of which were based on endosperm composition. This system was generally used without modification for over 40 years; little other interest and activity in classification were shown. N. I. Vavilov and his associates collected a large number of specimens from different areas of the world; they concluded the center of origin of maize was in Central America because of the variability of maize types in that area. As a follow-up of those collections, Kuleshov (1933) classified maize by endosperm types in the following groups:

1. *Zea mays* indurata--flint
2. *Zea mays* amylacea--floury
3. *Zea mays* indentata--dent
4. *Zea mays* everta--popcorn
5. *Zea mays* saccharata--sweet
6. *Zea mays* amylea saccharata--starchy-sugary
7. *Zea mays* ceratina--waxy
8. *Zea mays* tunicata--pod

The classification of Kuleshov was similar to Sturtevant's because in some instances only a one gene difference was needed to change the classification. The classification was satisfactory for kernel type, but it was not indicative of the germplasm's morphological and polygenic differences for other traits.

Geographical distribution of the eight systematic groups listed by Kuleshov follows. (1) Flint maize was distributed throughout the Western Hemisphere, but its greatest importance seemed to be in northern and southern frontiers of maize growing. (2) Flour-type maize occurred south of the northern range of flints in North America and in the southwestern states of the United States, and it predominated in the Andea valley of southern Colombia, Peru, and Bolivia; the greatest diversity occurred in Peru. (3) Dent varieties were predominant in what is called the U.S. Corn Belt and in some areas of Mexico. The dent group did not seem to occur in the aboriginal culture of South America, but the greatest diversity seemed to occur in the central and southern states of Mexico. (4) The popcorn types were collected in several countries and localities but, except for commercial production, had not moved into North America. (5) Sweet maize was collected principally in the central and northeastern regions of the United States and was nearly absent in the south and the tropics. (6) Nearly all the starchy-sugary types were collected in Bolivia and Peru with the greatest diversity found in Peru. (7) The waxy types seemed to be restricted to eastern Asia. (8) No fixed geographical area was identified for the tunicata types. The *Tu* allele is never found in any particular type, and the spontaneous occurrence of tunicata types can occur in different areas.

The method of classifying maize germplasm by endosperm type was not satisfactory, as noted by Sturtevant (1899). But it was not until Anderson and Cutler (1942) investigated the range of variability among collections of maize germplasm and developed the concept of "races of maize" that vigorous effort was given to classifying maize germplasm. The definition of a race by Anderson and Cutler was: "For the classification of *Zea mays* we shall define the word race as loosely as possible, and say that a race is a group of related individuals with enough characteristics in common to permit their recognition as a group." They continued by stating, "As Hooton (1926) has said in his discussion of racial analysis, 'Races are great groups and any analysis of racial elements must be primarily an analysis of groups, not of separate individuals. One must conceive of race not as the combination of features which gives each person his individual appearance, but rather as a vague physical background, usually more or less obscured or overlaid by individual variations in single subjects and realized best in a composite picture.'" Genetic as well as phenotypic differences were considered in their definition of a race. Anderson and Cutler thus state that "a race or subrace is defined as a number of varieties with enough characters in common to permit their recognition as a group; in genetical terms it is a group with a significant number of genes in common." Thus a more natural system rather than an artificial description was developed. Anderson and his colleagues used this descriptive definition for classifying maize germplasm and the concept has been used extensively. If used only for cataloging, inventorying, or storing, the method of classifying germplasm suggested by Sturtevant is adequate, but the classification suggested by Anderson and Cutler is certainly more useful in attempting to derive and trace the origins of different races.

The concept of race is not easily understood. Wellhausen et al. (1952) and Brieger et al. (1958) discussed the concept of a race and how races of maize may have originated. It is generally agreed that races exist and are characterized by complexes of traits that make one race distinguishable from another. Levels of differentiation among races are not always the same, but races seem to maintain themselves for many generations without losing their identity. Brieger et al. elaborated on the definition of a race as follows: "We may define as a race any group of populations having sufficient number of distinctive characters in common, maintaining itself through panmictic reproduction within populations, and occupying definite areas." This definition was not in conflict with the one given by Anderson and Cutler, but the concise definition does describe a race as a random mating population possessing definite phenotypic and genetic characteristics. The difficulty of precisely defining a race in general terms was equated with the problems of a precise definition for a species; i.e., there usually are exceptions to the general definition.

Races of maize have arisen, but it is not always clear what mechanisms were involved and how the races maintained their integrity through many generations of reproduction. Distinctive races have evolved in different regions following the distribution of primitive maize several thousands of years ago. Initially, frequent mutations and isolation mechanisms (geographical, flowering, and gametophyte factors) must have played a prominent role. Superimposed on the evolutionary trends were the activities of peoples by migration and thus the movement of germplasm to different geographical areas. Native Americans contributed to the maintenance of different races by isolating populations with special characteristics, particularly ear and grain traits (such as grain color) for use in their ceremonies or other purposes. Races of maize arose simply because of artificial and natural selection pressures that caused changes in gene frequencies in successive generations of propagation. Wellhausen et al. (1957) cited evidence that maize under domestication is potentially a self-improving species, as evidenced by the increasing size of the ear for the past 4000 years. The crossing of distinct races, probably unintentional initially, enhanced productivity, and new races evolved because of hybridization. Wellhausen et al. concluded that mutation and racial hybridization were the two important evolutionary factors for the development of races in Mexico and Central America.

Brieger et al. (1958) also emphasized that isolating mechanisms must exist for races to have maintained their racial characteristics for many generations of intermating, which, in some examples cited, were not obvious. Races of maize were considered to have arisen by selection of mutant genes accompanied by favorable modifier complexes and by a similar procedure after hybridization of previously existing races. Both ways of explaining the development of races were similar to those of Wellhausen et al. (1952), but Brieger et al. emphasized that the intermediacy of presumed synthetic races created by hybridization may not be valid for postulating the putative parent races. Synthetic populations formed by crossing two parents tend to be intermediate to the two parents, but Brieger et al. question selecting putative parents that possess characters differing equally in opposite directions of the assumed synthetic race; i.e., one should not expect that the synthetic race will be intermediate in all or even most of the characters.

The concept of race was an advancement in attempting to classify maize germplasm. But races were characterized by differences in quantitative characters that are often variable within races (Wellhausen et al. 1952). Use of quantitative traits rather than simply inherited traits was a more natural system of classification, but expressions of quantitative traits are subject to environmental biases and breeding information was usually lacking to assist in the classification. Because of the extent of germplasm available for study, other methods of classification are being investigated to assist in classifying races. Goodman and Paterniani (1969) listed three ways that environmental biases may be reduced: (a) evaluate the germplasm in several environments and use average values of traits over environments, (b) evaluate the germplasm in several environments and determine similarities of responses within each of the environments, and (c) limit comparisons to those characters that have the least environmental bias relative to size of differences among means. Goodman (1967, 1968), Goodman and Paterniani (1969), Bird and Goodman (1977), and Goodman and Bird (1977) have used numerical taxonomy techniques in an attempt to identify characters that fulfill condition (c) and to identify how the techniques relate to previous race classifications. Characters least affected by environmental factors and their interactions with environments were reproductive characters (e.g., ear and kernel characters) that had components of variance greater than the sum of corresponding components of variance for years and race by year interactions (Goodman and Paterniani 1969). Vegetative characters tended to have greater interactions with environments; tassel characters were intermediate to the reproductive and vegetative characters. Hence reproductive characters seemed to be better indicators of racial differences than vegetative characters.

Preliminary information on numerical methods of classification shows that multivariate methods (principal component analysis, factor analysis, canonical variate analysis, and cluster analysis using unweighted variables) can provide additional information in sorting out relations of races of maize. Use of the numerical methods of taxonomy, however, is very complex and only limited efforts have been directed to their use. The feasibility of different methods of classification, as well as more conventional classification procedures, depends on the choice of characters measured and how they are influenced by environmental factors. There has been a consistent relation in classification of races between multivariate analyses and conventional procedures. As classification procedures and techniques become developed, as additional data are collected for different morphological and physiological traits, and as additional breeding information becomes available, classification of races will become more refined (Paterniani and Goodman 1977). Adjustments will be made for duplications and classifications will become more subjective. Collectively the information will be valuable to breeders in the range of variability among maize races and will help them to recognize patterns of variation in choice of germplasm.

11.3 RACES OF MAIZE IN THE WESTERN HEMISPHERE

The first extensive race description of a comprehensive collection of maize germplasm was reported by Wellhausen et al. (1952) in Mexico. Similar types of programs were later initiated throughout the Western Hemisphere under the auspices of the Committee on the Preservation of Indigenous Strains of Maize within the National Academy of Sciences-National Research Council (NAS-NRC). Because of experiences in the United States where new hybrids and varieties had rapidly replaced open-pollinated varieties, the NAS-NRC was concerned with collection and preservation of maize varieties indigenous to the Western Hemisphere. Much of the original germplasm in the United States in the form of open-pollinated varieties was lost. Maize is one of the basic food plants in the Western Hemisphere; and its great diversity, resulting from thousands of years of domestication, was considered one of the important natural resources of the Western Hemisphere. It was felt that rapid development of communication, travel, and breeding programs would result in the same fate for indigenous varieties and races of other countries. Hence, an effort to collect, study, and preserve the vast reservoir of genetic variability was initiated.

The format of collecting, studying, and classifying used by Wellhausen et al. (1952) was followed in nearly all instances. Characters used in classifying the maize collections comprised four principal categories:

1. vegetative characters of the plant--response to altitude, height, total leaf number, number of leaves above the ear, width of ear-bearing leaf, venation index, and internode patterns
2. tassel characters--tassel length, peduncle length, length of branching space, percentage of branching space, percentage of secondary and tertiary branches, and total number of branches
3. external and internal ear characters--ear diameter, length, and row number; shank diameter and length; number of husks; kernel width, thickness, length, and denting; cob diameter; rachis diameter; cob/rachis index; rachilla length; rachilla/kernel index; glume/kernel index; cupule hairs; rachis flag; lower and upper glume traits; rachis induration; and teosinte introgression
4. physiological, genetical, and cytological characters--number of days from planting to anthesis, pubescence of leaf sheath, plant color, midcob color, chromosome knobs, and B-chromosomes

The characters measured described the entire genetic constitution of the germplasm under consideration rather than only Sturtevant's endosperm differences.

Studies conducted on the races of maize in the Western Hemisphere are summarized in Table 11.1. They are listed in chronological order of publication and show the number of races described in each country and the number of collections studied in arriving at the racial classifications. In addition, each report gives a description of ecological conditions, maize cultures, and methods of collection and classification. A wealth of information is included in the reports, and they probably represent the only detailed repository of information relative to the races of maize. A total of 285 races was described. There was some overlapping of races for different countries and regions, but classification of the collections differentiated distinctive races. It was fortunate that the studies were undertaken when they were (although

Table 11.1. Distribution of the races of maize described for Mexico, Central America, and South
America, Europe, and the United States.

Source	Areas	Number of collections	Number of races described	
			Total	Subdivisions
Wellhausen et al. (1952)	Mexico	2,000	32	
			Ancient indigenous	4
			Pre-Columbian exotic	4
			Prehistoric mestizos	13
			Modern incipient	4
			Poorly defined	7
Hathaway (1957)	Cuba	---	7	
			Commercial	4
			Domestic	3
Roberts et al. (1957)	Colombia	1,999	23	
			Primitive	2
			Probably introduced	9
			Colombia hybrid races	12
Wellhausen et al. (1957)	Central America	1,231	13	
			Primitive	2
			Exotic and derived	11
Brieger et al. (1958) Paterniani & Goodman (1977)	Brazil	3,000[+]	52	
			Argentina	11
			Eastern slopes of Andes	1
			Under the Capricorn	26
			Amazon Basin	14
Ramirez et al. (1960)	Bolivia	844	32	
Brown (1960)	West Indies	135	7	
Timothy et al. (1961)	Chile	39-114	19	
Grobman et al. (1961)	Peru	1,600	49	
			Primitive	5
			Anciently derived	19
			Lately derived	9
			Introduced	5
			Incipient	5
			Imperfectly defined	6
Timothy et al. (1963)	Ecuador	675	23	
Grant et al. (1963)	Venezuela	685	19	
Brown and Goodman (1977)	United States	---	9	
Brandolini (1969)	Europe	6,000	11[‡](33)	
			285	

[+] Obtained from Paterniani and Goodman (1977), who also describe 91 populations belonging to 19 races and 15 subraces.

[‡] Eleven races described by Leng et al. (1962) and 33 groups listed by Pavlicic (1971) not included in total of 285.

sooner would have been preferable) because considerable hybridization of races was occurring by interchange of germplasm. The classification identified where and when hybridization had occurred and the putative parents involved. Collection and classification were effective in developing the lineage of many races.

Brown and Goodman (1977) recently summarized the reports listed in Table

11.1. They provided what seemed to be well-defined racial groupings. The
groupings provide information on relationships among races and should be helpful to breeders in selecting germplasm. A reexamination of races described by
authors listed in Table 11.1 indicates that many of the races were duplicates
and that about 130 more or less distinct racial complexes make up the aggregate maize germplasm of the Western Hemisphere. Most of the reports listed in
Table 11.1 were intended to be only preliminary because little or no breeding
information (inbreeding and hybridization) was available to assist in the
classification of the races. Brown and Goodman also list and describe nine
races for the United States. According to reports listed in Table 11.1, variability of maize germplasm in the United States is considerably less than in
other areas (e.g., Wellhausen et al. 1952; Roberts et al. 1957; Brieger et al.
1958; Grobman et al. 1961). Fewer races were identified, but much of the
maize germplasm of the United States was lost before it became apparent that
it would be useful in breeding programs. Development of and interest in inbred lines and hybridization caused maize breeders to virtually ignore the basic germplasm sources in the United States. It is unfortunate that extensive
collections were not classified and preserved in the early 1900s.

Paterniani and Goodman (1977) summarized relative proportions of described races based on adaptation to elevation and endosperm type. About 50%
of the races was adapted to low altitudes (0 to 1000 m), about 10% was grown
at intermediate altitudes (1000 to 2000 m), and about 40% was adapted to higher elevations (greater than 2000 m). Based on endosperm type, the following
classifications of the races were made: about 40% floury, about 30% flint,
slightly more than 20% dent, about 10% popcorn, and about 3% sweet corns. No
classification was given relative to endosperm types adapted to low, intermediate, and high altitudes. Adaptation to elevation was considered mainly a
result of natural selection, whereas distribution of endosperm types was primarily due to human preferences.

11.4 EUROPEAN RACES OF MAIZE

Maize was introduced in Europe by Christopher Columbus and was first cultivated in fields near Seville, Spain, during 1492 (Brandolini 1971). During
the next four centuries, movement of maize germplasm from the Western Hemisphere to Europe continued intermittently and at different rates. Continued
influx of germplasm from the Western Hemisphere resulted in a range of variation available for use, but introduction of improved varieties and hybrids
from the U.S. Corn Belt rapidly led to the disappearance of many of the adapted varieties either by substitution or intercrossing with previously introduced germplasm. Consequently, a repetition of the U.S. experience occurred
in Europe; i.e., many of the varieties and races selected for environmental
conditions of Europe were lost because of interest in recently improved introductions from the U.S. Corn Belt.

Brandolini (1969, 1971) and Leng et al. (1962) have described briefly the
distribution of maize in Europe and surrounding regions after it was first
grown in 1492. Brown (1960) suggested that the two West Indies races, Coastal
Tropical Flint and Early Caribbean, may have been the original maize germplasm
introduced into Europe after the voyages of Columbus but Leng et al. reported
that maize types currently available in southeastern Europe bear little resemblance to these two races. Maize was poorly adapted to environments of Spain,
but because of repeated collections by the explorers of the Western Hemisphere, germplasm was continuously introduced to the European continent. Four
hundred years of selection developed varieties that were adaptable to the
broad spectrum of environmental conditions, from the arid conditions surround-

ing the Mediterranean Sea to the short growing seasons of northern Europe. Introductions of different germplasms had different impacts on the composition of the maize gene pools, and natural and artificial selection developed varieties with specific fitness to the new environments. Because of the necessity of surviving sea transport, it seems that flint and popcorn varieties played a prominent role in the early European germplasm--especially in southern Europe. Eventually the Northern Flint and Southern Dent races were introduced in the central part of Europe as a result of English and French explorations in North America. Later, about 1900, the Corn Belt Dent race became an important part of the European germplasm. The different stages of introduction, hybridization of introduced germplasm with previously introduced germplasm, and selection of types to meet the wide range of environmental conditions in Europe created a complex array of European germplasm.

Preliminary classifications of European germplasm have been given by Leng et al. (1962) for southeastern Europe and by Brandolini (1969) for the European continent. Leng et al. concluded from their studies that at least 11 maize races occur in southeastern Europe:

1. small-eared Montenegrin flints--probably a direct derivation of the introductions from the Western Hemisphere and with ear type similar to that of the Andean ràce, Amarillo de Ocho
2. small-kerneled flints--closely resembling South American pearl popcorns in plant and ear characteristics
3. eight-rowed (Northern) flints--typical of the Northern Flints described by Brown and Anderson (1947) and probably direct introductions from the United States
4. Mediterranean flints--relatively rare and seemingly unlike any of the races described in the Western Hemisphere
5. derived flints--seemingly resulted from hybridization of different races, e.g., hybridization of the Mediterranean flints with other flint varieties
6. many-rowed soft dents (Southern Dents)--probably direct introductions of the Old Southern Dent races described by Brown and Anderson (1948)
7. large-kerneled dents--may have resulted from direct introductions or evolved from hybridization between dent varieties and eight-rowed flints
8. beaked (Rostrata) dents--seemingly derived from the Southern dent races described by Brown and Anderson (1948)
9. Corn Belt dents--direct importations from the United States in 1890-1910, constituting the major source of germplasm in southeastern European breeding programs at the present
10. derivatives of hybrids between flint and dent races due to hybridization of the two races
11. modern hybrids--introduced after World War II and rapidly displacing older races

Leng et al. (1962) emphasized that European germplasm may be quite useful to breeders in temperate zone regions because it is adapted to temperate zone climatic conditions and day lengths. The most common problem encountered by breeders in the U.S. Corn Belt and other temperate zone regions from the introduction of Central and South American germplasm is its adaptability and photoperiod response. The introduced germplasm usually requires several (5 to 10) years to adapt to Corn Belt conditions before it can be seriously worked into the breeding programs. European germplasm would be more adaptable to the U.S. Corn Belt, but the gain from adaptability may be offset by the limited new genes introduced in the program. Development of the races in Europe has

spanned a much shorter time than development of the old races in the Western
Hemisphere. Use of European germplasm would probably have greater short-term
gains, but total genetic gain may be greater with use of germplasm from outside the temperate zones.

Sanchez-Monge (1962), Brandolini (1969, 1971), Brandolini and Avila
(1971), and Pavlicic (1971) have given preliminary reports on classification
of maize germplasm for southern Europe and Mediterranean areas. Collection,
classification, conservation, and exchange of germplasm for this area was conducted by the Southern Committee of EUCARPIA. About 6000 samples were collected and 3260 samples have been studied (Brandolini 1969). Major goals of
the collections were to determine similarities or differences among representative samples collected from the different countries and to obtain knowledge
about genetic mechanisms involved in adaptation to different areas. The collections were more extensive and included a greater area than those reported
by Leng et al. (1962); it was found that the types of maize collected and
studied have many traits in common, suggesting a common origin. For example,
the small-eared Montenegrin flints described by Leng et al. were similar to
the Poliota varieties of Italy. Pavlicic (1971) presented some preliminary
data on the collections that showed: (1) considerable similarity among countries for the flint types; (2) the greatest variability among varieties in Italy, with Yugoslavia second; (3) considerable overlap of types from Italy and
Yugoslavia; and (4) some similarities in the Yugoslavian types to the types
from Romania, Bulgaria, and USSR. From the same collection of germplasm,
Brandolini (1971) found that the frequency of chromosome knobs was consistently low.

Brandolini (1969) supported Leng et al. (1962) in that southern European
races seem to possess germplasm that is of value to breeding programs in temperate regions. Although contamination by hybridization with recently introduced germplasm has occurred, a large reservoir of germplasm is available in
Europe for breeding programs in temperate regions. Different selection pressures have been applied to different areas of Europe, so different races and
varieties include genes that contribute to disease and insect resistance,
drought tolerance, short growing seasons, spring cold tolerance, and low moisture at harvest. So far, however, movement of germplasm from Europe to the
U.S. Corn Belt has been minimal.

11.5 U.S. CORN BELT GERMPLASM

Although study of the maize germplasm of the United States was generally
initiated too late, descriptions of the array of varieties available before
acceleration of inbred line and hybridization breeding programs are available
in several places. Most of the descriptions of varietal germplasm were not
systematic or directed to origin or lineage of the varieties. Brief descriptions of many of the varieties used to study heterosis observed in variety
crosses listed in Table 10.1 were presented; observed heterosis of varietal
crosses would be a measure of genetic diversity among parental varieties.
Wallace and Bressman (1925), other similar types of references, and experiment
station bulletins (e.g., Atkinson and Wilson 1914) provide descriptive ac-
counts of some of the open-pollinated varieties but they are neither extensive
nor complete. Similar to racial descriptions of reports listed in Table 11.1,
there were many duplications of the same open-pollinated variety, which in
several instances were only slight modifications; e.g., many known versions of
the Reid open-pollinated variety existed that bore different names, depending
on individuals growing and practicing some mild selection within the variety.

Despite deficiencies in classification of germplasm in the United States,

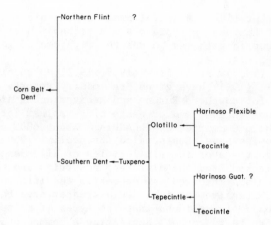

Fig. 11.1. Suggested probable origin of the Corn Belt Dents of the United
            States (Wellhausen et al. 1952).

much is known concerning lineage of the germplasm and how the invaluable Corn
Belt Dent race arose. Wellhausen et al. (1952) outlined the probable origin
of the U.S. maize germplasm that contributed to the Corn Belt Dent race (Fig.
11.1)--it is clear that the Corn Belt Dent varieties arose by repeated hybridization between the Northern Flints and the Southern Dents. Origin of the
Northern Flints is still unclear and the Southern Dents somewhat conjectural.
Brown and Anderson (1947) gave a detailed account of the Northern Flint racial
complex, described some of the flint varieties, and listed the traits (hard
kernels, low row number, cylindrical ears, and early maturity) that contributed to the Corn Belt Dent varieties. It has been suggested that the Northern
Flints were derived from the Mexican race Harinoso de Ocho (Galinat and Gunnerson 1963); from the southwestern United States (Mangelsdorf and Reeves
1939); and from the San Marcenô and Serrano races of the highlands of Guatemala (Brown and Anderson 1947). As Brown and Goodman (1977) point out, each of
these conjectures has some faults. More evidence is available on the ancestral origin of the Southern Dent race, the other half of the parentage of the
Corn Belt Dent race. Wellhausen et al. (1952) have involved several of the
Mexican races in the parentage of the Southern Dents (Fig. 11.1). Brown and
Anderson (1948) studied the Southern Dent complex in some detail and concluded
that it was almost certainly derived from certain Mexican varieties. Brown
and Goodman (1977) also concluded that many Southern Dent varieties are related to the dents of central Mexico and that they are simply northern counter-
parts of races still prevalent in Mexico. There is not total agreement of the
possible Mexican races involved in the Southern Dent race, but the Tuxpeno
race seems prominent. Contributions of Southern Dents to Corn Belt Dents include high number of kernel rows, tapering ears, softer textured kernels, and
pointed kernels (Brown and Anderson 1948). Other traits that Southern Dents
could have contributed are prolific tendencies and greater frequency of genes
for disease and insect resistance.
     Of the nine racial complexes described by Brown and Goodman (1977) for
the United States, five (Great Plains Flints and Flours, Pima-Papago, Southwestern Semidents, Southwestern 12 Row, and Derived Southern Dents) have had
little impact on U.S. maize breeding. Derived Southern Dents seem to have
arisen from Southern Dents hybridized with (perhaps) Southeastern Flints,
Northern Flints, and Corn Belt Dents. Derived Southern Dents tend to have

greater prolificacy than Southern Dents and include several varieties that
have played an important role in breeding programs in the southeastern United
States and as sources of prolificacy in Corn Belt Dents. The Great Plains
Flints and Flours racial complex includes several distinctive varieties; apparently this race was derived from the hybridization of Northern Flints and
varieties from southwestern United States. Collections of Great Plains Flints
and Flours are available but it does not seem that they have made a significant contribution to germplasm of the Corn Belt.

Maize breeders in the U.S. Corn Belt have been concerned about the genetic variability of their breeding materials and have attempted to maintain some
semblance of diversity in their breeding populations. Rapid development of
breeding programs emphasizing inbred line development for use in hybrids and
rapid acceptance of hybrids by farmers of the Corn Belt (Fig. 1.2) removed
Shull's proposal from theory to practicality. Empirical evidence showed that
the greatest expression of heterosis in hybrids was from use of inbred lines
derived from divergent sources. Initial samplings of plants for deriving inbred lines were from open-pollinated varieties, but this soon changed to pedigree selection in F_2 populations formed from inbred lines considered superior
to the average or possessing certain traits to correct known weaknesses of
otherwise desirable lines (Fig. 1.4).

It became apparent during the latter part of the 1940s that some order
was needed to maintain the heterotic pattern of hybrids produced from lines
derived from recycling of previously used lines. Although no attempt was made
to classify material, collection and storage of germplasm in the form of open-
pollinated varieties was emphasized in the 1930s. The minutes of the North
Central Corn Improvement Conferences frequently include reports by subcommit-
tees on the preservation of germplasm. The extent of the preservation depended on the interests of the individuals within the 12 states in regard to collecting, maintaining, and describing the collections. Each state was encouraged to maintain storage of germplasm locally and at some central storage
agency. Germplasm of open-pollinated varieties was collected for preservation, but the extent of its inclusion in breeding programs was usually minimal.

Renewed interest in maize germplasm occurred after the *Helminthosporium
maydis* outbreak on T-cytoplasm in 1970. In 1975 maize breeders in the 12
north central states appointed a new subcommittee on genetic vulnerability
that was charged with the responsibility to assess the stage of germplasm base
in the North Central Region of the United States. The subcommittee surveyed
all the public maize breeders in the North Central Region to determine the
scope of the germplasm base of their breeding programs. A summary of the subcommittee's report was included in the minutes of the U.S. North Central Corn
Breeding Research Committee (NCR-2) Report (1977) (Tables 11.2, 11.3, and
11.4), and a brief description of each population was given. Surprisingly,
the germplasm base of U.S. Corn Belt breeding programs was not as restricted
as originally feared; 246 populations were currently undergoing some form of
selection, and an additional 200 populations included in the breeders' inven-
tory were in storage and available for use. There were several duplications
of basic germplasm undergoing selection (Table 11.2); the most striking example was the 24 versions of Stiff Stalk Synthetic being used.

Improvement for grain yield was the most common single trait under selection (Table 11.3); disease resistance also had a high priority. The primary
traits of selection are listed in Table 11.3 and usually more than one trait
was included, particularly in selection programs for yield; e.g., selections
based on mechanically harvestable yield would include selection for stalk

Table 11.2.   Populations in the U.S. North Central Region
survey classified by origin and whether they
are currently undergoing selection.[†]

Origin	Active		Inactive	
	No.	%	No.	%
Krug	9	3.6	1	0.1
Lancaster	4	1.6	3	1.5
Hays Golden	7	2.8	0	0.0
Other open-pollinated varieties	7	2.8	38	19.0
Exotics[‡]	62	25.2	39	19.5
Stiff Stalk Synthetic	24	9.8	8	4.0
Other synthetic varieties Single-line[§]	2	0.8	10	5.0
Early (A.E.S. 400 to 600)	42	17.1	45	22.5
Late (A.E.S. 700 to 800)	89	36.2	56	28.0
Total	246		200	

[†]Report of the North Central Corn Breeding Research
Committee (NCR-2), 1977, p. 66.

[‡]Populations that include at least some exotic germplasm.

[§]Populations that include different versions of the same
inbred line.

Table 11.3.   Traits undergoing selection in
the populations included in the
U.S. North Central Region
survey.[†]

Trait	Number of populations
Grain yield	44
Grain yield and maturity	9
Agronomic traits	43
Insect resistance	14
Disease resistance	57
Chemical composition	28
Adaptation	15
Unidentified	36
Total	246

[†]Report of the North Central Corn Breeding
Research Committee (NCR-2), 1977, p. 67.

Table 11.4. Selection methods being used for
the populations included in the
U.S. North Central Region
survey.[†]

Method of selection	Number of populations
Mass	140
S_1 per se	36
Full-sib plus mass	11
S_2 per se	11
Inbred tester	11
Reciprocal full-sib	8
Full-sib	7
Reciprocal recurrent	4
Modified ear-to-row	4
Half-sib	3
Miscellaneous	11
Total	246

[†]Report of the North Central Corn Breeding
Research Committee (NCR-2), 1977, p. 68.

quality (disease resistance) and ear droppage (corn borer resistance). Mass
selection was the most commonly used breeding method (Table 11.4). In 36 instances S_1 progeny evaluation was used, with about equal use of the other
methods. Of the populations actively under selection, 62 (25.2%) included
germplasm that was considered exotic; by comparison, 54.1% of the populations
were synthesized from Corn Belt lines (excluding Stiff Stalk Synthetic) and
10.8% were open-pollinated varieties. The inactive populations were nearly
equal for the U.S. Corn Belt open-pollinated varieties (20.6%) and populations
that included exotic germplasm (19.5%).

The survey of germplasm included in the U.S. Corn Belt breeding programs
indicated that the problem of genetic vulnerability was less serious than previously assumed. A wide range of germplasm was undergoing selection for several different traits by use of several different selection procedures. How
could one reconcile the evidence that there was a dependence on only a few inbred lines for use in hybrids (Corn 1972, p. 105; Zuber 1975) and in the range
of germplasm included in the breeding program? In all instances the surveys
were for lines released from publicly supported breeding programs. It seems
reasonable to assume that a similar survey involving privately supported
breeding programs would have given similar results (Duvick 1975). It does not
seem that breeding programs are extremely limited in genetic variability, but
the economic restraints demanded by modern farmers guarantee that genetic variability within and among fields of maize will be much less than when open-
pollinated varieties were predominantly used. Modern farmers demand the use
of hybrids that have proven yield performance and uniform grain quality, plant
type, and maturity. A unique combination of two inbred lines that provides
the desired hybrid will be extensively used. The hybrid seed industry is very
competitive and will provide seed to meet the demands. In many cases, the

same or similar lines will be used. Extensive use of a few inbred lines of similar genetic background in the production of hybrids is a cause for concern of genetic vulnerability. Maize breeders are cognizant of the problem but they may not be able to do much about it. However they have the materials available if and when the need arises.

11.6 GERMPLASM PRESERVATION

Collection, classification, and maintenance of germplasm has been emphasized to maintain its availability for future use. Although original forms of races and varieties of maize are fast disappearing because of the change to hybrids, preservation of race collections is less than adequate because of limited interest, funds, and facilities. Races and varieties are populations of individuals, each with unique genotypes, that require sizable (e.g., 200 to 500) numbers to maintain genetic characteristics. Thousands of collections have been made, but the potential and future use of the materials is limited if the collections are not properly stored and maintained.

Three major centers established for the storage and preservation of collections in the Western Hemisphere are:

1. Maize Germplasm Bank--CIMMYT, Chapingo, Mexico, which serves Central America and the Caribbean area. It is supported by the Rockefeller Foundation Agricultural Program in Mexico and the Mexican government.
2. Seed Storage Center at Medellin, Colombia, which serves Colombia and the other Andean countries (Bolivia, Chile, Ecuador, Peru, and Venezuela). It is maintained jointly by the Colombian Ministry of Agriculture and the Rockefeller Foundation.
3. Brazilian Germplasm Bank at CNPMS-EMBRAPA (Sete Lagoas, M.G.) (formerly at Instituto de Genética, Escola Superior de Agricultura "Luiz de Queiroz," Piracicaba, Brazil). This center serves Argentina, Brazil, eastern Bolivia, Paraguay, Uruguay, and Guyana.

Additional collections are maintained at INIA, Peru la Molina; Argentina-INIA; USDA National Seed Storage Laboratory at Fort Collins, Colorado; and the USDA North Central Regional Plant Introduction Station at Ames, Iowa. Many of the collections at the different storage and preservation centers are duplicates, and the number of collections available probably exceeds those for most other crop species. Lonnquist (1974) indicated there are over 25,000 collections at the storage centers in the Western Hemisphere and about 6000 collections in the European-Mediterranean area. Most breeding projects also maintain a limited number of collections, which are included in the seed storage centers. Duplications of collections in the storage centers and breeding projects are desirable because of differences that can arise from their maintenance from genetic drift.

Paterniani and Goodman (1977) discussed problems associated with preservation of collections: labor and facilities required to maintain the collections; population sizes required to minimize the effects of inbreeding, loss of genes, and genetic drift; contamination; amount of seed required to meet the demands of researchers; and loss of materials that are poorly adapted to the area used for propagation. All problems associated with preservation of germplasm are related to the one important problem of population size.

Duplicate samples at storage centers probably become quite different for some traits because of the limited population sizes used in their maintenance. To overcome some of these problems, Paterniani and Goodman (1977) used a more

practical approach for preservation of their collections by compositing similar original samples. Use of this method provides more adequate maintenance
of a smaller representative number of populations.

Maize populations are usually maintained by hand sib-pollinations that
require extensive supplies and labor, and only a limited number of plants may
be included for each population; hence the problems of expenses, genetic
drift, seed supplies, and contamination can arise. Omolo and Russell (1971)
propagated by hand-controlled pollinations 500, 200, 80, 32, and 18 plant populations for five successive generations of the open-pollinated variety Krug.
They wanted to determine what size of population was required in order to
maintain itself without causing significant genetic changes. Significant
yield decrease was obtained as sample size decreased, and they concluded that
a sample of 200 plants was adequate to maintain a heterogeneous population by
hand sib-pollination. Only if some inbreeding could be tolerated and reproduction was infrequent would an 80 plant sample be adequate. The results of
Omolo and Russell emphasize that adequate population sizes are important to
retain the original genetic variability, but the size also imposes serious
problems in preservation of germplasm. Reproduction of the collections in
isolation fields would be preferable, but locating sites, field husbandry, and
adequate isolation make this approach unfeasible. Adequate isolation would
minimize all problems associated with germplasm maintenance except for cost of
labor and facilities.

The effective size of a population is an important aspect of germplasm
preservation. The degree of maintenance of genetic properties of a population
depends partly on the number N of seeds or individuals in the population (cen-
sus number) and primarily on the number of individuals intercrossed in previous generations, which reflects the effective number N_e. The effective number
depends also on how female and male gametes are sampled to originate the descendant population. In addition, if the population is treated as a dioecious
species, N_e will depend on the number of males and females taken for crosses.
If the population is crossed as a monoecious species each plant is potentially
assumed to participate as a source of male and female gametes.

The largest possible effective size of a population occurs in homozygous
self-pollinated species when one seed or an equal-size sample is taken from
each plant to originate the next generation. In cross-pollinated species,
such as maize, effective size depends on the crossing system and the manner in
which female and male gametes are sampled. Therefore, maintenance of germplasm stocks can be performed by any of the following procedures.

1. Population is treated as a monoecious species--Each plant in the population
    is a potential source of female and male gametes, which can be sampled
    with or without control on their relative number. Three sampling proce-
    dures are possible:
    (a) Control of the number of female and male gametes, which is only possi-
        ble under controlled hand pollination so that each male plant contrib-
        utes an equal number of gametes to the next generation. To obtain
        equal contributions of male gametes an equal number of seeds must be
        taken from each pollinated ear, thus resulting also in equal numbers
        of female gametes contributed to the next generation.
    (b) Control of the number of female gametes only. An equal number of con-
        tributed female gametes is obtained by taking an equal number of seeds
        from each ear (female parent), which is pollinated at random so that
        each male plant contributes unequally to the next generation.
    (c) No control of either female or male gametes. No control results when

pollination is at random (no control of the number of male gametes) and samples of varied sizes are taken from each ear (no control of female gametes). This procedure is common when all the ears of an open-pollinated field are harvested together (by hand or machine) and one sample is taken from the bulked seeds.

2. Population is treated as a dioecious species--A monoecious species like maize can be crossed as a dioecious species, provided that some plants are used as male and others as female parents. Separation of male and female plants in a field of maize can be accomplished through controlled hand pollination or by detasseling plants or plant rows. Control of the number of female and male gametes results in the following cases:

   (a) Control of both female and male gametes, which requires either controlled hand pollination in plant to plant (male × female) crosses or one plant used as a male to pollinate several plants as females. In the first case we have the same number of males N_m and females N_f; i.e., $N_m = N_f = N/2$, where N is the total number of plants. In the second case the number of females will be greater than the number of males; e.g., if one male is used to pollinate four female plants, then $N_m = N/5$ and $N_f = 4N/5$. In either case, equal-sized samples must be taken from each pollinated ear to control the number of female and male gametes.

   (b) Control of only the number of female gametes, which means that pollination is completely at random so that each male plant contributes unequally to the next generation. The easiest way to do this is by detasseling a proportion of rows in a field of maize; e.g., by detasseling alternate rows we have $N_m = N_f = N/2$, assuming a constant number of plants per row. To have control of the number of female gametes, samples of equal size must be taken from each pollinated ear in the female rows.

   (c) No control of either female or male gametes, which requires either hand pollination or random pollination in a detasseled field with unequal-size samples from each pollinated ear. This would result when a single sample of seeds is taken from a bulk of all ears from female plants.

Effective sizes for populations resulting from each of the given procedures are shown in Table 11.5. For calculation of the effective size of a population there are two different approaches--inbreeding effective size and variance effective size procedures. They are equivalent for a constant population size over generations.

With control of the number of female and male gametes when the population is crossed as a monoecious species [case 1(a)], the effective number can double in relation to the simplest procedure where there is no control on gametes of either sex [case 1(c)]. This is particularly important for small populations, which occur very frequently with introduced germplasm. On the other hand, seed sampling and controlled hand pollinations are not difficult in the maintenance of relatively small populations. When the population is pollinated as a dioecious species, the effective number is the same as when treated as a monoecious species if the same number of male and female parents are used. Otherwise, the effective number tends to decrease in the direction of the sex that participates with the smaller number, because N_e is proportional to the harmonic mean of N_m and N_f and the harmonic mean is more strongly influenced by the smaller values (Crow and Kimura 1970). Therefore, for $N_f > N_m$, the number of males is more important than the number of females in determining the effective size of the population and vice versa.

Table 11.5. Effective size[†] of a population under different procedures relative to crossing and sampling female and male gametes.

Crossing system[‡]	Gamete control Female	Male	No. of parents Female	Male	Effective number (N_e)[§]	Relative number[¶]
1. Monoecious						
a.	with	with	N	N	$2N$	200
b.	with	no	N	N	$4N/3$	133
c.	no	no	N	N	N	100
2. Dioecious						
a.	with	with	N_f	N_m	$\dfrac{8N_m N_f}{N_m + N_f}$	200
b.	with	no	N_f	N_m	$\dfrac{16N_m N_f}{3(N_m + N_f)}$	133
c.	no	no	N_f	N_m	$\dfrac{4N_m N_f}{N_m + N_f}$	100

[†]The authors are indebted to R. Vencovsky for derivation of the formulas.

[‡]See text for explanation; [§]indicates constant population number of each procedure, without random elimination of plants.

[¶]Relative to 1c ($N_e = N$) when monoecious and to 2c ($N_m = N_f$) when dioecious.

When considering several generations in a random mating population, the effective number may change with differences in sample sizes and procedures. In a random mating population where size fluctuates, it would be desirable to know what population of constant size would give the same effective number. Crow and Kimura (1970) showed that the effective population number is roughly the harmonic mean of the various values; i.e., $(1/N_e) = (1/t)\Sigma_i(1/N_i)$, where t is the number of generations considered. As an example, consider four generations where the effective numbers are 100, 100, 20, and 1000. The effective number of a corresponding population of constant size would be $N_e = 56.3$. If one wants to compensate for the bottleneck in the third generation and sample 10,000 individuals instead of 1000 in the fourth generation, little is gained; the effective population size would correspond to a population of constant size $N_e = 57.1$. This example shows that in maintaining germplasms care must be taken to keep the population size at least nearly constant over generations, because the decrease in population size in one generation is not easily recoverable in the next.

When there is a reduction in sample size in one generation after some generations of constant size, it is possible to recover the original trend of effective size by using one or more generations of larger sample sizes. Over t generations we have $t - k - 1$ generations of constant size (N_1) followed by one generation of reduced size ($N_2 < N_1$), and we want to recover the original trend (N_1) of effective size in the next k generations with effective size of N_3. It can be shown that this is possible only for $(k + 1)N_2 > N_1$ and that the required effective size in the last k generations must be $N_3 = (kN_1N_2)/[(k + 1)(N_2 - N_1)]$. Note that N_3 does not depend on the number of initial generations under constant size.

To recover the loss in effective size in just one generation after the bottleneck, then $k = 1$ and it is necessary that $N_2 > N_1/2$. In the example given, we had $N_1 = 100$, $N_2 = 20$, and $N_3 = 1000$. In this case we have $N_2 < N_1/2$, showing that it is not possible to recover in one generation the loss in the third generation. However, if $N_2 = 60$, it would be possible to recover the loss by taking $N_3 = 300$; i.e., effective sizes of 100, 100, 60, and 300 in four generations would be equivalent to a constant population size $N_e = 100$. For $N_2 = 50$, recovery would be possible by taking more than one generation of larger sample sizes. Thus, effective sizes of 100, 100, 50, 200, and 200 would give an effective population size equivalent to a population of constant size $N_e = 100$.

Race collections are irreplaceable and represent a source of germplasm that is either limited or not available at the present. Races of maize have evolved over hundreds of years of natural and artificial selection that developed unique populations adaptable to a wide range of environmental conditions. Seed storage banks are a repository of genes and gene combinations that otherwise would have been lost because of use of hybrids, increased cultivation of land, and decrease in numbers of people involved in agriculture. Genetic improvement of hybrids has been made from use of only a limited amount of the total germplasm available. Brown (1975) has stated that in the United States more than 90% of the breeding effort is devoted to germplasm whose origin is traced to not more than 3 of the 130 existing races. Hence, he continues, U.S. maize improvement programs have largely ignored 98% of the germplasm that makes up *Zea mays*. Galinat (1974) also has expressed concern about the genetic erosion of maize and the loss of natural genetic variation because of the pressures for developing uniform varieties that produce the greatest yields. In the future, seed repositories would become immediately useful in breeding programs for increasing the genetic variability of the materials undergoing selection. The only recourse for recapturing natural variation is to take advantage of the seed storage centers. They contain material that can be incorporated into long-range breeding programs--they are sources of genes for resistance to maize pests (to meet emergencies); sources of germplasm to further basic studies on heterosis, relationships, and origin of maize; and a wealth of genes and combinations of genes for basic genetic studies. The importance of indigenous maize germplasm is emphasized by all the authors listed in Table 11.1. It is imperative that efforts be expended to maintain the reservoir of maize germplasm for future breeding programs.

11.7 POTENTIAL AND USE OF EXOTIC GERMPLASM

The possibilities of including exotic germplasm in maize breeding programs has been emphasized (Brown 1953, 1975; Wellhausen 1956, 1965; Leng et al. 1962; Paterniani 1962; Brandolini 1969; Lonnquist 1974; Brown and Goodman 1977; reports listed in Table 11.1). In most instances, it also was emphasized that the immediate usefulness of exotic germplasm may be limited to single-gene transfers. The wealth of germplasm available is staggering if one considers the volume of the maize collections that have been assembled for possible use (Lonnquist 1974).

Exotic germplasm can have several connotations; for applied breeding programs, exotic germplasm includes all germplasm that does not have immediate usefulness without selection for adaptation for a given area. Paterniani (1962) distinguished two possible alternatives in exotic germplasm: (1) races or varieties having a broad genetic base and (2) inbred lines having a narrow genetic base. Races and varieties can be used either in population improvement programs or in intervarietal crosses to determine the hybrid vigor expressed in crosses. Inbred lines can be used either for conventional hybrids or to synthesize populations for special purposes. The choice between the two alternatives and its success depends on the level of maize improvement, the social and economic situation, and genetic variability and potential of the local germplasm.

In spite of the extreme variability within *Zea mays* the gross chromosome morphology is rather uniform among the many races, varieties, and strains (Galinat 1977). Similar chromosome numbers and morphology of exotic and adapted germplasm permit wide crossing. However the immediate effects of crossing exotic germplasm with native germplasm and initiating selection are often disappointing. Productivity and desirable segregates usually are limiting and the material is usually discarded as not being promising. To obtain gene combinations that have efficient biochemical functions within the genome, it is necessary to allow genetic recombination accompanied by mild selection. Lonnquist (1974) emphasized the necessity and importance of recombination and the possible problems of linkage. The only recourse for the use of exotic germplasm in, say, the U.S. Corn Belt is patience and adequate recombination after several generations of random mating with mild selection pressure (Brown 1953; Lonnquist 1974).

Troyer and Brown (1972) demonstrated the effectiveness of gradual introductions of exotic germplasm into adapted germplasm of the U.S. Corn Belt. They crossed Mexican germplasm with U.S. Corn Belt lines and grew the materials in isolated fields for 10 years to allow for recombination. Mass selection for recombinant genotypes that had desirable plant traits was carried out before intense selection was practiced for earliness.

Each race of germplasm has evolved over a period of time to develop a population of genotypes that are physiologically adapted to a particular ecological niche. The collection of genetic factors for each race has been assembled over time to develop genotypes that function efficiently to survive and propagate the race. Although the Corn Belt Dent race has not been subjected to selection for as long as most of the other races described in Table 11.1, it has been under relatively intense selection pressure for the factors that collectively formed the highly productive race. The crossing of races would disrupt the harmonious gene combinations of each race. It would be similar to the situation that existed in the past when a race was established by crossing two existing races.

An example from Wellhausen et al. (1952) shows that the Harinoso Flexible and Teocintle races were crossed to form the race Olotillo; the Harinoso de Guatemala and Teocintle races were crossed to form the race Tepecintle; finally the Olotillo and Tepecintle races were hybridized to form the important race Tuxpeno. Tuxpeno is one of the most productive and desirable modern races of Mexico and has been a source of germplasm for the Southern Dents of the United States. The Corn Belt Dent race, in turn, was formed by the hybridization of the Southern Dent and Northern Flint races. Successive hybridizations were followed by a period of intermating and selection to gradually evolve into the new races. Interracial hybridization seems to have been an important factor in the development of superior germplasm (Wellhausen et al. 1952).

The evolution of the formation of new races would be similar to the methods needed for incorporating exotic germplasm into adapted germplasm. The key issue seems to be allowing sufficient time for the integration of the genetic factors from the two sources of germplasm. If the evolution of the develop-

ment of the superior new races is to be mimicked in a breeding program, it
seems imperative to permit adequate recombination with only mild selection
pressure to sort out the desirable alleles in the new breeding population.
After the derived germplasm has developed to the point of seeming adaptability
to the particular environment, more intense selection accompanied by inbreeding may be initiated.

Hallauer (1978) summarized information available from the use of exotic
germplasm in the United States under the following categories: variability
among and within exotic germplasm, heterosis expressed among exotic varieties
and among exotic × adapted varieties, effects of selection within exotic germplasm and within populations formed by crossing exotic and adapted varieties,
and potential of exotic germplasm as sources for line development and genes
for disease and insect resistance. Although Wellhausen et al. (1952), Brown
(1953), Wellhausen (1956, 1965), Leng et al. (1962), and others have emphasized the importance and potential of the germplasm from outside the United
States, the amount of information from the use of this material is rather limited. Results reported generally are positive, but they may be misleading as
an indication of the effort that has been expended on screening and evaluating
exotic germplasm. It probably has been included in most applied breeding programs at some time and the results were either negative or not reported. The
survey of the North Central Region of the United States shows that 25.2% of
the populations actively undergoing selection included at least some exotic
germplasm (Table 11.2). The proportion of exotic germplasm in commercial applied breeding programs may be even greater.

Goodman (1965b) reported the only data that include a critical comparison
of the estimates of genetic variability for an adapted population (Corn Belt
Composite) and a population that included exotic germplasm (West Indian Composite). His results showed that genetic variability was greater for the population that included exotic germplasm and not at the expense of lower yields.
Predicted gain was greater in the West Indian Composite; and the opportunities
seemed good that material developed from the West Indian Composite would contribute to the heterosis of hybrids, which was substantiated by Eberhart
(1971) in a diallel series of variety crosses that included West Indian Composite. Shauman (1971) also reported greater estimates of additive genetic variance in the hybrid population of Krug × Taboncillo 13 Hi Synthetic 3 than in
the adapted variety Krug.

Heterosis among exotic varieties and among exotic × adapted varieties has
been determined for several races and varieties (Wellhausen 1956, 1965; Table
10.1). Evaluations of races and varieties (both adapted and exotic) have not
been as extensive as desirable. Widespread trials of races and varieties
would assist in identification of those most desirable for breeding programs
with added information on the manifestation of heterosis.

Wellhausen et al. (1952) and others (Table 11.1) have emphasized that
heterosis information would be a valuable additional aid in classification of
maize germplasm. Heterosis usually was observed for crosses that included exotic germplasm. The contribution of genetic diversity to manifestation of
heterosis, however, seems to have a limit. Moll et al. (1962) and Moll et al.
(1965) in studies of crosses among adapted and exotic varieties found that
heterosis increased as presumed genetic diversity increased but decreased in
variety crosses that were assumed to be most genetically diverse. The possible
explanation is that the combination of genetic factors from extremely diverse
germplasm was too great to allow compatible functioning of the physiological
mechanisms. Before extremely diverse exotic germplasm is introduced it would
be helpful to know these relationships in order to increase efficiency of a
breeding program. The problem may be alleviated to some extent because selec-

tion for adaptability of exotic germplasm is necessary before it is useful in a breeding program.

Population improvement programs conducted in exotic and semiexotic populations usually have been effective in the limited instances reported. Selection for improvement usually was conducted in populations formed by crossing adapted and exotic germplasm, but Hallauer and Sears (1972) selected directly in ETO Composite. Initial selection generally is for adaptability, i.e., maturity and reduced plant stature. Mass selection has proven effective for adaptability (Hallauer and Sears 1972; Troyer and Brown 1972). Heritability is rather high for adaptability traits, and mass selection techniques permit additional recombination with a mild selection pressure. Number of cycles of mass selection necessary to attain acceptable standards of adaptability depend on the germplasm included, selection intensity, and the range in latitude over which the germplasm is being transferred. Recurrent selection in populations that include exotic germplasm was initiated for yield improvement to compare rates of gain and genetic variability; S_2 progenies have been evaluated in replicated field trials (Hallauer 1978). Populations undergoing S_2 recurrent selection included different proportions of exotic germplasm: BS16 was developed by mass selection for early maturity from ETO Composite; BS2 was formed by crossing ETO Composite with six early inbred lines followed by five generations of intercrossing; BSTL was developed by crossing Lancaster Surecrop with Tuxpeno and backcrossing to Lancaster Surecrop; and Krug Hi I Synthetic 3 was an adapted Corn Belt variety. The relative proportions of exotic germplasm are 100%, 50%, 25%, and 0% for BS16, BS2, BSTL, and Krug Hi I Synthetic 3, respectively. One objective of the four concurrent selection programs is to determine what effect if any the different proportions of exotic germplasm have on gain and variability in successive cycles of selection. Wellhausen (1965) was of the opinion that only small doses (25% or less) of exotic germplasm should be incorporated initially into adapted populations. Preliminary results do not show any striking differences among the four populations (Hallauer 1978), but critical comparisons will not be available until after additional cycles of selection.

Evaluations of materials that included exotic germplasm have been reported for development of lines for use in hybrids (Griffing and Lindstrom 1954; Paterniani 1964; Efron and Everett 1969; Nelson 1972), sources of disease (Kramer and Ullstrup 1959) and insect (Sullivan et al. 1974) resistance, and silage production (Thompson 1968). Nelson (1972) has used exotic germplasm in an applied breeding program developing lines for use in hybrids in the southern part of the United States; he has developed lines that have played a prominent role in the production of hybrids grown in that area. Other maize breeders, public or private, undoubtedly have integrated exotic germplasm into their breeding programs; but the stage of development either does not permit their use in hybrids or the material is used in proprietary hybrids and the extent of their use is unknown. Considerable screening for sources of resistance to diseases and insects occurs in nearly all breeding programs and a significant amount of the material being screened includes some proportion of exotic germplasm. In Table 11.2, populations that include at least some exotic germplasm were one of the largest categories currently undergoing active (25.2%) or inactive (19.5%) selection, and the number undergoing selection for disease (57%) or insect (14%) resistance is substantial (Table 11.3). It seems that considerable effort is being expended on populations that include some exotic germplasm, but in most instances results are in the preliminary or initial stages of selection.

Wellhausen et al. (1952), Brown (1953), Wellhausen (1956, 1965), and Leng et al. (1962) have emphasized the importance of exotic germplasm and thought

U.S. maize breeders were in an excellent position to make use of exotic germplasm; however, a concerted effort has not been given to exploiting the wealth
of germplasm available for use (Brown 1975). Maize breeders in the United
States have been making significant genetic improvements in their hybrids
through use of conventional breeding procedures and adapted germplasm (Russell
1974; Duvick 1977). The status quo has been satisfactory on a short-term basis; incorporation of exotic germplasm into adapted populations or its direct
use would be included in long-range objectives of the breeding program (Fig.
1.3). Exotic germplasm must include useful genes, but they will not be available until they are incorporated with the highly productive adapted germplasm.
It will require time and patience. Immediate payoffs are not to be expected
but long-range payoffs seem likely. Most of the evidence reported from use of
exotic with adapted germplasm has been encouraging. The initiation of selfing
in recently hybridized exotic-adapted germplasm usually has been disastrous;
inbreeding depression is severe and few vigorous lines were obtained. Frequently the material was discarded and the use of exotic germplasm seemed
fruitless. A common error was not recombining the best progenies and initiating another cycle of recombination of the best material. Additional selection
and recombination would permit further integration of linkage blocks and
choice of genes desired for the particular environment.

Choice of exotic germplasm to include has to be considered. Wellhausen
(1965), Brown and Goodman (1977), and others have discussed some of the Mexican and Caribbean races and populations that seem promising. Lack of systematic information on the relative merits of different races in breeding programs prevents making precise recommendations for given situations; this was
emphasized by the authors in Table 11.1. Hence, choice of exotic germplasm
often depended on limited experience in the area of adaptation and limited data available.

One consistent theme is that the most productive races and populations
arose from hybridization of previous races. Older races developed in isolated
regions were brought together by the movement of people and natural hybridization occurred. It seems that a certain fraction of heterosis between the
crossing of two races persisted in future generations to produce more productive races (see Chap. 10). The highly productive U.S. Corn Belt Dents is a
good recent example of this phenomenon. Crosses of races or populations that
are considered promising will require time to allow for adequate recombination. The time required to evolve new races by hybridization will fit long-
term objectives and the payoff may not be obvious to the breeder who initiated
the program. The "hodge-podge" (Brown 1953) or "mess" (Wellhausen 1956, 1965)
of hybridization will require mild selection with time to meld the favorable
factors of parents in the population derived by hybridization.

Because it is believed that maize originated in the Western Hemisphere,
maize grown in other parts of the world is in a sense exotic. Hundreds of
years of natural and artificial selection in all parts of the world where
maize was grown, however, have developed many distinct local races and varieties. Only in the last 30 to 40 years, when it was realized that a great
amount of variability existed within the species *Zea mays*, were organized efforts made to collect, study, preserve, and use potential genetic resources in
all parts of the world (Table 11.1). Consequently, in most parts of the world
attempts have been made to introduce foreign materials to local breeding programs. The fate of the foreign introductions varied from little or no success
to the greatest possible success in replacing local germplasm.

Until World War II most of the European breeding programs developed varieties and hybrids from use of local germplasms. After World War II extensive
use of inbred lines from the United States was made in the production and

growing of hybrids. The most productive hybrids of Europe involve crossing U.S. dent lines with European flint lines. Recently, there has developed a greater interest in the European flint germplasms as sources of inbred lines. Improvement programs in the European germplasms are under way, and in the future U.S. inbred lines will be needed less.

Almost no maize breeding programs existed in Africa until about 20 years ago. Extensive evaluations of exotic races and varieties have been made in the past 15 to 20 years because local varieties were on the average very poor. Introduced varieties that showed a much better performance than local types were used effectively for population improvement. Breeding programs using local and exotic germplasm are currently active in most of Africa.

The majority of races and varieties of maize is found in Central America and South America. Practically all natural genetic variability for adaptations to all latitudes and altitudes, and germplasms for special and general purposes can be obtained in this area. Thus in this area the interchange of exotic material should result in the best success. Local habits and social conditions present some limitations for use of exotic material. In many regions maize is used for human consumption and needs to fit rigid local standards. For instance, the famous Cuzco material, with its eight-rowed ears, very large kernels, and very late maturity, is widely used and not easy to replace. People do not easily accept a different product even if it has proved to be more productive or has some other agronomic advantages. In these instances the use of exotic material to include some genetic variability without changing the plant and ear patterns would be a very long process.

In Argentina substantial progress has been made where most of the maize grown as hybrid. There has always been a policy, however, to produce the dark orange flint material to take advantage of the premium prices in the international market. As a consequence very little use has been made of exotic materials, although U.S. dent materials perform quite well and give higher yields than the local flint hybrids. Recently there has been a tendency to move to softer types of kernels--dents or semidents--in order to obtain greater productivity.

Chile has almost completely adopted U.S. hybrids with great success. Since most of the maize is grown for animal feeding, there was no objection to use of the higher yielding U.S. yellow dent hybrids.

During the Civil War many Americans came to Brazil bringing samples of U.S. yellow dent maize, which were crossed to the local orange flint Cateto and resulted in many types of dent maize. During 1910 to 1915, maize shows similar to those in the United States were organized in some areas of Brazil, and new introductions were made. These materials were used also in the natural formation of local dents. The first hybrids developed were of Cateto germplasm (orange flint kernels) and were not highly productive. Semident hybrids were subsequently obtained with a little increase in performance. Experience showed that local flint and dent germplasms were generally poor with low possibilities for inbred line development, especially the dent material. Only after the 1950s were organized introductions performed in Brazil. Tuxpeno, a race from Mexico (see p. 390), substantially increased productivity in population improvement programs and later was used in the development of inbred lines that resulted in superior commercial hybrids. Significant contributions were also achieved by use of flint germplasms, especially the ones from Colombia (ETO material) and Cuba. Brazil is one of the countries in which all local varieties had poor performance and exotic germplasm gave a substantial improvement. It is estimated that exotic germplasm is responsible for 100% improvement of yield in relation to the local germplasm.

Wellhausen (1978) reported yield data for 10 dent or semident varieties

crossed with each of 10 flint or semiflint varieties. The crosses were tested
at six sites in Latin America. From these trials and others, Wellhausen identified four racial complexes that he considered useful for immediate improvement of maize in the tropics: Tuxpeno and its related Caribbean and U.S.
dents, Cuban Flint, Coastal Tropical Flint, and ETO. Tuxpeno is the only pure
flint and it was considered to be outstanding because of its yield capacity,
exceptional vigor, and resistance to common maize diseases. Tuxpeno also
should be useful in U.S. breeding programs because it is generally considered
to be one of the putative parents of Corn Belt dents (Fig. 11.1). Crosses of
Tuxpeno with Cuban Flint, Coastal Tropical Flint, and ETO were high yielding
and expressed considerable heterosis. Cuban Flint, Coastal Tropical Flint,
and ETO are flint complexes that showed good disease resistance and had good
yield potential. Wellhausen stated that the four racial complexes had been
used extensively in maize breeding programs in the lowland tropics in the past
20 years.

REFERENCES

Anderson, E.   1945.   What is *Zea mays*? *Chron. Bot.* 9:88–92.

Anderson, E., and H. C. Cutler.   1942.   Races of *Zea mays*: I. Their recogni-
    tion and classification. *Ann. Missouri Bot. Gard.* 29:69–89.

Atkinson, A., and M. W. Wilson.   1914.   Corn in Montana: History, character-
    istics, and adaptation. Montana Agric. Exp. Stn. Bull. 107.

Beadle, G. W.   1939.   Teosinte and the origin of maize. *J. Hered.* 30:245–47.

———.   1972.   The mystery of maize. Field Mus. Nat. Hist. Bull. 44:2–11.

———.   1977.   The mystery of maize. *Proc. Annu. Corn Sorghum Res. Conf.* 32:
    1–5.

Bird, R. M., and M. M. Goodman.   1977.   The races of maize: V. Grouping maize
    races on the basis of ear morphology. *Econ. Bot.* 31:471–81.

Brandolini, A. G.   1969.   European races of maize. *Proc. Annu. Corn Sorghum
    Res. Conf.* 24:36–48.

Brandolini, A.   1971.   Preliminary report on south European and Mediterranean
    maize germplasm. *Proc. Fifth Meet. Maize Sorghum Sect*. EUCARPIA. Pp.
    108–16.

Brandolini, A., and G. Avila.   1971.   Effects of Bolivian maize germplasm in
    south European maize breeding. *Proc. Fifth Meet. Maize Sorghum Sect*.
    EUCARPIA. Pp. 117–35.

Brieger, F. G.; J. T. A. Gurgel; E. Paterniani; A. Blumenschein; and M. R. Al-
    leoni.   1958.   Races of maize in Brazil and other eastern South American
    countries. NAS–NRC Publ. 593.

Brown, W. L.   1953.   Maize of the West Indies. *Trop. Agric.* 30:141–70.

———.   1960.   Races of maize in the West Indies. NAS–NRC Publ. 792.

———.   1975.   Broader germplasm base in corn and sorghum. *Proc. Annu. Corn
    Sorghum Res. Conf.* 30:81–89.

Brown, W. L., and E. Anderson.   1947.   The northern flint corns. *Ann. Mis-
    souri Bot. Gard.* 34:1–28.

———.   1948.   The southern dent corns. *Ann. Missouri Bot. Gard.* 35:255–68.

Brown, W. L., and M. M. Goodman.   1977.   Races of maize. In *Corn and Corn Im-
    provement*, G. F. Sprague, ed., pp. 49–88. Am. Soc. Agron., Madison, Wis.

Corn.   1972.   In *Genetic Vulnerability of Major Crops*, pp. 97–118. Natl.
    Acad. Sci., Washington, D.C.

Crow, J. F., and M. Kimura.   1970.   *An Introduction to Population Genetics
    Theory*. Harper & Row, New York.

de Wet, J. M. J., and J. R. Harlan.   1971.   Origin and evolution of teosinte
    (*Zea mexicana* [Schrader] Kunte). *Euphytica* 20:255–65.

————. 1972. Origin of maize: The tripartite hypothesis. *Euphytica* 21: 271-79.

————. 1976. Cytogenetic evidence for the origin of teosinte (*Zea mays* ssp. mexicana). *Euphytica* 25:447-55.

Duvick, D. N. 1975. Using host resistance to manage pathogen populations. *Iowa State J. Res.* 49:505-12.

————. 1977. Genetic rates of gain in hybrid maize yields during the past 40 years. *Maydica* 22:187-96.

Eberhart, S. A. 1971. Regional maize diallels with U.S. and semiexotic varieties. *Crop Sci.* 11:911-14.

Efron, Y., and H. L. Everett. 1969. Evaluation of exotic germplasm for improving corn hybrids in northern United States. *Crop Sci.* 9:44-47.

Galinat, W. C. 1970. The cupule and its role in the evolution of maize. Univ. Massachusetts Agric. Exp. Stn. Bull. 585.

————. 1971. The origin of maize. *Annu. Rev. Genet.* 5:447-78.

————. 1974. The domestication and genetic erosion of maize. *Econ. Bot.* 28:31-37.

————. 1975. The evolutionary emergence of maize. Torrey Bot. Club Bull. 102:313-24.

————. 1977. The origin of corn. In *Corn and Corn Improvement*, G. F. Sprague, ed., pp. 1-47. Am. Soc. Agron., Madison, Wis.

Galinat, W. C., and J. H. Gunnerson. 1963. Spread of eight-rowed maize from the prehistoric Southwest. Bot. Mus. Leafl., Harvard Univ. 20:117-60.

Goodman, M. M. 1965a. The history and origin of maize. North Carolina Agric. Exp. Stn. Tech. Bull. 170.

————. 1965b. Estimates of genetic variance in adapted and exotic populations of maize. *Crop Sci.* 5:87-90.

————. 1967. The races of maize: I. The use of Mahalanobis' generalized distances to measure morphological similarity. *Fitotec. Latinoam.* 4:1-22.

————. 1968. *Corn: Its Origin, Evolution, and Improvement*. Harvard Univ. Press, Cambridge.

Goodman, M. M., and R. M. Bird. 1977. The races of maize. IV. Tentative grouping of 219 Latin American races. *Econ. Bot.* 31:204-21.

Goodman, M. M., and E. Paterniani. 1969. The races of maize. III. Choices of appropriate characters for racial classification. *Econ. Bot.* 23:265-73.

Grant, U. J.; W. H. Hathaway; D. H. Timothy; C. Cassalett D.; and L. M. Roberts. 1963. Races of maize in Venezuela. NAS-NRC Publ. 1136.

Griffing, B., and E. W. Lindstrom. 1954. A study of the combining abilities of corn inbreds having varying proportions of Corn Belt and non-Corn Belt germplasm. *Agron. J.* 46:545-52.

Grobman, A.; W. Salhauana; and R. Sevilla; with P. C. Mangelsdorf. 1961. Races of maize in Peru. NAS-NRC Publ. 915.

Hallauer, A. R. 1978. Potential of exotic germplasm for maize improvement. In *International Maize Symposium*, W. L. Walden, ed., pp. 229-47. McGraw-Hill, New York.

Hallauer, A. R., and J. H. Sears. 1972. Integrating exotic germplasm into Corn Belt breeding programs. *Crop Sci.* 12:203-6.

Hathaway, W. H. 1957. Races of maize in Cuba. NAS-NRC Publ. 453.

Hitchcock, A. S. 1935. Manual of the grasses of the United States. USGPO, Washington, D.C.

Hooton, E. A. 1926. Methods of racial analysis. *Science* 63:75-81.

Iltis, H. H. 1970. The maize mystique: A reappraisal of the origin of corn. Mimeogr. Dep. Bot., Univ. Wisconsin, Madison.

————. 1972. The taxonomy of *Zea mays* (Gramineae). *Phytologia* 23:248–49.

————. 1979. *Zea diploperennis* (Gramineae): A new teosinte from Mexico. *Science* 203:186–88.

Kato, Y., T. A. 1975. Cytological studies of maize (*Zea mays* L.) and teosinte (*Zea mexicana* [Schrader] Kuntze) in relation to their origin and evolution. Massachusetts Agric. Exp. Stn. Res. Bull. 635.

Kramer, H. H., and A. J. Ullstrup. 1959. Preliminary evaluations of exotic maize germplasm. *Agron. J.* 51:687–89.

Kuleshov, N. N. 1933. World's diversity of phenotypes of maize. *J. Am. Soc. Agron.* 25:688–700.

Leng, E.; R. A. Tavcar; and V. Trifunovic. 1962. Maize of southeastern Europe and its potential value in breeding programs elsewhere. *Euphytica* 11:263–72.

Lonnquist, J. H. 1974. Consideration and experiences with recombinations of exotic and Corn Belt maize germplasms. *Proc. Annu. Corn Sorghum Res. Conf.* 29:102–17.

Mangelsdorf, P. C. 1974. *Corn: Its Origin, Evolution, and Improvement.* Harvard Univ. Press, Cambridge.

Mangelsdorf, P. C., and R. G. Reeves. 1939. The origin of Indian corn and its relatives. Texas Agric. Exp. Stn. Bull. 574.

Moll, R. H.; W. S. Salhuayra; and H. F. Robinson. 1962. Heterosis and genetic diversity in variety crosses of maize. *Crop Sci.* 2:197–98.

Moll, R. H.; J. H. Lonnquist; J. V. Fortuno; and E. J. Johnson. 1965. The relationship of heterosis and genetic divergence in maize. *Genetics* 42:139–44.

Nelson, H. G. 1972. The use of exotic germplasm in practical corn breeding programs. *Annu. Corn Sorghum Ind. Res. Conf. Proc.* 27:115–18.

North Central Corn Breeding Research Committee (NCR-2) Report. Mimeogr. Library, Iowa State Univ., Ames.

Omolo, E., and W. A. Russell. 1971. Genetic effects of population size in the reproduction of two heterogeneous maize populations. *Iowa State J. Sci.* 45:499–512.

Paterniani, E. 1962. Evaluation of maize germplasm for the improvement of yield. UN Conf. Appl. Sci. Tech. Benefit Less Developed Areas. Pp. 1–3.

————. 1964. Value of exotic and local inbred lines of corn. *Fitotec. Latinoam.* 1:15–22.

Paterniani, E., and M. M. Goodman. 1977. Races of maize in Brazil and adjacent areas. CIMMYT.

Pavlicic, J. 1971. Contribution to a preliminary classification of European open-pollinated maize varieties. *Proc. Fifth Meet. Maize Sorghum Sect.* EUCARPIA. Pp. 93–107.

Ramirez, R.; D. H. Timothy; E. Diaz B.; and U. J. Grant, with G. E. N. Calle; E. Anderson; and W. L. Brown. 1960. Races of maize in Bolivia. NAS-NRC Publ. 747.

Recommended actions and policies for minimizing the genetic vulnerability of our major crops. 1973. USDA and Natl. Assoc. State Univ. Land-Grant Coll. Spec. Rep.

Roberts, L. M.; U. J. Grant; R. Ramirez E.; W. H. Hathaway; and D. L. Smith, with P. C. Mangelsdorf. 1957. Races of maize in Colombia. NAS-NRC Publ. 510.

Russell, W. A. 1974. Comparative performance for maize hybrids representing different eras of maize breeding. *Proc. Annu. Corn Sorghum Res. Conf.* 29:81–101.

Sanchez-Monge P., E. 1962. Razas de maiz en Espana. Publ. Minist. Agric., Madrid.

Shauman, W. L. 1971. Effect of incorporation of exotic germplasm on the genetic variance components of an adapted, open-pollinated corn variety at two plant population densities. Ph.D. dissertation, Univ. Nebraska, Lincoln.

Sprague, G. F. 1971. Genetic vulnerability in command sorghum. *Proc. Annu. Corn Sorghum Res. Conf.* 26:96–104.

Sturtevant, E. L. 1899. Varieties of corn. USDA Off. Exp. Stn. Bull. 57:1–108.

Sullivan, S. L.; V. E. Gracen; and A. Ortega. 1974. Resistance of exotic maize varieties to the European corn borer *Ostrinia nubilalis* (Hübner). *Environ. Entomol.* 3:718–20.

Thompson, D. L. 1968. Silage yield of exotic corn. *Agron. J.* 60:579–81.

Timothy, D. H.; B. Pena V.; and R. Ramirez E., with W. L. Brown, and E. Anderson. 1961. Races of maize in Chile. NAS-NRC Publ. 847.

Timothy, D. H.; W. H. Hathaway; U. J. Grant; M. Torregroza C.; D. Sarria V.; and D. Varela A. 1963. Races of maize in Ecuador. NAS-NRC Publ. 975.

Troyer, A. F., and W. L. Brown. 1972. Selection for early flowering in corn. *Crop Sci.* 12:301–4.

Wallace, H. A., and E. N. Bressman. 1925. *Corn and Corn Growing.* Wallace Publ., Des Moines, Iowa.

Weatherwax, P. 1955. History and origin of corn. I. Early history of corn and theories as to its origin. In *Corn and Corn Improvement*, G. F. Sprague, ed., pp. 1–16. Academic Press, New York.

Wellhausen, E. J. 1956. Improving American corn with exotic germplasm. *Proc. Annu. Hybrid Corn Ind. Res. Conf.* 11:85–96.

———. 1965. Exotic germplasm for improvement of Corn Belt maize. *Proc. Annu. Hybrid Corn Res. Conf.* 20:31–45.

———. 1978. Recent developments in maize breeding in the tropics. In *Maize Breeding and Genetics*, D. B. Walden, ed., pp. 59–84. Wiley, New York.

Wellhausen, E. J.; A. Fuentes O.; and A. Hernandez C., with P. C. Mangelsdorf. 1957. Races of maize in Central America. NAS-NRC Publ. 511.

Wellhausen, E. J.; L. M. Roberts; and E. Hernandez X., with P. C. Mangelsdorf. 1952. *Races of Maize in Mexico.* Bussey Inst. Harvard Univ. Press, Cambridge.

Zuber, M. S. 1975. Corn germplasm base in the United States: Is it narrowing, widening, or static? *Proc. Annu. Corn Sorghum Res. Conf.* 30:277–86.

12

Breeding Plans

Recurrent selection methods can contribute to meeting the goals of continuous genetic improvement--they will not replace other breeding methods, but should be integrated with them (Hallauer 1981, 1985).

Cultivated maize arose from its wild ancestors as a result of mutation, natural and artificial selection, and adaptation to the environment. Plant breeding is the art and science of improvement of crop plants to meet the needs of people; processes of evolution are involved, but they are directed by people to hasten the attainment of their projected goals. Plant breeding is a broad discipline that requires some level of competence in genetics, botany, statistics, pathology, entomology, and an appreciation of the forces of the environment on plant growth and development. The relative importance of the art and science of plant breeding is ill-defined, but plant breeding has played an important role in the development of crop plants since the hunting and gathering stage of humans because of their dependence on plants for survival. Although people have always attempted to direct the evolution of plants to meet their needs, not until the genetic laws of Mendel and the principles of randomization and replication were understood and developed did plant breeding as a science become prominent.

The principles of maize breeding as we understand them today have been developed during the twentieth century, beginning with the publications of Shull (1908, 1909, 1910). The change from use of open-pollinated varieties to use of double-cross hybrids was a significant advancement in developing maize plants with improved standability and greater yields. The impact of inbreeding and hybridization techniques on yields is reflected in Fig. 12.1. The national average for U.S. maize yields was consistently low until the latter part of the 1930s. Yields did not fluctuate greatly and average U.S. yields attained 18 q/ha in only two years before 1935. Small but consistent improvements have been made since the introduction of double-cross hybrids (Fig. 12.1) in the 1930s to meet the changes and challenges of the improvements of maize husbandry and mechanization. Yield increases shown in Fig. 12.1 reflect the acceptance of the use of hybrid maize shown in Fig. 1.2.

Evidence that genetic improvements were made in maize hybrids was reported by Duvick (1977) and Russell (1974, 1986) in the U.S. Corn Belt. Both authors included hybrids produced and grown in the decades since the introduction of hybrids in replicated yield trials. Each report provides evidence of significant genetic improvement in yield. Duvick included two sets of hybrids in his trials: (1) hybrids grown in the five decades, including double crosses grown in the 1930s, 1940s, and 1950s and single crosses grown in the 1960s and 1970s; and (2) single crosses produced from inbred lines included in the hybrids for the five decades. Russell included hybrids (double and single)

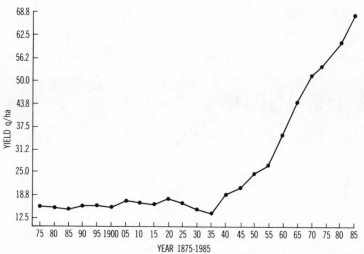

Fig. 12.1. National average maize yields in the United States, 1875-1985.

that were used for the five decades. Both studies included three plant densities, which were intended to simulate plant population densities used for
growing early hybrids, plant densities presently used, and plant densities
greater than generally used in the U.S. Corn Belt at the present. All comparisons showed that genetic improvement had been made in the development of hybrids to meet the conditions under which they were grown. Russell calculated
that gain in hybrid performance attributable to breeding was 63.2% and Duvick
determined that proportion of total gain due to breeding was 57% and 60% for
his two sets of hybrids. Gains were in good agreement for the three sets of
experiments. Russell also found that if we were still growing maize at plant
population densities commonly used for the first double crosses, we would realize little yield improvement over the first experiment station double crosses. Hence genetic improvement depended also on planting practices.

Response of hybrids for yield improvement during the five decades depended on correlated responses for improved roots and stalks, improved disease and
insect resistance, and resistance to barrenness at higher plant population
densities. Modern maize production requires hybrids that have acceptable
standards of root and stalks and ear attachment to permit retrieval by mechanical harvesters. Grain yield itself is the result of the total genotype of
the hybrid in modern maize growing; i.e., harvestable yield is the standard of
measure of a hybrid rather than the potential genotypic yield (which can be
measured by hand harvesting to collect all ears whether they are on standing
or broken plants). If there are indeed yield genes, they must be incorporated
within genotypes that include genes contributing to good roots and stalks,
strong ear attachment, acceptable maturity, and good health and vigor. Additional requirements for modern hybrids reduce selection intensity for the specific trait (e.g., yield); hence inclusion of additional traits broadens our
definition of yield. In spite of restrictions imposed for developing superior
yielding hybrids, significant progress has been made (Fig. 12.2). Plant
breeding is dictated by economic uses of crop species, and selection techniques have to be adapted to meet demands.

Although it seems slow, continuous genetic improvements have been made in
developing hybrids; the concern is whether the same progress can be achieved
in the future. Progress in the development of mechanized equipment has improved the timeliness of planting and harvesting; extensive applications of

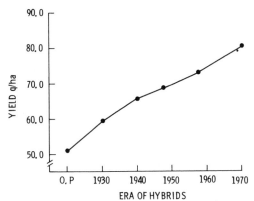

Fig. 12.2.   Predicted yields of hybrids at ten-year intervals based on actual
             yields of the two best hybrids in each double-cross group (Russell
             1974).

fertilizer (particularly nitrogen) have been common since 1950; plant population densities have been slowly increasing; development and use of pesticides
have increased dramatically to provide a better environment for maize growth
and development; and management abilities of farm operators have developed to
a high level of sophistication. All these developments and refinements in
concert with genetic improvement have contributed to higher yields. Many of
the factors, however, may have attained their plateaus so that dramatic changes may not be forthcoming. For instance, perhaps less fertilizer will be used
in the future than in the past or at the present. Development and use of pesticides may be restricted because of the concerns of environmental quality.
Potential limitations on further cultural and management changes and improvements emphasize that continued genetic improvements are imperative.
     It seems that genetic improvement of maize can be maintained by expansion
and refinement of breeding techniques, integration of cyclical selection programs with applied breeding programs, and greater usage of germplasm available. The information summarized in Chaps. 5, 7, 8, 9, 10, and 11 represents
about 60 years of research conducted since the suggestion of the pure line hybrid concept by Shull (1908, 1909, 1910). The precepts of Shull with some minor modifications have been essentially used in most maize breeding programs.
Population improvement schemes and maize germplasm available have not been
used extensively. Dudley and Moll (1969) and Moll and Stuber (1974) have discussed interpretation and uses of genetic variances for populations and relevance of results. Integration of these facets with those currently used is
required if we are to continue to have genetic improvement. None will result
in any spectacular leaps forward, but integration of all phases will ensure a
slow, steady rate of genetic improvement (Hallauer 1985).

12.1 DEVELOPING BREEDING PLANS
     Maize breeding programs include three important phases to meet the
short-, intermediate-, and long-term objectives: (1) choice of germplasm, (2)
cyclical improvement of germplasm chosen, and (3) development of lines for use
as parent stocks in production of single-cross hybrids (for areas requiring
hybrids) and development of improved varieties, synthetics, and composites
(for areas where hybrids are not currently practical). All phases are equally
important, but resources and time expended for each may vary considerably.

12.1.1 Germplasm

Germplasm to include in the breeding program may involve choosing from current sources of germplasm and information available or collection, development, and evaluation of germplasm before a choice is made. In the first instance the breeder depends on the information and opinions of others. This approach may be adequate if the evaluations were made in an environment similar to that in which the germplasm is to be used. If the information was obtained from a very different environment, the germplasm may not be adapted to the environment desired. Collection, development, and evaluation of germplasm requires several seasons, but the breeder has the opportunity to directly observe the response of the germplasm to the particular environment. Several collections may be obtained from germplasm banks and grown to determine which ones have the best plant and ear development and adaptability. Similar collections may be combined to form a composite or a few may be sufficiently distinct from one another and have sufficient genetic variation for direct use. Either approach may be used, but generally a combination is used. Some evaluation trials may be conducted and used in conjunction with previously available information that in many instances may be experiences of other breeders.

Choice of germplasm is a critical decision that requires considerable thought. Hasty decisions either to eliminate or to decrease number of growing seasons required may in the long run increase the number of growing seasons required to develop usable materials. Germplasm chosen forms the basic material of the breeding program. Germplasm that has a low frequency of genes for the traits desired may require either several additional growing seasons or greater samplings to develop desirable genotypes or populations. Unfortunately, information available to the breeder for the large collections of germplasm is limited--this emphasizes the importance of considering all the information available before choosing. In many instances the selected germplasm will be the basis of the breeding program for the lifetime of the breeder. Choice of germplasm will determine maximum potential improvement that can be attained via breeding; the breeding system used will determine how much of that maximum potential can be realized.

12.1.2 Recurrent Selection

After choice of germplasm, the next procedure is to use some type of cyclical selection program to improve the general level of performance in maturity, roots, stalks, resistance to pests, or yield. The breeder can proceed immediately to extract materials that are planned for use by the farmer; this is a natural, practical response. The important aspect is that superior material selected for the breeding nursery also should be recombined to resynthesize the basic population. If considerable thought, time, and expense have been devoted to choice of germplasm, the logical process is to continue using the germplasm in the breeding program. Presumably the superior progenies isolated from the basic population included the genetic factors that meet standards imposed by the breeder. Probably none will meet the desired standards of all traits, but at least the selected progenies are better than a random sample. Recombination of superior progenies will increase frequency of favorable alleles, which in future cycles will increase the opportunities of isolating a greater frequency of progenies that meet desired standards for most traits.

A simple example of the possibilities of recycling is illustrated in Fig. 12.3. In this idealized example for finite populations, expected variability of single crosses of original and improved populations is the same but the mean has increased, and the best possible hybrid extracted from the improved population is superior to that of the original population. The example in Fig. 12.3 usually is not realized after only one recycling. If the recycled

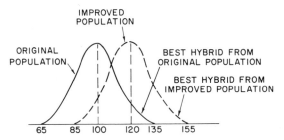

Fig. 12.3. Expected distribution of single crosses from the original and im-
            proved populations (Eberhart 1970).

population represented ten cycles of selection and recombination the improvement for yield would be striking. Instead of intermating superior progenies,
the breeder may be tempted to resample the original population. If sampling
was adequate, resampling would be expected to give progenies that have the
same relative distributions. Furthermore, changes in relative distribution
and mean of progenies extracted after only one to three cycles of selection
may not be detectable. It is essential that recycling be continued on a regular basis over a period of time.

     Results of recurrent selection programs conducted in maize are summarized
in Chap. 7. Effectiveness of recurrent selection depends on genetic variability and gene frequencies within original populations and heritabilities of
traits under selection. For example, Penny et al. (1967) reported that three
cycles of recurrent selection for first brood European corn borer resistance
were effective in developing populations that had an acceptable level of resistance (Fig. 12.4). Based on a rating scale of 1 (resistant) to 9 (susceptible), the mean of 484 S_1 progenies for the C0 population of five varieties
was a high intermediate (5.5), whereas the mean of the 484 S_1 progenies for
the five C3 populations was within the resistant range (2.5). Distributions

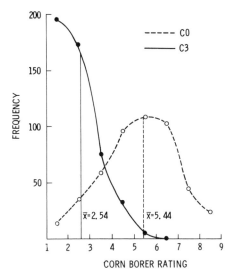

Fig. 12.4. Distributions of S_1 progenies for the original population (C0) and
            after three cycles of recurrent selection (C3) for first brood
            European corn borer resistance (Penny et al. 1967).

of S_1 progenies from the original and last cycles of selection also showed that three cycles of recurrent selection seemed adequate for establishing a high level of resistance. Frequency of resistant S_1 progenies for the C3 cycle was skewed to the resistance classes with none in the 7, 8, and 9 susceptible classes.

Similar recurrent selection procedures were used by Jinahyon and Russell (1969a) for developing resistance to *Diplodia zea* in the open-pollinated variety Lancaster Surecrop, which has a high level of susceptibility to stalk rot organisms. One of the methods of evaluating response to selection was use of 100 S_1 lines from the CO and C3 populations. Distributions and means of these lines are shown in Fig. 12.5; ratings were made on a scale of 0.5 (resistant) to 5.0 (susceptible). Average stalk rot ratings of the 100 S_1 lines were 4.1 for the CO population and 2.4 for the C3 population. An important feature of recurrent selection for stalk rot resistance was the change in distribution of S_1 progenies. Very little overlap occurred between the two distributions; the C3 population had very few S_1 progenies exceeding a rating of 3.0 and very few CO S_1 progenies had a rating less than 3.0.

In both instances, three cycles of recurrent selection based on S_1 progeny evaluation were effective in changing the mean and distribution of the populations. Heritability of first brood European corn borer and stalk rot resistance is greater than heritability for yield because techniques were developed for artificial infestation and infection of these two important pests. Development of artificial means for minimizing the frequency of escapes improves the heritability and, consequently, the effectiveness of selection--response to selection would be expected to be greater than for traits that depend on natural infestation or infection and the effects of the environment on the establishment and development of the pest. If the techniques are available, it seems that recurrent selection is an effective method for improving mean performance of a population and chances of obtaining superior progenies from improved populations. Results for first brood European corn borer and *Diplodia zea* stalk rot agree with the hypothetical example illustrated in Fig. 12.3. Three or four cycles of recurrent selection seem sufficient in most instances in developing populations that have an acceptable level of resistance (Hallauer 1973a).

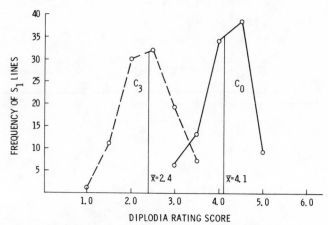

Fig. 12.5. Distributions of S_1 progenies for the original population (CO) and after three cycles of recurrent selection (C3) for *Diplodia zea* resistance (Jinahyon and Russell 1969a).

In contrast, Penny and Eberhart (1971) reported on 20 years of reciprocal recurrent selection (RRS) for yield in two synthetic populations, BSSS and BSCB1. Twenty years were required to complete four cycles of recurrent selection for yield whereas only 4 to 8 years would be required for first brood European corn borer and stalk rot resistance. Four cycles of RRS improved the population cross (1.18 ± 0.24 q/ha) and BSSS (1.38 ± 0.62 q/ha), but no improvement was realized in BSCB1 (-0.64 ± 0.62 q/ha). Although results were not striking and may seem discouraging, they were not totally unexpected because of limited opportunities (four in this instance) for selection and recombination. Russell and Eberhart (1975) tested the population cross and crosses of 5 selections from BSSS and BSCB1 after five cycles of RRS. The 5 selections from BSSS and BSCB1 were in the S_2 generation and were included among the 10 selections used to synthesize the sixth cycle in each population. Average yield of S_2 line crosses exceeded the population cross in all instances, and the best S_2 line cross exceeded the population cross by 35%. Two of the S_2 line crosses yielded significantly greater than B37 × Oh43, a single cross that had been used in commercially grown hybrids.

Moll et al. (1977) reported that six cycles of RRS improved chances of obtaining superior single crosses from the two source populations, Jarvis and Indian Chief. They examined frequency distributions of single crosses derived from unselected lines developed from original and improved populations, which had been improved by RRS (Fig. 12.6). Average yield of single crosses from the improved populations was 12.5% greater than those from the original populations. RRS improved the means of the hybrids but the distributions of the crosses were similar, which indicates that chances of obtaining a superior hybrid are greater in the improved populations than in the original populations. The 10 best single crosses from the selected populations also averaged 8.6% greater yield than the 10 best single crosses from the unimproved populations.

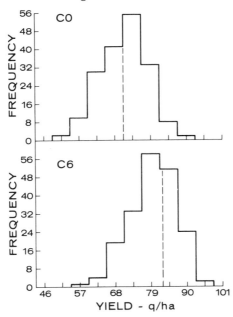

Fig. 12.6. Frequency distributions of the single crosses produced from unselected lines of the original (C0) and after six cycles of reciprocal recurrent selection (Moll et al. 1977). Dotted line indicates the average yield of all hybrids for C0 and C6 populations.

Hallauer (1984) reported on effectiveness of reciprocal full-sib selection for improving chances of obtaining superior crosses from improved populations. In Fig. 12.7 the distributions of full-sib progenies (S_0 x S_0 crosses) from the original populations (C0) and after seven cycles of reciprocal full-sib selection (C7) show the same trends reported by Moll et al. (1977). Relative to the mean of the six check hybrids included in the yield trials, yield distributions of full-sib progenies were transposed after seven cycles of selection. The mean of the C0 full-sibs was more than three standard errors below mean of the six check hybrids, whereas mean of C7 full-sibs was nearly three standard errors above the mean of the six check hybrids. Only 2 (1.4%) of the C0 full-sibs exceeded mean of checks, but only three (2.1%) of C7 full-sibs were less than the mean of the checks.

Available evidence suggests that recurrent selection procedures increase opportunities for obtaining superior progenies and crosses. In most instances only a limited number of cycles of recurrent selection have been completed, but the results are encouraging. Reported results tend to conform to the simplistic model illustrated in Fig. 12.3. The primary objective of recurrent selection procedures is to gradually increase frequency of desirable alleles for breeders' primary objectives. Constant selection pressure and recombination of progenies that possess genes meeting the standards of desirability established by breeders will develop improved breeding populations that will enhance chances of isolating superior lines and hybrids. If the maize program has not developed to the point of use of lines and hybrids, recurrent selection procedures can develop cultivars to be used directly by farmers. The relative success of recurrent selection depends on the complexity of the trait under selection, experimental techniques available for screening progenies, and effects of the environment. Comparisons of results show that significant improvements can be expected for most traits but the rate of improvement is greater for some traits than for others.

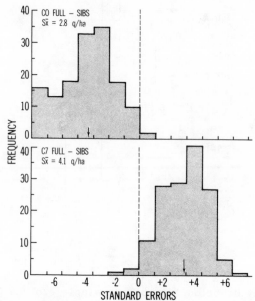

Fig. 12.7. Frequency distributions for yield of full-sib progenies for the C0 and C7 cycles of reciprocal full-sib selection relative to mean of six check hybrids (dotted vertical line). Arrows indicate means of full-sib progenies.

Yield is usually the most important economic trait considered in maize breeding programs. Use of recurrent selection techniques for yield improvement has not been as impressive and consistent as for traits that have a greater heritability and for which techniques are available to minimize the effects of the environment. Yield is a specific measurable trait that is a composite of the plant genotype in response to the environment. Good vigor and health in a high yield environment result in a high yield, but this situation is usually the exception. Good vigor and health, or freedom from maize pests, usually have greater heritabilities than yield. It seems that our breeding populations can be improved for these traits to the point that economic losses are not serious in a relatively short time, perhaps four to eight years. Then emphasis can be given to selection for improved yield. This suggests a two-stage selection program, but as described later, the selection of vigor and health traits can be combined with recurrent selection for yield improvement. Combining selection of as many traits as possible into one program will save time because the two-stage, or tandem, selection increases the time required.

Success from combination selection depends on level of genetic correlation among traits included for selection (Chap. 7). If genetic correlations are low, simultaneous selection for different traits will cause no difficulties. Low genetic correlations are an advantage to the breeder who wants to combine good vigor and health traits with improved yield. Strong positive correlations between desirable traits are of greater advantage. Experience and data seem to indicate that many of the vigor and health traits do not have a strong genetic correlation with grain yield. Jinahyon and Russell (1969b), for example, found that the selection imposed for improved stalk rot resistance (Fig. 12.5) did not reduce yield. The associated changes were greater plant vigor, later maturity, better disease resistance, greater stalk strength, and greater yields in hybrids. Selection for greater yield and vigor in temperate regions tends to develop materials having later maturity. Unless one emphasizes selection for early maturity with selection for greater yields, the populations tend to become later with associated changes of higher ear placement and taller plants.

Hallauer (1978), from use of reciprocal full-sib selection for harvestable yield in BS10 and BS11, found that grain yields were significantly increased in the populations themselves and in their crosses after three cycles of selection for yield. Associated changes showed that stalk lodging was significantly reduced, grain moisture was reduced, and no change occurred in root lodging. Selection pressure also was effective for other traits in desired directions. In reciprocal full-sib selection, however, selection was based on selfed progenies and full-sib crosses.

Selection pressure for only one trait, such as those traits reported by Penny et al. (1967) and Jinahyon and Russell (1969a), probably will be more effective than in combination with other traits if we consider the time interval for the one trait. Three cycles of recurrent selection for first brood European corn borer and stalk rot resistance developed populations having acceptable levels of resistance. If we combine selection for these traits with selection for yield, some trade-offs usually are made in the final selections for recombination. Instead of three cycles, it may require three to six cycles of selection to attain a comparable level of resistance. Progress will be made, but at a slower rate because of the compromises made in the selection process.

Selection for plant vigor and health traits often can be made before selection for yield. In some types of recurrent selection schemes it is convenient to include selection for other traits before conducting expensive yield trials. Selection occurs at different stages of plant and progeny development and undesirable plants and progenies can be discarded before they are considered for yield evaluation. Strong selection pressure can be imposed; if the genetic correlations between traits are low, effective selection can be made in conjunction with yield.

Because most breeding programs are concerned with simultaneous improvement of several traits, there has developed recently a greater interest in use of the selection index proposed by Smith (1936). The selection index has been shown to be the most efficient method for maximum aggregate genetic progress, provided that (1) reliable estimates of genotypic and phenotypic variances and covariances are available and (2) appropriate economic weights of each trait can be determined. Williams (1962) suggested the base index in which the traits are weighted only by their economic values; economic values, however, are not easily determined for many traits. Pesek and Baker (1969) recognized the difficulty of assigning relative economic weights and proposed a modified selection index using desired gains of the traits. Suwantaradon et al. (1975) compared the use of these three selection index methods (conventional, base, and modified) in recurrent selection programs for the simultaneous improvement of several traits. From their comparisons of S_1 testing, they recommended use of the modified selection index when relative economic values of traits are difficult to determine.

The main prerequisites for use of a selection index in improvement programs are reliability and simplicity. Subandi et al. (1973) examined gains expected from selection based on five selection indexes that included three agronomic traits (grain yield, stalk lodging, and dropped ears). The objective of their study was to develop a simple and useful selection index to determine machine harvestable yield. They found that use of selection indexes was more efficient in increasing machine harvestable yield than selection based only on yield. Although selection indexes have not had an important role in maize improvement, it seems that they will be studied more in the future, particularly when simultaneous selection for several traits is considered. Simultaneous selection for several traits is necessary if recurrent selection is used. Index selection needs to be examined and applied for increasing efficiency of selection. In most systems of recurrent selection, selection for other agronomic traits can be made to enhance the relative future usefulness of improved breeding populations.

Methods of recurrent selection include selection either within one or more populations (intrapopulation) or between two populations (interpopulation). The choice of which to use depends on traits under selection, objectives of selection, and capabilities of the breeding program. Intrapopulation selection is more appropriate in some instances, e.g., selection to improve stalk quality or adaptability of a particular population. If the primary objective is to improve the population cross, interpopulation selection would be more appropriate. Intrapopulation selection would emphasize selection for additive genetic effects, whereas interpopulation selection would enhance selection for nonadditive as well as additive genetic effects. Data by Moll and Stuber (1971) and Moll et al. (1977) show that interpopulation selection increases heterosis of the population crosses without changing the heterosis of crosses of populations subjected to intrapopulation selection. Intrapopulation can be used to improve two populations that express heterosis in their cross for some specific traits (e.g., stalk quality) without changing the heterosis expressed for yield, assuming no correlated response. Some intrapopulation methods of improvement are simpler to conduct than interpopulation methods; advantages and disadvantages must be considered relative to the primary objectives. Except for reciprocal full-sib selection, progress depends

on the increase in frequency of alleles with favorable additive effects for
both intra- and interpopulation recurrent selection.

12.1.3 Integrating Recurrent Selection with Applied Breeding

The third important phase of increasing effectiveness and efficiency of
maize breeding is to integrate recurrent selection with applied breeding programs. There has been a tendency in the past to consider recurrent selection
schemes as basic research studies that have little or no relevance to applied
breeding programs. This conflict was equally valid for those conducting research on recurrent selection and those conducting applied breeding improvement programs for varieties, lines, and hybrids. The first recurrent selection programs were initiated to obtain information on response of populations
to different selection schemes and to determine how the observed response was
related to possible types of gene action. Questions were asked concerning the
relative importance of overdominance in heterosis, adequate genetic variation
in populations for effective selection, appropriate choice of testers to use
for selection, and relative importance of general and specific combining ability. Answers were needed to assist in development and use of effective breeding procedures. Summaries of these studies suggested that adequate additive
genetic variance was present in maize populations to expect progress from selection (Chap. 5) and that response from selection was realized in most instances regardless of selection scheme used and trait under selection (Chap.
7). Summaries by Sprague and Eberhart (1977) also showed that response to selection for yield improvement was similar for the different intra- and interpopulation recurrent selection schemes.

An important feature of all cyclical selection schemes is recombination
of superior progenies. Progenies selected for recombination are determined
either from evaluation of progenies themselves, or from testcrosses in which
either remnant seed is recombined or (more commonly) the selfed seed of the
plants used in the testcrosses is recombined. The basis for selection is, in
all instances except for mass selection, yield of the progenies themselves or
their testcrosses. Early generation testing is conducted in all instances.
Because testing is limited in early generations, the relative ranking is not
exact but at least the poorest progenies are not consistently included. (Exact rankings are nullified because of limited testing.) Because progress from
selection usually is realized, limitations of early testing in a few environments is overcome to some extent by recombination and additional testing and
selection in subsequent cycles of selection. Most evidence indicates that
frequency of superior progenies is greater in advanced cycles of selection,
which is encouraging to the applied maize breeder.

Each recurrent selection scheme is discussed in reference to how it can
be used in conjunction with an applied breeding program. Recurrent selection
should not be considered as a separate phase of applied breeding programs.
Only when recurrent selection is considered an integral part of applied breeding programs will its real benefits be realized. Additional information can
be obtained from recurrent selection programs, and they should be considered
as one of the breeding tools available. If early generation yield evaluations
are considered a valid basis of selecting superior progenies for recombination
to form the next cycle of selection, superior progenies also should be included in the applied breeding program. Number of progenies extracted from recurrent selection programs will not be large for each selection cycle, but they
will not be random selections. Additional progenies that were borderline cases and not included for recombination may be included. In subsequent cycles
material will be continuously and regularly generated for the breeding nursery
for additional selection and testing. Greatest success will be obtained when

the frequency of favorable alleles is fairly high, and the frequency of desirable genotypes will be discouraging in the early cycles of selection.

To be most useful, recurrent selection must be an integral part of the
breeding program and not a minor part that is conducted only when it is convenient in certain seasons. All breeders have to plan their programs because
of the seasonal nature of plant breeding. Recurrent selection has to be included as one phase of the breeding program that is conducted on a continuous
and regular basis. Recurrent selection should be considered as applied breeding, not primarily as a basic research program to determine how realized response compares with predicted response.

12.2 INTRAPOPULATION IMPROVEMENT
12.2.1 Mass Selection

Mass selection is not as amenable as other recurrent selection schemes
for generating progenies in each cycle of selection. Mass selection is useful
for traits that have a relatively high heritability, e.g., maturity, adaptation, prolificacy. Because selection is based on individual plants that were
sib-mated by open pollination, the precision is not as great as for progenies
evaluated in replicated trials. Controlled pollinations can be made, but this
would reduce the advantages of mass selection as a relatively simple and inex-
pensive method of recurrent selection. Harris et al. (1972) have shown that
the frequency of superior testcrosses was greater in an advanced generation of
mass selection than in the original, unselected population. But it required
the development of progenies by controlled pollinations to evaluate mass selection. Ordinarily, half-sib seed of each ear in mass selection is not grown
in the breeding nursery because no inbreeding has occurred, and the large numbers (e.g., 100 sib-units, each with three plants) from each cycle may preclude it. Efficiency of mass selection is improved if tassels of undesirable
plants are removed before pollination. Parental control is doubled (Table
6.12), which increases the efficiency of mass selection. Gardner (1977,
1978) has summarized the response to mass selection for yield improvement in
Hays Golden.

12.2.2 Modified Ear-to-Row Selection

Modified ear-to-row selection is based on among and within half-sib family selection. Because the selection unit is a half-sib family, its theoretical aspects are discussed in Chaps. 6 and 7 under half-sib family selection.
Here these methods are discussed under the usual nomenclature because of their
peculiarities in methodology. Limited data of modified ear-to-row selection
(Lonnquist 1964) indicate this is an effective method of recurrent selection
(Paterniani 1967a, 1967b; Webel and Lonnquist 1967). Modified ear-to-row has
an advantage over mass selection and the original form of ear-to-row selection
(Hopkins 1899) because more than one environment is used to evaluate half-sib
families. The important modification of ear-to-row selection was to include
additional environments. Inclusion of more than one environment permits estimation of the genotype-environment interaction term and combining data across
environments to determine superior families for recombination. Bias due to
environmental forces is removed to a greater extent than for mass selection,
but greater resources are needed than for mass selection.

Modified ear-to-row includes growing a single replication of families in
two or more environments. One replication is grown in isolation to facilitate
recombination. A sample of ears is taken from the population chosen for improvement. If the half-sib families are to be evaluated in three environments, a single replication of each half-sib family is included in each envi-

ronment. Separate randomizations are conducted for each environment. The one
replication grown in isolation provides data on each half-sib family in addition to recombination. In addition to the half-sib families, the isolation
planting includes a bulk of the families that pollinate the half-sib families.
The half-sib families are detasseled and cross-pollination provides seed for
each of the progenies. If selection is only on the basis of the half-sib family means for the three environments, the component of variation among family
means is $(1/8)\sigma_A^2$ (Empig et al. 1972). If selections also are made among
plants within progeny rows, expected progress includes an additional component
of $(3/8)\sigma_A^2$. Each plant within the progeny rows, for example, can be infected
with stalk rot organisms and infested with first brood European corn borers.
Data from all environments are used to choose superior families, and individual plant selections are made within selected half-sib families.

Compton and Comstock (1976) offered a further modification using two seasons rather than one for conducting modified ear-to-row selection. All environments in season 1 are used to identify superior families with no isolation
planting for recombination. In season 2 only selected families are planted
ear-to-row, detasseled, and cross-pollinated with males consisting of a bulk
of previously selected half-sib families. Parental control is greater because
the selected families are mated with only males formed from a bulk of the selected families; in modified ear-to-row selection conducted in only one season
the selected families were mated with males of all families, selected and unselected. Hence, in the prediction equation we have $(1/4)\sigma_A^2$ for among-family
and $(3/8)\sigma_A^2$ for within-family selection. Compton and Comstock concluded that
in temperate zones with only one growing season the two-season scheme would
have about 75% as much gain per year as the one-season scheme of modified ear-
to-row selection.

Modified ear-to-row selection is an intrapopulation scheme that should be
effective in maize populations having adequate genetic variability. It may be
used initially to improve a population or after mass selection has been practiced for highly heritable traits. Because of the ever present environmental
effects, modified ear-to-row would seem more appropriate for improvement of
such traits as yield. Because modified ear-to-row is for intrapopulation improvement, Lonnquist (1964) suggested that two populations exhibiting heterosis in their cross should be submitted to modified ear-to-row selection. The
natural procedure would be to use some form of interpopulation recurrent selection after response to modified ear-to-row selection seems to be diminishing.

Modified ear-to-row selection does not evolve progenies that are inbred
to any extent. Consequently, selected progenies may not be amenable for the
applied breeding nursery. But modified ear-to-row is an improvement over mass
selection for identifying superior progenies. With the two-season scheme, a
smaller number of selected families is included in the isolated recombination
planting; selected plants within selected families would be logical candidates
for the breeding nursery. Otherwise, lines could be extracted from the populations improved by modified ear-to-row selection, which is similar to mass
selection.

12.2.3 S_1 Progeny Recurrent Selection

The S_1 progeny recurrent selection has been used for improving several
traits. Response to S_1 progeny recurrent selection usually has been positive
(Chap. 7; Figs. 12.4, 12.5). It lends itself to evaluation of most traits on

a progeny basis. The coefficient for σ_A^2 is 1 (Table 6.12) for a restricted genetic model (Chap. 2) and an estimate of σ_A^2 for the population under selection is determined from the component of variance for S_1 progenies. Usually supplies of S_1 seed produced on S_0 plants are sufficient to permit replicated trials in different environments. Selection among individual S_0 plants can be practiced as a basis for self-pollination. Because the heritability on an individual plant basis is usually lower than on a progeny basis in replicated trials (e.g., Table 7.22), effectiveness of selection will not be as good as estimates of heritabilities based on replicated progeny trials. Higher heritabilities of replicated progeny trials were the reason Penny et al. (1967) and Jinahyon and Russell (1969a) used S_1 progeny evaluation rather than S_0 plants for selecting individuals having resistance to first brood European corn borer and stalk rot organisms.

Although S_1 progeny evaluation lends itself to improvement of most traits of maize, it has not been used as extensively as the half-sib or testcross method for improvement of yield. Types of genetic effects affecting yield expression have caused some hesitation in the use of S_1 progeny recurrent selection for yield improvement. Selection based on inbred progenies (S_1, S_2, etc.) is theoretically more effective for changing frequencies of genes having additive effects than are testcross methods of selection (Table 6.12). Testcross methods would be expected to capitalize on those genes that have overdominant effects for the specific tester used.

Selection studies comparing S_1 and testcross progeny evaluation are not in complete agreement, but both methods have significantly improved populations under selection. Lonnquist and Lindsey (1964) and Horner et al. (1973) found that S_1 and S_2 selection were less effective than expected in comparison with testcross selection. Genter and Alexander (1962), Burton et al. (1971), and Genter (1973) found that both methods of selection (S_1 and testcross) improved populations for yield of S_1 lines, but improvement was greatest with S_1 progeny selection. In nearly all instances, inbreeding depression was less for S_1 lines extracted from populations improved by S_1 progeny selection than for S_1 progenies extracted from populations improved by testcross selection. The concerns are how the combining abilities of the populations are affected by the two methods of selection and particularly the possible implications of type of gene action. Present evidence is not conclusive, but it seems that the testcross selection method favors selection of dominant favorable alleles. Burton et al. (1971) found evidence that S_1 and testcross methods of selection either select for different genes and the difference in their frequencies in selected populations was great enough for the heterosis of the population crosses to increase or genetic drift was great enough to cause differences in gene frequency. Because S_1 and testcross selection emphasize selection for different alleles, differences in gene frequency are to be expected.

Combining ability is a concern, but S_1 progenies selected for recombination to form the next cycle of selection can be included in the breeding nursery for further inbreeding and initial testcross evaluations. Jinahyon and Russell (1969b) found that S_1 progeny selection for stalk rot resistance did not change combining ability. It increases average yield level of S_1 progenies, which is desirable for developing lines for use in hybrids. Selected S_1 entries chosen for recombination and included in the topcross nursery to produce testcross seed will be a selected sample of S_1 lines based on S_1 line per se evaluation. All S_1 lines may not be included in testcross trials because some may be discarded in the breeding nursery. Or before including selected S_1 entries in the topcross nursery they may be included only in the breeding nursery and not used in producing testcross seed until a later generation of inbreeding, e.g., S_3 or S_4. Number of selections is usually not great (20 to 40). Production of testcross seed in a topcross nursery would not require much additional space and separate seasons are not used for the two phases, thus saving time in collecting combining ability information on S_1 lines selected for recombination. Preferences on generation of inbreeding to testcross lines will influence when selected lines are included in the testcross nursery, i.e., S_1 or S_2 vs. S_4 or S_5 (Chap. 8).

Number of years needed to complete each cycle of S_1 progeny recurrent selection depends on number of growing seasons available and trait under selection (Table 12.1). If selections can be made before pollinations (e.g., first brood European corn borer), recombination of selected S_1 progenies can be done in the same season the evaluation trials are conducted. Recombination in the same season will require considerable walking and carrying of pollen to make the necessary crosses among the selected S_1 progenies, but one cycle of selection can be completed each year with the use of off-season nurseries. If it is not desirable to make the crosses the same season, the time required for

Table 12.1. Description of the use of different seasons for conducting S_1 progeny recurrent selection in temperate zones.

Season	Population improvement program	Breeding program
1 (Winter)	Self-pollinate: In the population under selection (C0), self-pollinate 300 to 600 S_0 plants. Practice selection among the S_0 plants at pollination and harvest. Harvest S_1 ears that have sufficient seed for replicated trials.	---
2 (Summer)	Evaluate S_1 progenies: Conduct replicated trials; e.g., if for yield use 2 replications at 3 or 4 locations in the same year. On the basis of the replicated trials, select 20 to 30 S_1 progenies for recombination. If selections can be made before flowering, recombination also can be made this season.	---[†]
3 (Winter)	Recombine: Recombine the selected S_1 progenies to form the C1 cycle population. If recombination can be made in season 2, plants can be self-pollinated to develop S_1 progenies for evaluation in season 4.	Include selected S_1 progenies in breeding and topcross nurseries for selection, inbreeding, and production of testcross seed.
4 (Summer)	Recombine or self-pollinate: In temperate zones, either recombine to form C1 Syn. 2 or self-pollinate S_0 plants in the C1 Syn. 1 to form S_1 progenies for testing in season 5.	Include selected S_1 progenies in breeding and topcross nurseries for selection, inbreeding, and production of testcross seed.

[†]S_1 progenies included in replicated trials also can be included in breeding nursery to advance to the S_2 generation.

each generation will be the same as that for traits evaluated later in the growing season (e.g., stalk rot resistance and yield). Two years per cycle is the shortest generation time for traits evaluated after flowering in temperate zones, provided off-season nurseries are available. Off-season or winter nurseries can be used for recombination and self-pollination to initiate the next cycle of selection. Self-pollinations can be made in season 4 (Table 12.1), but because off-season nurseries usually are not suitable for evaluation trials another generation of recombination can be made before selfing. A decision has to be made as to whether another recombination is more valuable than additional costs of producing S_1 progenies in the off-season. In areas with more than one growing season each year, adjustments can be made for the scheme in Table 12.1 to make maximum use of the available growing seasons (Table 7.30). One important way to increase efficiency of selection is to use the least possible number of years to complete each cycle. Length of each cycle depends on the trait under selection and the seasons available to produce progenies for test, evaluation of the progenies, and recombination of selected progenies (Chap. 7).

Because most breeders are anxious to have new materials in their breeding nurseries, S_1 progeny selection will provide progenies that have been evaluated in one generation. Selections used for recombination in each cycle are logical candidates to include in breeding nurseries for additional selection and testing. Although additional space and time would be required, all S_1 progenies included in evaluation trials also could be included in the breeding nursery for further selection and inbreeding. Some selections not included for recombination also may be included in the breeding nursery; they may be progenies that did not quite meet the standards for recombination but were within the range of the least significant difference of the progenies used. The one generation of S_1 progeny testing will screen out poorest progenies, but some that were marginal for including for recombination may be useful after further selection and testing.

12.2.4 S_2 Progeny Recurrent Selection

Mechanics of S_2 progeny recurrent selection are similar to those described for S_1 progeny selection except that another generation of inbreeding is accomplished before evaluation in replicated trials. The S_2 progeny selection has some definite advantages over S_1 progeny selection, but an additional year is required to complete each cycle in temperate zones. As in S_1 progeny recurrent selection, selection for additive genetic effects is emphasized.

The component of variance for among S_2 progenies is $(3/2)\sigma_A^2$ compared with σ_A^2 for among S_1 progenies for a restricted genetic model (Table 6.12). Because an extra year is required, strong selection pressure can be applied to S_1 progenies before conducting S_2 progeny trials. The S_2 progeny selection usually is used for yield improvement, whereas S_1 progeny selection also has been used for traits other than yield.

The sequence of operations for S_2 progeny recurrent selection is described in Table 12.2. The population chosen for improvement should have potential as a source population for inbred lines that have good general combining ability with other elite lines currently used or to be used in the future for the production of hybrids. The S_1 progenies in season 2 should be included in as many nurseries as seed, time, and facilities permit. The primary objective for growing the S_1 progeny generation is to eliminate progenies obviously sus-

Table 12.2. Description of the use of different seasons for conducting
S_2 progeny recurrent selection for temperate zones.

Season	Population improvement program	Breeding program
1 (Winter)	Self-pollinate: In the population under selection (C0), self-pollinate 300 to 600 S_0 plants. Practice selection among the S_0 plants at pollination and harvest.	---
2 (Summer)	Evaluate S_1 progenies: Grow the 300 to 600 S_1 progenies in the breeding and pest nurseries. It probably is not necessary to replicate. Make selections among and within S_1 progenies for disease and insect resistance and for agronomic traits. Advance selected plants to S_2 generation by self-pollination. Harvest S_2 seed on S_1 plants that have sufficient seed for replicated yield trials.	---
3 (Summer)	Evaluate S_2 progenies: Grow the 100 to 200 selected S_2 progenies in replicated yield trials (e.g., 2 replications at 3 or 4 locations in one year). On the basis of the replicated yield trials, select 20 to 30 progenies for recombination.	---[+]
4 (Winter)	Recombine: Recombine the selected progenies to form the C1 cycle population. Remnant S_1 or S_2 seed can be used for recombination (see discussion).	---
5 (Summer)	Recombine or self-pollinate: In temperate zones, either recombine to form C1 Syn. 2 or self-pollinate S_0 plants in the C1 Syn. 1 to develop S_1 progenies for selection in season 7.	Include selected S_2 progenies in the breeding and topcross nurseries for selection, inbreeding, and production of testcross seed.
6 (Winter)	Self-pollinate: The off-season nursery is used to produce S_1 progenies in C1 cycle population to initiate next cycle of selection.	---
7 (Summer)	Evaluate S_1 progenies: Same as season 2.	Include selected progenies in breeding nurseries for further selection and inbreeding. Conduct testcross trials in replicated tests.
	Repeat as described for seasons 2 through 5.	

[+]S_2 progenies that are in yield trials also can be included in the breeding nursery to advance to the S_3 generation.

ceptible to major pests of an area before conducting yield trials in the S_2
progeny generation. If sufficient numbers of S_1 progenies are available, selection pressure can be increased in discarding susceptible S_1 progenies. Use
of only one replication in each nursery will not eliminate the possibility of
escapes or environmental effects, but if continuous selection pressure is
maintained in subsequent cycles of S_2 progeny recurrent selection the general
level of resistance to major pests will increase. If selections can be made
before flowering the number of self-pollinations is reduced, but in most instances selection has to be made after flowering. Selection among and within
S_1 progenies can be made for several agronomic traits at flowering, e.g., barrenness or poor shoots, tassel type, pollen shed, and any other traits necessary for the line to be useful either as a seed or as a pollen parent in hybrid seed production.

Major emphasis on selection can be made at harvest for ear size, stalk
rot resistance, ear diseases, kernel type, etc. One of the needs of S_2 progeny recurrent selection is an ear that has sufficient seed for replicated yield
trials. Ear size and number of kernels on the ear also are important for use
of lines as parental seed stocks. Replicated yield trials of selected S_2
progenies are conducted in as many environments as seed and facilities permit.
Superior yielding progenies are selected for recombination based on the combined data. The S_2 progenies included in yield trials also can be included in
the breeding nursery for additional selection and selfing. By the time yield
trials are completed, S_3 seed of selected progenies will be available for the
applied breeding program. Seed supplies may be a limiting factor if remnant
S_2 seed is used for recombination.

Mechanics of recombination are the same as for S_1 progeny recurrent selection. The breeder has a choice, however, of which generation of seed to
use for recombination, i.e., remnant S_1 or S_2 generation seed. Effects of inbreeding are greater with S_2 seed than with S_1 seed (Table 9.2); but additional selection within S_1 progeny rows was imposed in selecting S_2 progenies for
yield traits and using remnant S_2 seed takes advantage of it. If sufficient
S_2 remnant seed is not available or one is concerned about effects of inbreeding, remnant S_1 seed can be used for recombination. As an example, if 20 S_1
progenies are recombined vs. 20 S_2 progenies, effects of inbreeding are 22%
for S_1 progenies vs. 31% for S_2 progenies in the fifth cycle of recurrent selection. Or it would require 18 S_2 progenies vs. 12 S_1 progenies to have the
same level of inbreeding. If remnant S_2 generation seed is used for recombination, it is an advantage to have another generation of recombination in season 4 (Table 12.2). Because season 6 will generally not be as satisfactory
for evaluating S_1 progenies for temperate zones, the additional generation of
recombination will not increase the time for each cycle of recurrent selection. The S_1 progenies can be obtained in the next off-season nursery (season
6, Table 12.2).

Practical usefulness of S_2 progeny recurrent selection is to include selected S_2 progenies in the breeding nurseries for selection, inbreeding, and
producing testcross seed. The S_2 progenies selected for recombination are the
survivors of intensive multiple trait selection in the S_1 and S_2 generations
and should have potential in developing new lines. They are an elite sample
of progenies that have to be yield tested to determine their combining ability
with other elite lines in the S_2 or later generations.

The S_2 progeny selection method requires at least three years for each
cycle. For an applied breeding program in search of new lines, S_2 progeny selection seems to be a good choice. The increased length of cycle has advantages because of increased opportunities to practice selection before yield
testing. Selection in the S_1 generation eliminates progenies that do not meet

accepted standards. The important point is to recombine selected progenies to continue selection for future cycles. A logical division of germplasm in the heterotic response of elite lines, such as dent vs. flint and Reid vs. Lancaster, suggests S_2 progeny recurrent selection in two populations to test combining ability of selected progenies, i.e., testing selections from a dent population against those from a flint population.

12.2.5 Half-sib Recurrent Selection

Half-sib recurrent selection includes several versions, depending on types of testers used to produce the half-sib (testcross) progenies for evaluation. (Pros and cons of different types of testers are discussed in Chaps. 7 and 8.) Half-sib recurrent selection can be useful for applied breeding programs. Mechanics of conducting half-sib recurrent selection are similar for different types of testers. The choice of testers depends on the breeder's current knowledge of relative importance of different types of gene action, materials available, and stage of breeding program. In contrast to S_1 and S_2 progeny recurrent selection based on performance of progenies themselves, half-sib recurrent selection of progenies for recombination is based primarily on their performance in crosses with other genotypes (combining ability). Recurrent selection on the basis of S_1 and S_2 progenies emphasizes selection that changes frequencies of genes having additive effects more than do recurrent selection methods involving testcrosses (Table 6.12). If overdominance is important at some loci, half-sib recurrent selection would favor selection for alleles that have overdominant effects more than does S_1 and S_2 progeny evaluation. Stage of testing also will influence the decision for use of half-sib recurrent selection, which emphasizes early testing for combining ability. Early testing is effective (Chap. 7), and selection among and within inbred progenies before conducting expensive yield trials is desired by many. Comparisons of inbred progeny evaluation and half-sib progeny evaluation are discussed in Chap. 8.

Half-sib progeny evaluation can be conducted by producing either S_0 plants or S_1 plant testcrosses, but it seems that use of S_1 plants would be preferable in applied breeding programs. Description of half-sib selection in Table 12.3 is based on testcrosses from use of S_1 plants. This procedure requires one additional season, but the number of years required to complete one cycle of recurrent selection does not change. Its major advantage is that S_1 progenies can be screened in choosing the S_1 plants used to produce the testcrosses. The S_1 progenies represent the S_0 plant genotype and may be included in different pest nurseries and selections made at either pollination or harvest for plants to include in testcross trials. Testcrosses of only selected plants in selected S_1 progenies are yield tested. Off-season nurseries can be used to produce S_1 progenies without increasing the cycle interval for half-sib recurrent selection.

Type of tester will influence amount of effort required to produce testcross seed. Use of either an inbred or a single-cross tester will require making only enough crosses to have sufficient seed supplies for replicated yield trials. For a single-cross tester, perhaps only two or three ears are necessary. Use of the parental population or some other type of population with a broad genetic base involves some sampling problems. To have the gametic array of the tester adequately represented in the testcross, 6 to 10 plants of the tester should be pollinated (Sprague 1939; Noble 1966). Number of plants pollinated with use of a broad genetic base tester, however, may not be any greater than with use of an inbred tester. We need adequate sampling when a broad genetic base tester is used, and we need to ensure an adequate quantity of seed when an inbred tester is used. Number of pollinations required,

Table 12.3. Description of the use of different seasons for conducting
half-sib recurrent selection for temperate zones.

Season	Population improvement program	Breeding program
1 (Winter)	Self-pollinate: In the population under selection (C0) self-pollinate 300 to 600 S_0 plants. Practice selection among the S_0 plants at pollination and harvest.	---
2 (Summer)	Produce testcrosses: Grow the 300 to 600 S_1 progenies in the breeding and pest nurseries. Make selections among S_1 progenies where possible (e.g., 1st brood European corn borer) before pollination. Produce testcrosses (or half-sibs) of selected S_1 plants in selected S_1 progenies. Self-pollinate the S_1 plants used to produce testcrosses.	---
3 (Summer)	Evaluate testcrosses: Grow the 100 to 200 test-crosses in replicated yield trials (e.g., 2 replications of 3 or 4 locations in one year). On the basis of the combined replicated yield trials, select 20 to 30 progenies for recombination.	---[†]
4 (Winter)	Recombine Recombine either selfed progeny or remnant half-sib seed (See Table 12.1). Remnant S_1 or S_2 seed can be used for recombination (see discussion).	---
5 (Summer)	Recombine or self-pollinate: In temperate zones, either recombine to form C1 Syn. 2 or self-pollinate S_0 plants to develop S_1 progenies for selection and testcrossing in season 7.	Include S_2 progenies of selected half-sibs in the breeding and topcross nurseries for selection, inbreeding, and production of testcross seed.
6 (Winter)	Self-pollinate: The off-season nursery used to produce S_1 progenies in the C1 Syn. 2-cycle population to initiate next cycle of selection.	---
7 (Summer)	Produce testcrosses: Same as season 2.	Include selected pro-genies in breeding nurseries for further selection and inbreeding. Conduct testcross trials in replicated tests.

Repeat as described for seasons 2 through 5.

[†] S_2 progenies of the S_1 plants used to produce testcrosses also can be included in the breeding nursery to advance to the S_3 generation.

therefore, may not be greatly different for different testers.

Careful planning of the nursery is necessary in half-sib recurrent selection. In pest nurseries the replicated or nonreplicated S_1 progenies may be grown in sequence. In breeding nurseries for half-sib recurrent selection, it is necessary to alternate S_1 progeny rows or S_1 progeny ranges with the tester parent. To minimize the mechanics and time for producing testcrosses, the tester parent should be close to the S_1 progenies. Planting the tester parent in ranges adjacent to each range of S_1 progenies minimizes time and labeling required to make the testcrosses. The tester parent and S_1 progenies may have poor timing at flowering because of inbreeding depression and maturity of S_1 progenies. It may be necessary to have one or two delay-plantings of the tester parent. The first planting is made at the same time as the S_1 progenies with subsequent plantings made 7 to 10 days later. Low planting densities are used for the first planting to permit row identification for delayed plantings. If an inbred tester is used it may be necessary to delay-plant the S_1 progenies, but this usually is not critical. If the tester and S_1 progenies are planted end-to-end in different ranges, harvesting also is simplified. Testcrosses produced on inbred and single-cross testers are shelled in bulk, but equal quantities of seed to meet the requirements for testing are shelled from each ear when a broad genetic base tester is used (e.g., if 500 seeds required for yield testing and eight testcrossed ears are available, 64 seeds are shelled from each ear). The half-sib progenies are evaluated in replicated trials (2 or 3) in three or four environments in season 3 (Table 12.3).

Selected progenies to be recombined to form the next cycle population are determined by performance of half-sib progenies. Remnant S_1 or S_2 seed of plants used to produce the testcrosses are used for recombination (see S_2 progeny recurrent selection for use of S_1 or S_2 seed for recombination). Recombination can be completed in off-season nurseries (winter for temperate zones), with either recombination or self-pollination the following season. Since another off-season nursery is available for temperate zones an additional generation of recombination (season 5, Table 12.3) seems desirable, particularly if S_2 seed is used for recombination.

Half-sib progeny recurrent selection also is amenable to generating material for the applied breeding nursery. The S_1 plants used in making the testcrosses also are self-pollinated to produce S_2 seed. Remnant S_2 seed can be included in the applied breeding nursery after the testcross information has become available. Selected S_2 progenies have had selection in the S_0 and S_1 generations and testcross information in the S_1 generation; thus they have had considerable selection pressure as inbred progenies and for combining ability with the tester used to produce the half-sib progenies. Or if S_2 generation seed was included in the breeding nursery in season 3, S_3 generation progenies are available. (The additional information on combining ability, of course, was not available for the S_1 and S_2 progeny recurrent selection methods discussed previously.) If the breeder considers early testing of S_1 plants an adequate test for combining ability, half-sib recurrent selection would be appealing. Again, some additional S_2 or S_3 progenies may be included in the breeding and topcross nurseries that were not included in the 20 to 30 selections recombined for the next cycle population. Additional inbreeding, selection, and testcrossing may reveal some superior lines that were not ranked correctly on the basis of initial testcrosses.

The sequence of methods for half-sib recurrent selection in Table 12.3 was described for only one population. An applied breeding program for development of new inbred lines and hybrids will include some populations that exhibit a greater heterotic response than others. A natural sequence is to pair populations that have the greatest heterotic response and potential usefulness as fruitful breeding populations and use one as the tester for the other, and vice versa. The description of half-sib recurrent selection in Table 12.3 would be applicable, but rather than one population there would be two populations under recurrent selection. To reduce the amount of testing in each season, it may be advisable to alternate the sequence of breeding activities for the two populations. Testing in alternate seasons would reduce the number of test plots, particularly when considered in reference to other testing priorities of the breeding program. If each population is the tester for the other, recurrent selection procedures would be similar to reciprocal recurrent selection described by Comstock et al. (1949). Use of an inbred line or a single-cross hybrid as tester would be equivalent to the method of recurrent selection described by Russell and Eberhart (1975).

Half-sib recurrent selection conducted concurrently in two breeding populations provides a natural partition of the derived lines for testing in hybrids. It is logical that lines extracted from a Reid or dent-type population would be tested with lines extracted from Lancaster Surecrop or flint-type populations. Selections obtained from the two half-sib recurrent selection programs would generate lines for the breeding nursery for both sides of the heterotic pedigree, which is highly desirable. If one highly promising population is used for half-sib recurrent selection, the lines extracted may not have a natural classification as to type of tester needed for further evaluation or how they should be used in currently produced hybrids. This dilemma is not serious if the population under half-sib recurrent selection has had previous testing and the heterotic response is known with reference to other breeding materials.

12.2.6 Full-sib Family Recurrent Selection

Full-sib family recurrent selection has not been used as extensively as some of the other intrapopulation recurrent selection schemes. Full-sib family recurrent selection in the Jarvis and Indian Chief varieties was reported by Moll and Stuber (1971). Comparisons were made between full-sib family recurrent selection (intrapopulation) and reciprocal recurrent selection (RRS) (interpopulation). Responses of both varieties to full-sib family recurrent selection were 2.1 times greater than their response to RRS; hence full-sib family recurrent selection was more effective than RRS for improvement of the populations themselves. In contrast RRS was 1.3 times more effective than full-sib family recurrent selection for improvement of the variety cross between Jarvis and Indian Chief. Results from this study seem to agree with the basic principles of the two recurrent selection schemes because RRS was designed primarily for improvement of the cross of two populations. Full-sib family recurrent selection was described by Mather (1949) as biparental crosses.

Full-sib families are established in populations chosen for improvement by crossing one plant with another. Reciprocal crosses can be made between the two plants included in a cross and the seed bulked for evaluation trials. The number of full-sib families deemed sufficient for adequate sampling (200) of the population under improvement would be produced. Selection of S_0 families to include for producing full-sib families can be made at pollination and harvest. Selection at pollination would be only for obvious phenotypic plant traits (e.g., pollen shed, shoot emergence, plant type), but selection at harvest could be made for ear type, seed set, and stalk quality. Adequate crosses would be needed to allow for discarding at harvest. Selection would not be as effective as on progenies that have been inbred and can be replicated

Table 12.4. Description of the different seasons for conducting full-
            sib recurrent selection in temperate zones.

Season	Population improvement program	Breeding program
1 (Winter)	Produce full-sib progenies: In the population under selection (C0), produce 150 to 200 full-sib crosses by reciprocal pollination of two selected plants. Practice selection among S_0 plants at pollination and harvest. Harvest full-sib crosses that have sufficient seed for replicated yield trials (see discussion for possible modifications of applied breeding program.).	---
2 (Summer)	Evaluate full-sib progenies: Conduct replicated trials of 100 to 200 full-sib progenies at 3 or 4 locations the same year with say 2 or 3 replications at each location.	---[†]
3 (Winter)	Recombine: Recombine the selected 20 to 30 full-sib progenies to form the C1 population.	---
4 (Summer)	Recombine or produce full-sib progenies: In temperate zones, additional recombination can be obtained to form C1 Syn. 2 population or full-sib progenies of the C1 Syn. 1 population can be produced for testing in season 6.	---
5 (Winter)	Produce full-sib progenies: The next cycle of selection is initiated in the C1 Syn. 2 population by producing the full-sib crosses.	---

Continue as for seasons 2 to 5.

[†]Full-sib progenies also can be included in the breeding nursery for
selfing. Selfed progenies can be used for recombination and advanced
in the breeding nursery (Sprague and Eberhart 1977).

(e.g., S_1 or S_2). Full-sib families are evaluated in replicated trials and
superior full-sib families identified. Remnant full-sib seed of superior families
are recombined to initiate the next cycle of selection (Table 12.4).
    At least three seasons are required to complete one cycle of full-sib
family recurrent selection. In temperate zones it requires two years to complete
one cycle: produce full-sib families in season 1; evaluate full-sib
families in season 2; and recombine selected full-sib families in season 3,
which is the off-season or winter nursery. If full-sib families are produced
in the winter nursery, five seasons in two years are used to complete each cycle
with two generations (or seasons) of recombination included. If intercrosses
produced during recombination are full-sibs, they could be included in

evaluation trials to complete one cycle per year.

Full-sib family recurrent selection does not lend itself to generating progenies for applied breeding nurseries. Remnant full-sib seed is available, but the families are noninbred at this stage. The selected full-sib families are produced from noninbred progenies with only mass selection of S_0 plants. Lines could be derived from selected full-sib families but it may be preferable to extract lines from the population improved by full-sib recurrent selection, which is a procedure similar to that described for mass selection and modified ear-to-row selection. Full-sib family recurrent selection can be modified to make it more amenable to generating lines for applied breeding nurseries. Either S_0 plants can be self-pollinated to produce S_1 progenies or S_0 plants that produce seed on more than one ear can be used to produce full-sib families for evaluation. Sprague and Eberhart (1977) suggested that full-sib families be included in the breeding nursery the same season of yield trials and selfed to generate S_1 progenies for further evaluation. Modifications are discussed in more detail in reciprocal full-sib selection.

Full-sib family recurrent selection is an intrapopulation improvement scheme, but two populations that have manifested heterosis in the population cross can be chosen for full-sib family recurrent selection. Although it seems that intrapopulation full-sib recurrent selection is not as effective as RRS for improvement of the population cross, comparison of improvements shows that full-sib family recurrent selection may be a logical choice for improvement of two populations as source populations for extraction of lines. Moll and Stuber's (1971) results showed that full-sib selection seems to be a viable recurrent selection scheme for improvement of breeding populations. Minor modifications to facilitate extraction of lines would increase potential of full-sib family recurrent selection.

12.3 INTERPOPULATION IMPROVEMENT
12.3.1 Reciprocal Recurrent Selection

RRS was recommended by Comstock et al. (1949) for improvement of performance of the cross between two populations that included selection for specific and general combining abilities. Hull (1945) believed that selection was not effective in maize because of the paucity of additive genetic effects and that overdominant loci played an important role in manifestation of heterosis in crosses of inbred lines. Hull suggested that to maximize effectiveness of selection for overdominant loci either an inbred line or a single-cross hybrid should be used as tester. For improvement of a breeding population the narrow genetic base tester should be used in each cycle of selection--hence, recurrent selection for specific combining ability. It was shown in Chap. 5, however, that adequate additive genetic variability was present in maize populations to expect progress from selection based on additive genetic effects with partial to complete dominance. Some earlier estimates obtained in F_2 populations suggested overdominance, but subsequent studies in random mated F_2 populations showed that estimates of overdominance were due largely to repulsion phase linkages (Table 5.3). Pseudooverdominance was detected because of linkage biases in the F_2 populations produced from inbred lines. Recurrent selection for specific combining ability was in contrast to selection for general combining ability, which was based primarily on additive genetic effects (Jenkins 1940). Because of divergence of opinion on whether selection should be emphasized for general or for specific combining ability, Comstock et al. (1949) proposed the use of RRS, which would include selection for both general and specific combining abilities. If the populations under RRS were to have value in breeding programs, either inbred lines extracted from the two populations or the population cross can be used to produce hybrids. Evidence summarized in Chap. 7 shows that RRS effectively improves the population cross with small or no changes in the population itself.

Comstock et al. (1949) showed that if overdominant loci are more important, recurrent selection for specific combining ability will be more effective than RRS. If only additive genetic effects with partial to complete dominance are more important, recurrent selection for general combining ability will be more effective than RRS. But if both types of gene effects are operative in expression of a trait, RRS will be the more effective method of selection. The advantages of recurrent selection for specific and general combining abilities over RRS for the specific cases were not great. Hence it seemed that RRS would be at least as effective as recurrent selection for the specific cases and more effective if both types of gene action were involved.

If additive genetic effects with partial to complete dominance are of primary importance, most of the recurrent selection schemes should be equally effective. This is evidenced by the summaries in Chap. 7 and by Sprague and Eberhart (1977). Recurrent selection for specific combining ability, as discussed previously, also seems to be effective for general combining ability (e.g., Horner et al. 1973; Walejko and Russell 1977). Hence the usefulness of RRS may seem limited. But it is a useful selection scheme for breeding programs, particularly if one is interested in developing new lines from two breeding populations that manifest heterosis in their crosses. Interpopulation selection also has an advantage over intrapopulation selection if there are multiple alleles unless one can accumulate all the alleles in one population. Accumulation of all alleles in one population, however, negates heterosis between populations; it may take several cycles of selection in the hybrid population to obtain intrapopulation hybrids that are equivalent to hybrids obtained from use of interpopulation selection (i.e., RRS). Inclusion of populations should be determined on the basis of the heterotic response of their crosses (Chap. 10). A natural method is provided for selecting and testing lines derived by RRS. Information reported for the North Carolina and Iowa RRS programs demonstrated that selected lines from improved populations performed better in hybrids than those from original populations (Russell and Eberhart 1975; Moll et al. 1977).

RRS requires three years to complete each cycle in temperate zones that have winter nurseries for making self-pollinations and recombination (Table 12.5). Description of RRS in Table 12.5 used S_1 plants to make the testcrosses. Either S_0 or S_1 plants can be used, but additional selection can be made among S_1 progenies for other traits that are important. The S_1 progenies can be developed in the winter breeding nursery and do not extend the length of the cycle. Inasmuch as possible, selection can be made among S_0 plants at pollination and harvest for advancing to the S_1 generation. The S_1 progenies are planted in the breeding and pest nurseries, and selection is applied for traits as discussed for intrapopulation schemes. Selected plants within the selected S_1 progenies are crossed to the opposite parental population, which is the tester population. That is, S_1 plants from population A (e.g., Iowa Stiff Stalk Synthetic) are crossed with population B (e.g., Iowa Corn Borer Synthetic No. 1), the tester population for population A. Also S_1 plants from population B are crossed with population A, the tester for population B. The number of plants that the S_1 plants are crossed with should be 6 to 10. Each S_1 plant used to produce a testcross also is self-pollinated to produce S_2 generation seed. All ears are harvested for each S_1 testcross. Equal quantities from each harvested ear are needed to form an adequate seed lot for yield trials.

Because two populations are undergoing selection, two sets of testcrosses

Table 12.5. Description of the use of different seasons for conducting
             reciprocal recurrent selection in temperate zones.

Season	Population improvement program	Breeding program
1 (Winter)	Self-pollinate: In the two populations (CO) under selection, self-pollinate 300 to 600 S_0 plants in each population. Practice selection among the S_0 plants and pollination and harvest.	---
2 (Summer)	Produce testcrosses (half-sibs): Grow the 300 to 600 S_1 progenies in the breeding and pest nurseries. Use selected S_1 plants within selected S_1 progenies and make testcrosses (half-sibs) with the opposite population. Cross each selected S_1 plant with 6 to 10 plants. Self-pollinate each S_1 plant to make testcrosses, say 100 to 200	---
3 (Summer)	Evaluate testcrosses: Grow the 100 to 200 testcrosses in replicated yield trials (e.g., 2 or 3 replications at 3 or 4 locations). On the basis of the replicated yield trials select 20 to 30 selfed progenies for recombination. There will be two sets of testcrosses, one for each of the two populations under selection.	---[†]
4 (Winter)	Recombine: Recombine the selfed progenies of the selected testcrosses (half-sibs) (e.g., 20 to 30) to form Cl Syn 1.	---
5 (Summer)	Recombine: Recombine Cl Syn. 1 to form Cl Syn. 2.	Grow S_2 progenies of selfed testcrosses in breeding and topcross nurseries. Practice additional selection within and among S_2 progenies in the breeding nursery.
6 (Winter)	Self-pollinate: In the two Cl populations, self-pollinate 300 to 600 S_0 plants in each population. Practice selection among S_0 plants at pollination and harvest.	---
7	Repeat as in seasons 2 to 5.	Grow selected S_3 progenies in breeding nursery and testcrosses of selected S_2 progenies in replicated yield trials.

[†]The S_2 generation progenies can be grown in the breeding nursery the
same season the testcrosses are grown. Selection among and within S_2
progenies can be made and advanced to the S_3 generation.

are included in the yield trials. Number of entries for yield trials will be double that for intrapopulation improvement, but if two populations are undergoing intrapopulation recurrent selection the number of yield test plots will not be different. The two sets of testcrosses are grown in separate experiments. The S_2 generation progenies of the S_1 testcrosses also can be included in the breeding nursery to advance to the S_3 generation. After harvesting, summarize the data from all locations and identify highest yielding testcrosses for each of the two sets of testcrosses. Because intrapopulation half-sib progenies are evaluated, S_1 or S_2 seed is used for recombination. Parental control is greater with S_1 and S_2 seed than with remnant half-sib seed (Table 6.12). (Advantages and disadvantages for use of S_1 and S_2 seed are discussed for S_2 recurrent selection.) Two sets of recombinations can be made in the winter nursery following harvest. The S_1 progenies can be developed the following summer (season 5, Table 12.5), but another generation of recombination is desirable--particularly if S_2 seed is used for recombination. The S_1 progenies can be developed the following winter season (season 6, Table 12.5) and not increase the cycle time. After obtaining S_1 progenies from the C1 cycle, the process is repeated in subsequent cycles.

Two sets of selected S_2 progenies (or S_3 progenies if S_2 progenies were grown in season 3) are available to include in the breeding nursery in season 5. The S_2 progenies are the survivors of plants from S_1 progenies that were selected to make the testcrosses and had the highest yielding testcrosses. Hence the S_2 progenies were selected from S_1 progenies that had desirable agronomic traits and above average general combining ability. Additional selection among S_2 progenies can be made either before or at the same time the lines are included in the topcross nursery. The stage of additional testing depends on the preference of the breeder, but topcross seed can be produced the same season without much additional expense. Whether topcross seed is included in testcross trials depends on whether S_2 progenies were deemed desirable for continuing in the breeding nursery.

The logical tester for the two sets of testcrosses again depends on the previous heterotic pattern of the germplasm. For example, the tester for the lines from Iowa Stiff Stalk Synthetic will be one of Lancaster Surecrop origin (e.g., Mo17), and the tester for the lines from Iowa Corn Borer Synthetic No. 1 will be one of Iowa Stiff Stalk Synthetic (Reid) origin (e.g., B73). Rather than using a common tester for each set of lines, testcrosses also can be produced by use of the cross-classification (design II) mating scheme. If, for example, 20 lines are selected from the A and B populations, 5 lines from A can be crossed with 5 lines from B to produce 25 testcrosses. This procedure would require four sets of design II crosses, or 100 testcrosses. Use of design II requires additional effort at pollination and more testcrosses are included in testcross trials. Use of a common tester for the two sets of selected S_2 lines would result in only 40 entries, whereas the example for the design II scheme would have 100 entries. The design II scheme is used only when no logical tester is available for selected lines. Otherwise it is easier and cheaper to include selected lines in the topcross nurseries.

Although RRS was the first interpopulation scheme proposed for the cyclic improvement of two breeding populations, the scheme has not been tested as much as desirable. At the present only two RRS programs are being conducted in the United States: (1) in North Carolina for the two open-pollinated varieties, Jarvis and Indian Chief; and (2) in Iowa for the two synthetic varieties, Iowa Stiff Stalk Synthetic and Iowa Corn Borer Synthetic No. 1. Both programs have been conducted for nearly 30 years and significant improvements have been made in the population cross for both programs with no evidence that

genetic variability has decreased (Moll et al. 1977; Tables 5.7, 5.8). Two
programs also are being conducted in Africa: (1) at Kitale, Kenya, in KII and
Ec573 (Darrah et al. 1978); and (2) at Peitermaritzburg, South Africa, in Teko
Yellow and Natal Yellow Horsetooth (Gevers 1975). Paterniani and Vencovsky
(1978) reported 3.5% gain from modified RRS in Dent Composite and Flint Composite in Brazil. Three cycles of selection have been completed for the African and the Brazilian programs with use of modified RRS. RRS seems to be a
logical recurrent selection scheme for the development of new lines, provided
the selection program is combined with the applied breeding program: and it
provides for the development of new lines for both sides of the hybrid pedigree,
e.g., Reid vs. Lancaster Surecrop and flint vs. dent. Its limited use may be
due to its seeming more complex than intrapopulation schemes; but if one
conducts intrapopulation improvement in two populations as suggested previously,
RRS is no more complex than half-sib selection in two populations.

12.3.2 RRS Based on Testcrosses of Half-sib Families

RRS based on testcrosses of half-sib families, a modification of the
original RRS, was suggested by Paterniani (1967b) and reduces effort for making testcrosses because they are obtained by open pollination in isolated
fields. The main genetic difference from the original procedure is the type
of parentage among individuals in testcrosses (selection units); i.e., individuals are related as half-sibs and cousins, whereas in the original RRS individuals are related as half-sibs among female-male progenies and as full-
sibs within female progenies.

A cycle of the modified method starts with 200 or more open-pollinated
ears (half-sib families) from population A that are planted ear-to-row as females in an isolation block. Male rows are planted with seeds of population
B with a ratio of 3:1, 2:1, or 4:2 (female:male). In a separate isolation
block a similar number (200 or more) of open-pollinated ears (half-sib families) of population B is planted as females and population A is used as male.
It may be desirable to plant a greater number of half-sib families and discard
undesirable ones at harvest. The second phase is evaluation of testcrosses
(half-sib family A × population B and half-sib family B × population A) in
replicated yield trials. Selection is on the basis of yield means and agronomic traits. Recombination is performed in populations A and B for the selected half-sib families, using remnant seeds from the selected half-sib families (see Table 12.6).

One cycle of selection was evaluated by Paterniani and Vencovsky (1977).
The population cross increased 7.5%, although (as emphasized by the authors)
genetic properties of the populations may have been an important factor in the
response to selection. Simplicity of the scheme and nonlimited seed supplies
were cited as advantages. In addition, the scheme may be more appropriate for
long-term objectives because of the larger number of progenies tested and the
nature of the recombination unit. The method uses a smaller portion of the
genetic variability but it permits a higher selection pressure because effective population size is about four times the number of families tested.
Therefore, the breeder can easily protect the populations against low effective size or depletion of genetic variability due to genetic drift during selection (Paterniani and Vencovsky 1977).

12.3.3 RRS Based on Half-sib Progenies of Prolific Plants

RRS based on half-sib progenies of prolific plants is another modification of the original RRS. The main difference between this modification and
RRS is that individuals within the selection units (testcrosses) are related
only as half-sibs and the recombination unit is a half-sib family instead of

Table 12.6.   Description of the method of reciprocal recurrent selection
based on testcross of half-sib families.

Season	Population improvement program	Breeding program
1 (Summer)	Ears from isolated open pollinated fields (A and B) are planted ear-to-row as females in detasseling blocks where the opposite population is used as males to produce the testcrosses.	---
2 (Summer)	Evaluate testcross (seeds from each female row constitute one entry): 100 to 200 testcrosses half-sib (A) × B and same number of testcrosses half-sib (B) × A are evaluated in replicated trials (e.g., 2 or 3 replications at 3 or 4 locations). Identify the best testcrosses half-sib (A) × B and half-sib (B) × A.	---
3 (Winter)	Recombine: Remnant seeds of half-sib families (A) corresponding to the best test-crosses half-sib (A) × B are planted for recombination. The diallel system of recombination can be used but a simpler procedure is the use of an isolation block where the selected progenies are used as females (detasseled) and a bulk of them as males. Same procedure is used to recombine half-sibs (B). At the harvest time, select 100 to 200 open pollinated ears, getting about the same number from each female row.	---
4 (Summer)	Repeat as in seasons 1 to 3.	---

an S_1 family. In the first phase testcross progenies are obtained in isolated
detasseling blocks. In one isolation block, population A is planted as females (detasseled rows) and population B as males, and vice versa in the other. Female rows in each isolation are planted either as bulked seed or ear-
to-row. The second (lower) ear shoots of prolific plants in the female rows
in each block are covered. The first (upper) ear of each plant is open pollinated by the opposite population (male) in each block. When the tassels start
pollen shedding, a sample of pollen is collected each day from the male rows
in each field. The bulk pollen samples are used to pollinate the second ears
of plants from the same population (female rows) in the other field; i.e.,
pollen from male A plants in the first field is used to pollinate female A
plants in the second field, and vice versa. Pollen from male rows is collected each day from at least 50 plants and passed through a sieve to eliminate
anthers. A pollen gun can be used to pollinate each female plant with the
bulk of pollen. For bulking of pollen no special care is needed to avoid contamination because all pollen grains of the particular isolated field belong
to the same population. For better control pollinations can be performed al-
ternately; i.e., in one day pollination is in one direction (pollen collected
from field A) and in the next day pollination is in the opposite direction
(pollen collected from field B).

Table 12.7. Description of the method of reciprocal recurrent selection
based on half-sib progenies of prolific plants.

Season	Population improvement program	Breeding program
1 (Winter)	Population A is planted as females in a detasseling isolated block (Field 1), where population B is planted as male rows with a ratio (male:female) of 1:2, 1:3, or 2:4. Population B is planted as female rows in another isolation block (Field 2) where population A is used as males. Protect the second (lower) ear shoots of prolific plants in female rows in both fields. Pollen is bulked, after passing through a sieve to eliminate anthers, and a pollen gun is used to pollinate the second ears in the opposite field. Alternate pollination can be used, i.e., from Field 1 to Field 2 in one day and vice versa in the next day. The first (upper) ears in each field are open pollinated. Harvest both ears together with an appropriate identification. The upper ear is an interpopulation half-sib family, whereas the lower ear is an intrapopulation half-sib family.	---
2 (Summer)	Evaluate testcross: 100 to 200 testcrosses from each field, i.e., half sibs (A × B) and half-sibs (B × A) are evaluated in replicated trials (2 to 3 replications at 3 or 4 locations). Identify the best testcrosses A × B and B × A.	---
3 (Winter)	Remnant seeds of lower (hand pollinated) ears of population A harvested in season 1, Field 1, corresponding to the best testcrosses A × B, are planted ear-to-row as females in a detasseling block, as in season 1. Male rows are a bulk of remnant half-sib seeds of population B that correspond to the selected testcross B × A. Remnant half-sibs B related to the best testcrosses B × A are also planted ear-to-row in another isolation field where selected half-sibs A are bulked and planted as males. Repeat procedures as described for seasons 1 and 2.	---

Each female plant in each isolation field provides: (1) the interpopulation half-sib family, obtained from the upper ear that was open pollinated by the opposite population; and (2) the intrapopulation half-sib family, obtained by controlled pollination using a bulk of pollen from the same population (males in the other field). Interpopulation half-sib families (inter-HS) are evaluated in replicated yield trials and intrapopulation half-sib families (intra-HS) are used for recombination. Intra-HS from each population, corresponding to the selected inter-HS, are planted ear-to-row as females in one field; at the same time a bulk of seeds of the same families is planted as male rows in the other field. The second cycle starts by repeating the procedure of the first season. This technique provides simultaneously for recombination of selected progenies and for testcross material. Therefore seeds obtained from the control hand-pollinated (lower) ear in the female rows in the first phase of the second cycle will correspond to the improved populations of the first cycle, A_I and B_I (see Table 12.7).

One cycle is completed every two seasons, which increases gain per year. Selection also is practiced in every generation, i.e., selection within female rows and selection for prolificacy in one year and selection based on testcrosses in the other year. Prolificacy is under a very strong selection intensity because when the second ear is protected its development is impaired in favor of the first ear. This reduces the number of plants with good seed sets in the first cycles, but in subsequent cycles the success in obtaining good seed set on both ears increases.

Variations of the procedure are possible that can make it more applicable to the breeder's facilities and available materials. For materials with a low level of prolificacy, both ear shoots could be protected and the second ear pollinated a few days before removing the bag of the first ear. Another variation may be to plant each population in an isolation block. The first ears are protected and the second ears are open pollinated with pollen from the same field (same population). The first ears are then pollinated with pollen collected from the other field. Such a modification may result in a greater number of plants with good seed sets but also in some disadvantages: (1) risk of contamination on the second ear is increased; and (2) outcrossed ears will have a poorer seed set, limiting the number of replications in yield trials.

The method also can be modified to complete one cycle per season. In each year outcrossed ears are evaluated in yield trials and two isolation blocks are planted for detasseling. In one field half-sibs obtained by hand pollination of A plants (intrapopulation half-sibs) are planted ear-to-row as females and a bulk of seeds from intra- half-sibs (B) as males. In the other field population B is planted ear-to-row as females (half-sib family) with population A as male. When data of yield trials are available, only the best (selected) female rows are considered in each field. In the selected female rows both open-pollinated (inter-HS) and hand-pollinated (intra-HS) ears are harvested to initiate the next cycle (Paterniani and Vencovsky 1978). Parental control is the same in both (one season/cycle and two seasons/cycle) schemes.

12.3.4 Reciprocal Full-sib Selection

Reciprocal full-sib selection was suggested by Hallauer and Eberhart (1970) as another method of interpopulation improvement and for development of new lines from the two populations under selection. The operational procedures of reciprocal full-sib selection are similar to those proposed by Comstock et al. (1949) for RRS; the main difference is that full-sib progenies rather than half-sib progenies are evaluated. Reciprocal full-sib selection includes two breeding populations for cyclical improvement; the bases of choosing breeding populations for reciprocal full-sib selection are the same as those for RRS. Because full-sib progenies are evaluated, only one set of progenies is tested in reciprocal full-sib selection (in RRS two sets of half-sib progenies are evaluated). Hence, in reciprocal full-sib selection we can either sample twice as many plants in each of the two populations as in RRS or we can sample the same number of plants in each scheme and have half as many yield test plots. For example, if we sample 100 plants in each population in

each selection scheme, we will have 200 half-sib progenies to test with RRS
(100 for A tested with B and 100 for B tested with A) and 100 full-sib progenies for reciprocal full-sib selection. We will have equal numbers of yield
test plots if we sample 200 plants from each population for reciprocal full-
sib selection and 100 plants from each population for RRS. In the first case,
we will have 50% fewer test plots with reciprocal full-sib selection and equal
sampling of populations by both schemes. In the second case, we will have the
same number of test plots for each scheme but twice as many plants are sampled
with reciprocal full-sib selection.

Hallauer and Eberhart (1970) described reciprocal full-sib selection for
populations including plants that produced seed on at least two ears (Hallauer
1967a). Use of plants that produce seed on two ears will permit completion of
one cycle of selection in two years (Table 12.8). Each cycle of selection is
initiated in the summer season rather than in the winter season, as described
for most of the other recurrent selection schemes. Unless the off-season

Table 12.8. Description of the use of different seasons for conducting
reciprocal full-sib selection with prolific plants in temperate
zones.

Season	Population improvement program	Breeding program
1 (Summer)	Produce S_1 progenies and full-sib crosses: Plant the two populations (C0) under selection in alternate rows in the breeding nursery. Self-pollinate one ear and cross the other ear. Practice selection among S_0 plants at pollination and harvest.	---
2 (Summer)	Evaluate full-sib crosses: Grow the full-sib crosses in replicated yield trials (e.g., 2 or 3 replications at 3 or 4 locations). It is optional, but pairs of S_1 progenies can be grown in breeding nursery to initiate line and hybrid development (see discussion).	Grow S_1 progenies to produce S_2 progenies and full-sib crosses between pairs of S_1 progenies.
3 (Winter)	Recombine: Recombine the selected S_1 progenies of the selected full-sib crosses (e.g., 20 to 30) to form C1 Syn. 1 for each of the two populations.	---
4 (Summer)	Produce S_1 progenies and full-sib crosses: Plant the C1 Syn. 1 populations in alternate rows in the breeding nursery. Self-pollinate one ear and cross the other ear. Practice selection among S_0 plants at pollination and harvest. Repeat as in seasons 2 and 3.	Grow S_2 progenies of selected full-sib crosses in breeding and topcross nurseries for additional selection. Grow selected S_3 progenies in breeding nursery and repli-cated topcross trials.

nursery is in another temperate zone, the quantity of seed set on the second ear is usually not satisfactory. If the off-season nursery can be used to produce S_1 progenies and full-sib crosses each cycle of selection can be completed in two years, but an additional cycle of recombination can be completed between each cycle of selection.

If we assume the summer season is used to produce S_1 and full-sib progenies, plant the two base populations in alternate rows (say 25 plants per row) in the breeding nursery. Plant density used will depend on frequency of two-ear development in the breeding nurseries. In the Iowa program 40,000 plants/ ha have been used for the two populations undergoing reciprocal full-sib selection. Frequency of two-ear development was consistently high in both populations; but it will depend to some extent on environmental conditions, such as moisture and temperature. Selected S_0 plants of the two populations, planted in alternate rows, are chosen at pollination. The most important trait usually is the time of flowering. The S_0 plants that have the same flowering dates are prepared (by covering silks on both ear shoots) for pollination with the first pollination made on the second ears. If the cross-pollinations are made first, the tassels are covered and used to pollinate the second ears the next day. The same day the second ears are pollinated the tassels are covered again, and the top ears are self-pollinated the second day. It has been found that if the second ears are pollinated first (rather than the top ears first), more successful seed sets are obtained. It is necessary to accurately label the bags of the two plants used to make the cross-pollinations in order to bulk the two ears included in producing the full-sib progenies. The primary reason the second ears were used to produce the full-sib progenies was to increase the frequency that selfed seed was available for each plant included in the crosses. If one of the two second ears is lost, we usually have sufficient seed on one of the second ears to conduct a two-replicate yield trial at three locations (about 300 seeds). Saving seed from only plants that produce seed on both ears exerts strong selection pressure for two ears. If the plants do not have seed on two ears, they are discarded. Continued selection fixes a high frequency of two-ear plants. After pollination inoculate all pollinated plants for stalk rot organisms.

Harvest the S_1 and full-sib progeny seed from plants that have sound stalks. All bags are labeled and one can check to determine if the particular pairs of S_0 plant crosses were successful. Bulk the two (or in some instances only one) full-sib ears for yield trials in season 2 (Table 12.8). Each pair of S_1 ears is identified for each cross so that we harvest four ears for each pair of S_0 plant crosses. The full-sib progenies, say 150 to 200, are grown in replicated yield trials of two replications at three or four locations. Data of the full-sib progenies are summarized for all locations, and the highest yielding full-sib progenies are identified. The S_1 seed of the full-sib progenies is planted in the winter nursery for recombination. As for RRS, there will be two sets of S_1 progenies for recombination--one for each of the two populations undergoing reciprocal full-sib selection. After recombination, the next cycle of reciprocal full-sib selection is initiated in the Cl Synthetic 1 generation, which is season 4 (Table 12.8). The same season that another cycle of reciprocal full-sib selection is initiated, the S_1 progenies included for recombination are included in the breeding and topcross nurseries. The S_1 progenies have not had any previous selection except for combining ability in the full-sib progeny tests. Choice of testers and stage of topcrossing would be the same as described for RRS.

Reciprocal full-sib selection was designed to maximize selection for specific combining ability in the development of single-cross hybrids (Hallauer 1967a, 1967b). Instead of testing for general combining ability and then

searching for a specific single-cross combination, emphasis was given to selecting for specific combining ability in each generation of inbreeding. Selected progenies were not included in the topcross nursery but only in breeding and pest nurseries. The S_1 progenies of selected full-sib progenies are
planted ear-to-row in pairs that correspond to the respective S_0 plants included in full-sib progenies tested in season 2 (Table 12.8). Each pair of S_1
progenies is planted in 16 to 25 plant rows in the breeding nursery. Within
each pair of progeny rows, three to five pairs of S_1 plants are self- and
cross-pollinated as described for season 1. Some of the pairs are discarded
before pollination for certain agronomic traits and first brood European corn
borer. Many pollinations will not be used because data for full-sib progeny
yield trials are not available at pollination and harvest. All plants pollinated in the S_1 generation are inoculated for stalk rot organisms. Seed is
harvested only from plants that have good stalks and produce seed on both
ears; one ear is S_2 seed and the second ear is the full-sib seed produced between S_1 plants. After full-sib progeny yield trial data are available, many
progenies are discarded; only those that exhibit good yield potential as full-
sib progenies are continued for additional inbreeding, selection, and testing.
In some instances a pair of S_1 progenies may have been discarded for some
trait but had high yield in the full-sib progeny yield tests. If the S_1 progenies were marginal for some traits but high yielding, they may be included
for recombination but not in the breeding nursery in the S_2 generation. Stalk
lodging is the most serious problem, but machine harvesting of full-sib progeny trials will eliminate progenies with a high frequency of stalk breakage and
usually confirm the data on poor stalk quality of S_1 progenies in the breeding
nursery. Opportunities exist for selecting for good stalk quality in the
breeding nursery and the full-sib progeny trials. Inbred development phase of
reciprocal full-sib selection is continued as described by Hallauer (1973b),
Fig. 12.8.

Although specific combining ability was emphasized with the use of reciprocal full-sib selection, Hoegemeyer and Hallauer (1976) found that elite in-

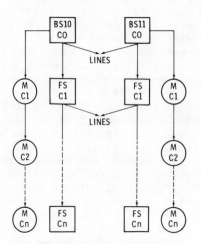

RECIPROCAL FULL-SIB SELECTION

Fig. 12.8. Integration of selection methods in two populations for contribut-
            ing to short-term (line development), intermediate-term (recipro-
            cal full-sib selection), and long-term (mass selection) breeding
            goals.

breds developed by reciprocal full-sib selection also had good general combining ability with other elite lines. There was concern that new inbred lines developed by testing only particular pairs of lines may have limited usefulness because they may not be acceptable for use as either male or female parent seed stock in production of the single-cross hybrid. Hoegemeyer and Hallauer used the design II mating scheme to test single-cross hybrids of inbred lines tested through the S_4 generation via reciprocal full-sib selection. Twenty-four pairs of selected lines in the S_7 generation were included in 96 single crosses. Primary interest was to determine how the diagonal single crosses of the design II mating scheme compared with the off-diagonal single crosses. The diagonal single crosses were those that had been tested and selected by reciprocal full-sib crosses, whereas the off-diagonal single crosses (not previously tested) were produced among the selected lines. The mean yield of the diagonal crosses (92 q/ha) for the six sets was significantly greater than the mean yield of the off-diagonal crosses (88 q/ha). Single-crosses selected by reciprocal full-sib selection, therefore, were on the average greater yielding than the previously untested off-diagonal crosses. But in each set at least one of the off-diagonal crosses was greater yielding than the diagonal single crosses.

Analysis of variance of the design II mating scheme permits the estimation of general and specific combining ability effects and tests of mean squares for general and specific combining ability sources of variation. Hoegemeyer and Hallauer (1976) found that specific combining ability mean squares were significant in 4 of the 6 sets of single crosses. General combining ability mean squares were significant in all instances. Estimates of specific combining ability effects for the 24 selected diagonal single crosses, however, were predominantly positive and significant (16 of 24); whereas estimates of specific combining effects for the 72 off-diagonal untested single crosses were significantly positive in 21 instances, significantly negative in 38 instances, and not significantly different from zero in 13 instances. It seems that reciprocal full-sib selection was effective in selecting for specific combining ability, but that specific effects were relatively small compared with general effects. Hence elite lines developed by reciprocal full-sib selection that emphasized selection for specific combining ability also had good general combining ability with other elite lines. Concern that inbred lines developed by reciprocal full-sib selection would combine well only with the particular line with which it had been tested was not valid. Conclusions were that although reciprocal full-sib selection could effectively select for specific combining ability (nonadditive effects), the procedure also was effective for general combining ability (primarily additive effects) and general combining ability was generally more important than specific combining ability.

Techniques of reciprocal full-sib selection were initially described with the use of two-eared plants (Hallauer 1967a). Reciprocal full-sib selection also is usable if either two-eared germplasm is not available or the maize breeder does not want to use a breeding procedure that involves multiple pollinations on the same plant (Hallauer 1967b, 1973b). The chronological description of reciprocal full-sib selection with use of primarily one-eared maize plants is given in Table 12.9. Use of one-eared rather than two-eared plants increases cycle time from two to three years. Effect of increased cycle time is compensated for by the additional selection that can be practiced for agronomic and pest traits. Theoretically, increasing cycle time from two to three years will decrease predicted gain on a per year basis, but additional selection will result in improved material for the applied breeding program.

Table 12.9. Description of the use of different seasons for conducting
reciprocal full-sib selection with use of primarily one-
ear plants in temperate zones.

Season	Population improvement program	Breeding program
1 (Winter)	Self-pollination: Self pollinate S_0 plants in the two populations (C0) under selection. Practice selection among S_0 plants at pollination and harvest. Note date of pollination of S_0 plants.	---
2 (Summer)	Produce full-sib crosses: Plant the S_1 progenies of the two C0 populations in alternate rows in the breeding and pest nurseries. Plant the S_1 progenies in pairs according to the date of pollination of S_0 plants. Produce full-sib progenies on a portion of the S_1 plants and self-pollinate selected S_1 plants (see discussion).	---
3 (Summer)	Evaluate full-sib progenies: Grow the full-sib progenies in replicated trials (e.g., 2 or 3 replications at 3 or 4 locations). It is optional, but S_2 progenies also can be grown in the breeding nursery.	S_2 progenies also can be included for additional self and cross pollinations.
4 (Winter)	Recombination: Recombine remnant seed of 20 to 30 S_1 or S_2 progenies to produce C1 Syn. 1 cycle.	---
5 (Summer)	Recombination: Recombine C1 Syn. 1 cycle to produce the C1 Syn. 2 cycle population.	Grow the S_2 progenies of the selected full-sib crosses in the breeding and topcross nurseries for additional selection.
6 (Winter)	Self-pollination: Self-pollinate S_0 plants in the two C1 Syn. 2 populations. Practice selection among S_0 plants at pollination and harvest. Note date of pollination of each S_0 plant.	---
7	Continue as in seasons 2 to 5.	Grow S_3 progenies in breeding nursery for additional selection and topcrosses in replicated yield trials to determine combining ability.

Reciprocal full-sib selection with use of one-eared plants is initiated in the winter nursery by selfing selected S_0 plants in the two broad genetic base populations. Date of pollination should be recorded on the pollination bag. At harvest, selected ears with their dates of pollination are saved and prepared to plant ear-to-row in the summer nursery (season 2, Table 12.9). The S_1 progenies will need to be planted in paired ear rows: one for population A paired with one for population B. The purpose for recording dates of pollination of the S_0 plants is to pair the S_1 progenies that have similar dates of flowering. If dates of pollination are not recorded, split plantings for each row could be made to ensure timing of the tassels and silks at pollination. Paired S_1 rows also should be planted in the pest nurseries to eliminate those that are obviously susceptible to the primary pests. For some traits (e.g., first brood European corn borer), some of the progenies can be discarded before flowering.

Because only one-eared plants are available, the same plants cannot be used to produce the self (S_2) and full-sib seed and different S_1 plants within each pair of S_1 progenies will be self-pollinated and cross-pollinated. Thus there will be some sampling differences of genotypes used for the self- and cross-pollinations. But the sampling should not be too great because the genotypic array of the S_1 progeny represents the genotype of the S_0 plant that was self-pollinated. Self-pollinate selected plants in each row. Similarly, cross-pollinate other selected plants between each pair of S_1 progeny rows. Or (preferably) self-pollinated plants can be used to cross-pollinate other selected plants. All pollinations can be made the same day. It will minimize errors if one portion of the S_1 progeny rows is self-pollinated and the other portion is cross-pollinated. Clearly mark each pollination to indicate self- or cross-pollination. Another option would be to use plain pollination bags for self-pollinations and striped bags for cross-pollinations. After pollination, inoculate all pollinated plants within each row with stalk rot organisms.

Harvest self- and cross-pollinated ears in pairs of S_1 progenies that have acceptable stalks, ears, kernel type, etc. Self-pollinated ears need labeling for each progeny row. Harvest cross-pollinated ears and shell in bulk in preparation for yield trials. One advantage of this procedure with one-eared plants is that adequate pollinations can be made to ensure adequate seed for the yield trials. If three to five plants are cross-pollinated for each plant self-pollinated, adequate quantities of seed should be available for yield testing and the differences due to sampling should not be serious. The full-sib progenies of S_1 progenies are tested in season 3 (Table 12.9).

Other procedures of reciprocal full-sib selection with use of one-eared plants are the same as those described with use of two-eared plants. The primary difference is the self-pollination in the first season to initiate each cycle of recurrent selection. One additional season is required but effective selection can be made among S_1 progenies, whereas only selection among S_0 plants was possible with use of two-eared plants. Labeling is reduced with use of one-eared plants. Also, less opportunities for errors exist at pollination, harvest, and shelling. If the winter nursery is used to initiate the next cycle of selection, another recombination can be made in the summer season to produce the Synthetic 2 generation. Reciprocal full-sib selection with use of one-eared plants may be amenable to more applied breeding programs than with use of two-eared plants.

Table 12.10. Observed comparisons of BS10 and BS11 maize populations and crosses for four traits after three cycles of reciprocal full-sib selection (Obilana et al. 1979a, 1979b).

Trait	BS10			BS11			BS10 x BS11			Heterosis, %	
	C0	C3	$\bar{d}$	C0	C3	$\bar{d}$	C0	C3	$\bar{d}$	C0×C0	C3×C3
Yield, q/ha	53.9	62.8	8.9 ± 3.6	56.6	66.8	10.2 ± 2.6	61.2	75.5	14.3 ± 4.5	10.8	16.5
Grain moisture, %	22.1	19.9	-2.2 ± 1.0	24.4	22.0	-2.4 ± 1.0	22.0	21.5	-0.5 ± 1.0	-5.4	2.6
Root lodging, %	5.6	8.3	2.7 ± 3.6	7.5	5.0	-2.0 ± 3.6	4.3	8.5	4.2 ± 3.6	-34.4	27.8
Stalk lodging, %	18.1	12.8	-5.3 ± 3.3	19.4	12.3	-7.1 ± 3.3	20.7	15.5	-5.2 ± 3.3	10.4	23.5

    Hallauer (1978) and Obilana et al. (1979b) have reported on the effec-
tiveness of reciprocal full-sib selection in Iowa Two-Ear Synthetic (BS10) and
Pioneer Two-Ear Composite (BS11).   Selection was effective in significantly
improving yield of the populations themselves and the cross of the two populations (Table 12.10).   Observed progress in the population cross was in close
agreement with predicted yield, which was based on components of variance estimated for the interpopulation cross of the CO populations.   Progress for
stalk and root lodging and grain moisture at harvest also was encouraging.
Although differences between the original and selected cycles were not all
significant, the changes were in the desired direction; stalk lodging was decreasing and root lodging and grain moisture at harvest tended to decrease, but
not significantly in all instances.   Selection was effective in changing populations for the primary trait, yield, and correlated agronomic traits that are
important in applied breeding programs.
    Figures 12.7 and 12.9 show distributions for yield and stalk lodging of
full-sib progenies produced and tested in CO and C7 cycles.   There were 144
full-sib progenies tested in 1964 for the CO cycle and 145 full-sib progenies
tested in 1984 for the C7 cycle.   The same six hybrid checks were included in
all trials for a point of reference.   Frequencies of full-sib progenies were
plotted relative to the mean of the six hybrid checks.   Standard errors were
calculated by use of the genotype-environment component of variance.   Distributions for yield from CO to C7 were essentially transposed relative to the
mean of the six check hybrids (Fig. 12.7).   Reciprocal full-sib selection was
effective increasing the performance of full-sib progenies.   Also yields of CO
full-sib progenies were obtained by hand harvesting with retrieval of all ears,
whereas yields of C7 full-sibs were obtained by mechanical harvesting with no
retrieval of dropped ears.   Theoretical effects of selection illustrated in

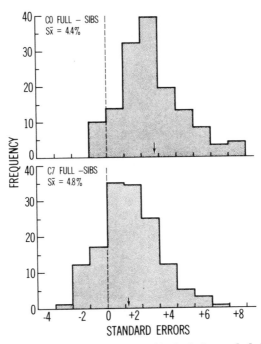

Fig. 12.9.   Frequency distributions for stalk lodging of full-sib progenies for
the CO and C7 cycles of reciprocal full-sib selection relative to the
mean of six check hybrids (dotted verticle line) (Hallauer 1984).

Fig. 12.3 are supported by data shown in Fig. 12.7. Only 24.1% of C0 full-sib progenies exceeded the mean of the checks, but only 10.5% of C3 full-sib progenies were below the mean of the checks.

Good stalk quality is essential if populations are to be useful as breeding material and if inbred lines are to be useful in hybrids. Table 12.10 shows that stalk lodging decreased in the populations themselves and in the population cross. Figure 12.9 shows progress has been made for improved stalk quality of full-sib progenies. The means of C0 and C7 full-sib progenies exceeded the mean of the six check hybrids, but change in the distributions is more evident. Only 6.9% of C0 full-sib progenies had less stalk lodging than the mean of the six check hybrids, but 20.6% of C7 full-sib progenies had less stalk lodging than the mean of the hybrid checks. The mean of the 144 C0 full-sib progenies was nearly four standard errors above the mean of the checks, whereas the mean of the 145 C7 full-sibs was nearly one standard error higher. Selection, and testing in each generation of inbreeding provide good opportunities for developing improved stalk quality: (1) inoculation and selection among and within progenies in the breeding nursery and (2) taking notes and measuring harvestable yield from mechanically harvested yield trials. In the breeding nursery information is obtained on progenies themselves and in yield trials information is obtained on the combining ability for stalk quality in full-sib progeny crosses. Use of plants that produce seed on two ears puts additional stress on plants for good stalk quality. It seems that the physiological limits of increased kernels and improved stalk quality can be combined.

Reciprocal full-sib selection (like other recurrent selection schemes) can be used to meet short-, intermediate-, and long-term objectives of a breeding program (Fig. 12.8). Mass selection in BS10 and BS11 has been conducted in Iowa to provide backup populations for the reciprocal full-sib program. Mass selection is conducted in isolated plantings of about 0.4 ha and ears are selected from plants that produce two ears on standing plants. If in the future it seems that genetic variation is being reduced by reciprocal full-sib selection, mass selected populations can be crossed with populations under reciprocal full-sib selection. The mass selection portion of the program satisfies long-term objectives. Long-term objectives are improvement by mass selection, intermediate-term objectives are improvement of populations by reciprocal full-sib selection, and short-term objectives are line development from each cycle of reciprocal full-sib selection.

Empirical data comparing RRS and full-sib reciprocal selection are limited to those shown in Table 12.10. Realized response for three cycles of reciprocal full-sib selection in BS10 and BS11 was 5% per cycle, which compares favorably with the response of 7% and 6% reported by Darrah et al. (1978) and Gevers (1975) for three cycles of RRS. Some theoretical guidelines were provided by Jones et al. (1971) for response to selection expected by RRS and full-sib recurrent selection. For the situation of equal additive and dominance variances, the phenotypic variation among half-sib families was one-third the phenotypic variation among full-sib families. To compensate for the lower phenotypic variation, the selection differential would need to be 1.7 times greater with full-sib than with half-sib RRS. Comparisons with actual estimates of components of variance showed that the selection differential should be 1.2 times greater for reciprocal full-sib selection. Interpopulation estimates of additive genetic and dominance components of variance and their interactions with environments were reported by Obilana et al. (1979a) for BS10 and BS11. Their estimates were used to calculate the phenotypic variance for half-sib and full-sib progenies; for the situation of two replications in each of three environments the full-sib phenotypic standard deviation was 1.24 times greater than the half-sib phenotypic standard deviation. Jones et al. (1971) concluded that reciprocal full-sib selection was favored at lower selection intensities and for large environmental variance relative to total genetic variance but declined when selection intensity increased.

12.3.5 RRS with Use of Inbred Lines

Because it seems that specific combining ability is of minor importance, Russell and Eberhart (1975) and Walejko and Russell (1977) suggested that RRS be conducted with use of inbred lines as testers. Choice of inbred lines to use as testers should be the same as choice of two populations to include for RRS; i.e., tester lines should conform to the heterotic pattern expressed in hybrids. Description of RRS with use of inbred testers is the same as shown in Table 12.6 except inbred lines rather than opposing populations are used as respective testers. Instead of crossing plants from population A with plants from population B (the tester), cross plants from population A with an inbred tester extracted from population B. The reverse situation would be used for population B. If the two populations are of Reid and Lancaster Surecrop origin, the Reid population would be crossed with an inbred line tester (e.g., Mol7 or Oh43) of Lancaster Surecrop derivation, and the Lancaster Surecrop population would be crossed with an inbred line tester (e.g., B73 or A632) of Reid derivation. The same conditions hold if flint and dent populations are undergoing selection.

After choices of populations and inbred line testers are decided, mechanics of the breeding operation follow the description given in Table 12.6. Self selected S_0 plants in both populations in the off-season (winter nursery) in season 1. Select among and within S_1 progenies of each population in season 2. Cross selected S_1 plants to the inbred tester. Also self each S_1 plant used to make the testcrosses. Grow the two sets of testcrosses in replicated yield trials. The S_2 progenies of S_1 plants used to make the testcrosses also can be included in the breeding nursery for another generation of selection and inbreeding. By the time the testcross data are available, S_3 generation progenies are available for the breeding nursery. On the basis of testcross trials, select S_1 or S_2 progenies for recombination to form the next cycle. Choice of progenies (S_1 or S_2) to recombine are the same as discussed for RRS. Another generation of recombination can be made in the summer season before initiating selfing of S_0 plants in the recombined population. Each cycle of RRS with use of inbred testers would require a minimum of three years per cycle, which includes two generations of recombination (seasons 4 and 5).

Use of inbred line testers in RRS provides S_2 lines for the breeding nursery that have had selection among and within S_1 progenies and combining ability tests with the inbred line tester. Choice of tester is important. Use of inbred parental lines of a highly heterotic hybrid (e.g., B73 × Mol7 or A619 × A632) seems appropriate. Although inbred line testers were suggested by Walejko and Russell (1977), single crosses of the appropriate pedigree also can be used to reduce the concern of having adequate seed for the yield trials. Because of the seemingly small importance of specific combining ability, testers can be changed to meet the current use of specific lines in hybrids. A particular line or lines may be suitable for two or three cycles of selection and then changed if the line is not useful or being used in hybrids. Since general combining ability seems of greater importance than specific combining ability (e.g., Horner et al. 1973; Horner et al. 1976) even with use of a specific inbred line, the change in lines used as testers should not affect continued selection for improvement of populations.

RRS with use of an inbred line was designed for applied breeding programs (Russell and Eberhart 1975). Accumulated information of relative importance of specific vs. general combining ability with different types of testers was instrumental in developing the recurrent selection scheme. Elite lines are required as parental stocks in production of single-cross hybrids; consequently, an important objective of most applied breeding programs is the improvement of lines for one or both sides of the hybrid pedigree. Most pedigree selection programs emphasize improvement of one or both lines of the pedigree. RRS by use of an inbred line tester is a useful recurrent selection scheme for supplementing pedigree selection programs. Advantages and disadvantages for use of an inbred line as tester for recurrent selection are discussed in Chap. 8. Comparisons of advantages and disadvantages of the method relative to RRS were discussed by Comstock (1979).

12.4 ADDITIONAL CONSIDERATIONS

Choice of population(s) to include for improvement and choice of recurrent selection scheme(s) to use for the improvement of the population(s) are decisions that must be made before the improvement program is initiated. Populations to include for improvement should have adequate genetic variability, have high mean, and manifest heterosis in crosses, particularly if some form of interpopulation recurrent selection is proposed. Information may not be explicitly available for populations available, but a combination of experimental data, personal experience, and the experience of others may provide leads to the most potentially useful populations to include for improvement. It seems that adequate genetic variability is present in most populations to expect progress from selection, but some populations may have more useful genetic variability than others; e.g., Iowa Stiff Stalk Synthetic seems to be a more useful population for developing new inbred lines than Iowa Corn Borer Synthetic No. 1, although the populations seem to have similar genetic variability. A population that has high mean yield performance, other traits being equivalent, would be preferable to one that has low mean yield. Equivalent rates of improvement may be made for both low and high mean yield populations, but the low mean yield population would not be equivalent to the high mean yield population. Even if the rate of improvement was somewhat greater for the low mean yield population, it probably would not yield as many potentially useful lines as the higher mean yield population. If the breeding program is not committed to use of inbred lines in hybrids, use of populations that have high mean performance also would be more feasible. It is important to capitalize on the expression of heterosis either in crosses of inbred lines extracted from populations undergoing recurrent selection or in population crosses for breeding programs not committed to use of inbred lines. Manifestation of heterosis should be considered for both intra- and interpopulation improvement programs. If two populations are undergoing recurrent selection, it is important that heterosis is expressed in their cross. Chapter 5 emphasizes importance of selection among populations to initiate a recurrent selection program; and formulas are given in Chap. 10 that provide the basis for prediction of composite variety means or the mean of a cross between two composites, using data from variety diallel crosses. Use of prediction procedures permits selection not only among existing varieties but also among populations that would result from crosses involving such a set of varieties.

Choice of recurrent selection scheme to use for improvement of populations depends on the breeder's objectives. It seems that all recurrent selection schemes effectively improve populations chosen for improvement. Based on realized gain per cycle, all recurrent selection schemes are nearly equally effective. A summary of selection results for those schemes included in Chap. 7 and those included by Sprague and Eberhart (1977) is presented in Table

12.11. Rate of realized gain is about 3% to 5% per cycle for intra- and interpopulation recurrent selection schemes. Different populations were included in different recurrent selection schemes conducted in different areas, but
realized gain per cycle on the average is not greatly different. Rate of gain
per year would be different because different numbers of years are required
for different recurrent selection schemes. Comparisons for different methods
of recurrent selection listed in Table 12.11 are averages for recurrent selection in different populations.

Darrah et al. (1978) reported on a comprehensive series of selection
studies conducted for three populations, Kitale Composite, Ec573, and KII
(Eberhart et al. 1967). Because different methods of recurrent selection were
conducted in the same populations, direct comparisons could be made for realized gain (Table 12.12). Darrah et al. (1978) presented their results on a
per year basis, but the data were converted to gain per cycle to make them
equivalent to the average realized gains given in Table 12.11. Except for
RRS, realized gains per cycle were similar to average realized gains given in
Table 12.10 for different methods of recurrent selection. Realized gain for
RRS was 7%, which was greater than for the programs in the United States but
similar to that reported by Gevers (1975) for Teko Yellow and Natal Yellow

Table 12.11. Summary of the average gain per cycle for several
different recurrent selection schemes.

Selection scheme	Average gain per cycle for different recurrent selection schemes	
	Chapter 7	Sprague and Eberhart (1977)
Intrapopulation		
Mass	2.9	3.4
Ear-to-row	6.5	3.8
Full-sib	3.8	3.1
Testcross	5.4	2.8
S_1	6.4	4.6
S_2	---	2.0
Average	5.0	3.3
Interpopulation		
Reciprocal	4.2	2.9
Darrah et al. (1978)[+]	7.0	---
Gevers (1975)[+]	6.0	---
Paterniani & Vencovsky (1978)[+]	3.5	---
Testcross	4.2	3.5
Reciprocal full-sib[+]	5.0	---
Average	5.3	3.2

[+]Not included by Sprague and Eberhart (1977).

Table 12.12. Response to selection for yield in a breeding
              methods study in Kenya, East Africa (Darrah
              et al. 1978).

Selection methods	Gain per cycle	
	q/ha	%
Mass: KCA-M9C6	0.38	0.8
M10C6	0.93**	1.6
M17C6	-0.70	---
Ear-to-row: KII	0.83**	1.8
Ec573	2.59**	6.6
KCA-E3C6	0.98**	2.1
E4C6	0.82**	1.4
E5C6	1.04**	2.2
E6C6	0.56**	1.2
E7C6	1.40**	2.5
Half-sib[†]: KCA-H14C3	1.26	2.0
HL15C3	0.62	1.0
HI16C3	-3.26**	---
S_1[‡]: KCAC3	-0.50	---
Full-sib: KCAC3	1.60*	2.4
Reciprocal: KIIC3	-0.04	---
Ec573C3	1.94**	2.5
(KII × Ec573)C3	4.18**	7.0

* and ** indicate significance at P = 0.05 and 0.01, respectively.

[†]Population by tester crosses and not populations themselves.

[‡]Based on bulk S_1 lines rather than random mated populations.

Horsetooth in South Africa (6%). Hence the choice of recurrent selection
scheme depends on how the particular scheme complements other phases of the
breeding program.

Initial selection among and within progenies for agronomic traits may be
preferable before obtaining a measure of combining ability for either noninbred or S_1 progenies; for this instance, S_1 or S_2 recurrent selection may be
preferable to some type of half-sib recurrent selection. Choice of intra- or
interpopulation recurrent selection for two populations also depends on the
breeder's preference. The S_1 or S_2 recurrent selection would not be appropriate for generally accepted schemes of interpopulation recurrent selection but
it could be used for two populations that meet the same criteria. The logical
sequel would be either to cross improved populations to produce the population
cross or to extract lines from improved populations for use in hybrids regardless of whether intra- or interpopulation recurrent selection was used.

Decisions of choice of populations and recurrent selection schemes are
not always obvious, but adjustments can be made during the course of the program if the breeder feels the wrong choices were made initially. Changes in a
recurrent selection scheme can be made from say half-sib recurrent selection
to S_1 recurrent selection in subsequent cycles if the breeder feels that the
scheme currently used is not meeting objectives. For example, a broad genetic
base population can be effectively improved through half-sib family selection
if the objective is to use the improved population directly for grain production. However, if the population was never submitted to a strong inbreeding
pressure it would not have immediate value as a source of vigorous inbred

lines, since greater inbreeding depression is expected to occur after selfing.
Should the breeder change objectives during the program and want to use the
improved population as a source of inbred lines for hybrids, one or more cycles of S_1 family selection would provide an opportunity for selection against
deleterious recessive genes and make the population more suitable for extraction of inbred lines. If the base population originated from inbred material,
such as synthetics from inbred lines, use of inbred family selection would
have less pronounced results than if it had never been submitted to strong inbreeding. Additional germplasm can be included in populations under recurrent
selection to rectify some obvious weaknesses, i.e., keep the populations open
and not closed.

Neither suggestion for modifying populations for recurrent selection
schemes would be appropriate if one is interested primarily in obtaining basic
information on breeding methods, but we are primarily interested in using recurrent selection schemes as adjuncts to the applied breeding program. To
stay current with applied breeding objectives, it is imperative that necessary
adjustments be made as indicated with time. Adjustments should not be made
frivolously to cause disruptions in genetic gain from selection. Adjustments
made in a haphazard manner would tend to put the improvement program on a
treadmill, causing either inconsistent or no genetic improvement for the germplasm chosen to provide materials for the applied breeding program.

Except for mass selection, each recurrent selection scheme includes three
operational stages for each cycle of selection: development of progenies for
evaluation, evaluation of progenies in replicated trials, and recombination of
superior progenies chosen on the basis of replicated trials. Contrary to
choices of populations for improvement and of recurrent selection schemes, the
decisions are not easy. Choices of populations and recurrent selection
schemes are determined to a large extent on how they relate to the applied
breeding program and experience. Very little experience and theory are available to assist in making decisions regarding operational steps involved in recurrent selection. For instance, how many progenies should be tested to adequately represent genetic variability of the population under selection? How
many progenies should be selected for recombination? How many generations of
recombination should be conducted before initiating the next cycle of selection? And what method of recombination should be used when 20 to 30 selections are chosen for recombination? Compromises usually are necessary. Obviously, the greater the number of progenies available for test, the greater the
opportunities for having a more representative sample of the population. But
testing of progenies in replicated yield trials prevents including an exorbi-
tant number. Most recurrent selection schemes have included a minimum of 100
progenies, but some programs have included 500 or more progenies. Evidence
from estimates of components of genetic variances in Chap. 5 show that reasonable standard errors are obtained with 100 progenies; the greater the number
of progenies, however, the more valid the estimates. Marquez-Sanchez and Hallauer (1970a, 1970b) found from empirical studies that 180 to 200 individuals
should be included for stabilizing standard errors on estimates of components
of variance from use of the nested (design I) mating design.

Sample size to be used for recurrent selection schemes is determined to
some extent by selection intensity, i.e., how many progenies are selected for
recombination. If a 10% selection intensity is used and 100 progenies are
tested, only 10 progenies are recombined to form the next cycle population.
Use of only 10 progenies for recombination, however, results in an appreciable
level of inbreeding after several cycles of selection (Table 9.2). Empirical
evidence agrees with projected effects of inbreeding because of limited number
of progenies recombined (Smith 1979). Some theoretical studies also provide

some guidelines for sample sizes to use in plant selection programs. Robertson (1960) showed that expected total advance and half-life of recurrent selection programs were proportional to effective population size; hence effective population size should be as large as possible for long-term selection programs. Baker and Curnow (1969) and Rawlings (1970) concluded that with a selection intensity of 10%, effective population sizes of 30 to 45 would be reasonable compromises for both short- and long-term objectives of plant selection programs. Number of progenies to include for recombination depends on the type of progeny for the minimum effective size based on theory; it seems 20 to 30 progenies are a minimum, but greater numbers are desirable.

Types of progenies recombined also have an effect, e.g., recombining S_1 vs. S_2 progenies (Table 9.2). Changing number of progenies for recombination changes selection intensity. If 100 progenies are tested, a 10% selection intensity includes 10 for recombination. If it is desired to increase the number of progenies to say 20 for recombination for 100 progenies tested, selection intensity is reduced to 20%. To maintain selection intensity of 10% by recombining 20 progenies, number of progenies tested needs to be increased to 200. It is desirable to have an adequate sample of progenies for testing and also to have an acceptable selection intensity to realize progress from selection. Obviously some compromises must be made to conduct a manageable selection program. A high selection intensity is desirable for short-term programs and for certain traits but would not be desirable for long-term programs because the effects of inbreeding and genetic drift would hinder future progress. It seems that at least 20 to 30 progenies should be included for recombination; hence, selection intensity is reduced if a smaller number of progenies is tested or increased if a greater number of progenies is tested.

The inclusion of 20 to 30 selections for recombination poses problems of the most appropriate method of recombination. If 10 selections are recombined, a diallel mating design frequently has been used to form the next cycle population. Making diallel crosses among 10 selections is not formidable (45 possible crosses), but as number of selections increases to 20 to 30, number of crosses increases rapidly (Table 4.8). For example, diallel crosses of 20 selections result in 190 crosses (or possibly 380 nursery rows to make the crosses) as well as the mechanics of making all the crosses. Diallel crossing is feasible if number of selections is 10 or less but becomes increasingly difficult for a greater number of selections. The primary objective of recombination in cyclical selection programs is to intermate superior selections to form the next cycle population for continued selection. Hence, an effective method of recombination is imperative.

Different recombination methods are available; choice depends on number of selections included, seasons available, and resources available. (1) Diallel mating scheme is acceptable if number of selections included for recombination is not too large. Diallel crossing ensures mating each selection with all other selections. (2) Chain crossing the selections requires less space than the diallel mating scheme, but equal mating may not be made among selected progenies. Chain crossing 20 selections would require 20 rows in the nursery. (3) Bulk entry method entails planting selections in alternate rows; remaining rows are planted to a bulk of equal number of seeds of all progenies except the one they are to be crossed with. That is, if row 1 involves selection A, row 2 includes a bulk of all selections except selection A. All plants in row 2 are crossed to row 1. This method of recombination permits intercrossing among all entries unless some of the plants in the rows that include the bulks are missing. Reciprocal crosses between pairs of rows will minimize the possibility of not making crosses among all selections. If 20 selections are intercrossed, 40 rows are needed to complete the recombination.

(4) Equal quantities of seeds of the 20 selections could be bulked and inter-matings made among plants of the bulk plantings either by open pollination in isolation plantings or by hand pollination in the breeding nursery. Space would be minimized, but if one or more selections had late flowering, poor germination, or poor competitive ability, some of the selections may not be included equally in the intercrosses. All four methods of recombination have some disadvantages and advantages, and it is necessary to make a choice that one feels ensures adequate intercrossing among selected progenies. Choice of method used for recombination is not as critical as some other steps in recurrent selection. Different methods of recombination may be used between different cycles of selection without seriously jeopardizing progress of selection.

After initial intercrosses, it is necessary to include a representative sample of plants for additional intercrossing. It is important that duplicate samples be obtained to minimize chances of losing the intercrossed seed because of a crop failure. In some recurrent selection schemes, only one intercrossing is made between cycles of selection. It is imperative that a representative sample be obtained from intercrossed selected progenies to ensure that each one is equally included in bulk samples. The best procedure is to have an equal quantity of seed for each cross by counting the number of ears available and taking an equal number of seeds from each ear to form the two bulks. For example, if 10 ears are available for each of 20 crosses and two 1000 seed samples are needed, five seeds are sampled from each ear for each of the two bulks. If each selection crossed with a bulk planting of the other selections is used for recombination, one needs to count the total number of ears to determine how many seeds per ear should be included in each of the two bulk samples.

For recurrent selection schemes that permit two recombinations between cycles of selection, the same procedures are used to form the bulk samples for the next generation of recombination. One bulk sample is planted either in an isolation field or the nursery for intercrossing and the other bulk sample is placed in suitable cold storage conditions for either emergency or future use. Adequate isolation (200-500 m) is necessary to minimize the chances of contamination from other maize plants, and it is easier to obtain in the field than in a nursery planting. If a nursery planting is used for recombination, it is desirable that each plant be included as either a male or female; this requires checking the planting each day to cover silks and prepare tassels and ear shoots for pollination. The usual procedure is to pollinate each ear shoot and not use pollen from each tassel for more than two pollinations. Pollen also can be bulked and used to pollinate all ear shoots that were prepared for pollination the previous day. Regardless of the system used for the second recombination, equal quantities of seed should be taken from each ear to form two bulk samples of the Synthetic 2 seed--one for cold storage and the second to initiate the next cycle of selection.

When recombination is done through controlled hand pollination, there is an opportunity to control the number of female and male gametes that are contributed by each female and male parent to the next generation. Therefore, taking a sample of equal size from each hand-pollinated ear we will get the highest effective number for a given number of progenies in the recombination block (Table 11.5).

A simpler procedure that can be used to recombine selected progenies does not involve controlled hand pollination. Seeds from each selected progeny are planted ear-to-row, and rows are detasseled before flowering. A bulk of seeds of all progenies is planted in alternate rows as males, which will provide pollen for all detasseled rows so that each female row is pollinated with a bulk of pollen of all progenies. The ratio male:female rows may vary from 1:1 to 1:4, and male rows can be split-planted to reduce assortative mating. Because pollination is random there is no control on the number of male gametes contributed by each male plant, but a control is possible on the number of female gametes by taking a sample of equal size from each pollinated ear. This sampling procedure assures a larger effective population size than would result if a single sample were taken from a bulk of the whole set of seeds. On the other hand, control of the number of female gametes only will result in a smaller effective population size as compared with other recombination procedures previously discussed. Small differences in effective population size make no great difference when number of progenies saved for recombination is not too small.

Effects of linkage are a concern when considering number of recombinations or intermatings between cycles of selection. Effects of linkage in recurrent selection programs are not as obvious as effects of inbreeding. Linkage effects have been examined in F_2 populations of two inbred lines, and they did affect estimates of genetic components of variance (Table 5.3). Progenies were formed from individuals in F_2 populations for estimating genetic parameters. Broad based populations that have had several opportunities for recombination would be expected to be near linkage equilibrium. Difficulty of linkage effects arises from recombination of selected progenies. Because linkage is known to exist, the question is how many generations of recombination are needed for the selected population to approach linkage equilibrium. Hanson (1959) concluded that four generations of intermating seemed to be adequate. In temperate zones four generations of intermating would require two years to complete, provided winter nurseries were available. A minimum number (one or two) of intermating generations is used in most recurrent selection programs. In some recurrent selection schemes (e.g., Table 12.3), an additional generation of recombination can be included without increasing cycle time if efficient use is made of winter nurseries. Additional recombination of selected progenies is usually acknowledged to be useful; it usually is carried out between cycles of selection to reduce cycle time. Without any empirical evidence, it seems that greater gains can be made by increasing number of cycles of selection than by increasing number of recombination generations between cycles of selection.

Latter (1965, 1966a, 1966b) examined effects of linkage on limits of selection. He concluded that reduction in total response was minor unless recombination probability among loci was less than 0.1. As interpreted by Rawlings (1970), if recombination probability was 0.1 reduction was about 2.5% and if recombination probability was 0.05 reduction was about 6%. Linkage had very little effect on rate of response in early generations of selection, particularly if the populations were in linkage equilibrium when selection was started. Rawlings concluded that if linkage effects are a concern, effective population size should be increased to allow for linkage depression. It seems that linkage effects will not seriously affect response to selection.

The basic relation between factors involved in operational stages of recurrent selection was given by Eberhart (1970) and Sprague and Eberhart (1977). They expressed predicted gain Δ_G in the following relation:

$$\Delta_G = \frac{ck\sigma_{g'}^2}{y[\sigma^2/(re) + \sigma_{ge}^2/e + \sigma_g^2]^{1/2}}$$

where c is parental control (Table 6.12), k is the standardized selection differential, $\sigma_{g'}^2$ is the portion of genetic variance among progenies due to addi-

tive effects, y is number of years per cycle, and the expression within the
brackets is the phenotypic variance. This formula differs slightly from those
in Chap. 6, where k in the numerator is selection intensity (which shows if
selection is for one or both sexes, or parental control) and c is a coefficient that takes into account the crossing system (which dictates the covariance between relatives in different generations); k and c together multiply σ_A^2
(additive genetic variance) or one of its homologues. On the other hand, in
the formula presented by Sprague and Eberhart (1977), k has the same meaning
and c is a coefficient that takes into account only parental control. The coefficient that results from the covariance between relatives is inherently
contained in $\sigma_{g'}^2$, which is the additive portion of variance expressed among
selection units or the portion of covariance between relatives in the same
generation. For example, the numerator for expected gain from selection among
half-sib families for only one sex (ear-to-row procedure) is shown in Chap. 6
as being $k(1/8)\sigma_A^2$, where 1/8 is the coefficient for the covariance between
half uncle-nephew. On the other hand, Sprague and Eberhart (1977) give the
same coefficient expressed as $k(1/2)(1/4)\sigma_A^2$ where 1/2 is a coefficient due to
parental control (selection for one sex) and $(1/4)\sigma_A^2$ is the additive portion
of variance among half-sib families or covariance between half-sibs (same generation).

Although both formulations give essentially the same results, a discussion of differences between them follows. Genetic gain in the formula is expressed on a per year basis. Relation for predicted gain can assist in making
decisions in each operational stage of recurrent selection. Table 7.30 shows
that the prediction formula provides guidelines in the choice of recurrent selection procedure that maximizes rate of gain per year. We want to obtain
maximum gain possible on either a per year or a per cycle basis. If the
choice among recurrent selection schemes is on a genetic gain per year basis,
the choice may be different if different numbers of years are required to complete each cycle of selection (Table 7.30). Prediction formulas for different
methods of recurrent selection are listed in Table 12.13.

The predicted gain formula shows that genetic gain can be increased if
parental control is increased, years per cycle are reduced, and additive genetic variance among progenies is increased (Table 6.12). Two other important
factors are selection intensity (Table 6.2) and phenotypic variance. If information is available, various combinations of these five factors can be determined to provide some guidelines for the potentially most efficient recurrent selection scheme to use (Empig et al. 1972). The formula does not provide any guidelines on choice of population. The three operational stages of
recurrent selection can be adjusted to affect predicted gain; number of progenies to test and recombine affects selection intensity and number of generations of recombination determines how many years are required to complete each
cycle of selection. How the various factors affect efficiency of selection is
illustrated in Chap. 7.

Although the prediction formula can provide assistance in determining the
most efficient method for conducting recurrent selection, the difference in
predicted gain per cycle or year may often be too small to be detectable in
experimental trials. Then the appropriate course is to use the recurrent selection scheme that satisfies objectives of the applied breeding program.
Other factors that enhance efficiency of recurrent selection schemes include
such items as use of mechanical planting and harvesting equipment to permit
timely handling of a greater number of plots, high-speed computers to analyze
large volumes of data, adequate test sites to estimate the genotype-environ-

Table 12.13. Expected genetic gain per cycle (Δ_g) for different intra- and interpopulation methods of recurrent selection.

Selection method	Expected genetic gain (Δ_g)[†]	Years per cycle
1. Mass[‡]		
a. One sex	$k(1/2)\sigma_A^2 \Big/ \sqrt{\sigma_w^2 + \sigma_{DE}^2 + \sigma_{AE}^2 + \sigma_D^2 + \sigma_A^2}$	1
b. Both sexes	$k\sigma_A^2 \Big/ \sqrt{\sigma_w^2 + \sigma_{DE}^2 + \sigma_{AE}^2 + \sigma_D^2 + \sigma_A^2}$	1 or 2
2. Modified ear-to-row[‡§]		
a. One sex	$k(1/8)\sigma_A^2 \Big/ \sqrt{\dfrac{\sigma^2}{re} + \dfrac{(1/4)\sigma_{AE}^2}{e} + (1/4)\sigma_A^2}$	1
b. Both sexes	$k(1/4)\sigma_A^2 \Big/ \sqrt{\dfrac{\sigma^2}{re} + \dfrac{(1/4)\sigma_{AE}^2}{e} + (1/4)\sigma_A^2}$	2
3. Half-sib[¶]		
a. Remnant half-sib seed	$k(1/4)\sigma_A^2 \Big/ \sqrt{\dfrac{\sigma^2}{re} + \dfrac{(1/4)\sigma_{AE}^2}{e} + (1/4)\sigma_A^2}$	2
b. Self-seed	$k(1/2)\sigma_A^2 \Big/ \sqrt{\dfrac{\sigma^2}{re} + \dfrac{(1/4)\sigma_{AE}^2}{e} + (1/4)\sigma_A^2}$	3
4. Full-sib	$k(1/2)\sigma_A^2 \Big/ \sqrt{\dfrac{\sigma^2}{re} + \dfrac{[(1/2)\sigma_{AE}^2 + (1/4)\sigma_{DE}^2]}{e} + [(1/2)\sigma_A^2 + (1/4)\sigma_D^2]}$	2
5. S_1[#]	$k\sigma_A^2 \Big/ \sqrt{\dfrac{\sigma^2}{re} + \dfrac{[\sigma_{AE}^2 + (1/4)\sigma_{DE}^2]}{e} + [\sigma_A^2 + (1/4)\sigma_D^2]}$	2
6. S_2[#]	$k(3/2)\sigma_A^2 \Big/ \sqrt{\dfrac{\sigma^2}{re} + \dfrac{[(3/2)\sigma_{AE}^2 + (3/16)\sigma_{DE}^2]}{e} + [(3/2)\sigma_A^2 + (3/16)\sigma_D^2]}$	3
7. Reciprocal recurrent[††]	$\dfrac{k_1(1/4)\sigma_{A_{12}}^2}{\sqrt{\dfrac{\sigma_{12}^2}{re} + \dfrac{(1/4)\sigma_{AE_{12}}^2}{e} + (1/4)\sigma_{A_{12}}^2}} + \dfrac{k_2(1/4)\sigma_{A_{21}}^2}{\sqrt{\dfrac{\sigma_{21}^2}{re} + \dfrac{(1/4)\sigma_{AE_{21}}^2}{e} + (1/4)\sigma_{A_{21}}^2}}$	3

8. Reciprocal full-sib

$$\frac{k(1/2)\sigma^2_{A(12)}}{\sqrt{\dfrac{\sigma^2}{re} + \dfrac{(1/2)\sigma^2_{AE(12)} + (1/4)\sigma^2_{DE_{12}}}{e} + [(1/2)\sigma^2_{AE(12)} + (1/4)\sigma^2_{D_{12}}]}} \qquad 2$$

9. RRS based on testcross of half-sib families[††]

$$\frac{k_1(1/16)\sigma^2_{A_{12}}}{\sqrt{\dfrac{\sigma^2_{12}}{re} + \dfrac{(1/16)\sigma^2_{AE_{12}}}{e} + (1/16)\sigma^2_{A_{12}}}} + \frac{k_2(1/16)\sigma^2_{A_{21}}}{\sqrt{\dfrac{\sigma^2_{21}}{re} + \dfrac{(1/16)\sigma^2_{AE_{21}}}{e} + (1/16)\sigma^2_{A_{21}}}} \qquad 3$$

10. RRS based on half-sib families of prolific plants[††]

$$\frac{k_1(1/8)\sigma^2_{A_{12}}}{\sqrt{\dfrac{\sigma^2_{12}}{re} + \dfrac{(1/8)\sigma^2_{AE_{12}}}{e} + (1/8)\sigma^2_{A_{12}}}} + \frac{k_2(1/8)\sigma^2_{A_{21}}}{\sqrt{\dfrac{\sigma^2_{21}}{re} + \dfrac{(1/8)\sigma^2_{AE_{21}}}{e} + (1/8)\sigma^2_{A_{21}}}} \qquad 1 \text{ or } 2$$

[†]k is the standardized selection differential; σ^2_A and σ^2_D are the additive and dominance variances; σ^2_{AE} and σ^2_{DE} are the additive and dominance by environment interactions; r is the number of replications in each of e environments; σ^2_w is within-plot environmental variation; and σ^2 is experimental error.

[‡]Parental control is only for female and one sex and both parents for both sexes. Add σ^2 if no blocking.

[§]If mass selection is practiced within ear rows for the primary trait, an additional component should be added to the expected gain:

$$\frac{k(3/8)\sigma^2_A}{\sqrt{\sigma^2_w + (3/4)\sigma^2_{AE} + \sigma^2_{DE} + (3/4)\sigma^2_A + \sigma^2_D}}$$

[¶]a: Use σ^2_A for parental population tester; b: use $\sigma^2_A = 2pq\alpha_1\alpha_2$ for unrelated tester (see p. 274).

[#]Definitions of additive genetic variances change slightly with inbreeding and dominance variances are difficult to define unless p = q = 0.5.

[††]$\sigma^2_{A_{12}}$ and $\sigma^2_{A_{21}}$ are homologues of σ^2_A (additive genetic variance); $\sigma^2_{A(12)} = (1/2)(\sigma^2_{A_{12}} + \sigma^2_{A_{21}})$ and is the additive genetic variance in the crossed population; see Chap. 2.

ment interaction, and availability of off-season nurseries. These items are
not included directly in the formula, but they contribute greatly to predicted
gain on a per year basis. Inclusion of these items permits better sampling to
get a better estimate of $\sigma_{g'}^2$, reduces number of years per cycle, and reduces
confounding effects of genotype-environment interactions. It is necessary to
conduct cyclical selection programs on a regular basis without interruptions.
If the population improvement program via some form of recurrent selection
does not contribute to breeding program objectives, it should not be conducted. Basic selection studies are conducted to provide information for applied
breeders, but they also provide breeding material for applied programs (e.g.,
Hallauer 1973a, 1973b; Suwantaradon and Eberhart 1974; Russell and Eberhart
1975).

Recurrent selection will not, and is not intended to, replace classical
breeding methods that have been so successful. The progress in maize yields
illustrated in Figs. 1.1 and 1.2 resulted from use of classical breeding methods. Pedigree and backcross methods of breeding will continue to have a prominent role. Recurrent selection methods should supplement other breeding
methods and be considered as only one other type of breeding method in the
breeder's arsenal available for developing improved varieties and hybrids.
One will not replace the other. The time and effort expended on recurrent selection in applied breeding programs are justified only when they contribute
to them. Integration of recurrent selection methods with other selection and
breeding methods is feasible in nearly all instances.

Comparisons of recurrent selection and classical pedigree methods of
breeding are limited. Sprague (1952) reported comparisons for selection for
oil content of the kernel and Duvick (1977) for yield. Depending on the comparison, Sprague reported that recurrent selection was two to five times more
effective than pedigree selection and the recurrent selection program still
had genetic variability present for continued response to selection. Duvick
found that rates of gain were nearly equal for the two distinct methods of
breeding; 0.68 q/ha per year of pedigree selection (13.3 years per cycle) vs.
0.71 q/ha per year for recurrent selection (3 years per cycle). The comparisons of Sprague and Duvick do not resolve which method of selection is more
effective, and it is doubtful whether any arguments can show that one method
(pedigree or recurrent) is more important or more efficient.

Pedigree and recurrent selection methods are two different methods of selection with different objectives. Pedigree selection usually is restricted
to breeding populations having a narrow genetic base (often developed from a
cross of two inbred lines) that emphasizes inbreeding, while recurrent selection methods are applied to populations having a broad genetic base that minimizes effects of inbreeding. In addition, the breeding methods are usually
quite distinct. Effective recurrent selection methods contribute selections
that can be used either in pedigree selection programs or directly as parental
stocks in hybrids or as improved varieties; that is, recurrent selection sup-
plements other methods of selection.

Jenkins (1978) emphasized the importance of pedigree and backcrossing
methods in maize breeding programs for developing inbred lines. In 1936, 350
lines developed by state and federal agencies were listed; only 7 (2%) of the
lines listed were second-cycle lines. Subsequent summaries of lines released
by state and federal agencies show that the percentage of second-cycle lines
was 20% in 1948, 26% in 1952, 40% in 1956, and 50% in 1960. Since 1960 most
of the new inbred lines released by state and federal agencies have been second-cycle lines developed by pedigree and backcrossing methods of selection.
The same trend is probably occurring in proprietary lines. The genetic base

of breeding materials, therefore, is becoming more restricted. Recurrent selection methods can contribute to broadening the genetic base of breeding programs, but only when all methods of selection are integrated in all phases of breeding programs will the projected goals of continuous genetic improvement be realized.

REFERENCES

Baker, L. H., and R. N. Curnow. 1969. Choice of population size and use of variation between replicate populations in plant breeding selection programs. *Crop Sci.* 9:555-60.

Burton, J. W.; L. H. Penny; A. R. Hallauer; and S. A. Eberhart. 1971. Evaluation of synthetic populations developed from a maize population (BSK) by two methods of recurrent selection. *Crop Sci.* 11:361-67.

Compton, W. A., and R. E. Comstock. 1976. More on modified ear-to-row selection in corn. *Crop Sci.* 16:122.

Comstock, R. E. 1979. Use of inbred lines in reciprocal recurrent selection. *Crop Sci.* 19:881-86.

Comstock, R. E.; H. F. Robinson; and P. H. Harvey. 1949. A breeding procedure designed to make maximum use of both general and specific combining ability. *Agron. J.* 41:360-67.

Darrah, L. L.; S. A. Eberhart; and L. H. Penny. 1978. Six years of maize selection in Kitale Synthetic II, Ecuador 573, and Kitale Composite, using methods of the comprehensive breeding system. *Euphytica* 27:191-204.

Dudley, J. W., and R. H. Moll. 1969. Interpretation and uses of estimates of heritability and genetic variances in plant breeding. *Crop Sci.* 9:257-62.

Duvick, D. N. 1977. Genetic rates of gain in hybrid maize yields during the past 40 years. *Maydica* 22:187-96.

Eberhart, S. A. 1970. Factors affecting efficiencies of breeding methods. *African Soils* 15:669-80.

Eberhart, S. A.; M. N. Harrison; and F. Ogada. 1967. A comprehensive breeding system. *Züchter* 37:169-74.

Empig, L. T.; C. O. Gardner; and W. A. Compton. 1972. Theoretical gains for different population improvement procedures. Nebraska Agric. Exp. Stn. Bull. MP26 (revised).

Gardner, C. O. 1977. Quantitative genetic studies and population improvement in maize and sorghum. In *Proceedings of the International Conference on Quantitative Genetics,* E. Pollak, O. Kempthorne, and T. B. Bailey, Jr., eds., pp. 475-89. Iowa State Univ. Press, Ames.

————. 1978. Population improvement in maize. In *Maize Breeding and Genetics,* D. B. Walden, ed., pp. 207-28. Wiley, New York.

Genter, C. F. 1973. Comparison of S_1 and testcross evaluation after two cycles of recurrent selection in maize. *Crop Sci.* 13:524-27.

Genter, C. F., and M. W. Alexander. 1962. Comparative performance of S_1 progenies and test-crosses of corn. *Crop Sci.* 2:516-19.

Gevers, H. O. 1975. Three cycles of reciprocal recurrent selection in maize under two systems of parent selection. *Agroplantae* 7:107-8.

Hallauer, A. R. 1967a. Development of single-cross hybrids from two-eared maize populations. *Crop Sci.* 7:192-95.

————. 1967b. Performance of single-cross hybrids developed from two-ear varieties. *Proc. Annu. Hybrid Corn Ind. Res. Conf.* 22:74-81.

————. 1973a. Recurrent selection for polygenic resistance. Report of workshop on the downy mildews of sorghum and corn. Texas Agric. Exp.

————. Stn. Tech. Bull. 74-1:32-40.

————. 1973b. Hybrid development and population improvement in maize by reciprocal full-sib selection. *Egyptian J. Genet. Cytol.* 1:84-101.

————. 1978. Recurrent selection programs. Illinois Corn Breeders Sch. 14: 28-45.

————. 1981. Selection and breeding methods. In *Plant Breeding II*, K. J. Frey, ed., pp. 3-55. Iowa State Univ. Press, Ames.

————. 1984. Reciprocal full sib selection in maize. *Crop Sci.* 7:55-59.

————. 1985. Compendium of recurrent selection methods and their application. *Critical Rev. Plant Sci.* 3:1-33. CRC Press, Inc., Boca Raton, FL.

Hallauer, A. R., and S. A. Eberhart. 1970. Reciprocal full-sib selection. *Crop Sci.* 10:315-16.

Hanson, W. D. 1959. The breakup of initial linkage blocks under selected mating systems. *Genetics* 44:857-68.

Harris, R. E.; C. O. Gardner; and W. A. Compton. 1972. Effects of mass selection and irradiation in corn measured by random S_1 lines and their testcrosses. *Crop Sci.* 12:594-98.

Hoegemeyer, T. C., and A. R. Hallauer. 1976. Selection among and within full-sib families to develop single crosses of maize. *Crop Sci.* 16:76-81.

Hopkins, C. G. 1899. Improvement in the chemical composition of the corn kernel. Illinois Agric. Exp. Stn. Bull. 55.

Horner, E. S.; H. W. Lundy; M. C. Lutrick; and W. H. Chapman. 1973. Comparison of three methods of recurrent selection in maize. *Crop Sci.* 13:485-89.

Horner, E. S.; M. C. Lutrick; W. H. Chapman; and F. G. Martin. 1976. Effect of recurrent selection for combining ability with a single-cross tester in maize. *Crop Sci.* 16:5-8.

Hull, F. G. 1945. Recurrent selection and specific combining ability in corn. *J. Am. Soc. Agron.* 37:134-45.

Jenkins, M. T. 1940. Segregation of genes affecting yield of grain in maize. *J. Am. Soc. Agron.* 32:55-63.

————. 1978. Maize breeding during the development and early years of hybrid maize. In *Maize Breeding and Genetics*, D. B. Walden, ed., pp. 13-28. Wiley, New York.

Jinahyon, S., and W. A. Russell. 1969a. Evaluation of recurrent selection for stalk-rot resistance in an open-pollinated variety of maize. *Iowa State J. Sci.* 43:229-37.

————. 1969b. Effects of recurrent selection for stalk-rot resistance on other agronomic characters in an open-pollinated variety of maize. *Iowa State J. Sci.* 43:239-51.

Jones, L. P.; W. A. Compton; and C. O. Gardner. 1971. Comparison of full- and half-sib reciprocal recurrent selection. *Theor. Appl. Genet.* 41:36-39.

Latter, B. D. H. 1965. The response to artificial selection due to autosomal genes of large effect. II. The effects of linkage on limits to selection in finite populations. *Australian J. Biol. Sci.* 18:1009-23.

————. 1966a. The response to artificial selection due to autosomal genes of large effect. III. The effects of linkage on the rate of advance and approach to fixation in finite populations. *Australian J. Biol. Sci.* 19: 131-46.

————. 1966b. The interaction between effective population size and linkage intensity under artificial selection. *Genet. Res. Comb.* 7:313-23.

Lonnquist, J. H. 1964. A modification of the ear-to-row procedure for the improvement of maize populations. *Crop Sci.* 4:227-28.

Lonnquist, J. H., and M. F. Lindsey. 1964. Topcross versus S_1 line performance in corn (*Zea mays* L.). *Crop Sci.* 8:50-53.

Marquez-Sanchez, F., and A. R. Hallauer. 1970a. Influence of sample size on the estimation of genetic variances in a synthetic variety of maize. I. Grain yield. *Crop Sci.* 10:357-61.

————. 1970b. Influence of sample size on the estimation of genetic variance in a synthetic variety of maize. II. Plant and ear characters. *Iowa State J. Sci.* 44:423-36.

Mather, K. 1949. *Biometrical Genetics.* Methuen, London.

Moll, R. H., and C. W. Stuber. 1971. Comparisons of response to alternative selection procedures initiated with two populations of maize (*Zea mays* L.). *Crop Sci.* 11:706-11.

————. 1974. Quantitative genetics. Empirical results relevant to plant breeding. *Adv. Agron.* 26:277-313.

Moll, R. H.; A. Bari; and C. W. Stuber. 1977. Frequency distributions of maize yield before and after reciprocal recurrent selection. *Crop Sci.* 17:794-96.

Noble, S. W. 1966. Sampling of heterogeneous testers and the comparison of a double cross with parental and non-parental single crosses as testers for the evaluation of maize lines. Ph.D. dissertation, Iowa State Univ., Ames.

Obilana, A. T.; A. R. Hallauer; and O. S. Smith. 1979a. Estimation of genetic components of variance in the interpopulation formed by crossing two maize synthetics, BS10 and BS11. *J. Hered.* 70:127-32.

————. 1979b. Predicted and observed response to reciprocal full-sib selection in maize. *Egyptian J. Genet. Cytol.* 8:269-82.

Paterniani, E. 1967a. Selection among and within half-sib families in a Brazilian population of maize (*Zea mays* L.). *Crop Sci.* 7:212-15.

————. 1967b. Interpopulation improvement: Reciprocal recurrent selection variations. Maize 8, CIMMYT.

Paterniani, E., and R. Vencovsky. 1977. Reciprocal recurrent selection in maize (*Zea mays* L.) based on testcrosses of half-sib families. *Maydica* 22:141-52.

————. 1978. Reciprocal recurrent selection based on half-sib progenies and prolific plants in maize (*Zea mays* L.). *Maydica* 23:209-19.

Penny, L. H., and S. A. Eberhart. 1971. Twenty years of reciprocal recurrent selection with two synthetic varieties of maize. *Crop Sci.* 11:900-903.

Penny, L. H.; G. E. Scott; and W. D. Guthrie. 1967. Recurrent selection for European corn borer resistance. *Crop Sci.* 7:407-9.

Pesek, J., and R. J. Baker. 1969. Desired improvement in relation to selection indices. *Canadian J. Plant Sci.* 49:803-4.

Rawlings, J. O. 1970. Present status of research on long- and short-term recurrent selection in finite populations--choice of population size. *Proc. Second Meet. Work. Group Quant. Genet.*, sect. 22. IUFRO, Raleigh, N.C. Pp. 1-15.

Robertson, A. 1960. A theory of limits in artificial selection. *Proc. R. Soc. London* B153:234-49.

Russell, W. A. 1974. Comparative performance for maize hybrids representing different eras of maize breeding. *Proc. Annu. Corn Sorghum Res. Conf.* 29:81-101.

————. 1986. Contribution of breeding to maize improvement in the United States, 1920's-1980's. *Iowa State J. Res.* 61:5-34.

Russell, W. A., and S. A. Eberhart. 1975. Hybrid performance of selected
  maize lines from reciprocal recurrent and testcross selection programs.
  Crop Sci. 15:1-4.
Shull, G. H. 1908. The composition of a field of maize. Am. Breeders'
  Assoc. Rep. 4:296-301.
————. 1909. A pure line method of corn breeding. Am. Breeders' Assoc.
  Rep. 5:51-59.
————. 1910. Hybridization methods in corn breeding. *Am. Breeders' Mag.* 1:
  98-107.
Smith, H. F. 1936. A discriminant function for plant selection. *Ann. Eugen.*
  7:240-50.
Smith, O. S. 1979. A model for evaluating progress from recurrent selection.
  Crop Sci. 19:223-26.
Sprague, G. F. 1939. An estimation of the number of top-crossed plants required for adequate representation of a corn variety. *J. Am. Soc. Agron.*
  31:11-16.
————. 1952. Additional studies of the relative effectiveness of two systems of selection for oil content of the corn kernel. *Agron. J.* 44:329-
  31.
Sprague, G. F., and S. A. Eberhart. 1977. Corn breeding. In *Corn and Corn
  Improvement*, G. F. Sprague, ed., pp. 305-62. Am. Soc. Agron., Madison,
  Wis.
Subandi, W.; W. A. Compton; and L. T. Empig. 1973. Comparison of the efficiencies of selection indices for three traits in two variety crosses of
  corn. *Crop Sci.* 13:184-86.
Suwantaradon, K., and S. A. Eberhart. 1974. Developing hybrids from two improved maize populations. *Theor. Appl. Genet.* 44:206-10.
Suwantaradon, K.; S. A. Eberhart; J. J. Mock; J. C. Owens; and W. D. Guthrie.
  1975. Index selection for several agronomic traits in the BSSS 2 maize
  population. *Crop Sci.* 15:827-33.
Walejko, R. N., and W. A. Russell. 1977. Evaluation of recurrent selection
  for specific combining ability in two open-pollinated maize cultivars.
  Crop Sci. 17:647-51.
Webel, O. D., and J. H. Lonnquist. 1967. An evaluation of modified ear-to-
  row selection in a population of corn (*Zea mays* L.). *Crop Sci.* 7:651-55.
Williams, J. S. 1962. The evaluation of a selection index. *Biometrics* 18:
  375-93.

INDEX